EARTH'S MAGNETOSPHERE

EARTH'S MAGNETOSPHERE

Formed by the Low-Latitude Boundary Layer

Second Edition

WAYNE KEITH

WALTER HEIKKILA

ACADEMIC PRESS
An imprint of Elsevier

ELSEVIER

Library of Congress Cataloging-in-Publication Data
A catalog record for this book is available from the Library of Congress

British Library Cataloguing-in-Publication Data
A catalogue record for this book is available from the British Library

ISBN 978-0-12-818160-7

For information on all Academic Press publications
visit our website at https://www.elsevier.com/books-and-journals

Cover image courtesy: C.J. Kale
Publisher: Candice Janco
Acquisitions Editor: Amy Shapiro
Editorial Project Manager: Andrea Dulberger
Production Project Manager: Kumar Anbazhagan
Cover Designer: Miles Hitchen

Typeset by SPi Global, India

Working together
to grow libraries in
developing countries

www.elsevier.com • www.bookaid.org

Contents

Dedication

To my ever-supportive family. My wife Melinda, daughters Amy and Ella, and my parents, Donald and Lanita Keith.

Wayne Keith

To Jeanine, my wife, and sons Eric and Richard, and my father Kustaa Heikkilä:

From his lonely perch he let his mind roam the cosmos.

Walter Heikkila

Preface to the first edition

The magnetosphere is the vast region of space magnetically connected to the Earth and responsible for that magnificent spectacle, the northern lights, the aurora borealis (northern dawn). In northern Ontario, a long time ago, I was impressed with the rays of light of various colors dancing and prancing along east-west arcs (see the frontispiece drawn for me by Sheré Chamness). My mother said ominously that war would come. I would spend most of my working life trying to find an explanation for this strange and intriguing phenomenon.

Earth with its magnetic field is embedded in a high-velocity stream of plasma of solar origin, the solar wind. This situation can be viewed in two ways. First, it constitutes a variable input of plasma, momentum, and energy to the magnetosphere; by observational techniques now firmly in place, we can see the consequences for this intricate system. Second, the solar wind is one result of processes that operate on the Sun, especially in the corona. These are somewhat different than in the Earth's magnetosphere, partly because of scale, but it is generally thought that the two are essentially alike. For solar research, we have some additional observational techniques that we can use to our advantage; for one, we can look from afar to witness radiation of all kinds, including plasmoids, leaving the system. Solar flares, coronal mass ejections, heating of coronal plasma, changes in magnetic topology, collisionless shock waves, and solar energetic particles are observed.

The magnetosphere is readily accessible for direct observations, even experiments to test what is the result of a known input; as such, it is a giant laboratory for solar and astrophysical processes. Here we have the auroras, both quiescent and active. With the latter, we have rapid particle energization, like solar flares. We have bursty bulk flows, plasmoids and flux ropes, and much more. It would be easy to overload our account with observational facts, for they exist in profusion.

Yet, despite half a century of space research, there is still no definitive conclusion about the rapid particle energization, about substorms, or even the obvious correlation between geomagnetic activity and the solar wind parameters, including the orientation of the interplanetary magnetic field (IMF). Theoretical arguments are complicated for many reasons. Thus, for example, Phan et al. (1996, p. 7827) concluded that "crucial questions about the interplay between the upstream magnetosheath conditions and the structure and dynamics of the magnetopause boundary [remain] unanswered." Although Baker et al. (1999) "concluded that the global magnetospheric substorm problem has now largely been solved ..." they nevertheless noted that "fundamental issues remain to be resolved ... Why, for example, is the magnetosphere stable most of the time, and why do substorms occur just when they do? What allows the violation of the frozen-flux constraint necessary for an efficient energy release by reconnection in the course of substorms?" More recently, Donovan (2006) stated, "At this ICS, we somehow managed to avoid debates that are fundamentally unresolvable, but still consider the dominant paradigms."

Perhaps the main shortcoming of all previous substorm theories is their difficulty of explaining the initial localization in space, the rapidity of the onset of the expansion phase of the substorm. "Explaining the sudden onset of the expansion phase of magnetospheric substorms has proved to be one of the most intractable problems in magnetospheric physics to date" (Vasyliunas, 1998).

One semantic difficulty is that the language in general use in space plasma physics comes in large part from analogies with fluid theory. It is highly suggestive; unfortunately, it often gives the impression of providing an explanation when it can be partly, or even totally, wrong. Birkhoff (1960) has noted that "it is easy for mathematicians to become convinced that rational hydrodynamics is, in principle, infallible." He concluded that the failures of fluid theory could generally be laid upon the use of various plausible hypotheses, sometimes made explicitly, sometimes implicitly, that are not fully appropriate.

It is arguable that, in our case, the plausible hypothesis is the use of the B,V paradigm; here the magnetic field **B** and the plasma velocity **V** are regarded as the primary independent variables (Parker, 1996). The plasma is described by a fluid theory, by magnetohydrodynamics (MHD) invented by Hannes Alfvén (1950), and the so-called principle of frozen-field convection. According to that, if we follow the fluid initially on the surface S as it moves through the system, the flux through the surface will remain constant even as the surface changes its location and shape. Furthermore, the frozen-flux condition implies that all particles initially on a flux tube will remain along a single tube as they convect through space (Kivelson, 1995). "The key point here is that the particle and fluid pictures give us equivalent answers" (Hughes, 1995).

However, that statement is only partly true; the use of the B,V paradigm requires an equation of state to get a complete set of equations so that the problem is well posed; that can, and does, lead to difficulties (Heikkila, 1997). An equation of state is local, by definition, while global forces by the principle of superposition (such as due to the electric field) are important in a plasma. Even Alfvén (1986) was opposed to this extreme use of MHD, without considering its limitations.

The use of a different paradigm, the E,J paradigm where the electric field **E** and current density **J** are regarded as primary, is sometimes a necessity. Energization of plasma particles is due to an electric field, with a force $\mathbf{F} = q\mathbf{E}$ acting on particles with a charge q. In contrast, the force due to a magnetic field is transverse to the particle motion, and so, by a vector identity, $q\mathbf{V} \cdot (\mathbf{V} \times \mathbf{B}) = 0$: the particles move without energization. The two paradigms are not equivalent; this can be forced through an equation of state but it has limitations of its own. I believe that the E,J paradigm is the key to solving the "intractable problems in magnetospheric physics."

The equivalence of the two paradigms has been questioned for some time. On the energization of particles within the plasma sheet Hines said:

It has been shown that energization of particles in the presence of an electric field and an inhomogeneous magnetic field may be described equivalently by means of a hydromagnetic–thermodynamic approach or … a particle-drift approach, insofar as drifts transverse to the magnetic field are concerned, and if attention is confined to differential variations [emphasis added]. If collisions maintain an equipartition of the energy amongst the three translational modes, then this conclusion can be extended to integrated variations and the only important drifts are those transverse to the magnetic field. … The equivalence might be restored for integrated variations if suitable definitions of the 'energization' are introduced and suitable limiting cases are considered, but in general there seems to be little advantage in the further pursuit of the hydromagnetic–thermodynamic approach when a precise description is required. The particle approach is then necessary for at least some aspects of the problem, and it might as well be adopted for all.

Hines (1963)

This has been repeated many times (e.g., Kivelson, 1995, p. 41), regrettably to no avail. Schmidt expressed it as follows:

It is rather fortunate, therefore, to have found that these fictitious quantities obey some quasi-hydrodynamic equations: an equation of continuity and an equation of motion. This is about as far as the analogy can be stretched. Unfortunately these equations do not suffice to provide solutions for the great number of unknown quantities.

Schmidt (1979, p. 57)

I will not go so far as did Colin Hines (my then boss!); there is much work based on MHD theory that is truly worthwhile. Rather, my task will be to separate the good from the bad. Examples of the good are the search for the de Hoffman-Teller frame by Sonnerup et al. (2004) and the MHD Gumics theory of Laitinen (2007). An example of the bad is the model used for reconnection (in two dimensions?), for example, in the summary by Birn et al. (2001).

As Richmond (1985) has recognized, "The solar and magnetospheric plasmas are represented as ohmic media in which the current and electric field are linearly related by a conductivity parameter even though collisionless plasma comes nowhere near to satisfying Ohm's law [even the generalized version]. Unfortunately, in many cases the parameters that are used in the models [even the models] tend to conceal or distort some of the key physical processes that occur, a high price to pay for simplified analytical [or computer] solutions."

When David Winningham and I discovered the cusp (cleft) (Heikkila et al., 1970; Heikkila and Winningham, 1971) using ISIS-1 data, I thought we had verified Dungey's concept of magnetic reconnection. It was Ian McDiarmid using energetic particle data on ISIS-1 who convinced me otherwise: something was wrong with this simplistic interpretation. He showed that trapped particles were present regularly in the cleft, implying closed field lines. At the AGU fall meeting in 1972, Bill Olson, Juan Roederer, and I shook hands to fight reconnection as presently conceived. For me, the case was proved in their paper by McDiarmid et al. (1976) "Particle properties in the day side cleft." As Ian said, particles would go away on open field lines without doing any work!

My interest in the low-latitude boundary layer (LLBL) was a direct result, initially at the Royal Institute of Technology, where I had the privilege of using Hannes Alfvén's office. This research was published 10 years later (Heikkila, 1984), but hardly anyone noticed (but see Lundin and Evans, 1985; Lundin, 1988).

Another person who has influenced me greatly is Joseph Lemaire. His work on the patchiness of the solar wind in the late 1970s (Lemaire and Roth, 1978) opened my eye to what might be happening. He noted that the solar wind plasma must be patchy, and blobs with more momentum than the adjacent plasma could penetrate through the magnetopause current sheet by an impulsive penetration mechanism (IP). Schmidt (1960, 1979) had demonstrated a mechanism that could be involved, a polarization current to form an electrostatic field that draws on electric energy. My ideas were advanced six years later, the same principle but a complementary idea of a plasma transfer event (PTE) drawing on magnetic energy.

Chapter 1 is more than a historical introduction. It is organized by material on spatial location starting with the solar wind, with Figure 1.1 showing the different regions of the magnetosphere, in three dimensions, as a guide. A secondary theme is to present a historical approach wherever possible, at the same time covering the simple material of the early investigations. A third objective is to cover the assumptions made in each case, to test for applicability of the results, quite often in hindsight.

Kinetic statistical theory of plasma problems may provide for a complete analysis, but it is difficult at best. Chapter 2 discusses various approximate methods, such as circuit theory, plasma fluid theory, and computer simulation. Each has its own problems and benefits. Observations are excellent, leaving little doubt about what happens.

Chapter 3 treats the electric field quite differently from the usual in plasma physics. The implications of Helmholtz's theorem on any vector field **V** are discussed first. The *sources* of any general vector field **V**, the inhomogeneous terms in the equations of the governing differential equations, are the *divergence* and *curl*; div **V** and curl **V** are not simply characteristics of the vector **V**. The same holds true for the electric field **E** where div **E** is the charge density, the source for the electrostatic field, and curl **E**, source for the field due to induction to a perturbation current.

The pivotal concept is to divide the electric field into parallel and transverse components with reference to the magnetic field. Since the electrostatic field has zero curl, it cannot modify the curl of the inductive component, nor the electromotive force; if the parallel component of the total electric field is reduced by field-aligned charge separation (even to zero), the transverse component will be increased to preserve the curl (Heikkila and Pellinen, 1977; Heikkila et al., 1979). This has many consequences, for example, solar wind plasma transfer across the magnetopause, current thinning event during the growth phase, tailward-moving plasmoids, bursty bulk flows, omega band auroras, and prompt particle acceleration to high energies.

Chapter 4 (Chapter 5 in the second edition) dissects Poynting's theorem at some length. Perhaps the main objection to the current practice in space plasma physics, including theoretical analysis of magnetic reconnection, is that the time-dependent volume integral concerned with magnetic energy was not used. By this simple, yet fundamental, argument it can be concluded that magnetic reconnection, as presently understood and practiced, is flawed. There is no doubt that there can be changes in the state of interconnection between the geomagnetic and the IMFs, but the present theory for that is not appropriate. For that, we must use 3D reconnection with the full electromagnetic field as in Generalized Magnetic Reconnection (GMR) (Schindler et al., 1988).

The next five chapters use the latest data on the magnetopause, the cusp/cleft, LLBL, auroral phenomena, and substorms, partly from the 4-spacecraft Cluster mission. This part, over half the book, attempts to demonstrate how these ideas explain complex, time-varying, magnetospheric situations better than the fluid-based theories do, giving clear physical reasons for the deficiencies in the conventionally accepted views. A direct comparison of this type should take the subject forward in a useful way.

Finally, the Epilogue discusses some outstanding problems as reviewed by a NASA panel. It is also a brief review of the book, by chapters.

This book presents a new way of looking at problems of solar wind interaction with the magnetosphere, and in space plasma physics in general. The PTE into the LLBL allows the chain of cause and effect relationships to be followed throughout the cycle of solar wind interaction with the magnetosphere. New ideas are appraised, and reminders to students on well-known principles that have apparently been ignored are provided.

Throughout the book, I limit discussion of my own work (which some have described as controversial) for the Discussion sections. The material in the rest of the book, including the Introduction, the key results of past research, and the Summaries, can be backed up by fundamental ideas on mathematics, physics, and space physics.

This is not a review book; many excellent papers that have formed my knowledge of the subject have not been mentioned. It is not difficult; a beginning graduate student should have no challenges if he understands, for example, the Delta function in Chapter 3. I will rely on the textbook *Introduction to Space Physics* as a primer (denoted by Kivelson and Russell (1995) throughout the text); 14 authors cover a vast amount of material in 568 pages, with Kivelson and Russell serving as editors. Most of the book is excellent, with the most notable exception being magnetic reconnection (GMR is not even mentioned!). Additional material is given by several texts I have used in my lectures. A few references not mentioned in the text have been included in the list.

Walter Heikkila

References

Alfvén, H., 1950. Cosmical Electrodynamics. Oxford University Press, London and New York.

Alfvén, H., 1986. Double layers and circuits in astrophysics. IEEE Trans. Plasma Sci. PS-14, 779–793.

Baker, D.N., Pulkkinen, T.I., Büchner, J., Klimas, A.J., 1999. Substorms: a global instability of the magnetosphere-ionosphere system. J. Geophys. Res. 104, 14601–14611.

Birkhoff, G., 1960. Hydrodynamics: A Study in logic Fact and Similitude, first ed. Princeton University Press, Princeton, NJ.

Birn, J., Drake, J.F., Shay, M.A., Rogers, B.N., Denton, R.E., Hesse, M., et al., 2001. Geospace environmental modeling (GEM) magnetic reconnection challenge. J. Geophys. Res. 106, 3715.

Donovan, E., 2006. Preface. In: Syrjäsuo, M., Donovan, E. (Eds.), Proceedings of the Eighth International Conference on Substorms. University of Calgary, Alberta, CA, Institute for Space Research, p. iii.

Heikkila, W.J., 1997. Comment on "The alternative paradigm for magnetospheric physics" by E. N. Parker. J. Geophys. Res. 102 (A5), 9,651–9,656.

Heikkila, W.J., Pellinen, R.J., 1977. Localized induced electric field within the magnetotail. J. Geophys. Res. 82, 1610–1614.

Heikkila, W.J., Winningham, J.D., 1971. Penetration of magnetosheath plasma to low altitudes through the dayside magnetospheric cusps. J. Geophys. Res. 76, 883–891.

Heikkila, W.J., Smith, J.B., Tarstrup, J., Winningham, J.D., 1970. The Soft particle spectrometer in the ISIS-I satellite. Rev. Sci. Instrum. 41, 1393–1402.

Heikkila, W.J., Pellinen, R.J., Fälthammar, C.-G., Block, L.P., 1979. Potential and induction electric fields in the magnetosphere during auroras. Planet. Space Sci. 27, 1383.

Heikkila, W.J., 1984. Magnetospheric topology of fields and currents. In: Potemra, T.A. (Ed.), Magnetospheric Currents. AGU Geophysical Monograph 28. American Geophysical Union, Washington, DC, pp. 208–222.

Hines, C.O., 1963. The energization of plasma in the magnetosphere: hydromagnetic and particle-drift approaches. Planet. Space Sci. 10, 239.

Hughes, W.J., 1995. The magnetopause, magnetotail, and magnetic reconnection. In: Kivelson, M.G., Russell, C.T. (Eds.), Introduction to Space Physics. Cambridge University Press, New York, NY.

Kivelson, M.G., 1995. Physics of space plasmas (p. 27-55), and pulsations and magnetohydrodynamic waves (p. 330-353). In: Kivelson, M.G., Russell, C.T. (Eds.), Introduction to Space Physics. Cambridge University Press, New York, NY.

Kivelson, M.G., Russell, C.T., 1995. Introduction to Space Physics. Cambridge University Press, New York, NY.

Laitinen, T.V., 2007. Rekonnektio Maan magnetosfäärissä—Reconnection in Earth's Magnetosphere. Finnish Meteorological Institute, Ph.D. thesis.

Lemaire, J., Roth, M., 1978. Penetration of solar wind plasma elements into the magnetosphere. J. Atmos. Terr. Phys. 40 (3), 331–335.

Lundin, R., 1988. On the magnetospheric boundary layer and solar wind energy transfer into the magnetosphere. Space Sci. Rev. 48, 263–320.

Lundin, R., Evans, D.S., 1985. Boundary layer plasmas as a source for high-latitude, early afternoon, auroral arcs. Planet. Space Sci. 33, 1389–1406.

McDiarmid, I., Burrows, J., Budzinski, E., 1976. Particle properties in the day side cleft. J. Geophys. Res. 81 (1), 221–226.

Parker, E.N., 1996. The alternative paradigm for magnetospheric physics. J. Geophys. Res. 10, 10.587.

Phan, T.-D., Paschmann, G., Sonnerup, B.U.Ö., 1996. The low-latitude dayside magnetopause and boundary layer for high magnetic shear: 2. Occurrence of magnetic reconnection. J. Geophys. Res. 101 (A4), 7817–7828. https://doi.org/10.1029/95JA03751.

Richmond, A.D., 1985. Atmospheric physics: atmospheric electrodynamics. Science 228 (4699), 572–573.

Schindler, K., Hesse, M., Birn, J., 1988. General magnetic reconnection, parallel electric fields, and helicity. J. Geophys. Res. 93, 5547–5557. https://doi.org/10.1029/JA093iA06p05547.

Schmidt, G., 1960. Plasma motion across magnetic fields. Phys. Fluids 3, 961.

Schmidt, G., 1979. Physics of High Temperature Plasmas, second ed. Academic Press, New York, NY.

Sonnerup, B.U.Ö., Hasegawa, H., Paschmann, G., 2004. Anatomy of a flux transfer event seen by cluster. Geophys. Res. Lett. 31. https://doi.org/10.1029/2004GL020134.

Vasyliunas, V.M., 1998. Theoretical considerations on where a substorm begins. In: Kokubun, S., Kamide, Y. (Eds.), Substorms-4. Terra Science, Lake Hamana, Japan, pp. 9–14.

Preface to the second edition

My journey to coauthoring this edition began in the late 1990s, while I was a graduate student working for David Winningham at Southwest Research Institute (SwRI) in San Antonio. David was visited on occasion by his mentor and friend Walter Heikkila, and so I was introduced to my graduate advisor's advisor, my "grand-advisor." The topic of my PhD Thesis was the Earth's magnetospheric cusps, so I was impressed to be collaborating with the authors of Heikkila and Winningham (1971) who had discovered the region. On my first visit to Kiruna, Sweden in 1997, I learned that magnetospheric scientists living at high latitudes have a term for those of us who had never actually seen its most striking and beautiful interaction with the atmosphere. I was a practitioner of a purely abstract phenomenon they called "Texas Aurora." Thankfully, during that same visit I was able to see some excellent REAL aurora, an experience that really did change my perspective, despite remaining in Texas for most of my career.

Walter and I continued to collaborate on occasion during and after my time at SwRI, mostly by me supplying him with data and other figures for his publications, including for the first edition of this book. In October of 2017, almost exactly three years before the publication of this edition, I was at a meeting at UT Dallas and arranged for a brief visit with Walter at his home. We met for an hour or so, talking about our past work together, including my contributions to his book. I was surprised to learn that he was considering a second edition, and shocked to be asked to be the lead author for it. Upon my return to McMurry University, I immediately began working to secure a sabbatical for the project, the deadline of which was only days away at the time. The formal proposal to the publisher for the second edition was finalized in the summer of 2018, and work on the book began in earnest in the spring of 2019 with my sabbatical semester.

This book is not a traditional textbook, or a review of facts and theories for a particular idea. Rather it is a reflection of our 80+ combined years of research starting (for Walter) with the launch of Sputnik in 1957. There has been an amazing advance of experimental techniques in this time, especially in the use of multiple spacecraft, and this trend is likely to continue into the future. Multiple spacecraft offer investigators a way to separate temporal changes due to the changing dynamics of the magnetosphere with spatial changes due to the spacecraft traversing into different regions along its orbit.

This second edition has been improved and corrected throughout the text, with high-resolution full color figures now integrated in the chapters rather than being grouped at the back. It has been updated to include recent missions such as MMS and THEMIS, 3D reconnection, space weather, and ionospheric outflow and coupling. Chapter 4 on Magnetohydrodynamic Equations has been split off from Chapter 2 and expanded. There is also an entirely new Chapter 8 on the Inner Magnetosphere, featuring recent results from the Van Allen Probes (Radiation Belt Storm Probes). End of chapter problems have been added to assist in using this text in a classroom setting.

Despite the many updates to include the latest research in the field, plenty of fundamental problems remain to be solved. Vasyliunas (2015), introduced his chapter of a book on Magnetotails in the Solar System by saying, "There is as yet (to my knowledge) no predictive first principles theory [for the structure of the magnetotail], comparable to the Chapman-Ferraro model which is the basis for understanding the dayside magnetosphere." Phan et al. (2018) found that in the turbulent magnetosheath downstream of the bow shock, "Contrary to the standard model of reconnection, the thin reconnecting current sheet was not embedded in a wider ion-scale current layer and no ion jets were detected." While no theory of the magnetosphere will ever be complete, we hope that the techniques and interpretations offered in this text will contribute to the continued progress of the field.

Wayne Keith

References

Heikkila, W.J., Winningham, J.D., 1971. Penetration of magnetosheath plasma to low altitudes through the dayside magnetospheric cusps. J. Geophys. Res. 76, 883–891.

Phan, T.-D., Eastwood, J.P., Shay, M.A., Drake, J.F., Sonnerup, B.U.O., Fujimoto, M., Cassak, P.A., Oieroset, M., Burch, J.L., Torbert, R.B., Rager, A.C., Dorelli, J.C., Gerchman, D.J., Pollock, C., Pyakurel, P.S., Haggerty, C.C., Khotyaintsev, Y., Lavraud, B., Saito, Y., Oka, M., Ergun, R.E., Rentino, A., LeContel, O., Argall, M.R., Giles, B.L., Moore, T.E., Wilder, F.D., Strangeway, R.J., Russell, C.T., Lindqvist, P.A., Magnes, W., 2018. Electron magnetic reconnection without ion coupling in Earth's turbulent magnetosheath. Nature 557 (2), 202–206.

Vasyliunas, V.M., 2015. Magnetotail: unsolved fundamental problem of magnetospheric physics. In: Keiling, A., Jackman, C.M., Delamere, P.A. (Eds.), Magnetotails in the Solar System., https://doi.org/10.1002/9781118842324.ch1.

Acknowledgments

We are indebted to C.J. Kale for the beautiful auroral photograph on the cover; to Sheré Chamness, the artist of the auroral fox; and to Tom Eklund for the auroral photograph of the 911 Plasmoid. Eric Heikkila provided invaluable records, memories, and advice from the production of the first edition. Risto Pellinen, Mark Hairston, Phillip Anderson, Qiugang Zong, and Robin Coley provided new high-resolution figures.

The 2002 Magnetic Reconnection Meeting in Kiruna, Sweden, was a significant milestone on the journey to the first edition of this book, published in 2011. In attendance were Zuyin Pu, Yan Song, Tsugunobu, Rickard Lundin, Christian Jacquey, Tony Lui, Walter Heikkila, Michel Roth, and Masatoshi Yamauchi. The goal of the meeting was to confront and discuss all of the issues and progress in the understanding of magnetic reconnection.

The Sixth Plasma Physics of the Magnetosphere Conference in Pollenzo, Italy in 2019 was important in the development of the second edition. Especially useful were the discussions, ideas, and later correspondence with Gian Luca Delzanno, Melvyn Goldstein, Steve Fuselier, Efyhia Zesta, Seth Claudepierre, Roderick Heelis, and Geoff Reeves.

We would like to thank everyone at McMurry University and the University of Texas at Dallas who supported this work, especially Keith Waddle, Reference Librarian at McMurry, and Tikhon Bykov, Physics Chair at McMurry.

Thank you to our editors Marisa LaFleur, and later Andrea Dulberger at Elsevier for keeping us on track. Thanks also to project manager Kumar Anbazhagan, whose push for improved quality figures has greatly enhanced the look and usefulness of this edition, and copyrights coordinator Swapna Praveen for helping us navigate the huge number of figure permissions needed for a book of this kind.

Finally, we are eternally grateful for the support and encouragement of our families, extended families and friends during the years that went into creating this book. It is not a journey to be undertaken lightly, and we could not have done it without you.

1

Historical introduction

Chapter outline

Earth's Magnetosphere. https://doi.org/10.1016/B978-0-12-818160-7.00001-6

We have found it is of paramount importance that in order to progress we must recognize our ignorance and leave room for doubt. Scientific knowledge is a body of statements of varying degrees of certainty—some most unsure, some nearly sure, but none absolutely certain.

Richard Feynmann, Engineering & Science (December 1955)

1.1 Early history

According to Lapp legend, a fox could create the northern lights with a swish of his tail across the snow, as suggested by the Frontispiece. There is an element of truth to this: the swish would create an electric field **E**, a necessary part of the process.

The year 1600 saw the publication of De Magnete by William Gilbert, chief physician to Queen Elizabeth; he noted that the globe of the Earth is a magnet. Thus another field was introduced, the magnetic field **B**, also essential to the creation of the auroras.

Another great pioneer of geomagnetism, Edmund Halley, had the distinction to organize the first purely scientific expeditions, in 1698, to the North Atlantic Ocean and, in 1700, to the south Atlantic. These investigations led to geomagnetic charts necessary for accurate use of the marine compass.

In the 17th century, Galileo Galilei in Florence used the new invention of the telescope to study sunspots. Richard Carrington's sighting of a solar flare on September 1, 1859, was followed by a strong magnetic storm some 18h later; auroras were seen at low latitudes, even in Puerto Rico. Apparently, the information traveled at a finite velocity **v** from the Sun to Earth, anticipating the solar wind velocity.

The emerging discipline of solar-terrestrial physics was prompted by the discovery in 1722 by George Graham that the compass needle is always in motion. His discovery was confirmed by Anders Celsius in 1741, and by O. Hiorter in 1741, both Swedish. The latter discovered that the diurnal variation of the geomagnetic field was caused by the current systems **J** in the upper atmosphere. Furthermore, on April 5, 1741, Hiortor discovered that geomagnetic and auroral activities were correlated. The early 19th century saw a network of magnetometers, advanced recording forms of the compass. The eminent German mathematician C. F. Gauss was a leader in the analysis.

The First Polar Year 1882–83 was truly interdisciplinary; beyond the planned auroral, magnetic, and meteorological observations, it included botany, ethnology, geology, and zoology. The records, now available in digital format, offer the arctic environment in historical perspective as it existed prior to the present era of environmental change.

Solar-terrestrial physics began in earnest with campaigns using magnetometers and auroral photographs. The study of the relation between the aurora and magnetic disturbances in 1882–83 inspired the Danish physicist Adam Paulsen to suggest that field-aligned currents exist, a milestone in auroral research (Jørgensen and Rasmussen, 2006). The Norwegian physicist Birkeland (1908) in his third campaign in 1902–03 observed magnetic perturbations to be associated with active auroral forms; he also concluded that large electric currents $\delta\mathbf{J}_{\parallel}$ flow during disturbances in what he called *elementary storms*.

The qualification of delta is vital; it is a perturbation current connected with magnetospheric dynamics. Only a transient current will induce an electric field, as Faraday (1832) first discovered. Birkeland invented a laboratory experiment in a device he called a *terrella*, a dipole inside a model Earth, attempting to prove his theories. There was much opposition to this concept, notably by Sydney Chapman from England.

The auroral zone is a narrow band roughly 20–25 degrees around each magnetic pole. As a result, auroral research is a global endeavor, especially when time variations spanning hours or days are involved. At first, the research involved the upper atmosphere and ionosphere. The ionosonde (Breit and Tuve (1925, 1926) in United States; Appleton and Barnett (1925) in United Kingdom) was a major instrumental advance, anticipating the first radar by three British scientists, Sir Robert Watson-Watt, A. F. Wilkins, and H. E. Wimperis, along with Sir Henry Tizard in 1935 in World War II.

The Second Polar Year (1932–33) program studied how much observations in the polar regions could improve weather forecasts and help transport by air and sea. Forty-four nations participated, and a vast amount of data was collected. By most accounts, the privations of these two early operations were extreme, with the men spending less than 10% of their time on science, the rest of the time devoted to survival. A world data center was created under the organization that eventually came to be called the International Meteorological Organization. This was affected by the worldwide economic depression, but nevertheless it benefitted from developments such as radiosondes (meteorological balloons with instruments and radio transmitters) and ionospheric sounders. It was becoming clear that the ionospheric layers are produced by ultraviolet light emitted from the Sun. The data that Appleton and his colleagues in England were starting to amass follows the theory of Chapman and Ferraro (1931). The regular behavior of these layers was, however, disrupted during magnetic storms; energetic charged particles precipitate into the atmosphere after being guided there by the Earth's magnetic field. The resulting data are therefore much more complex. This trend reached an apex with the International Geophysical Year (IGY) 1957–58 and the International Magnetospheric Study (IMS) 1976–79. The coming of rockets in the late 1940s and satellites in the late 1950s opened the field enormously.

1.2 International Geophysical Year

The IGY of 1957–58 was an audacious plan launched by a small committee of prominent scientists; it was an organized campaign that would involve planes, ships, and rockets as well as ground-based instruments. Sullivan's (1961) thorough account of the IGY is called, appropriately, "Assault on the Unknown." Korsmo (2007) has discussed it in *Physics Today* (July 2007):

After World War II the challenge of mobilizing U.S. science in service to national security was handed to Vannevar Bush, president of the Carnegie Institution of Washington. Bush was now asked to lead ... the newly developed U.S. Research and Development Board (R&D) in 1947. The R&D Board consisted of a civilian chairman and two representatives from each service: the Army, the Navy, and the newly independent Air Force. The board reported to the Secretary of Defense. Its primary duties were to prepare an integrated military R&D program, coordinate R&D among the services, and allocate responsibilities for programs. The R&D Board conducted its work through committees, which formed panels and working groups. Each committee, panel, and working group comprised military and civilian members; together they provided hundreds of forums for civilian–military interactions that encompassed all the physical, medical, biological, and geophysical sciences.

The voluble and expansive executive secretary for the R&D Board, Lloyd Berkner, did much to enliven the discussions. ... Berkner wanted government to rely on scientists rather than the other way around. Although he did much of his consulting for the Truman administration, his politics of engagement—for example, using international scientific meetings to gather information about the state of knowledge in the Eastern bloc—probably found a more comfortable home with the Eisenhower administration's policies of negotiation and accommodation with the Soviet Union. ... Under Bush and Berkner, the R&D Board became an incubator. Every topic that would be studied during the IGY had a board panel or working group that had looked into it 5 to 10 years earlier. ...

The R&D Board could not operate without foreign expertise, although that collaboration did not extend to the Eastern bloc. Working with colleagues across the Atlantic, on the other hand, was seen as beneficial not only to the U.S. but also to the postwar reconstruction of Western Europe. In this spirit, Caltech invited British geophysicist Sydney Chapman to Pasadena in 1950. Caltech had organized a meeting, funded by the armed services, concerning the upper atmosphere.

Fae Korsmo, Physics Today (July 2007)

Some international societies preferred a worldwide study, rather than a polar. Chapman agreed, suggested renaming the program the International Geophysical Year. A special IGY committee known as CSAGI had the initials taken from the French name Comité Spécial de l'Année Géophysique Internationale.

The scientific program would include 26 countries and practically the whole of Earth, ocean, and atmospheric sciences.

The launch of Sputnik 1 on October 4, 1957, just when the working group on rockets and satellites was convened in Washington shocked the Western attendees. Two months after Sputnik, the U.S. attempted to launch a 2 kg object with a Vanguard rocket. The rocket exploded on the launch pad. Chapman, head of the IGY, summarized "Thus is settled the identity of the first winner in this grand cooperative race to enrich geophysical knowledge by means of earth satellites."

In other ways the cooperative mechanisms of the IGY continued. SCOSTEP, the Scientific Committee on Solar-Terrestrial Physics, was established in 1966. The Committee on Space Research, the Scientific Committee on Oceanic Research, and the Scientific Committee on Antarctic Research all emerged from the IGY as international coordinating bodies. Legacies of the IGY include the launch of the first artificial Earth-orbiting satellites, the Antarctic Treaty, the World Data Center system, the discovery of the Van Allen belts, and the monitoring of atmospheric carbon dioxide and glacial dynamics. The IGY also led to the establishment of Earth sciences programs in many developing countries. Good material is provided by Ratcliffe (1960), Odishaw (1964), and Hines et al. (1965).

1.3   International Magnetospheric Study

Now let us advance to the 1980s; the field of space physics was being formed, and rapidly. There were a multitude of observations from instruments on the ground and in space, and our interpretations of these data (although sometimes not entirely correct). The International Magnetospheric Study (IMS) in 1976–79 had a decisive influence, including launch of the spacecraft pair International Sun-Earth Explorer (ISEE) 1 and 2 in 1977. Several books describe the situation that then existed; these include especially Egeland et al. (1973) and Carovillano and Forbes (1983), but also other conference proceedings such as Nishida (1982), Potemra (1984), Kamide and Slavin (1986), Lui (1987), and Meng et al. (1991).

It became clear that the explanation of the aurora rested upon the magnetosphere, a name coined by Gold (1959). "The region above the ionosphere in which the magnetic field of the Earth has a dominant control over the motions of gas and fast-charged particles is known to extend out to a distance of the order of 10 Earth radii; it may appropriately be called the magnetosphere."

Fig. 1.1 shows several important features of the magnetosphere explored in this chapter. The actual history of space research began from the ground upwards, from the neutral upper atmosphere, to the auroras, to the ionosphere, to the radiation belts, and beyond. We will reverse the order, starting with the solar wind, following its motion outwards from the Sun to its encounter with the magnetosphere.

1.3.1 Importance of three dimensions

It is essential to use three dimensions as in Fig. 1.1, not two as was common before (see, e.g., Fig. 1.15); the 3D figure created by Heikkila was first published nearly four decades ago, and in slightly revised form 5 years later by the Finnish National

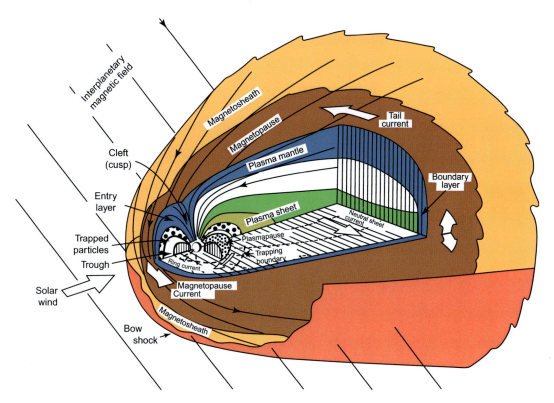

Fig. 1.1 The grand view of the magnetosphere within the bow shock. The solar wind first enters into the magnetosheath before its interaction with the magnetopause. Tailward flows of the higher density plasma mantle and low-latitude boundary layer are in contrast to the earthward flow of the low-density plasma sheet. The Van Allen belts are on lower *L*-shells around the plasmapause. From Heikkila, W.J., 1973, Aurora. EOS Trans. AGU, 54(8), 765.

Academy. Different regions are shown, involving quite different processes, requiring further definitions and explanations.

To the far left is the swift solar wind; solar wind plasma is responsible for the geomagnetic activity, the auroras. Its first interaction with the Earth is the Earth's bow shock. The shock is *collisionless* as the mean free path is much larger ($\sim$1 AU) than the shock and magnetopause dimensions ($<$1 Earth radius); the interaction is electric and magnetic in nature, not collisional. In the shock, the plasma is heated, and it slows down at the same time so that the speeds of various wave modes in the plasma (defined later) become higher. Now the plasma flow is subsonic; solar wind flow can go around the obstacle, the magnetosphere. Over 90% of it is diverted, the rest encounters the magnetopause; a still smaller fraction gets into the low-latitude boundary layer (LLBL). Some gets into the plasma sheet, and it is this small fraction ($<$1%) that is responsible for all auroral phenomena.

1.3.2   ISEE spacecraft

The IMS was an international program under the auspices of SCOSTEP in which a coordinated effort was made to understand magnetospheric processes. The active phase of the IMS, during which data were gathered, ran from 1976 to 1979 with the different data gathering efforts coming on line at various times during this period.

The Explorer-class spacecraft, ISEE 1 and 2 were launched in 1977 into Earth orbit; ISEE 3 was placed at the Lagrangian L_1 point—where the gravitational force of the Earth and the Sun balance—in 1978. Information is available at the National Space Science Data Center (NSSDC), National Aeronautics and Space Administration (NASA) permanent archive for space science mission data. The purposes of the mission were to:

- Investigate solar-terrestrial relationships at the outermost boundaries of the Earth's magnetosphere.
- Examine in detail the structure of the solar wind near the Earth and the shock wave that forms the interface between the solar wind and the Earth's magnetosphere.
- Investigate motions of, and mechanisms operating in, the plasma sheet.
- Continue the investigation of cosmic rays and solar flare effects in the interplanetary region near 1 AU.

The three spacecraft carried a number of complementary instruments for making measurements of plasmas, energetic particles, waves, and fields. The mother/daughter portion of the mission consisted of two spacecraft (ISEE 1 and ISEE 2) with station-keeping capability in the same highly eccentric geocentric orbit with an apogee of 23 R_E (Earth radii). During the course of the mission, the ISEE 1 and ISEE 2 orbit parameters underwent short-term and long-term variations due to solar and lunar perturbations. These two spacecraft maintained a small separation distance and made simultaneous coordinated measurements to permit separation of spatial from temporal irregularities in the near-Earth solar wind, the bow shock, and inside the magnetosphere. By maneuvering ISEE 2, the interspacecraft separation as measured near the Earth's bow shock was allowed to vary between 10 and 5000 km; its value is accurately known as a function of time and orbital position. The spacecraft were spin stabilized, with the spin vectors maintained nominally within 1 degree of perpendicular to the ecliptic plane, pointing north. The spin rates were nominally 19.75 rpm for ISEE 1 and 19.8 rpm for ISEE 2, so that there was a slow differential rotation between the two spacecraft. ISEE 1's body-mounted solar array provided approximately 175 watts initially and 131 watts after 3 years, at 28 V during normal operation. The ISEE 1 data rate was 4096 bits per second most of the time and 16,384 bps during one orbit out of every five (with some exceptions). Both ISEE 1 and ISEE 2 reentered the Earth's atmosphere during orbit 1518 on September 26, 1987. Seventeen of twenty-one on-board experiments were operational at the end.

Initially, ISEE 3 was kept at the Lagrangian point between the Earth and the Sun, 0.01 AU from the Earth for observations of the solar wind. To avoid telemetry interference with the Sun, ISEE 3 was kept slightly ($\sim$4°) off the Earth-Sun line (and moving slightly above and below the ecliptic). As seen from the Earth, it appeared to be circling the Sun, giving rise to the term "halo orbit." ISEE 3 was moved from this orbit and made transits through the Earth's geomagnetic tail by a series of complicated maneuvers from September 1982 until December 22, 1983 as shown in Fig. 1.2.

ISEE 3 made a very close swing-by of the Moon to begin its cometary mission (and was given the new name ICE, for International Cometary Explorer) to the comet 21P/Giacobini-Zinner. It flew through the plasma tail of G-Z on September 11, 1985, and was upstream in the solar wind from comet 1P/Halley when several other spacecraft flew close to Halley in March 1986. ICE remains in the heliospheric orbit at approximately 1 AU. Termination of ICE/ISEE 3 operations was authorized on May 5, 1997. Contact with ISEE 3 was reestablished briefly by a private group in 2014, but it is unknown if any further contacts are possible or will be attempted.

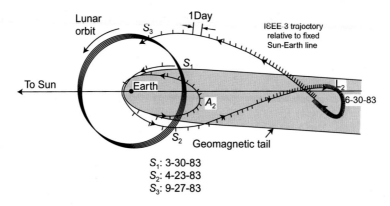

Fig. 1.2 The ISEE-3 orbit in 1983, approaching Earth at S_3 on its way to cometary encounters. *Newton would have been amazed at what he had wrought.* From von Rosenvinge, T., 1983. New observations from ISEE-3. EOS Trans. AGU 64(46), 929.

The ISEE program was an international cooperative program between NASA and European Space Agency (ESA) to study the interaction of the solar wind with the Earth's magnetosphere. For instrument descriptions written by the investigators, see Anderson et al. (1978).

1.4 Electric and magnetic fields in space

If an electric field **E** and a magnetic induction **B** act on a particle with charge q and velocity **v**, the particle experiences a force $\mathbf{F}_L$, the Lorentz force:

$$\mathbf{F}_L = q\mathbf{E} + q\mathbf{v} \times \mathbf{B} \text{ (Lorentz force)} \tag{1.1}$$

E is in volts per meter, **B** in tesla (T), and q in coulombs. This is expressed in the International System (SI) units expressing mass, length, and time in kilograms, meters, and seconds, respectively; this equation *defines* both **E** and **B** in terms of the other variables. Note that we take **v** for the velocity of an individual particle, reserving **u** for a fluid plasma velocity, to be used shortly. From Newton's laws, the rate of change of momentum ($m\mathbf{v}$) is given by

$$m\frac{d\mathbf{v}}{dt} = q\mathbf{E} + q\mathbf{v} \times \mathbf{B} + \mathbf{F}_g \tag{1.2}$$

where m is the mass of the particle, and $\mathbf{F}_g$ represents nonelectromagnetic forces such as gravity. The velocity has been assumed to be sufficiently small that relativistic corrections need not be taken into account.

1.4.1 Effect of an electric field

The effect of an electric field is easy to understand, following Coulomb (1736–1806): like charges repel, unlike charges attract. Positive charge is accelerated by the electric field in the same sense as **E** (Faraday, 1791–1867), while negative charge is accelerated in the opposite sense. Both gain energy, very much like a stone falling under gravity. In fact, this is the only way to change the energy of charged particles, by $\mathbf{F} = q\mathbf{E}$.

However, there is a serious problem here: the *real* electromagnetic electric field has two sources, not one, explored in this book:

$$\mathbf{E} = \mathbf{E}^{\text{es}} + \mathbf{E}^{\text{ind}} = -\nabla\phi - \frac{\partial\mathbf{A}}{\partial t} \tag{1.3}$$

1.4.2 Effect of a magnetic field

The magnetic field gets complicated, often counterintuitive. In a uniform magnetic field ($\mathbf{E}=0$, $\mathbf{F}_g=0$), a charged particle moves in a circle. The circular motion is in the left-hand sense for a positive particle, and in the right-hand sense for negative with an angular frequency (the cyclotron frequency or the gyrofrequency) ω_c:

$$\omega_c = 2\pi f_c = \frac{|q|B}{m} \tag{1.4}$$

Denoting $\mathbf{v} = \mathbf{v}_\| + \mathbf{v}_\perp$ (to **B**), the cyclotron, or Larmor radius, r_c is

$$r_c = \frac{v_\perp}{\omega_c} = \frac{mv_\perp}{|q|B} \tag{1.5}$$

As the magnetic force is perpendicular to the velocity, no work is done; the velocity remains constant. We can see that very simply:

$$m\mathbf{v} \cdot \frac{d\mathbf{v}}{dt} = \frac{d}{dt}\left(\frac{1}{2}mv^2\right) = q\mathbf{v} \cdot \mathbf{v} \times \mathbf{B} = 0 \tag{1.6}$$

The expression vanishes using a vector identity. A listing of the values of the gyroradius as a function of energy for protons and electrons in the dipole equator at different distances from the Earth is given in Kivelson and Russell (1995) Table 10.1. In that table, and elsewhere in their book and in this book, the particle energy is given in electron volts (eV). This is an energy unit that represents the energy that a particle carrying a charge e gains or loses in falling through a potential drop of 1 V: $1\,\text{eV} = 1.6022 \times 10^{-19}$ joules (J). Typical velocities and energies of interest to space plasma physics are presented in the Nomogram of Fig. 1.3 (a practicable nomogram).

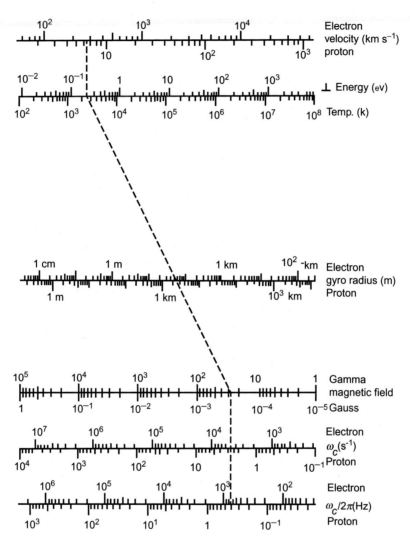

Fig. 1.3 Nomogram for frequencies and distances of interest for a space plasma. From Fälthammar, C.-G., 1973. Motions of charged particles in the magnetosphere. In: Egeland, H., Omholt (Eds.), Cosmical Geophysics. Oslo, Universitetsforlaget, Norway, p. 125.

1.4.3 Nonelectromagnetic forces

It may be that the gravitational forces will be important, as, for example, in planetary ionospheres and the solar corona. Eq. (1.1) shows that the magnetic field acts to change the motion of a charged particle only in directions perpendicular to that motion. Let us add that general force **F**:

$$\frac{d\mathbf{v}}{dt} = \frac{q}{m}(\mathbf{v} \times \mathbf{B}) + \frac{\mathbf{F}}{m} \qquad (1.7)$$

This splits up into two component equations

$$\frac{d\mathbf{v}_{\parallel}}{dt} = \frac{\mathbf{F}_{\parallel}}{m} \qquad (1.8)$$

$$\frac{d\mathbf{v}_{\perp}}{dt} = \frac{q}{m}(\mathbf{v}_{\perp} \times \mathbf{B}) + \frac{\mathbf{F}_{\perp}}{m} \qquad (1.9)$$

This can be solved by introducing a drift velocity $\mathbf{w}^D$ plus a time-dependent gyrating velocity $\mathbf{u}$, so that

$$\mathbf{v}_{\perp} = \mathbf{w}^D + \mathbf{u} \qquad (1.10)$$

$$\mathbf{w}^D = \frac{1}{q}\frac{\mathbf{F}_{\perp} \times \mathbf{B}}{B^2} \qquad (1.11)$$

Negative and positive particles go in opposite directions so there is an electrical current involved, a restoring force against gravity.

1.4.4 Combined fields: *E*-cross-*B* drift

If there is an electric field $\mathbf{E}$, it will accelerate both the electrons and ions from their initial velocity (which can be assumed zero). With 2D, $\mathbf{E} = \mathbf{E}_{\perp}$, as is shown by Fig. 1.4. As the particles gain energy they experience an increasing velocity $\mathbf{v} = \mathbf{v}_{\perp}$ and will now be affected by the magnetic field. We can use the previous result by taking the force $\mathbf{F} = q\mathbf{E}$ in Eq. (1.11). Then

$$\mathbf{v}^E = \frac{\mathbf{E} \times \mathbf{B}}{B^2} \qquad (1.12)$$

The drift is independent of q and m; there is no steady-state current. This electric drift is very important for any plasma, as the other drifts (magnetic gradient and curvature drift, discussed later) are generally smaller. However, the electric drift is sensitive to the reference frame as discussed in the next section.

1.4.5 Particle motion in a magnetic field

A realistic magnetic field such as in the magnetosphere has gradients across and along magnetic lines. On time scales of gyro-periods, the particle spirals about the magnetic field lines (see illustration of gyromotion in Fig. 1.5). If there is a field-aligned gradient of the field strength, the component of velocity parallel to the field decreases as the particle moves into regions of

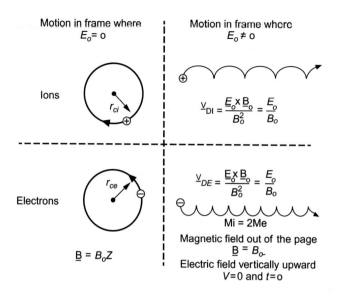

Fig. 1.4 Schematic showing the motions of ions (charge e) and electrons (charge $-e$) with $m_i/m_E = 2$ in a uniform magnetic field **B** in the presence of an electric field **E** perpendicular to **B**. The left diagram shows the spiraling motion with no electric field, while on the right represents the motion in a plane perpendicular to the magnetic field. As the particle gains velocity (due to the presence of the electric field), its radius of gyration increases. From Heikkila and Brown.

Particle motion in a dipolar magnetic field

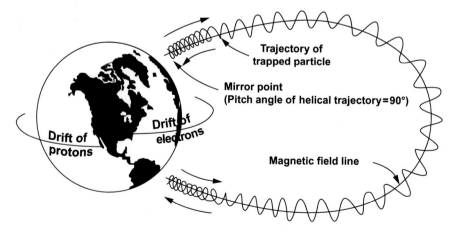

Fig. 1.5 Particle motion in a magnetic field. The fastest is the gyromotion (fraction of seconds), followed by the bounce motion between hemispheres (seconds to minutes) due to magnetic mirroring. The longitudinal drift can be hours long depending on the energy. From Heikkila and Brown.

increasing field magnitude, although the total velocity is conserved. Eventually, the parallel velocity reverses. The reflection of the parallel motion is called "magnetic mirroring." Motion between mirror points is called bounce motion. As the particle bounces, it also drifts about the source of the field, tracing out a drift shell.

1.4.6 Magnetic moment: First invariant

In the foregoing discussion the magnetic field was assumed to be uniform; this is not realistic. The Earth's magnetic field, to lowest order, is a dipole field that changes in direction and magnitude, both along and across the field lines. Charged particles will drift with a nonuniform magnetic field due to their inertia, for example, to produce a ring current around the Earth. The real magnetosphere field has higher-order terms.

If there is an electric field present, the particles will gain or lose kinetic energy. A remarkable feature of the motion of charged particles in collisionless plasmas is that even though the energy changes, there is a quantity that will remain constant if the field changes slowly enough. By "slowly enough" we mean that the field changes encountered by the particle within a single gyration orbit will be small compared with the initial field. If this condition is satisfied, then the particle's "magnetic moment" will remain constant:

$$\mu = \frac{\frac{1}{2}mv_\perp^2}{B} \qquad (1.13)$$

Note that if μ remains constant as the particle moves across the field into regions of different field magnitudes, some acceleration is required.

The quantity is also called the first adiabatic invariant. "Adiabatic" refers to the requirement that it may not remain invariant or unchanged unless the parameters of the system, such as its field strength and direction, change slowly. The definition of an adiabatic process is one for which no heat is gained or lost. A slowly changing pendulum arm will cause a slowly changing period such as to keep the energy constant.

The force on a charged particle due to an inhomogeneous magnetic field is given by

$$\mathbf{F} = \mu \nabla B \qquad (1.14)$$

We can use this expression as the force $\mathbf{F}$ in Eq. (1.11). (The short article by Fälthammar (1973) is highly recommended.)

1.4.7 Bounce motion: Second invariant

On time scales of gyroperiods, the particle spirals about the magnetic field between the northern and southern hemispheres as can be seen by the illustration in Fig. 1.5. There is always a field-aligned gradient (and thus a force, by Eq. (1.14)) of the field strength in a realistic magnetic field (including a simple dipole and the geomagnetic field), leading to convergence or divergence. The component of velocity parallel to the field decreases as the particle moves into regions of increasing field magnitude, although the total velocity is conserved if $\mathbf{E} = 0$. Eventually, the parallel velocity reverses at the magnetic mirror point. Motion between mirror points is called bounce motion, with a second adiabatic invariant being conserved. The angle between the magnetic field vector $\mathbf{B}$ and the particle velocity $\mathbf{v}$ is called the pitch angle and is defined in terms of the velocity components as:

$$\alpha = \tan^{-1}\left(\frac{v_{\perp}}{v_{\parallel}}\right) \tag{1.15}$$

The set of particles with equatorial pitch angles too small to mirror before reaching the Earth's atmosphere will be lost and are referred to as being in the "loss cone." As a direct result of the bounce motion, energetic electrons above 10 keV can be used to trace closed field lines between the northern and southern hemispheres in a matter of seconds. The loss cones would become empty (north and south) but other particles can bounce between mirror points. Ions would do likewise, but taking considerably longer.

As the particle bounces, it also drifts about the source of the field, tracing out a drift shell conserving a third adiabatic invariant (if everything else stays constant).

1.4.8 Gradient $\mathbf{B}$ drift

A gradient in the field strength in the direction perpendicular to $\mathbf{B}$ will produce a drift velocity transverse to $\mathbf{B}$ and $\nabla \mathbf{B}$. If a particle gyrates in a field whose strength changes from one side of its gyration orbit to the other, the instantaneous radius of curvature of the orbit will become alternately smaller and larger. Averaged over several gyrations, the particle will drift, this time in a direction perpendicular to both the magnetic field and the direction in which the strength of the field changes. The drift velocity, which we shall call $\mathbf{v}_g$ (for gradient drift velocity, see Chen, 1984, p. 27), is given by

$$\mathbf{v}_g = \frac{mv^2 \mathbf{B} \times \nabla B}{2qB^3} \tag{1.16}$$

Eq. (1.16) shows that the drift velocities produced by a gradient in the magnetic field will depend on the sign of the particle's charge, and so this drift will cause electric currents to flow across the magnetic field.

1.4.9   Curvature drift

The curvature of magnetic field lines will introduce additional drifts because, as the particles move along the field direction, they will experience a centrifugal acceleration.

$$\mathbf{F}_c = -\frac{mv_\parallel^2}{R}\hat{\mathbf{n}}_1 \qquad (1.17)$$

where R denotes the local radius of curvature of the magnetic field line and $v_\parallel$ is the particle instantaneous longitudinal speed (Chen, 1984, p. 29).

Fig. 1.6 shows another view indicating that the particle overshoots the field line moving in the $\mathbf{E} \times \mathbf{B}$ frame. The curvature drift is

$$\mathbf{v}_c = \frac{\mathbf{F}_c \times \mathbf{B}}{qB^2} = -\frac{mv_\parallel^2}{qB^2}\frac{\mathbf{R} \times \mathbf{B}}{R^2} \qquad (1.18)$$

To express the unit vector $\hat{\mathbf{n}}_1$ in terms of the unit vector $\hat{\mathbf{B}}$ along the magnetic field line, see the article by Fälthammar (1973). This is further treated in great detail by Shen et al. (2003), as discussed in Chapter 10.

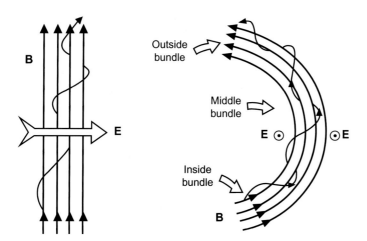

Fig. 1.6 Particles that cause the auroras have small pitch angles in the plasma sheet where **B** is very small, so their energy gain is by curvature drift. The particles overshoot the field line in the $\mathbf{E} \times \mathbf{B}$ drifting frame as shown here, gaining energy by the cross-tail electric drift if that is duskward. From Heikkila and Brown.

1.4.10 Collections of particles

Up to this point, we have described the motions of individual particles. Plasmas consist of collections of particles, some with positive and some with negative charges, with various energies. Let us next consider how to deal with collections of particles, particularly large numbers of particles that have different velocities. We describe the properties of a large number of particles by saying how many of them there are per unit volume of a 6D space that is called phase space. The phase-space density $f(\mathbf{r}, \mathbf{v}, t)$ is also called the single-particle distribution function. It provides a count of the number of particles per unit volume of ordinary, or *configuration*, space that also fall into a particular unit volume in a *velocity* space whose axes are labeled v_x, v_y, and v_z.

"Volume" in phase space needs to be specified. Interest focuses on a differential element of *phase-space* volume, which can be denoted as

$$d\mathbf{v}d\mathbf{r} = dv_x dv_y dv_z dx dy dz \qquad (1.19)$$

at phase position $(\mathbf{v}, \mathbf{r})$ where the number of particles in the differential phase-space volume

$$= f(\mathbf{r}, \mathbf{v}, t)d\mathbf{v}d\mathbf{r} \qquad (1.20)$$

The following discussion makes a distinction between the terms *volume* (which refers to configuration space only) and *phase-space volume* (which refers to a 6D space).

If there are particles of different masses or charges, it may be necessary to keep track of each type (designated by a subscript) separately. The symbol that is used for this density is $f_s(\mathbf{r}, \mathbf{v}, t)$, the phase-space density for species s particles. In order to determine how many particles of type s there are per unit volume of ordinary space, one must count all the particles in the phase-space volume, regardless of their velocities; that is, we integrate over all possible velocities

$$n_s(\mathbf{r}, t) = \int d\mathbf{v} f_s(\mathbf{r}, \mathbf{v}, t) \qquad (1.21)$$

Here, $n_s(\mathbf{r}, t)$ is the number density (number per unit volume) of type s particles, and it is related to the mass density ρ_s by $\rho_s(\mathbf{r}, \mathbf{v}, t)/m_s$. The number density is the zeroth-order moment of the distribution. Other quantities of interest are higher-order moments of the distribution obtained by integrating powers of the velocity and closely related quantities over the phase-space distribution function. For example, it is often important to determine the average velocity $\mathbf{u}_s$ of a collection of particles. The first moment of the distribution

$$\mathbf{u}_s(\mathbf{r}, t) = \int d\mathbf{v}\mathbf{v}f_s(\mathbf{r}, \mathbf{v}, t)/d\mathbf{v}f_s(\mathbf{r}, \mathbf{v}, t)a \qquad (1.22)$$

If this flow or "bulk" velocity is different for ions and electrons, then electric currents will flow in the system. Notice the notation that has been used to distinguish between the velocity of an individual particle, v, and the average velocity of a collection of particles, u_s. Often the velocities of the individual particles are orders of magnitude larger than the flow velocity, but if the velocities are random, they contribute little to the average velocity. More discussion is presented by Kivelson and Russell (1995).

1.4.11 Potential functions

ϕ, the scalar potential, and $\mathbf{A}$, the vector potential are often useful to express the fields $\mathbf{E}$ and $\mathbf{B}$ in terms of potential functions. The only source of a magnetic field is a current $\mathbf{J}$ by Ampere's law; therefore, to study changes in the magnetic field we should consider perturbation electric currents $\delta\mathbf{J}$, the source of $\delta\mathbf{B}$. We define a vector potential $\mathbf{A}$ with the requirement that:

$$\mathbf{B} = \nabla \times \mathbf{A} \qquad (1.23)$$

The electric field has two sources: charge separation and induction.

$$\mathbf{E} = \mathbf{E}^{es} + \mathbf{E}^{ind} = -\nabla\phi - \frac{\partial \mathbf{A}}{\partial t} \qquad (1.24)$$

By components:

$$\mathbf{E} = \mathbf{E}^{es}_{\perp} + \mathbf{E}^{es}_{\parallel} + \mathbf{E}^{ind}_{\perp} + \mathbf{E}^{ind}_{\parallel} \qquad (1.25)$$

This last equation is very important; the four terms on the right are associated with different physics and can have different signs. Charge separation is easy to understand since it has the same form as gravity, which we know well. Induction is somewhat more difficult as it depends explicitly on time dependence. For this reason alone it is better to use the $\mathbf{E},\mathbf{J}$ paradigm rather than the $\mathbf{B},\mathbf{v}$ (notwithstanding Parker (1996)).

With the $\mathbf{B},\mathbf{v}$ paradigm there is only one electric field, the convection electric field:

$$\mathbf{E}_{conv} = -\mathbf{v} \times \mathbf{B} + \text{other terms} \qquad (1.26)$$

Its shortcomings are discussed later in the book, especially Sections 5.5.1 and 6.8.1.

1.5 Reference frames and frozen fields

Each particle gyrates about the magnetic field lines while drifting at a constant velocity. This brings in a new concept: the magnetic field lines are drifting at the same velocity $\mathbf{v}^E$, and the particles are gyrating about the drifting field lines.

Maxwell's equations are covariant under the Lorentz transformation, as will be noted in Chapter 3. Let the velocity of the new frame be $\mathbf{v}$; the assumption $v \ll c$ is generally valid for plasma velocities in space plasmas, and the equations simplify to

$$\mathbf{E}' = \mathbf{E} + \mathbf{v} \times \mathbf{B} \tag{1.27}$$

$$\mathbf{B}' = \mathbf{B} \tag{1.28}$$

We can think of two kinds of electric field. $\mathbf{E}$ is the applied electric field in our reference frame, and $\mathbf{v} \times \mathbf{B}$ is the motional electric field due to motion across $\mathbf{B}$. The current $\mathbf{J}$ is given by Ohm's law, following the engineers (as we discuss in the following chapters). Thus the total electric field times the conductivity σ gives a relation for the current $\mathbf{J}$

$$\mathbf{J} = \sigma(\mathbf{E} + \mathbf{v} \times \mathbf{B}) \tag{1.29}$$

In a plasma the conductivity may be so high that $\mathbf{J}/\sigma$ is negligible, so

$$\mathbf{E} + \mathbf{v} \times \mathbf{B} = 0 \tag{1.30}$$

This is Ohm's law for ideal magnetohydrodynamic (MHD) plasma. In the rest frame of the drift, the electric field vanishes because of the high conductivity, and $\mathbf{E}'$ is zero.

We can use this to derive the concept of frozen-in magnetic field; the magnetic field lines are also drifting at the velocity $\mathbf{v}_B$:

$$\mathbf{v}_B = \mathbf{E} \times \mathbf{B}/B^2 \tag{1.31}$$

Eq. (1.30) becomes

$$\mathbf{E} + \frac{\mathbf{E} \times \mathbf{B}}{B^2} \times \mathbf{B} = \mathbf{E} - \mathbf{B} \times \frac{\mathbf{E} \times \mathbf{B}}{B^2} \tag{1.32}$$

$$= \mathbf{E} - \mathbf{E}\frac{\mathbf{B} \cdot \mathbf{B}}{B^2} + \frac{\mathbf{B}(\mathbf{E} \cdot \mathbf{B})}{B^2} \tag{1.33}$$

with the use of a vector identity in the last step. To be equal to 0 (by Eq. 1.30) we see that $\mathbf{E} \cdot \mathbf{B} = 0$; $\mathbf{E}$ must be entirely transverse to $\mathbf{B}$. No parallel component of the electric field is allowed in ideal MHD:

$$\mathbf{E} \equiv \mathbf{E}_\perp \tag{1.34}$$

General magnetic reconnection (GMR) of Schindler et al. (1988), treated in Sections 5.3 and 6.8, implies the violation of ideal Ohm's law based on "nonideal" plasma processes.

$$\mathbf{E} + \mathbf{v} \times \mathbf{B} = \mathbf{R} \qquad (1.35)$$

If $\mathbf{R}$ vanishes everywhere, the plasma is ideal; if it does not, the plasma is nonideal. The function $\mathbf{R}$ is complex. The physical processes that lead to a finite vector $\mathbf{R}$ can be very different, with collisions, fluctuations, and particle inertia being the major causes.

1.6 Coronal expansion

Now let us get to the physics of space plasmas. The gravitational potential decreases outward from the Sun as $1/r$. The temperature cannot decrease so fast, hence, sufficiently far from the Sun, the gas is free to escape. Consequently, we can turn away from the hydrostatic barometric equation (very common for analysis of the upper atmosphere and ionosphere) to the hydrodynamic equation.

Parker (1964) did that, creating a theory for the supersonic *solar wind*. He recognized that the physics of the expansion of the solar corona could be understood from MHD principles (Alfvén, 1942, 1950). The weak magnetic fields, and their associated instabilities, would maintain the pressure in an approximately isotropic state. The effect is much the same as a high collision rate, so that the usual hydrodynamic equations would be a valid approximation for large-scale motion of the gas. A magnetic pressure can be defined as

$$p_B = \frac{B^2}{2\mu_o} \qquad (1.36)$$

and the ratio between the plasma pressure and magnetic pressure is referred to as the beta value.

$$\beta = \frac{p}{p_B} \qquad (1.37)$$

Since this magnetohydrodynamic treatment of the plasma requires an imbedded magnetic field, wave propagation is more complex than a simple compression wave as in hydrodynamics. In fact, there are three distinct modes of MHD wave propagation. With a fast mode wave, the variations in particle pressure and magnetic pressure are in phase, this mode is the closest to a conventional "sound speed," and is typically implied when flow is said to be "supersonic." A slow mode wave is similar, except that the particle and magnetic pressures are out of phase with each other. The intermediate mode wave, as its name implies, has a speed between the fast and slow modes; however, it does not involve pressure variations at all. This wave, also called an Alfvén wave, is a shear wave that only causes the magnetic field lines to bend. Since the field lines do not get closer together, Alfvén waves cannot propagate perpendicular to the field.

Let us consider the simple case of a solar corona with spherical symmetry about the center of the Sun, composed solely of ionized hydrogen. For steady-state conditions the outward velocity $v(r)$, the numbers of ions/cm^3 $N(r)$, and the temperature $T(r)$ are related by the equation

$$v\frac{dv}{dr} = -\frac{1}{NM}\frac{d}{dr}2NkT - \frac{GM_\odot}{r^2} \tag{1.38}$$

where M is the mass of a hydrogen atom. Here the acceleration term on the left is equal to the thermal pressure and solar gravitational attraction on the right. Conservation of matter requires that

$$N(r)v(r)r^2 = N_0 v_0 a^2 \tag{1.39}$$

where $r=a$ is the reference level, taken to lie in the lower corona where the initial parameters are indicated by N_0 and v_0 at $r=a$.

The gravitational field of the Sun plays the same role in coronal expansion as the throat in a de Laval nozzle. In both cases, the gas attains supersonic velocity passing from a high pressure into a vacuum, due to constriction in the flow. Eq. (1.39) is the equation for conservation of mass in the solar corona. In this case, $N(r)$ decreases rapidly with height in the corona because of the strong solar gravitational field. Consequently, $v(r)$ must increase rapidly with height to maintain the net flow of matter, with the result that the velocity reaches the speed of sound at a distance of a few solar radii. Beyond that point, the velocity increases further as a consequence of expansion into the vacuum of interstellar space, just as in the nozzle shown in Fig. 1.7.

The hydrodynamic equations for the corona possess a critical point across which continues the solution of physical interest for expansion into a vacuum. The critical point may be demonstrated by using Eq. (1.39) to eliminate $N(r)$ from Eq. (1.38):

$$\frac{dv}{dr}\left(v - \frac{2kT}{Mv}\right) = -\frac{2kr^2}{M}\frac{d}{dr}\left(\frac{T}{r^2}\right) - \frac{GM_\odot}{r^2} \tag{1.40}$$

The critical point (r_c, v_c) is the point where the right-hand side of this equation and the coefficient of dv/dr both vanish. Equating

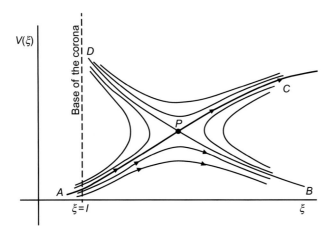

Fig. 1.7 Sketch of the topology of the solutions $v(r)$ of the momentum equation when $T(r)$ declines less rapidly than $1/r$. The critical point is labeled P. The *heavy line* represents the solution of physical interest for expansion of corona into vacuum. From Parker, E.N., 1964. In: Odishaw (Ed.), Research in Geophysics. MIT Press, Cambridge, MA, p. 108.

the right-hand side to zero leads to an expression that yields r_c as soon as the function $T(r)$ is specified. The velocity v_c at the critical point follows from the coefficient of dv/dr as $v_c^2 = 2kT(r_c)/M$. The critical point corresponds to the point at which the fluid velocity becomes supersonic. For expansion through a nozzle, the critical point lies in or near the throat. For the expanding corona it defines the position of the fictitious "gravitational throat." This shows that the corona not only expands, but that it expands with supersonic velocities of several hundred km/s.

1.7 Solar wind

The Sun loses mass in three different ways: the Solar Wind, Coronal Mass Ejections (CME), and Solar Energetic Particles (SEP) (see, e.g., Gopalswamy et al., 2006). These phenomena are signatures of solar variability from the matter point of view; electromagnetic radiation in the form of quiescent and flare emissions represents the other major variability. Fast solar wind originates from the open field regions on the Sun known as coronal holes, while CMEs originate from the closed field regions such as active regions and filament regions. The interplanetary plasma is the solar wind, and the propagation of CMEs through this plasma represents interaction between them (Fig. 1.8). CMEs often attain super-Alfvénic speeds in the magnetized coronal and

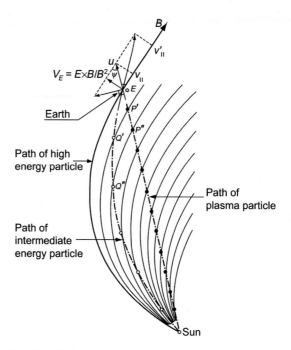

Fig. 1.8 Drift motion of charged particles of different energies in an ideal spiral magnetic field. The kinetic energy density in the solar wind bulk motion is much larger than the magnetic energy density so it determines the frozen-in magnetic field. From Fälthammar, C.-G., 1973. Motions of charged particles in the magnetosphere. In: Egeland, H., Omholt (Eds.), Cosmical Geophysics. Oslo, Universitetsforlaget, Norway, p. 100.

interplanetary plasmas, hence are bound to drive fast mode shocks. Almost all CMEs are associated with flares, which accelerate electrons and ions that flow toward the Sun and away from the Sun producing electromagnetic radiation at various wavelengths and injecting flare-heated plasmas into the CMEs.

The solar wind is a mixture of electrons and ions of solar origin, mostly protons and a few percent of helium doubly charged ions (alpha particles); the important parameters are listed by Hundhausen (1995). His Table 1 lists properties of the solar wind, Table 2 gives flux densities near the orbit of Earth, and Table 3 derives properties near Earth. Solar wind average velocity is $\mathbf{u} \sim 400\,\text{km/s}$, taking about 3–4 days to reach from the Sun to the Earth; occasionally it more than doubles. The solar wind is hot, tenuous, and fast by terrestrial standards.

Pressure in an ionized gas is

$$p_{\text{gas}} = nk(T_i + T_e) \tag{1.41}$$

where k is Boltzmann's constant and $T_{i,e}$ are the ion and electron temperatures. Thus

$$p_{\mathrm{gas}} = 3 \times 10^{-10} \mathrm{dyn} \times \mathrm{cm}^{-2} = 30 \, \text{pico pascals (pPa)} \qquad (1.42)$$

Sound waves travel with a speed c_S where γ is the ratio of specific heats at constant pressure and constant volume

$$c_S = \left\{ \frac{\gamma p}{\rho} \right\}^{\frac{1}{2}} \qquad (1.43)$$

and where γ is the ratio of specific heats at constant pressure and constant volume, and c_s is the speed of sound. Using $\gamma = 5/3$ for an ionized hydrogen gas and temperatures from Table 1 in Hundhausen (1995), gives

$$c_S \approx 60 \, \mathrm{km \, s}^{-1} \qquad (1.44)$$

The Alfvén velocity is of the order of $v_A \sim 100 \, \mathrm{km/s}$.

1.7.1 Interplanetary magnetic field

The expansion of the solar corona on a quiet day carries with it the general solar magnetic field of $1 \, T$ as a consequence of the high electrical conductivity of the coronal gas. If the Sun did not rotate, the result of the radial expansion of the corona would be to stretch out the magnetic lines of force in the radial direction, giving a magnetic field that declines outward as

$$B_r(r) = B_0 \left(\frac{a}{r} \right)^2 \qquad (1.45)$$

The sense of the radial field in any direction from the Sun would be determined by the sense of the field B_0 at the base of the corona in that direction. The rotation of the Sun, with a period of about 25 days, has little or no effect on the radial motion of the coronal gas, but it introduces a pronounced spiral in the magnetic field configuration.

The kinetic energy density in the solar wind bulk motion is much larger than the magnetic energy density. Therefore the frozen-in magnetic field is carried by the plasma without exerting any appreciable influence on the plasma flow. Combined with the rotational motion of the Sun, the radial flow it creates is a spiral structure in the IMF. In an idealized symmetric and steady state, the magnetic field lines would be shaped into Archimedes spirals as shown in Fig. 1.8. The spiral angle is sometimes referred to as the "garden hose angle." At the orbit of the Earth, the value of the angle is about 45 degrees. There are large spatial and temporal variations, and the instantaneous magnetic field can have any direction.

The Solar Wind conference series was started over 40 years ago after the first in-situ measurements of the solar wind plasma made by the Mariner 2 spacecraft. These conferences discuss the most recent findings on the Solar Corona, the origin and acceleration of the Solar Wind, the dynamical interactions throughout the heliosphere, and the interstellar medium. The 15th conference was held in Belgium in June 2018.

As the total magnetic flux through any closed surface around the Sun must be zero, inward and outward fluxes must balance each other. As spatial magnetic structures are swept past an observing spacecraft with the solar wind speed, these structures manifest themselves as time variations. The fluxes of energetic particles emitted from the Sun are generally anisotropic with a preferred direction aligned with the local magnetic field. These particles appear to be guided along filamentary magnetic flux tubes, which are often interwoven in a complicated way.

The interplanetary plasma also contains a small number of higher-energy particles representing a small energy density; these are controlled by electric and magnetic fields that result from the aggregate motion of all the particles. This motion of particles of different energy is illustrated in Fig. 1.8 for an ideal spiral magnetic field. Particles that reach the vicinity of the Earth at a given moment will have left the vicinity of the Sun at widely different times depending on their energy.

1.7.2 Collisionless shock waves

Earth's bow shock was proposed independently by Axford (1962) and Kellogg (1962). This suggestion, a *collisionless* shock wave, was met with considerable skepticism. An ordinary gas will respond to sound waves from the obstacle, a warning of the approaching collision. When the flow becomes supersonic, the sound speed is not fast enough to do that; instead, the gas particles will pile up, creating another obstacle, a shock. Could a similar feature happen in a collisionless plasma? The reality of a bow shock was confirmed in 1963 by spacecraft observations on IMP-1 (Wolfe and Intrilligator, 1970). The bow shock thickness is only 100–1000 km, much smaller than the mean free path in the solar wind, on the order of 1 AU.

In the case of a collisionless plasma, the conservation relations (also known as the shock-jump conditions or Rankine-Hugoniot relations) do not provide a unique prescription for the downstream state in terms of the upstream parameters, mainly because energy conservation gives information only about the total pressure (and hence temperature), not about how it is divided between the different types of particles in the plasma. In other words, we need to know about the shock structure, about how the shock works, in order to know how much the ions and electrons heat in passing through the shock.

Burgess, KR 137

A convenient way to describe the direction of the upstream field is the angle θ_B between the magnetic field and the shock normal. Depending on θ_B, the shock can have dramatically different behaviors. The terms quasiparallel and quasiperpendicular are used to divide the range of possible θ_B values, with the actual dividing line usually chosen as 45 degrees. When $\theta_B = 0$ the shock is called parallel, and when $\theta_B = 90°$ it is perpendicular. An oblique shock is one neither exactly perpendicular nor parallel. The oblique shocks are divided into three categories: the fast, slow, and intermediate (Alfvén), which correspond to the three modes of small-amplitude waves in MHD discussed in Section 1.6. The fast and slow shocks have the same behavior in terms of plasma pressure and magnetic field strength as do the corresponding MHD waves. (See Burgess (1995) for more treatment.)

1.7.3   De Hoffman-Teller frame

A key concept is the de Hoffman-Teller frame (HTF). In this frame, the electric field has been transformed away, as much as possible, both upstream as well as downstream. Particles move along the magnetic field; they have a gyrational motion as well, upstream and downstream of the shock. The following discussion follows Appendix 5A of Kivelson and Russell (1995).

In order to transfer from one shock frame to another that is in uniform relative motion, one should apply the appropriate relativistic transformations. However, for almost all work we can use a nonrelativistic transformation in which the velocity transformation is Galilean. The magnetic field is unchanged, and the electric field is obtained from the ideal MHD equation $\mathbf{E} = -\mathbf{u} \times \mathbf{B}$ in the appropriate frame. In Fig. 1.9A we consider a 1D shock; there is the shock normal $\hat{\mathbf{n}}$ and the upstream field $\mathbf{B}_u$, the upstream flow velocity $\mathbf{u}_u$ is in the same plane. We shall be considering only a 1D shock. In the normal incidence frame (NIF) the upstream flow is parallel to $\hat{\mathbf{n}}$, and the origin has been marked *0*. In this frame there is a motional electric field $\mathbf{E}_u = -\mathbf{u}_u \times \mathbf{B}_u$ that is perpendicular to both $\mathbf{u}_u$ and $\mathbf{B}_u$, and thus is parallel to the plane of the shock. This motional electric field can be made zero by

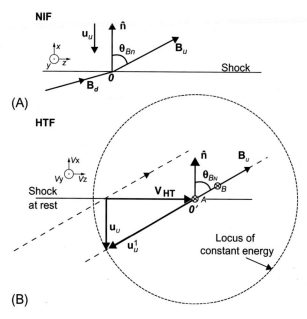

Fig. 1.9 Configurations for (A) the normal incidence frame and (B) the de Hoffmann-Teller frame with a typical coordinate system. The HTF is shown in velocity coordinates. The *dashed circle* represents the locus of all particles with energy equal to the energy of the incident flow. Marker *A* represents a particle with zero velocity in the HTF; marker *B* represents a particle with enough field-aligned velocity to escape upstream from the shock. From Burgess, D., 1995. Collisionless shocks. In: Kivelson, M.G., Russell, C.T. (Eds.), Introduction to Space Physics. Cambridge University Press, New York, NY, p. 156.

transforming to the frame where there is no flow velocity (the upstream flow frame), but in this frame the shock is moving.

The alternative is to find a frame in which the upstream flow and magnetic field are parallel and the shock is stationary. This is obtained from any shock frame by adding a transformation velocity parallel to the shock plane. This is shown in Fig. 1.9B where the transformation velocity from the NIF is marked $\mathbf{v}_{HT}$, and $0'$ is the origin in the new frame; this is called the HTF.

In the HTF, the upstream flow velocity is $\mathbf{u}_u'$, parallel to $\mathbf{B}_u$. ($\mathbf{B}_u$ is the same in all frames because we are not using the full relativistic transformation.) From Fig. 1.9B we see that

$$\mathbf{v}_{HT} = \mathbf{u}_u \tan\theta_{Bn} \tag{1.46}$$

so that the required transformation velocity increases rapidly as θ_{Bn} approaches 90 degrees. Also, the assumption about a nonrelativistic transformation will break down if θ_{Bn} is sufficiently close to 90 degrees, and when θ_{Bn} is exactly equal to 90 degrees it is not possible to find an HTF.

Burgess gets a more general expression for $\mathbf{v}_{HT}$ starting from the transformation of a velocity to a new frame in which all quantities are indicated by a prime (except $\mathbf{B}_u$):

$$\mathbf{u}'_u = \mathbf{u}_u - \mathbf{v}_{HT} \tag{1.47}$$

The final expression (Burgess, 1995, p. 157) is then

$$\mathbf{v}_{HT} = \frac{\hat{\mathbf{n}} \times (\mathbf{u}_u \times \mathbf{B}_u)}{\hat{\mathbf{n}} \cdot \mathbf{B}_u} \tag{1.48}$$

Because there is no electric field, the particles upstream have a very simple motion, with two parts: motion parallel to the magnetic field direction, and gyrational motion around it. A consequence of $\mathbf{E} = 0$ is that the energy of a particle is constant, and surfaces of constant particle energy are spheres centered on the HTF origin.

Although we have not marked the downstream field and flow velocity, there is another useful property of the HTF. From Maxwell's equation $\nabla \cdot \mathbf{B} = 0$, the normal component of the magnetic field is continuous ($B_n = $ constant):

$$[B_n] = 0 \tag{1.49}$$

From $\nabla \times \mathbf{E} = -\partial \mathbf{B}/\partial t$, with the assumption that $\partial \mathbf{B}/\partial t = 0$, the tangential component of the electric field must be continuous under *steady state*. Using $\mathbf{E} = -\mathbf{u} \times \mathbf{B}$ this becomes

$$[u_n \mathbf{B} - B_n \mathbf{u}] = 0 \tag{1.50}$$

Eqs. (1.49) and (1.50) show that $\mathbf{B}_n$, the normal component of $\mathbf{B}$, and the transverse component of $\mathbf{u} \times \mathbf{B}$ are continuous across the shock. From Eq. (1.48), we see that $\mathbf{v}_{HT}$ depends on these components, and consequently, the de Hoffman-Teller transformation velocity is the same downstream as it is upstream.

On the other hand, at a perpendicular shock, the field lines are parallel to the shock surface, and so particle motion along the magnetic field does not let the particle pass away from the shock. Indeed, the particle gyration at a perpendicular shock brings the particle back to the shock.

1.7.4   The quasiperpendicular shock

The shock angle θ_{Bn} is the most important factor in controlling the shock type, but almost every plasma parameter can have an effect: temperature, composition (i.e., what types of ions present), and shock Mach number M_A. The Mach number is defined as the ratio of the flow speed to the local wave propagation speed, and indicates the strength of the shock. It can be used as a measure of the amount of energy being processed by the shock. As might be expected, the higher the Mach number, the more dramatic the behavior of the shock. In the solar system, shocks can be found with Mach numbers between (almost) 1 and perhaps 20. In astrophysical objects, the Mach number could be much higher.

In the category of quasiperpendicular shocks there is a clear distinction, at least theoretically, between a type of low-Mach-number shock, the subcritical shock, and a higher-Mach-number shock, the supercritical shock. For most of the time, the Earth's bow shock is supercritical, and to find subcritical shocks we usually have to wait for a suitable interplanetary shock. The critical Mach number is at $M_A \sim 2.7$ for $\theta_{Bn} = 90°$, and it decreases as θ_{Bn} decreases. The Earth's bow shock has values for M_A in the range of approximately 1.5–10. For an exactly perpendicular shock, there is no particle escape, because the magnetic field keeps the particle motion parallel to the shock front. Quasiparallel shocks remain a topic of debate not understood.

1.7.5 Energetic particles and foreshocks

The existence of a "foreshock" is a unique property of collisionless plasma shocks. This is a region upstream of the shock that contains particles and waves associated with the shock (Fig. 1.10). In a collisionless plasma, a particle can have an

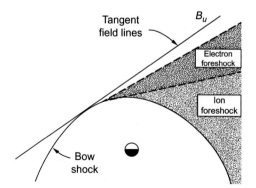

Fig. 1.10 Schematic of the Earth's bow shock, the tangent field line, and the electron and ion foreshocks. The diagram is drawn in the ecliptic plane. Although not shown, there will be a similar foreshock structure on the other side of the point of contact of the tangent field line. From Burgess, D., 1995. Collisionless shocks. In: Kivelson, M.G., Russell, C.T. (Eds.), Introduction to Space Physics. Cambridge University Press, New York, NY, p. 160.

arbitrarily high speed and thus not part of the plasma medium. A fast particle that escapes into the foreshock carries energy, and therefore it generally produces waves. Observationally, the Earth's foreshock is a rich "zoo" of different types of particles and the different waves they produce. Generally it is possible, and useful, to consider the shock and foreshock as separate aspects of the plasma.

There are two important issues. How do the particles gain these high energies? What is the structure within a foreshock? The subject of particle acceleration at shocks has a vast literature. Shocks in the heliosphere seem to be good accelerators of both electrons and ions. At the Earth's bow shock, ions are measured with energies up to several hundred kiloelectron-volts. The energy in these particles comes from the kinetic energy of the solar wind flow, but explanations of the various theories and their observational support are far from complete.

1.8 Magnetosheath

The magnetosheath lies between the bow shock and the magnetopause (Fig. 1.1). The observed magnetosheath plasma parameters show large-scale spatial ordering imposed by the shape of the magnetopause, but also great variability dependent on the solar wind input. The plasma properties of the magnetosheath depend on properties of the upstream solar wind, including density, velocity, and β. The properties depend also on whether the shock is quasiperpendicular or quasiparallel. In general, the magnetosheath tends to be in a more turbulent state behind the spatially extended quasiparallel bow shock than it is behind the quasiperpendicular shock.

The use of fluid equations for the collisionless gas, in particular the consideration of discontinuities, is justified by the presence of a magnetic field, which decreases the basic interaction lengths to something of the order of the gyroradii, at least in the transverse direction. Plasma waves excited by anisotropy driven instabilities will have the tendency to restore a near isotropic pressure tensor, as Parker found out previously. For modeling the flow around the magnetopause, an assumption of $\gamma = 5/3$ appears to be the best possible one within the limitations of the one-fluid approach.

1.8.1 Gas-dynamic model

Spreiter et al. (1966) have simplified the task of calculating the flow around the magnetosphere considerably by dropping all terms containing the magnetic field **B** from the fluid equations

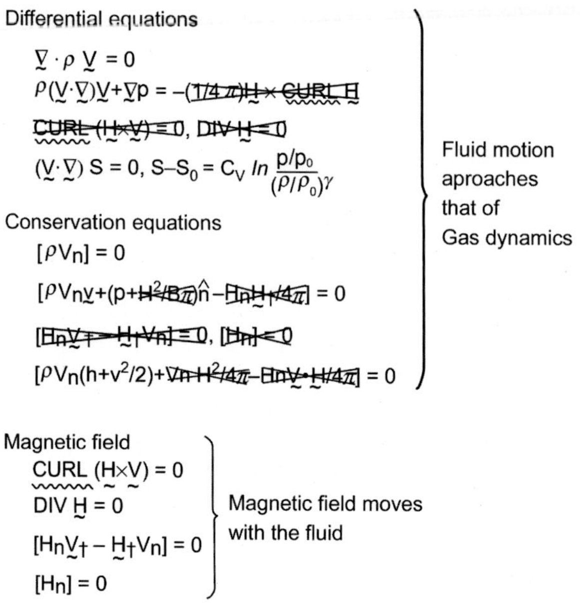

Differential equations

$$\nabla \cdot \rho \, \underline{V} = 0$$

$$\rho(\underline{V}\cdot\underline{\nabla})\underline{V} + \nabla p = -(1/4\pi)\underline{H} \times \text{CURL } \underline{H}$$

$$\text{CURL } (\underline{H}\times\underline{V}) = 0, \quad \text{DIV } \underline{H} = 0$$

$$(\underline{V}\cdot\underline{\nabla})\,S = 0, \quad S - S_0 = C_V \, ln\, \frac{p/p_0}{(\rho/\rho_0)^\gamma}$$

Conservation equations

$$[\rho V_n] = 0$$

$$[\rho V_n \underline{V} + (p + H^2/8\pi)\hat{n} - H_n \underline{H}/4\pi] = 0$$

$$[H_n\underline{V}_t - H_t V_n] = 0, \quad [H_n] = 0$$

$$[\rho V_n(h + v^2/2) + V_n H^2/4\pi - H_n \underline{V}\cdot\underline{H}/4\pi] = 0$$

Fluid motion approaches that of Gas dynamics

Magnetic field

$$\text{CURL } (\underline{H}\times\underline{V}) = 0$$

$$\text{DIV } \underline{H} = 0$$

$$[H_n\underline{V}_t - \underline{H}_t V_n] = 0$$

$$[H_n] = 0$$

Magnetic field moves with the fluid

Fig. 1.11 Equations of magnetohydrodynamics and their implication for large Alfvén Mach number $M_A = (4\pi\rho v^2/H^2)^{1/2}$. From Spreiter, J.R., Alksne, A.Y., 1969. Plasma flow around the magnetosphere. Rev. Geophys. 7(1, 2), 21.

and solving Maxwell's equations separately (Fig. 1.11). This approach is justified by the observation that the flow velocities, except near the stagnation point, are largely super-Alfvénic with little influence by the weak magnetic field. Once **v** has been found, the frozen-in magnetic field (provided it is passively transported by the fluid) can be calculated. The Alfvénic Mach number is given by:

$$M_A = \frac{v}{v_A} > 1 \tag{1.51}$$

with $v_A = B/(u)^{1/2}$. A typical value for v_A with $B = 10\,\text{nT}$ and $n = 10\,\text{cm}^{-3}$ is 70 km/s; even behind the bow shock the plasma flow velocity is typically >100 km/s. The magnetic stresses are related to the inertial term in the equation of momentum conservation by $M_A^2 \ll 1$.

Another simplification in solving the gas dynamic equations is to adopt cylindrical symmetry for the magnetopause. The method of calculations is described in detail by Spreiter and Alksne (1968). They use an iterative procedure of finding the shape of the shock. In their model, the shock is a normal hydrodynamic one. Its position and shape probably do not depend very much on the simplifications. It appears that this result does have considerable merit in spite of the oversimplification.

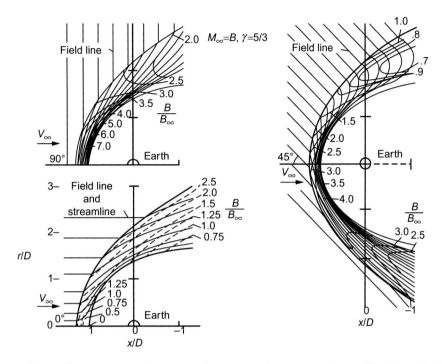

Fig. 1.12 Streamlines and wave patterns for supersonic flow past the magnetosphere; $M_\infty = 8$; $\gamma = 5/3$. From Spreiter, J.R., Summers, A.L., Alksne, A.Y., 1966. Hydromagnetic flow around the magnetosphere. Planet. Space Sci. 14, 243.

A few results are presented in Fig. 1.12. The figure shows the streamlines for a Mach number of 8 and $\gamma = 5/3$. In addition, characteristic or Mach lines are shown (by dashed lines), which can be thought of as standing compression or expansion waves.

1.8.2 Average properties

The average properties of the magnetosheath have been documented based on data from several missions including ISEE 1 and 2, AMPTE-IRM, and Wind. Magnetosheath plasma is characterized by the following:

- Its average density and magnetic field strength are higher than in the upstream solar wind by a factor consistent on average with the Rankine-Hugoniot relation for the fast mode shock.
- The average flow direction deviates from the antisolar direction such that the plasma flows around the blunt magnetosphere.
- The velocity downstream of the bow shock is lower than the local fast magnetosonic speed.
- The flow velocity increases again to supersonic speeds around the magnetopause flanks.

- The ion temperature of the sheath is higher than in the solar wind while the electron temperature does not increase very much over its upstream value.
- The plasma β shows large variations from the order of unity to values much greater than one.
- The magnetosheath plasma develops a noticeable temperature anisotropy behind the bow shock that increases toward the magnetopause and is more pronounced in the ions than in the electrons.

As a consequence of this, the magnetosheath seems to develop two regions of different turbulent behavior: one behind the bow shock and the other closer to the magnetopause.

1.9 Magnetopause

The magnetopause is the boundary of the magnetosphere (analogous to terminology for the lower atmosphere). It separates the solar wind plasma and the IMF from the geomagnetic field and plasma of primarily terrestrial origin. This boundary was first proposed by Chapman and Ferraro (1931).

The first clear identification of the magnetopause was made by Cahill and Amazeen (1963) with a magnetometer on the Explorer XII satellite. One radial pass of this experiment near the subsolar point is shown in Fig. 1.13. Inside 8 R_E the field is generally what is expected for a dipole but at 8.2 R_E an abrupt change occurs.

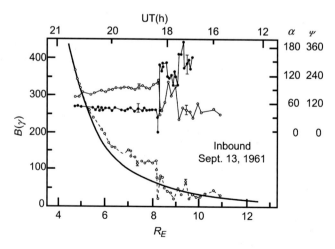

Fig. 1.13 A magnetometer record of the geomagnetic field measured on the inbound pass of Explorer XII on September 13, 1961. It shows the magnetopause at 8.2 R_E. From Cahill, L.J., Amazeen, P.G., 1963. The boundary of the geomagnetic field. J. Geophys. Res. 68, 1835.

Outside this, the field is variable in both magnitude and direction. This clearly is the magnetopause. Just inside this boundary, the field is roughly double what the dipole field would be; this is as expected as shown in Eq. (1.53).

Ions and electrons in the solar wind will penetrate into the boundary layer, be bent in opposite directions by the magnetic field, and create a surface current (Fig. 1.14). It is this surface current that produces the magnetic field change across the boundary.

If the magnetopause were a plane, these surface currents would just double the magnetic field inside the boundary and make it zero outside. Chapman and Ferraro (1931) considered a sheet of solar wind pushing the magnetic field in to make a plane surface. This field is exactly like that produced by an image magnetic dipole of the same strength as the Earth at a distance of twice the plane surface. The field inside the surface is given in this image dipole model by

$$B_{\text{total}} = B_{\text{geo}} + B_{\text{image}} = \frac{B}{r^3} + \frac{B}{(2R-r)^3} \qquad (1.52)$$

and at the boundary the field is

$$B_{\text{total}} = \frac{2B}{R^3} \qquad (1.53)$$

just twice that due to the Earth.

The solar wind dynamic pressure (or momentum flux), now subsonic in the magnetosheath, presses on the outer reaches of the geomagnetic field, confining it to a magnetospheric cavity. When the solar wind blows harder, the magnetosphere shrinks; when the solar wind abates, the magnetosphere expands. For supersonic flow, this wave is the fast magnetosonic wave. In the simplest approximation, the magnetopause can be considered as a boundary separating a vacuum geomagnetic field from solar wind plasma. A current is involved, often called the Chapman and Ferraro current.

A schematic illustration was advanced by Willis (1975) of the trajectories of magnetosheath ions and electrons incident normally on a plane boundary layer, shown in Fig. 1.14. In (a) the case assumes that the polarization electric field due to charge separation is present; the electrons gain energy and consequently they become the main current carriers. In (b) the polarization electric field is completely neutralized by ambient magnetospheric charged particles, accentuating the role played by the ions. The MP current has been discussed by Longmire (1963) and simulated by Berchem and Okuda (1990). There is no doubt about the mechanism involved.

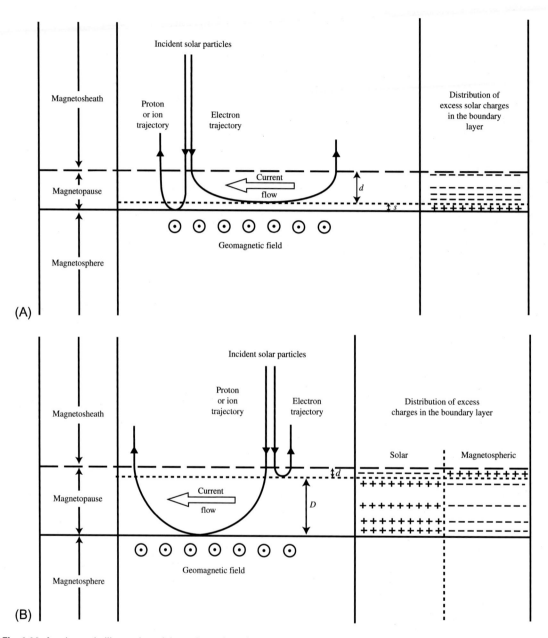

Fig. 1.14 A schematic illustration of the trajectories of magnetosheath ions and electrons incident normally on a plane boundary layer (A) when the polarization electric field due to charge separation is present, and (B) when it is completely neutralized by ambient magnetospheric charged particles. From Willis, D.M., 1975. The microstructure of the magnetopause. Geophys. J. Royal Astron. Soc. 41(3), 359–363.

The location of this MP boundary can be calculated from Ampere's law by requiring the total pressure on the two sides of the boundary to be equal. A current sheet can be defined as a thin surface across which the magnetic field strength and/or direction can change substantially as in the MP or in the magnetotail, as shown by Fig. 1.1.

"Thin" is a relative term, but here we mean that the sheet thickness is very much smaller than the other dimensions of the sheet or than the sheet radius of curvature. This also means that the sheet can be described locally as a plane. The thickness of the magnetopause current sheet is sometimes as little as a few hundred kilometers.

1.10   Cause and effect at the magnetopause

It is at the magnetopause that we must begin the sequence of events for interaction of the solar wind plasma with the magnetosphere, ending up with auroras. Two processes were proposed, both in 1961: magnetic reconnection (MR) by Dungey and viscous interaction by Axford and Hines. These have been the mainspring of research ever since. Another article on a dynamo powered by solar wind plasma was published in the same year by Cole (1961), and a year before that by Schmidt (1960, 1979) on the polarization current involved in a dynamo. These two significant papers were not noticed by a majority of researchers.

1.10.1   Dungey model

In ideal MHD, collisionless plasmas do not mix easily; instead, they tend to form cells of relatively uniform plasma permeated by a magnetic field. Dungey (1961) proposed that the required momentum transfer could be attributed to magnetic coupling if high-latitude geomagnetic field lines extend into interplanetary space, that is, "interconnect" with the IMF. His suggestion, illustrated in Fig. 1.15A, is the process called magnetic reconnection, that is, a process to establish interconnection.

This diagram shows three types of field lines for a strictly southward component of IMF. Closed field lines go from the South Pole, through the equatorial plane with positive B_z, and into the North Pole entering with negative B_z. Interplanetary field lines to the far left, and also right, do not connect to the geomagnetic field.

At higher latitudes the IMF connects to the geomagnetic field at both poles; these field lines are called *open*. Maxwell's equation

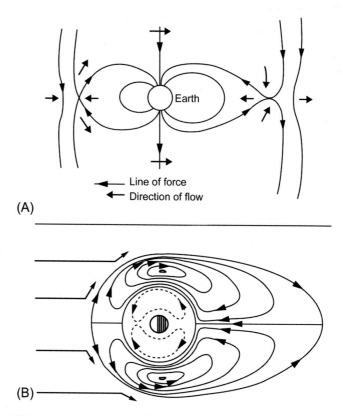

Line of force
Direction of flow

(A)

(B)

Fig. 1.15 Two models of solar wind interaction with the magnetosphere: (A) Dungey (1961) proposed that "Interplanetary magnetic field and the auroral zones" involving magnetic reconnection was a viable alternative to the model (B) proposed by Axford and Hines (1961), "A unifying theory of high-latitude geophysical phenomena and geomagnetic storms" with viscous-like interaction. From Slavin, J.A., Daly, P.W., Smith, E.J., Sanderson, T.R., Wenzel, K.-P., Lepping, R.P., et al., 1987. Magnetic configuration of the distant plasma sheet-ISEE 3 observations. In: Lui, A.T.Y. (Ed.), Magnetotail Physics. Johns Hopkins University Press, Baltimore, MD, p. 60.

$\nabla \cdot \mathbf{B} = 0$ is satisfied by having equal amounts of open field lines at both poles. The dividing lines meet at an *X-line* shown by its intersection with the meridian plane called the *separatrix* surfaces. In general, with an arbitrary orientation of the IMF, there will be a guide field along the *X*-line.

In this MHD model, as viewed in the meridian plane, magnetic tension transfers solar wind momentum to the magnetosphere and ionosphere. The solar wind $-\mathbf{V} \times \mathbf{B}$ electric field is correspondingly "mapped" along interconnected field lines to the magnetosphere and high-latitude ionosphere. Dungey assumed that the available magnetic stress is sufficient to drive magnetospheric convection. An "interconnection rate" (ratio of interconnected

IMF flux to total unperturbed IMF flux intersecting the magneto-sphere's cross section) of the order of 0.1 is evidently adequate in this fluid picture.

For other IMF directions, the field geometry becomes more complicated but the same topology prevails; that is, the polar caps remain magnetically connected to interplanetary space and the solar wind $-\mathbf{V} \times \mathbf{B}$ electric field maps to a generally dawn-to-dusk polar-cap electric field. The interconnection geometry changes in two important ways as the IMF direction becomes less southward. First, the amount of interconnected flux, and hence the size of the interconnected polar caps, decreases. Second, any east-west component of the IMF produces an east-west component of magnetic tension, producing east-west flow asymmetries in the magneto-sphere. This did sound very convincing.

Dungey (1961) proposed magnetic reconnection as a key plasma process operating in the magnetosphere, it was as a qualitative principle that deserved serious consideration. The magnetosphere had just been discovered, and everyone was groping to understand its interaction with the solar wind. Magnetic reconnection had been proposed in solar flares over a decade earlier; although it had difficulties in explaining certain features, such as its rapid onset, it was a promising enterprise to explore.

If somehow this electric field could be maintained along the X-line then the local plasma should be energized lending credence to reconnection. Sonnerup et al. (1995) summarized the prevailing view:

A large amount of indirect evidence in support of an open, or reconnecting magnetosphere has accumulated since that time, among the most important being the observed dependence of geomagnetic activity on interplanetary magnetic field (IMF) direction (Fairfield and Cahill, 1966), the observed inward motion of the magnetopause following southward turning of the IMF (Aubry et al., 1970), the observed direct access of energetic solar protons to the polar cap regions (Morfill and Scholer, 1973; Fennell, 1973), and the ion dispersion effects observed in the cusp regions, e.g., from the spacecraft HEOS-2 (Rosenbauer et al., 1975) or AE-C (Reiff et al., 1977). ... This uncomfortable situation persisted until the late '70s when the ISEE 1/2 mission provided, for the first time, plasma measurements of sufficient time resolution and precision to allow detailed study of plasma dynamics in the narrow inward-outward-moving magnetopause layer. Such information was needed in order to identify what has been called the "smoking-gun" evidence of reconnection at the sub-solar magnetopause: jetting of plasma

along the magnetopause away from the reconnection site, as
illustrated in (Fig. 1.16). The jetting should be strongest in a thin
layer adjacent to the earthward edge of the magnetopause.

Sonnerup et al. (1995, p. 167)

Sonnerup et al. (1981), in a classic paper, discussed this jetting,
the "smoking-gun" evidence of reconnection (Fig. 1.17). Four
aspects of this process will be discussed now.

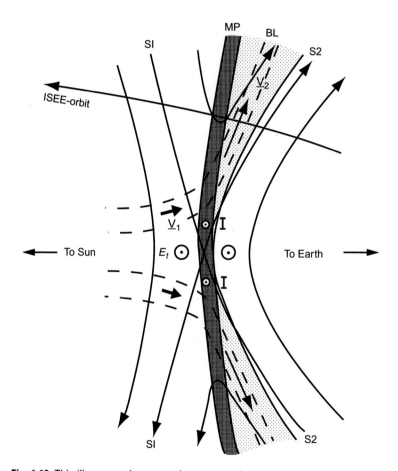

Fig. 1.16 This illustrates the magnetic reconnection process on the dayside
magnetopause with an X-line. The Chapman-Ferraro (magnetopause) current
flows from dawn to dusk. With a southward IMF, the electric field is also in this with
antisunward plasma flow; hence **E · J** is positive, indicating dissipation. This
diagram and Fig. 1.17 are perhaps the most persuasive in support of MR, but there
are pitfalls, as discussed in this book. From Sonnerup, B.U.Ö., Paschmann, G.,
Papamastorakis, I., Sckopke, N., Haerendel, G., Bame, S.J., et al., 1981. Evidence for magnetic
field reconnection at the Earth's magnetopause. J. Geophys. Res. 86, 10050.

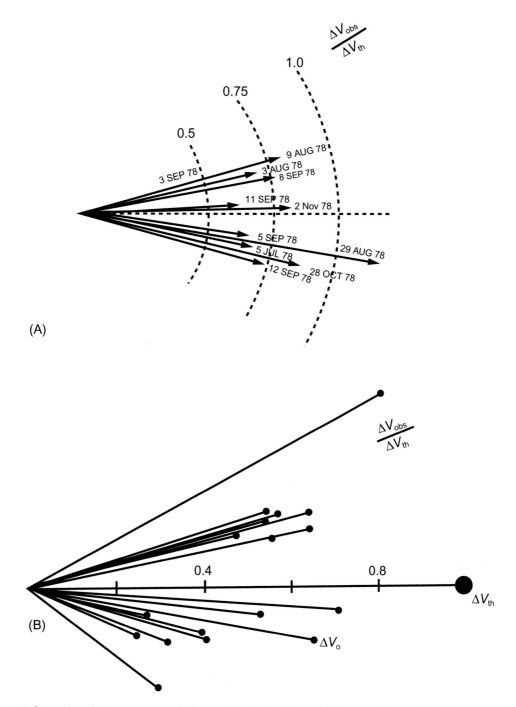

Fig. 1.17 Comparison between measured changes in velocity, Δv_{obs} and changes, Δv_A, predicted from tangential stress balance (the Walén relation). All vector lengths are normalized and all vector orientations by the orientation of Δv_A which is taken to be horizontal. (A) Summary of 11 ISEE crossings of the dayside magnetopause (Sonnerup et al., 1981); (B) summary of 17 flow reversal events observed by ISEE in the same region (Gosling et al., 1990). From Sonnerup, B.U.Ö., Paschmann, G., Song, P., 1995. Fluid aspects of reconnection at the magnetopause: in situ observations. In: Song, P., Sonnerup, B.U.Ö., Thomsen, M.F. (Eds.), Physics of the Magnetopause. AGU Geophysical Monograph 90. American Geophysical Union, Washington, DC, p. 171.

1.10.1.1 De Hoffmann-Teller frame

The plasma flow across a 1D, steady-state current sheet having field component $B_n \neq 0$ is most conveniently described in a 2D frame of reference (HTF) moving with the magnetopause as described in Section 1.7.3. The flow appears field aligned on both sides of the discontinuity (Sonnerup et al., 1995). The HTF slides along the magnetopause with the "field line velocity," that is, the velocity with which field-line intersection points with the inner and outer surfaces of the current layer move along that layer. This frame also participates in the (sometimes rapid) inward-outward oscillatory motion of the magnetopause. In the HTF, the electric field, $\mathbf{E}'$, vanishes on both sides of the current layer but a normal component E_n' may remain as part of its intrinsic electromagnetic structure. If accurate plasma velocity measurements, $\mathbf{v}_i$, are available rather than electric field data, the HTF velocity can be determined by least squares fitting of $\mathbf{E}_{HT}$ to the convection electric field

$$\mathbf{E}_C^i = -\mathbf{v}^i \times \mathbf{B}^i \qquad (1.54)$$

A convenient matrix method was given by Sonnerup et al. (1987). De Hoffmann-Teller frames were determined from measured electric and magnetic fields by Aggson et al. (1983) for five dayside magnetopause crossings by ISEE 1, the result being frame velocities in the range 118–285 km/s. The ISEE plasma data had insufficient directional and temporal resolution and insufficient accuracy to permit HTF determinations from measured plasma velocities and magnetic fields. Six successful and three unsuccessful HTF determinations from AMPTE/IRM plasma and magnetic field data, again taken in the dayside magnetopause, have been reported, yielding frame velocities in the range 265–600 km/s and frame accelerations in the range 0–7 km/s^2 with normal acceleration components in the range 0–1.3 km/s^2. It was found that a single HTF velocity was sometimes capable of reproducing measured electric vectors during, as well as for, substantial time intervals surrounding a magnetopause crossing. This result is consistent with the occurrence of reconnection, or, more precisely, with the presence of a nonzero normal magnetic field component B_n at the magnetopause. It does not prove that such a component must be present. In other words, for a simple locally 1D magnetopause structure, the existence of an HTF is a necessary but not a sufficient condition for ongoing reconnection.

1.10.1.2 Walén test

As magnetosheath plasma moves across a magnetopause layer in which $B_n \neq 0$, its momentum components tangential to that layer change as a result of the tangential Maxwell stresses, that is, in response to the force $\mathbf{J} \times \mathbf{B}_n$. The simplest description of this process pertains to an ideal MHD rotational discontinuity (RD) for which the following two relations are valid (Hudson, 1970)

$$\rho(1-\alpha) = \text{const} \tag{1.55}$$

$$\mathbf{v}' = \pm\mathbf{v}_A \tag{1.56}$$

with

$$\alpha = (p_{\parallel} - p_{\perp})\mu_0/B^2 \tag{1.57}$$

Here α is the pressure anisotropy factor, $\mathbf{v}' = (\mathbf{v} - \mathbf{v}_{HT})$ is the (field-aligned) plasma velocity observed in the HTF, and

$$\mathbf{v}_A = \mathbf{B}[(1-\alpha)/\mu_0\rho]^{1/2} \tag{1.58}$$

is the intermediate-mode wave speed. The second equation is called the Walén relation; in the spacecraft frame it is of the form

$$\Delta\mathbf{v} = \pm\Delta\mathbf{v}_A \tag{1.59}$$

where the symbol $\triangle$ refers to changes, say, relative to the upstream state. The choice of sign depends on whether the flow is parallel or antiparallel to $\mathbf{B}$. Detailed comparisons of ISEE plasma and magnetic field data with the Walén relation had been carried out for a total of nearly 50 crossings (Sonnerup et al., 1995).

> *It is clear from the overall evidence accumulated to date that the most common result for potential reconnection cases where the test has been carried out, is that the agreement is rather imperfect: the observed velocity changes are less in magnitude than predicted and have angular deviations of the order of $\pm 20°$ from predicted directions.*
>
> **Sonnerup et al. (1995, p. 172)**

1.10.1.3 Energy balance

Concerns in the pre-ISEE era about the relevance of reconnection at the dayside magnetopause (Heikkila, 1975) were centered, not on the Walén relation, that is, not on tangential momentum balance, but on the question of the electromagnetic power conversion. The value $\mathbf{E} \cdot \mathbf{J}$ associated with the standard reconnection model, specifically the deposition of this energy in plasma kinetic and thermal energy, had not been observed. The first attempt to check the energy balance was made by Paschmann et al. (1985) using ISEE data. They were able to cast the total energy equation into the form where enthalpy-like terms are collected on the left and terms containing the magnetic field on the right. The scatter is substantially larger, but otherwise the results appear generally consistent with the interpretation of the magnetopause as an approximately dissipation-free RD.

1.10.1.4   Reconnection rate

The most fundamental measure of the reconnection rate is in terms of the electric field along the reconnection line or X-line. However, most magnetopause crossings do not occur at the reconnection site; for this reason, it has become standard practice to use the local Alfvén Mach number M_{An} of the plasma flow into the magnetopause as a proxy for the true reconnection rate (Sonnerup et al., 1995). This practice is justified, provided the distance from the observation site to the X-line is not too large. Because of the normal flow velocity in an RD, we may write

$$M_{An} \equiv |\mathbf{v}_n| / |\mathbf{v}_A| \cong |B_n| / B \qquad (1.60)$$

Thus the size of the normal magnetic field component (or the normal flow component) serves to measure the reconnection rate. It is difficult to obtain the instantaneous reconnection rate from observations. The rate is proportional to $\mathbf{v}_n$, B_n, or $E_t = |v_{HT} B_n|$, all of which are usually small compared to the total vector magnitudes, and all of which are therefore highly sensitive to errors in the vector, $\mathbf{n}$, normal to the magnetopause.

1.10.2   Axford-Hines model

On the basis of ground-based geomagnetic and auroral observations, Axford and Hines (1961) noted the existence of a systematic convection pattern, this time as viewed in the equatorial plane. Subsequent ground-based and spacecraft observations have thoroughly confirmed its existence and elucidated many of its detailed characteristics. Axford and Hines introduced the concept of a frictional interaction between the solar wind and magnetosphere; they proposed a circulation pattern illustrated in Fig. 1.15B. Solar wind momentum is transmitted to the magnetosphere in a viscous boundary layer just inside the magnetopause, producing the required antisunward high-latitude flow and the implied low-latitude return flow. The corresponding "convection electric field" has a dawn-to-dusk orientation in the polar cap (region of antisunward flow in the ionosphere) and a dusk-to-dawn orientation in the lower-latitude return flow region of the ionosphere; these orientations are reversed in the magnetospheric equatorial plane.

Ordinary collisional viscosity fails by many orders of magnitude to account for this "viscous" boundary layer, just as in the earlier discussion of bow shock thickness. Axford and Hines recognized this discrepancy and suggested that wave-particle interactions might provide the required increase in effective collision frequency. Granted a sufficient effective viscosity, the Axford-Hines model provides a coherent phenomenological description of the basic solar wind-magnetosphere coupling. It has provided an invaluable paradigm for much subsequent work. The model, however, has several unresolved difficulties. It requires no IMF, and hence, presumably, predicts no first-order correlation between the coupling efficiency and the IMF direction, whereas significant such correlations have been clearly observed, and the mechanism to produce sufficient effective viscosity has not yet been found. The finding of the LLBL has resolved these and other difficulties.

1.10.3 Dynamic processes

Only when we bring about an explicit time dependence into the equations can we talk about dynamic processes. The same goes for considerations in three dimensions; only then can we talk about a near collision, deflection of flow, draping of flow around the magnetosphere, and so on.

1.10.3.1 Impulsive penetration

The idea that solar wind plasma-field irregularities with an excess momentum density penetrate deeper into the geomagnetic field was introduced by Lemaire and Roth (1978). It was based on the observation that the solar wind is generally patchy over distances smaller than the diameter of the magnetosphere. It is indeed unusual to find magnetogram periods of more than 30s when the IMF does not change at least by a few percent. The presence of these plasma irregularities indicates that the solar wind momentum density is not uniform, and that the dynamic pressure that the solar wind inflicts upon the magnetosphere must be patchy, nonuniform, and changing in time as shown in Fig. 1.18. The magnetopause itself might be in motion, locally, perhaps as a surface wave due to some instabilities.

Let us consider one of these many plasma density enhancements with an increase of the density ($dn > 0$) moving toward the magnetopause with the background speed v_{sw}. If this plasmoid corresponds to an excess density but has the same bulk velocity as the surrounding solar wind background, its

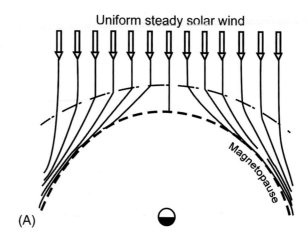

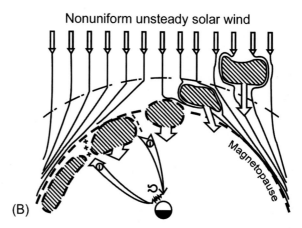

Fig. 1.18 Equatorial view of the magnetopause. (A) When the solar wind is steady and uniform, the MP is a smooth surface along which the solar wind slips without penetration. (B) When the solar wind is nonuniform and unsteady, plasma density irregularities carried in the solar wind will be able to penetrate deeper in the geomagnetic field provided they have an excess momentum density. From Lemaire, J., Roth, M., 1978. Penetration of solar wind plasma elements into the magnetosphere. J. Atmos. Terres. Phys. 40(3), 332.

momentum density $(n+dn)mv_{sw}$ is then necessarily larger than the average (nmv_{sw}) (Lemaire, 1985).

This plasma element must conserve its excess momentum. Therefore it can plow its way through the magnetosheath toward the magnetopause, with less deflection than the average flow. Unlike the other plasmoids, which have a lower momentum density than average, the former one will reach the position of the mean magnetopause with an excess momentum and an excess kinetic energy. At the mean magnetopause position, where the normal component of background magnetosheath plasma velocity becomes equal to zero, the plasmoid has a residual velocity (GC) given by

$$v_e = v_{sw} dn/(n + dn) \qquad (1.61)$$

The penetration mechanism of the plasmoid into the region of the magnetospheric field lines lies between the following two extreme cases: weakly diamagnetic, low-β plasmoids on one hand, and strongly diamagnetic, high-β plasmoids on the other hand.

It is *not proper to assume that these effects* are not important; nevertheless, only one paper in Kivelson and Russell (1995) is mentioned (Lemaire et al., 1979), and impulsive penetration is not listed in the index. This subject is covered in Sections 5.5 and 6.5 on the magnetopause when it becomes apparent that *they control the physics.*

1.10.3.2 Plasma transfer event

The only source of a magnetic field is a current **J**, by Ampere's law. Therefore to study changes in the magnetic field we must consider perturbation electric currents δ**J**, the source of δ**B**. A changing current will create an induction electric field by Faraday's and Lenz's laws.

The inferred immediate cause of a plasma transfer event (PTE) is a localized pressure pulse from the magnetosheath, an inward push by solar wind plasma associated with erosion (Aubry et al., 1970). The pressure pulse is likely to be in some small region, not extending to infinity in the dawn-dusk direction. A localized induction electric field, $\mathbf{E}^{ind} = -\partial \mathbf{A}/\partial t$, is forced upon the plasma. It is entirely local, opposed to the current perturbation; the perturbation current is solenoidal with div δ**J** $= 0$. It must have a component normal to the magnetopause at each edge of the perturbation, with the electric field with opposite polarities. We need to consider two cases regarding B_n.

If $B_n = 0$, the plasma cannot respond by charge separation and no electrostatic field is created (Fig. 1.19). Due to the collisionless nature of this region, the Pedersen conductivity is effectively zero. The induction electric field must be just the field to keep the motion of the magnetopause in step with the solar wind plasma.

The plasma response changes dramatically with an open magnetosphere. In this case, a rotational discontinuity will be present, with a finite B_n. Electron and ion mobilities are high along the magnetic field. Now we can use the very large direct conductivity σ_0; the plasma can polarize along the magnetic field lines, top and

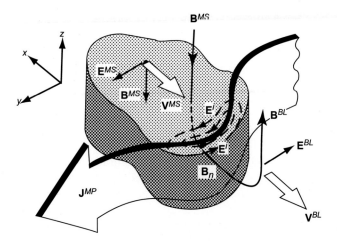

Fig. 1.19 Model of a localized diversion, or meander, of the magnetopause current due to some change in the solar wind or the interplanetary magnetic field. Any normal component of the magnetic field at the magnetopause indicates a rotational discontinuity, while $B_n = 0$ indicates a tangential discontinuity. In any case, an inductive electric field will exist, by Lenz's law, as shown (measured in a fixed frame) everywhere opposing the current perturbation that is involved with the change in magnetic topology. This cartoon is only a statement of the problem of localized erosion of the magnetopause, not the solution. It does not include the plasma response to the inductive electric field. From Heikkila, W.J., 1997. Interpretation of recent AMPTE data at the magnetopause. J. Geophys. Res. 102(A5), 2119.

bottom, in different senses. This causes an electrostatic field tangential to the MP, reversing as indicated. Thus we see that this E^{es} will drive the solar wind plasma into the current sheet, in the reconnection frame. On the other side, since both **B** and **E** reverse, the electric drift $\mathbf{E} \times \mathbf{B}$ will be also earthward. A PTE is produced. This will be treated at length in Sections 5.5 and 6.5.

1.11 Low-latitude boundary layer

The processes of plasma penetration through the magnetopause and the corresponding transfer of mass, momentum, and energy became a subject of intense discussion in the 1970s. A symposium dedicated to this subject was held at The University of Texas at Dallas in November 1973 with the title, "Magnetospheric Cleft Symposium." A second conference was held at St Jovite, Quebec, Canada in October 1976 and another conference was held in June 1979 at Alpbach, Austria, a Chapman Conference entitled "Magnetospheric Boundary Layers." Rather than focus only on the cusp and cleft regions, it was felt that the magnetospheric boundary layer as a whole should be the central subject. A total of 58 papers appeared in ESA SP-148 (August 1979). Eastman (2003) summarized the field until 1979 in another Chapman Conference "Earth's Low-Latitude Boundary Layer" in New Orleans in 2001 (AGU Monograph 133) with Newell and Onsager (2003) serving as editors.

The name "Low-Latitude Boundary Layer" (due to Gerhard Haerendel) is appropriate. Because the separatrix between open and closed field lines is, in general, distinct from the magnetopause current layer, the LLBL on closed field lines should be regarded strictly as a layer earthward of the magnetic separatrix. Fig. 1.20 shows a possible configuration for southward IMF conditions where the solar wind plasma convects through the magnetopause current layer and into the LLBL (shown exaggerated). If solar wind plasma (for a southward IMF) can get into the LLBL, the electric field must reverse its sense in the process if the electric drift is important.

The main distinguishing feature of the LLBL, whether on open or closed field lines, is the presence of a plasma population earthward of and adjacent to the magnetopause, at low to moderate geomagnetic latitudes, with properties intermediate between those of the neighboring magnetosheath and magnetospheric (plasma sheet) regions. The bulk density and flow velocity in the LLBL have values somewhat less than those in the adjacent magnetosheath but greater than those in the adjacent magnetosphere, while the plasma thermal energy in the LLBL exceeds that in magnetosheath and is less than that in magnetosphere. At times relatively cold ionospheric plasma components are also present in the LLBL. This is the subject of Chapter 9.

1.12 Discovery of the radiation belt

Analysis of the Explorer 1 data by the Van Allen group produced count rate-versus-altitude curves of energetic particles (Van Allen, 1959). When the data were replotted in terms of B, the magnetic field strength at different locations, the curves were identical. This demonstrates conclusively that the magnetic field controls the particles. The data can be understood in terms of the motion of charged particles in a dipole field. Trapped particles bounce back and forth, mirroring at the same value of B at all longitudes.

It is interesting to note that the Soviet satellite Sputnik II, launched on November 3, 1957 before Explorer I, had Geiger counters on board, and might have discovered the radiation belt; its

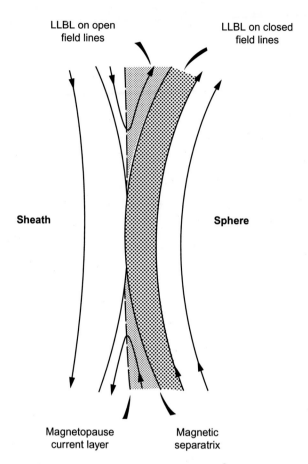

LLBL on open
field lines

LLBL on closed
field lines

Sheath

Sphere

Magnetopause
current layer

Magnetic
separatrix

Fig. 1.20 A possible configuration of the magnetospheric boundary region for IMF $B_z < 0$ conditions. The *heavy solid line* demarcates the magnetic separatrix between open and closed magnetic field lines. The *dashed line* indicates the magnetopause current layer, which coincides with the separatrix at low latitudes. From Lotko, W., Sonnerup, B.U.Ö., 1995. The low-latitude boundary layer on closed field lines. In: Song, P., Sonnerup, B.U.Ö., Thomsen, M.F. (Eds.), Physics of the Magnetopause. AGU Geophysical Monograph 90, Washington, DC, p. 373.

perigee was in the north so that it was underneath most of the radiation belt when it was monitored in the U.S.S.R. (it carried no tape recorder). The first paper published on Sputnik II showed the detector count rate increasing from 400 to 700 km, but this increase was only about 50% and it was not interpreted as being anything unusual. Sputnik III confirmed the existence of the radiation belt as already documented by Van Allen (Gringauz et al., 1960, 1962).

Explorer IV, launched July 26, 1958, was instrumented by Van Allen's group to study both the natural radiation belt and the artificial belts made by the Argus explosions. Its detectors were designed to handle large fluxes. There were four detectors with different thresholds designed to give an integral range spectrum of, but not to identify, the particles counted. Explorer IV provided a map over an extended region and also did give an integral range spectrum. Energy fluxes of larger than 50 erg/cm²-s-ster were detected through $1\,mg/cm^2$ shielding at about 1500 km. It is uncertain whether the particles were protons or electrons. It also monitored the three Argus nuclear explosions. These explosions, carried out during the summer of 1958, all produced belts of trapped electrons resulting from the β-decay of the fission fragments from the explosions. The decay of these artificial belts was followed for several weeks until they were no longer distinguishable.

On December 6, 1958, the Pioneer III spacecraft was sent out to a distance of 107,400 km from the Earth; it was intended as a lunar probe but failed. This carried two Geiger tubes to measure the spatial extent of the trapped radiation discovered on Explorer I. Combining of the data from Explorer IV and Pioneer III led to the first complete map of the trapped radiation. This picture has led to the concept of an inner and outer radiation zone. The inner high-intensity zone was that seen by Explorer I and III. The lower ends of the outer zone were seen by Explorer IV. The concept of inner and outer zones is somewhat useful qualitatively but is of limited merit.

1.12.1 Magnetic field lines and the L parameter

The spherical coordinate representation of a dipole magnetic field allows us to calculate readily the equation of a magnetic field line. A field line is everywhere tangent to the magnetic field direction.

$$r d\theta / B_\theta = dr / B_r \qquad (1.62)$$

$$d\varphi = 0 \qquad (1.63)$$

Integrating we obtain for the equation of a field line

$$r = r_0 \sin^2\theta \qquad (1.64)$$

where r_0 is the distance to the equatorial crossing of the field line. The equation in terms of L (with distance measured in planetary radii) and the magnetic latitude is:

$$r = L\cos^2(\lambda) \qquad (1.65)$$

A related parameter that is frequently used to organize low magnetospheric observations is the invariant latitude. It is the latitude where a field line reaches the surface of the Earth and is given by

$$\Lambda = \cos^{-1}(1/L)^{\frac{1}{2}} \qquad (1.66)$$

$$\cos^2\Lambda = \frac{1}{L} \qquad (1.67)$$

Thus a dipole field line that extends to 4 R_E in the equatorial plane of the magnetosphere maps to an *invariant* latitude of 60 degrees at the Earth's surface. A dipole field line that extends to 10 R_E in the equatorial plane maps to a latitude of 71.6 degrees. The invariant latitude refers to the L-value of an elongated magnetotail.

Although the dipole approximation is quite useful, often it is inadequate to express the significant complexities of magnetic fields in a planetary magnetosphere. In such situations, it is usual to express the scalar potential φ as a sum of internal and external contributions to the field using associated Legendre polynomials (Walker and Russell, 1995).

It is important to note that the coefficients are functions of time and have sizable secular or temporal variations. The dipole moment, the geographic colatitude (tilt) of the dipole axis, and the westward drift all vary. Although it is common to draw the Earth's magnetosphere with its dipole axis perpendicular to the solar wind flow, it is seldom in this configuration. In addition to the 10.8 degrees dipole tilt, the Earth's rotation axis is inclined 23.5 degrees to the ecliptic pole. Thus in the course of its daily rotation and its annual journey around the Sun, the angle between the direction of the dipole and the direction of the solar wind flow varies between 90 and 56 degrees. Because the interplanetary magnetic field (IMF) is ordered in the ecliptic plane (or, more properly, the Sun's equatorial plane), and because IMF that are opposite in direction to the Earth's field more strongly interact with it, there are annual and semiannual variations in geomagnetic activity.

1.12.2 Pseudo-trapping regions

Using realistic geomagnetic field models, calculations have revealed the concepts of L-shell splitting and pseudotrapping (Roederer, 1967). L-shell splitting refers to the imposition of a strong pitch angle dependence on invariant drift shells caused by the removal of longitudinal symmetry in the field due to boundary and tail currents. Pseudo-trapping follows immediately

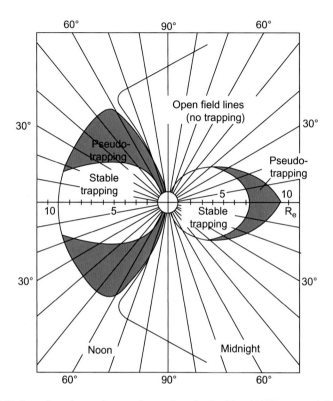

Fig. 1.21 Location of pseudo-trapping regions in the Mead-Williams model. Particles mirroring inside these regions are unable to complete a drift around the Earth. From Roederer, J.G., 1967. On the adiabatic motion of energetic particles in a model magnetosphere. J. Geophys. Res. 72(3), 987.

from L-shell splitting and refers to particles conserving μ and J, whose longitudinal drift carries them out of the magnetosphere. This is called *shell degeneracy* (see Fig. 1.21).

In the general case, particles starting on the same field line at a given longitude will populate different shells, according to their initial mirror point fields or equivalently, according to their initial equatorial pitch angles. Particles, starting on a common line in the noon meridian, do indeed drift on different shells, which intersect the midnight meridian along different field lines (further treated in Section 7.6).

1.13   The ionosphere

The ingredients for a planetary ionosphere are a neutral atmosphere and a source of ionization for the gases. Sources of ionization include photons primarily from the Sun, and energetic-particle precipitation from the galaxy (cosmic rays), the Sun, the magnetosphere. The process involving the former is referred to as photoionization, and the latter is often labeled impact ionization. The only requirement on the ionizing photons and particles is that their energies ($h\nu$ in the case of photons, and kinetic energy in the case of particles) exceed the ionization potential or binding energy of a neutral-atmosphere atomic or molecular electron. Solar photons in the extreme ultraviolet (EUV) and ultraviolet (UV) wavelength range of approximately 10–100 nm typically produce at least the dayside ionospheres of planets.

1.13.1 The upper atmosphere

The density n_n of a constituent of the upper (neutral) atmosphere usually obeys a hydrostatic equation:

$$n_n m_n g = \frac{dp}{dh} = -\frac{d}{dh}(n_n k T_n) \qquad (1.68)$$

which expresses a balance between the vertical gravitational force and the thermal-pressure-gradient force on the atmospheric gas. Here, m_n is the molecular or atomic mass, g is the acceleration due to gravity, h is an altitude variable, and p is the thermal pressure $n_n k T_n$ ($k=$ Boltzmann's constant, $T_n=$ temperature) of the neutral gas under consideration. If T_n is assumed independent of h, this equation has the exponential solution

$$n_n = n_0 \exp\frac{-(h-h_0)}{H_0} \qquad (1.69)$$

where H_0 defines the *scale height* of the gas,

$$H_0 = kT/m_n g \qquad (1.70)$$

and n_0 is the density at the reference altitude h_0. The scale-height dependence on particle mass is such that the lightest molecules and atoms have the largest scale heights. Most planetary atmospheres are dominated at high altitude by hydrogen and helium. The temperature may depend on h, so that this simple exponential distribution will not always provide an accurate description.

1.13.2 Photoionization

To model an ionosphere one must first calculate the altitude profile of the rate of ion production Q. For photoionization, this entails a consideration of the radiative transfer of photons through the neutral gas. Some simplifying assumptions can be made that together will allow an analytical approach to ionosphere modeling known as Chapman theory. The altitude profile of ion production has a peak at some altitude, because the rate of ionization depends on both the neutral density (which decreases with height) and the incoming solar-radiation intensity (which increases with height). The goal is to describe ion production as a function of height for the simple case in which the details of photon absorption are hidden in a radiation absorption cross section σ, and in which ion production is assumed to depend only on the amount of radiative energy absorbed.

The ionosphere consists of three main parts: (a) the D-region situated between about 50 and 90 km, (b) the E-region continuing upwards to about 150 km, and (c) the F-region extending further to about 500 km. Limit the discussion to the normal midlatitude ionosphere and the electron density N_e shows an increasing trend with altitude through the ionosphere up to the peak electron density of the F-region occurring at about 250 km, then falls off gradually at greater heights. The photoionization of a molecule or atom is written as

$$hv + X \rightarrow X^+ + e \tag{1.71}$$

where hv is the energy of the responsible photon, X is target, X^+ is the resulting positive ion, and e is the newly created free electron. For photoionization to occur, the photon energy hv must exceed the ionization potential of the species X. Landmark (1973) gives in his Table 5.1 the ionization potential for the most important ionospheric constituents with values $\sim$10–20 eV. For the process of ion production to be important, both the cross section for ionization σ, and the photon density must be large enough. The solar electromagnetic radiation is attenuated in its passage through the atmosphere because of this ionization, dissociation, and excitation of the atmospheric constituents. The complex structure around 120 nm implies penetration of some wavelengths to levels of the atmosphere as low as 75 km.

Let q be the number of ion pairs produced in a unit volume per unit time, and N_e the number density of free electrons. The rate of change of free electrons must then be

$$\frac{dN_e}{dt} = q - \text{loss terms} \tag{1.72}$$

In many cases the predominant loss term is due to electron-ion dissociative recombination,

$$XY^+ + e \rightarrow XY \rightarrow X^* + Y^* \tag{1.73}$$

The symbol * indicates excitation. This process is faster than other types of recombination because in the final state two bodies are present to enable momentum and energy to be conserved. For this chemical reaction, rate theory states

$$\frac{dN_e}{dt} = q - \alpha[XY^+]N_e = q - \alpha N_e^2 \tag{1.74}$$

The second equation is true because charge neutrality must hold. If there are several ion species present, an effective recombination coefficient α_{eff} must be used. This equation is believed to be valid at E-region levels, that is, in the height range between about 90 and 150 km. At F-region levels, that is, heights above about 150 km, charge transfer from primary to secondary ions becomes important, and also loss terms due to movements.

Chapman was the first to consider the case of ion production by monochromatic radiation. He assumed that there are present: (a) one gaseous type only, (b) planar stratification, (c) parallel beam of radiation, and (d) an isothermal atmosphere. The energy absorbed in thickness dh is

$$dI = -I\alpha N_n(h)dh\sec\chi \tag{1.75}$$

because the length is $dh\sec\chi$. The absorption cross section is σ, and $N_n(h)$ is the neutral gas density. Upon integration, one obtains

$$I = I_\infty \exp(-\tau\sec\chi) \tag{1.76}$$

Where

$$\tau = -\int_\infty^h \sigma N_n(h)dh \tag{1.77}$$

and is known as the *optical depth* of the atmosphere down to the height h. (The negative sign is due to positive direction of h upwards.) Then q pairs of electrons and ions will be produced per unit volume per second equal to

$$q(\chi, h) = N_0 \sigma I_\infty n \exp(-\tau\sec\chi) \tag{1.78}$$

with these assumptions

$$N_n(h) = N_0 \exp\left(-\frac{h}{H}\right) \tag{1.79}$$

where H is the scale height. The normalized Chapman production function q/q_m as a function of reduced height and zenith angle χ shows the same shape but with a reduced amplitude by a factor $\cos\chi$, and its peak shifted in height. The actual shape will of course vary, because the scale height may not be constant.

The various constituents of the upper atmosphere are shown by Fig. 1.22; they vary greatly, over 10 orders of magnitude. The

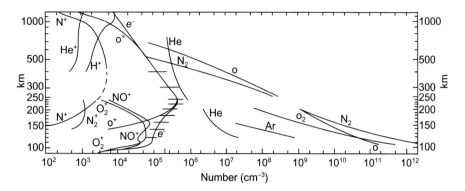

Fig. 1.22 IQSY daytime ionospheric and atmospheric composition based on rocket measurements above White Sands, New Mexico (32°N, 106°W). These data illustrate average daytime conditions for sunspot minimum. They demonstrate the great preponderance of neutral gases over ionic constituents in the upper atmosphere. From Rishbeth, H., Garriott, O.K., 1969. Introduction to Ionospheric Physics. Academic Press, New York, NY.

percentage of the ionized constituents is minuscule even at 300 km.

The diffusion of ions is interesting if several species are present. If only one species is present, the effective scale height is twice that of the neutral. However, with a mixture the lighter ions actually have a negative scale height, all because of the buoyant electrons.

1.14 High-frequency wave propagation

Electromagnetic waves are treated by Goertz and Strangeway in Chapter 12 of Kivelson and Russell (1995). We discuss some essentials here, with reference to ionospheric sounding. We first assume that the unperturbed plasma has no magnetic field. Electromagnetic waves are transverse waves, with propagation along the x-axis so that $\mathbf{k} \cdot \mathbf{E}_1 = 0$. In this case, we do not need Poisson's equation. The remaining equations in their linearized form are:

$$\nabla \times \mathbf{E}_1 = -\frac{\partial \mathbf{B}_1}{\partial t} \tag{1.80}$$

$$\frac{1}{\mu_0} \nabla \times \mathbf{B}_1 = \mathbf{j}_1 + \varepsilon_0 \frac{\partial \mathbf{E}_1}{\partial t} \tag{1.81}$$

$$m_e n_0 \frac{\partial \mathbf{u}_{e1}}{\partial t} = -\nabla p_{e1} - e n_0 \mathbf{E}_1 \tag{1.82}$$

We assume a cold plasma, and that the ions are immobile for high frequencies. With the assumed exponential dependence on x and

t of the wave properties, the dispersion relation is in linearized form:

$$\omega^2 = \omega_{pe}^2 + k^2 c^2 \tag{1.83}$$

In this case, $\mathbf{k} = k\hat{\mathbf{x}}$ without loss of generality. If the plasma frequency is much smaller than the wave frequency, the wave becomes a free-space light wave, with $\omega = kc$. It is useful to introduce an index of refraction:

$$n = c/v_{ph} = ck/\omega \tag{1.84}$$

For these waves,

$$n = \sqrt{1 - \omega_{pe}^2/\omega^2} \tag{1.85}$$

We see that for $\omega < \omega_{pe}$, the index of refraction becomes imaginary. For a real frequency, k is purely imaginary, and such a wave would decay in space and would not propagate.

When a magnetic field exists in the unperturbed plasma, the motion of charged particles is modified. Particles can move freely along the magnetic field, but not perpendicular. The Lorentz force $q(\mathbf{v} \times \mathbf{B})$ acts perpendicular to the magnetic field and causes the charged particles to gyrate about the magnetic field at their respective gyrofrequencies.

We are still dealing with linear waves, which involve the first-order electric field E_1 and the first-order magnetic field B_1. However, we must also include the zeroth-order (unperturbed) magnetic field B_0. Thus the Lorentz force contains a first-order term $\mathbf{u}_1 \times \mathbf{B}_0$ that did not appear in an unmagnetized plasma. Because of the presence of B_0, there is now also a unique direction; namely, the direction of this field. Waves, of course, can propagate in all directions. It is convenient to discuss special cases characterized by the direction of the wave vector k relative to the magnetic field and the perturbation electric field. We can distinguish among six different cases:

1. Parallel propagating waves have

$$\mathbf{k} \times \mathbf{B}_0 \tag{1.86}$$

2. Perpendicular waves have

$$\mathbf{k} \cdot \mathbf{B}_0 = 0 \tag{1.87}$$

3. Longitudinal waves have

$$\mathbf{k} \times \mathbf{E}_1 = 0 \tag{1.88}$$

4. Transverse waves have

$$\mathbf{k} \cdot \mathbf{E}_1 = 0 \qquad (1.89)$$

Any wave with $\mathbf{E}_1$ neither exactly parallel to $\mathbf{k}$ nor perpendicular to $\mathbf{k}$ can be constructed as a superposition of these two waves. There are two distinct physical modes distinguished by whether or not they have a wave magnetic field:

5. Electrostatic waves have no magnetic field perturbation

$$\mathbf{B}_1 = 0 \qquad (1.90)$$

6. Electromagnetic waves involve a magnetic field perturbation

$$\mathbf{B}_1 \neq 0 \qquad (1.91)$$

Waves can be classified according to this scheme as, for example, parallel propagating electrostatic waves, perpendicular electromagnetic waves, and so forth. If $\mathbf{k} \times \mathbf{E}_1 = 0$ (longitudinal waves), $\partial \mathbf{B}/\partial t = 0$, and hence $\mathbf{B}_1 = 0$. Thus longitudinal waves are electrostatic, and vice versa. All transverse waves are electromagnetic waves, but not all electromagnetic waves are transverse waves.

1.14.1 Perpendicular propagation

Eq. (1.83) also describes the *ordinary* (*O*) wave, which propagates perpendicular to the field $\mathbf{B}_0$ as if it did not exist. It is identical to the equation of the unmagnetized plasma only if the electron pressure vanishes.

We now consider a wave with $\mathbf{E}_1$ in the $\hat{\mathbf{x}} - \hat{\mathbf{y}}$ plane. Of course, any wave may be thought of as a superposition of this wave and the wave with $\mathbf{E}_1$ along $\mathbf{B}_0$. The wave with $\mathbf{E}_1$ perpendicular to $\mathbf{B}_0$ is called an *extraordinary* (*X*) wave. The dispersion relation can be written in 2×2 matrix form (see Kivelson and Russell, 1995, p. 382). It is instructive to write it in terms of the refractive index

$$n = c/v_{ph} = ck/\omega \qquad (1.92)$$

$$n^2 = \frac{k^2 c^2}{\omega^2} = 1 - \frac{\omega_{pe}^2}{\omega^2} \frac{\omega^2 - \omega_{pe}^2}{\omega^2 - \omega_{UH}^2} \qquad (1.93)$$

$$\omega^2 = \omega_{pe}^2 + \Omega_{ce}^2 \equiv \omega_{UH}^2 \qquad (1.94)$$

The angular frequency ω_{UH} is the upper hybrid frequency, which demonstrates the extra force due to the magnetic field.

1.14.2 Appleton-Hartree dispersion

For a general angle of propagation, a dispersion relation known as the Appleton-Hartree equation (neglecting ions and collisions) has been used for over 50 years. It is common to define X related to plasma density

$$X = \omega_{pe}^2 / \omega^2 \tag{1.95}$$

and Y related to the magnetic field

$$Y = |\Omega_{ce}| / \omega \tag{1.96}$$

to get

$$n^2 = 1 - \cfrac{X}{1 - \cfrac{\frac{1}{2} Y^2 \sin^2\theta}{1 - X} \pm \left\{ \left(\cfrac{\frac{1}{2} Y^2 \sin^2\theta}{1 - X} \right)^2 + Y^2 \cos^2\theta \right\}^{\frac{1}{2}}} \tag{1.97}$$

This looks impressive, but we cannot do it justice here; it is included for reference only. The interested reader is referred to Chapter 12 of Kivelson and Russell (1995).

1.14.3 Clemmow-Mullaly-Allis diagram

For a plasma containing only a single species of ions, the Clemmow-Mullaly-Allis (CMA) diagram may be drawn as in Fig. 1.23 in a parameter space of just two dimensions (this diagram is not used in Kivelson and Russell (1995)). The dimensionless quantities $|\Omega_e|/\omega$ and ω_{pe}^2/ω^2 can serve as ordinate and abscissa related to magnetic field and plasma density, respectively. The CMA bounding surfaces are formed by the surfaces for cutoff and for principal resonance. Ion-to-electron mass ratio is chosen to be an unrealistic 2.5.

1.14.4 The ionosonde

The fact that electromagnetic waves propagate or decay, depending on the relation between ω and ω_{pe}, can be used to diagnose the plasma. By sweeping through frequency, we can measure the value of the plasma frequency and hence infer the plasma density. This principle allows us to measure remotely the plasma density in the ionosphere. This was first used Breit and Tuve

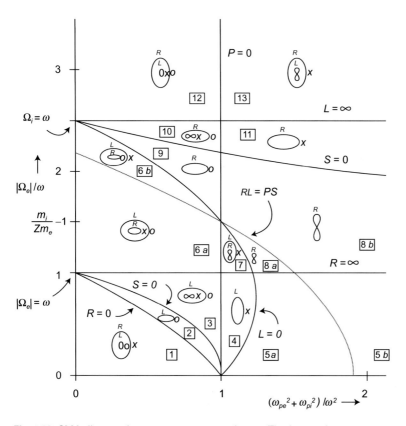

Fig. 1.23 CMA diagram for a two-component plasma. The ion-to-electron mass ratio is chosen to be 2.5. Bounding surfaces appear as lines in this 2D parameter space. Cross sections of wave-normal surfaces are sketched and labeled for each region. For these sketches the direction of the magnetic field is vertical. The small mass ratio can be misleading. The bounding surfaces divide the parameter space into 13 volumes (areas), which are numbered in an arbitrary fashion. In each numbered region, cross sections for the appropriate topological figure are sketched for the wave normal surfaces. Indicated on each wave normal surface representation are the labels R and L. The division into fast (F) and slow (S) waves is clear from the relative appearance (outside and inside) of the wave normal surfaces. This is included here because of its usefulness. From Stix, T.H., 1992. Waves in plasmas. American Institute of Physics Press, New York, NY, p. 27.

(1925) in United States and by Appleton and Barnett (1925) in England.

A transmitter on the ground (or a low-altitude spacecraft) emits a short pulse of radiation at a frequency ω. The wave will propagate through the atmosphere into the ionosphere, either upwards from the ground or downwards from a spacecraft. The pulse may reach a height where the plasma frequency ω_{pe} will equal ω. At that height, the wave will be reflected, and after a certain time delay Δt, it will return to the ground (or spacecraft), where it can be detected. Because the time delay is equal to twice the wave travel time from the ground to the reflection height h, we can calculate the height h where the electron density is n_e. By using different frequencies, we can deduce a height profile for the ionospheric electron density $n_e(h)$. The device of transmitter and receiver is called an ionosonde. An example of the output of such an ionosonde, called an ionogram, is shown in Fig. 1.24.

Benson and Bilitza (2009) have recently reviewed efforts to save a unique data set of Alouette 2, ISIS 1, and ISIS 2 digital topside ionograms. This project was initiated to preserve a significant portion of 60 satellite years of analog data, collected between 1962 and 1990, in digital form before the tapes were discarded. Digital topside ionograms are now available for downloading at http://nssdcftp.gsfc.nasa.gov and for browsing and plotting at http://cdaweb.gsfc.nasa.gov. The scientific results include evidence of extremely low-altitude ionospheric peak densities at high latitudes, improved and new ionospheric models including one connecting the F2 topside ionosphere and the plasmasphere, transionospheric HF propagation investigations, and new interpretations of sounder-stimulated plasma emissions that have challenged theorists for decades.

Fig. 1.24 (top) illustrates an ISIS 2 ionogram. The Z, O, and X ionospheric reflection traces are also labeled. There is also a wideband noise signal between the plasma frequency f_N and upper hybrid frequency f_T. The following passive ionogram (bottom) indicated that the former had a component due to natural emissions at f_N. It is this combination of active and passive operations that enables such labels to be made with confidence.

The question mark at the left is very interesting. The topside sounder transmitter pulses have a definite characteristic in terms of polarity, which in general is not orthogonal to the magnetic field; there must, in general, be a finite current. The plasma tries to have that equal to zero as given by Eq. (1.30). That sets up a counter emf as measured by the response shown by the question mark (treated later in Section 3.11.4).

Because the ionospheric plasma is magnetized, there are several different electromagnetic waves, all propagating at slightly different velocities. A single transmitter can emit a mixture of these waves that will return to the receiver after different time delays. Hence, an ionogram often contains multiple traces that track each other relatively closely.

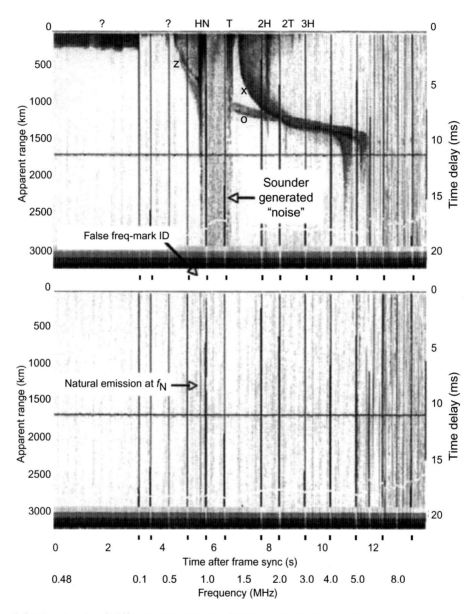

Fig. 1.24 ISIS 2 Resolute Bay (RES), year 1971, day 231, digital topside fixed/swept-frequency ionograms with normal 0.1–10.0 MHz sweeps and operating in the D mode, (Top) Transmitter ON (0146:17 UT). (Bottom) Transmitter OFF (0146:31 UT). Plasma resonances at the electron plasma frequency f_N, the electron gyrofrequency f_N and harmonics, the upper hybrid frequency $f_T(f_T^2 = f_N^2 + f_H^2)$ and at $2f_T$ are identified as N, H. From Benson, R. F., Bilitza, D., 2009. New satellite mission with old data: rescuing a unique data set. Radio Sci. 44, 6, RS0A04. doi:10.1029/2008RS004036.

1.14.5 Coherent and incoherent scatter radars

The radar aurora detects the ionization associated with auroral activity. These echoes come from the field-aligned irregularities, which are therefore strongly aspect sensitive, perpendicular to the magnetic field. Hargreaves (1992, p. 80) has discussed this, and his Fig. 3.16 shows the installation in northern Europe in the 1980s. When two radars look at a signal at the same location, they each produce the line-of-sight velocity from the Doppler shift of the echo. Thus 2D maps are produced, a great advance for auroral research.

SuperDARN auroral radar (Greenwald et al., 1995) is a global-scale network of HF radars capable of sensing backscatter from ionospheric irregularities in the F region of the high-latitude ionosphere. Currently, the network consists of a northern-hemisphere, longitudinal chain of 23 HF radars, and a southern-hemisphere chain of 12 radars. They primarily operate in pairs with common viewing areas so that the Doppler information contained in the backscattered signals may be combined to yield maps of high-latitude plasma convection and the convection electric field. The network is particularly suited to studies of large-scale dynamical processes in the magnetosphere-ionosphere system, such as the evolution of the global configuration of the convection electric field under changing IMF conditions and the development and global extent of large-scale MHD waves in the magnetosphere-ionosphere cavity.

Incoherent scatter radar is a relatively recent technique for observing the ionosphere (see Hargreaves, 1992). It is a remote sensing technique, which has been developed into a powerful and flexible tool. It is not restricted to the region below the level of peak electron density but can observe both sides of the peak simultaneously. Also, because the antenna has to be large relative to the radio wavelength, it produces a narrow beam and achieves far better spatial resolution. The principal disadvantage is that an incoherent scatter radar has to work with a very weak signal. It therefore requires a transmitter of high power, a large antenna, and the most sensitive receiver and sophisticated data processing available, all of which add up to a major facility.

European Incoherent Scatter Radar System (EISCAT) has been very successful in exploiting its two radars and three receiving stations located on the Scandinavian mainland since 1981 (one operating at 931 MHz and the second at 231 MHz). This success led directly to the design, construction, and operation of the EISCAT Svalbard Radar nearly 1000 km further north in 1996. EISCAT-3D, currently under construction with a completion date expected around 2021, will consist of five phased-array antenna fields located in northern Finland, Norway, and Sweden. One site will transmit at 233 MHz, with all five receiving. This new system will allow volumetric imaging, allowing geophysical events to be seen in their full spatial context, and to distinguish between spatial and temporal variations. Other radars exist at Irkutsk (Russia), Kharkov (Ukraine), Kyoto (Japan), Sondrestrom (Greenland), Millstone Hill (United States), Arecibo (Puerto Rico, United States), Jicamarca (Peru), and in Indonesia.

1.15 Polar caps

The polar caps are defined as the regions of the ionosphere around the magnetic poles with field lines that do not close back on the geomagnetic field, but are connected to the IMF. These regions are thus open to the solar wind and are prime regions for particle and energy exchange between the ionosphere, magnetosphere, and solar wind.

1.15.1 Polar cap during southward IMF

When the IMF has a southward component, the high-latitude ionospheric convection pattern shows a familiar two-cell geometry: antisunward flow at highest latitudes, sunward flow at lower latitudes, both on the dawn and dusk sides. This situation is associated with a dawn-dusk electric field within the plasma sheet, the earthward convection of PS plasma, and the energization of plasma sheet particles. A strictly southward IMF, $\mathbf{B}(0, 0, -B_z)$, although rare in practice, is worth special attention since it is conducive to geomagnetic activity. Phan et al. (1994) have shown that similar conditions are held until the IMF reaches to 60 degrees northward. Both the Standard Magnetic Reconnection (SMR) and the PTE are easier to understand if the x and y components are assumed to vanish. The dusk-to-dawn distribution of the equipotential is affected by the direction of the y component of the IMF.

1.15.2 Polar cap during northward IMF

One of the important questions is: What happens when IMF becomes northward for an extended period? As the power of the solar wind/magnetosphere generator decreases, the aurora becomes dim and the auroral electrojets become weak. However, an unexpectedly interesting auroral phenomenon takes place in this situation. Some auroral activity does occur on the polar cap during northward IMF, the theta aurora reported by Frank et al.

(1986). Observations show that theta aurora can form during strictly northward IMF with its motion consistent with a change in sign of IMF B_y. No sign change in B_z is needed for theta aurora formation.

This phenomenon and many others associated with it (field-aligned currents, convection, etc.) cannot be simply understood in terms of the decreasing power of the generator. Frank et al. (1986) showed that the azimuthal angle (or the east-west component) of the IMF plays an important role in determining the distribution of the aurora when the hour-angle θ becomes small. The convection pattern also becomes asymmetric with respect to the noon-midnight meridian; we also get sunward plasma convection. This is covered in Sections 7.3.1 and 9.11, 9.12, and Figs. 9.31 and 9.44.

1.16 The aurora and substorms

The driving force behind the effort of analysis of magnetic records was Kristian Birkeland. In his day, only the simplest ground instruments were available for auroral investigations. Today, auroral research is conducted mainly through the use of sophisticated instruments onboard rockets and satellites, as well as advanced balloon and ground-based equipment. Even artificial auroras have been produced in the Earth's atmosphere (e.g., Winckler et al., 1975). Some of the mysteries of the northern lights have been solved, partly or fully, but new problems have appeared, and the study of the auroras continues to engross many scientists.

Field-aligned currents (FAC), often called Birkeland currents, had been inferred to exist in the polar ionosphere quite some time ago. There has been a tremendous effort to synthesize all aspects of high-latitude observations of FAC, auroral displays, electric fields, and neutral winds into a coherent global description of the electrodynamics. The ultimate goal is to understand the nature of the solar wind interaction with the Earth's magnetosphere.

1.16.1 Akasofu's study of substorm aurora

It was the all-sky camera operation and the subsequent analyses during the IGY (1957/58) that revealed systematic auroral activity over the entire polar region, called the auroral substorm (Akasofu, 1964), shown in Fig. 1.25. *This figure is repeatedly used at conferences*! The first indication of an auroral substorm is a sudden brightening of the auroral curtain in the midnight or late evening sector. This brightening spreads rapidly so that

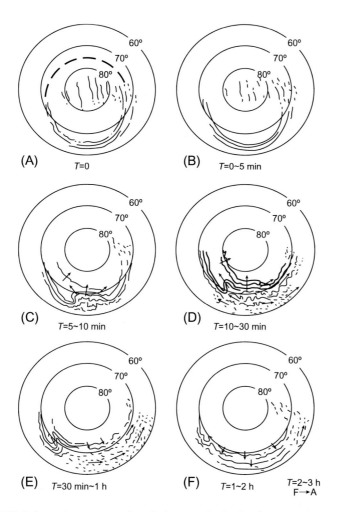

Fig. 1.25 Schematic representation of six stages in the development of an auroral substorm as determined from all-sky camera data during the IGY 1957. (A) Quiet state with multiple arcs drifting equatorward. (B) Sudden brightening at onset. (C) Rapid expansion westward and poleward. (D) Westward traveling surge and omega bands. (E) End of expansion phase, beginning of recovery. (F) Recovery phase. From Akasofu, S.-I., 1964. The development of the auroral substorm. Planet. Space Sci. 12(4), 275. doi:10.1016/0032-0633(64)90151-5.

in a matter of minutes the entire section of the curtain in the dark hemisphere becomes bright, with a speed of a few hundred meters per second. A large-scale wavy motion, called the westward traveling surge, propagates westward (toward the dusk-sunset line) with a speed of about 1 km/s. In the morning sector, auroral curtains appear to disintegrate into many patches. The poleward motion in the midnight sector lasts typically for about 30 min to 1 h. After this poleward advancing curtain reaches its highest latitude, auroral activity begins to subside. We continue this debate in Chapter 11 starting with Fig. 11.2 and ending with Fig. 11.41.

1.16.2 Iijima and Potemra: Field-aligned currents

Fig. 1.26 summarizes field-aligned current observations made by the Triad satellite (Iijima and Potemra, 1976, 1978). *This figure is repeatedly used at conferences!* The principal feature of this

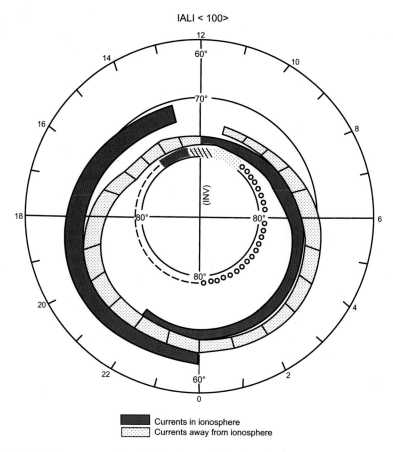

Fig. 1.26 A summary of the distribution and flow directions of large-scale field-aligned currents determined from data obtained from 439 passes of Triad during weakly disturbed conditions ($|AL| < 100\ \gamma$). From Akasofu, S.-I, 2007. Exploring the secrets of the aurora. Springer, New York, NY.

diagram is the unambiguous separation of the current systems into three components: the circumpolar region 1 current, which lies poleward of the neighboring region 2 current, and the cusp current system, which occurs in the region spanning 10–14 h local time. It took quite a while for the FAC dream by Birkeland to be accepted due to opposition by Chapman. Alfvén finally won; read the account by Dessler (1984).

Characteristics of field-aligned currents have been determined during a large number of substorms from the magnetic field observations acquired with the Triad satellite. The statistical features of field-aligned currents include the following:

- The large-scale regions of field-aligned currents persist during all phases of substorm activity.
- During active periods the average latitude width of regions 1 and 2 increases by 20%–30%, and the centers of these regions shift equatorward by 2–3 degrees with respect to the quiet time values.
- The average total amount of field-aligned current flowing into the ionosphere always equals the current flow away from the ionosphere during a wide range of quiet and disturbed conditions. The average total current during quiet periods is $\sim 2.7 \times 10^6$ A and during disturbed periods is $\sim 5.2 \times 10^6$ A.

1.16.3 Clauer and McPherron's current diversion

A substorm is the ordered sequence of events that occurs in the magnetosphere and ionosphere at auroral breakup (McPherron, 1979; Akasofu, 1979; Rostoker et al., 1980). During a substorm, quiet auroral arcs suddenly explode into brilliance; they become intensely active and colored. Over a period of about an hour, they develop through an ordered sequence that depends on time and location. Magnetic disturbances also accompany the aurora (Fig. 1.27). On the ground beneath the aurora, a magnetometer will record intense disturbances caused by electric currents in the ionosphere. Stations located in the afternoon-to-evening sector record positive disturbances, whereas stations near and past midnight record negative disturbances relative to the field measured on quiet days. Applying the right-hand rule to currents assumed to be overhead leads to the conclusion that the currents are, respectively, eastward and westward toward midnight. These currents are called electrojets, because the currents flow in concentrated channels of high conductivity by the particles that generate the auroral light. Typical disturbances have amplitudes in the range of 200–2000 nT, and durations of 1–3 h. *This figure is also repeatedly used at conferences*!

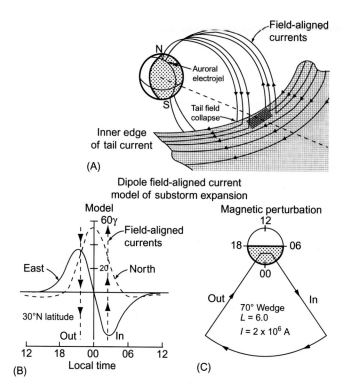

Fig. 1.27 Schematic illustration of the 3D current system that is responsible for the DP-1 current system during the expansion phase of a polar magnetic substorm. (A) Diversion of the cross-tail current through the midnight ionosphere. (B) Magnetic perturbations caused by this current system along a chain of northern midlatitude magnetic observatories. (C) The wedge shape of the projected equivalent current accounts for the name substorm current wedge. From Clauer, C.R., McPherron, R.L., 1974. Mapping the local time-universal time development of magnetospheric substorms using mid-latitude magnetic observations. J. Geophys. Res., 79, 2816.

Many 3D models of the magnetosphere-ionosphere current system include field-aligned or Birkeland currents as a principal element. Birkeland (1908) suggested that field-aligned currents were coupled to the auroral electrojets, which produced large ground level magnetic disturbances during what is now referred to as substorms. A large number of substorm studies have been, and are still, based upon the associated magnetic disturbances measured on the Earth's surface (e.g., Akasofu, 1968). As reviewed by Fukushima (1969) and Rostoker (2007), it is not possible to distinguish unambiguously between current systems that are field aligned and those which are completely 2D and confined to the ionosphere, from a study of only surface magnetic field measurements.

Observations of field-aligned currents with rockets and satellites have contributed significantly to an understanding of the
magnetosphere-ionosphere coupling processes as discussed in
the theoretical studies of Boström (1974) and Wolf (1983). The
rocket-borne experiments have provided detailed observations
of the fine structure of field-aligned currents and their relationship to energetic particles and auroral images, which are confined
to a limited range of space and time (Arnoldy, 1974). The satellite
experiments have provided observations of the large-scale features of field-aligned currents over a much larger spatial range.

1.16.4 Geomagnetic indices

Magnetic activity at the Earth's surface is produced by electric
currents in the magnetosphere and ionosphere. Over the past two
centuries, a large network of more than 200 permanent magnetic
observatories has been established. Data from these observatories
and from temporary stations are frequently used to study magnetospheric phenomena, often in conjunction with in-situ observations by spacecraft. Several organizations routinely generate
indices of magnetic activity. Ideally, an index is simple to generate,
but useful, as it varies monotonically with some meaningful
physical quantity, such as the total current flowing in a system.
New indices were defined, and older ones abandoned. Some indices are continued simply for historical reasons. They provide an
ever-lengthening sequence of measurements that can be used
for long-term studies of such phenomena as solar-cycle effects.
Reviews of magnetic indices are discussed by Davis and Sugiura
(1966), Lincoln (1967), Mayaud (1980), Rostoker (1972),
Troshichev (1988), Menvielle and Berthelier (1991), and
McPherron (1995).

1.16.5 Closed field line in the dayside cleft

McDiarmid et al. (1976) established with the Alouette and ISIS
satellites over four decades ago that there are closed field lines
within the cusp (cleft) precipitation. They were able to detect
the anisotropic pancake pitch angle distribution that is a mark
of closed field lines well into the cleft. Fig. 1.28 shows that "On
the average, anisotropic distributions extends about half way
across the cleft region; however, it is significant that in some cases
the pitch angle distribution is anisotropic throughout the entire
cleft region." The current carriers that are responsible for such
work as that in Fig. 1.26 must be on closed field lines.

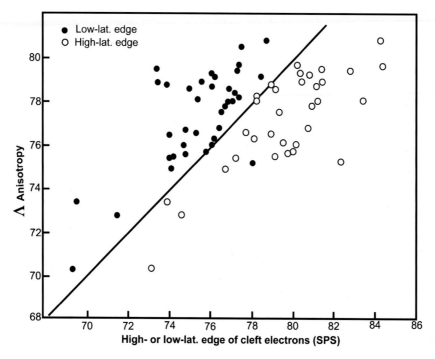

Fig. 1.28 Plot of the highest latitude where anisotropic pitch angle distributions are observed against the high- and low-latitude edges of the cleft determined from the soft particle detector SPS. From McDiarmid, I., Burrows, J., Budzinski, E., 1976. Particle properties in the day side cleft. J. Geophys. Res. 81(1), 225.

1.16.6 Impulsive injections by rocket

The first clear data on temporal structure came from the rocket flight in Greenland before noon on December 18, 1974 (Carlson and Torbert, 1980). Particle data is shown as spectrograms in Fig. 1.29. The upper panel displays the electron counting rate from energies of 50–5000 eV. The lower panel shows the ion counting rate with clear evidence of time dispersion, with the higher-energy ion flux preceding the lower. The authors concluded that the data are consistent with simultaneous injection at all energies at a distance of 12 R_E, placing the injection point near the magnetopause.

By the previous result, these data might be on closed field lines. Similar events are recorded regularly on spacecraft; injection events observed on a Viking pass are shown in Chapter 6 simultaneous with trapped electrons at higher energy. As already noted, the idea that solar wind plasma-field irregularities with an excess momentum density penetrate deeper into the geomagnetic field was introduced by Lemaire and Roth (1978). The presence of these plasma irregularities indicates that the solar wind momentum density is not uniform, and that the dynamic pressure that the solar

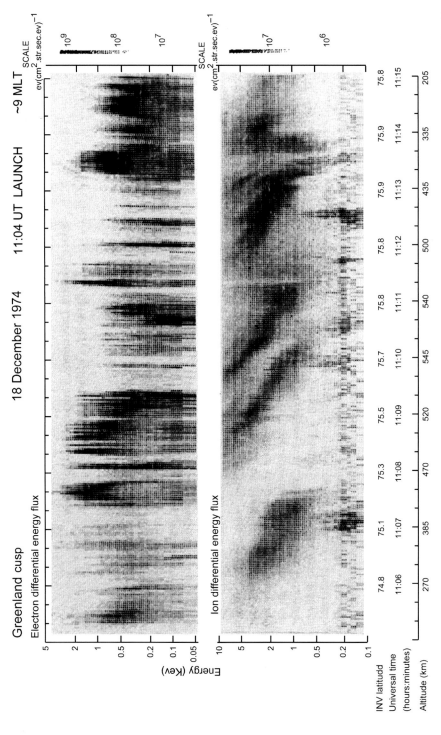

Fig. 1.29 An early rocket data showing ion time dispersion. The data are consistent with simultaneous injection of all energies at a distance of 12 R_E and time-of-flight dispersion. From Carlson, C.W., Torbert, R.B., 1980. Solar wind ion injections in the morning auroral oval. J. Geophys. Res. 85(A6), 2904.

wind inflicts upon the magnetosphere must be patchy, nonuniform, and changing in time. The magnetopause itself might be in motion, locally, perhaps as a surface wave due to some instabilities.

It was not until 1982 that Heikkila proposed another idea making use of magnetic energy. The electric field is not the convection electric field but the electromagnetic field with its emphasis on induction.

1.16.7 International conferences on substorms

The causes of the magnetospheric substorm have been one of the major topics among magnetospheric physicists during the past several decades. Now there is a biannual series of conferences on the subject, the first one at Stockholm in March 1992. ICS-14 was held September 2019 in Norway.

1.16.8 Introduction to current and future missions

The magnetosphere has been the target of direct in-situ measurements since the beginning of the space age. Gone are the days of single-spacecraft measurements plagued by the inability of a single observer to distinguish spatial from temporal changes. Recent multispacecraft missions flying in 3D configurations at variable separation distances represent a giant step forward. The following is not intended to be an exhaustive list, see Section 12.1 for a more complete listing of missions, but rather to highlight missions representative of the new era in the field.

Determining the manner by which plasma is transferred from one bounded region to another, together with the associated system response, has been one of space plasma physics' primary objectives since the birth of the discipline.

The Earth's magnetosphere, in particular, provides an excellent laboratory in which to investigate these plasma processes and as such we have benefited from a number of space missions examining them both in situ (e.g., ISEE, Interball, Geotail) and remotely (e.g., Polar, IMAGE). It is only recently however, with the successful launch and operation of multi-spacecraft missions (Cluster, Double Star and THEMIS), that we have been able to distinguish between temporal and spatial variations, and to probe small and meso-scale structures, leading to a step change in our understanding of the local plasma processes and the dynamics of boundaries of the magnetosphere. The recent observations provide glimpses of the micro- and multiscale processes that will be the focus of future missions like MMS and SCOPE.

Matthew Taylor, Malcolm Dunlop, C.-Philippe Escoubet, European Geosciences Union General Assembly 2011, April 3–8, 2011, Vienna, Austria.

1.16.8.1 Cluster

Cluster's history goes back to 1982 when scientists responded to a call for proposals by ESA. In February 1986, ESA chose Cluster and the Solar and Heliospheric Observatory (SOHO) as the first "Cornerstone" in its Horizons 2000 Science Program. Ready for launch in the fall of 1995, the four spacecraft were destroyed when the Ariane 5 rocket exploded on June 4, 1996. The spacecraft and instruments were rebuilt and then successfully launched pairwise on two Soyuz-Fregat rockets from Baikonur on July 9 and August 12, 2000, respectively. After extensive commissioning, the science phase officially began on February 1, 2001.

The Cluster spacecraft were placed in nearly identical, highly eccentric polar orbits, with an apogee of 19.6 R_E and a perigee of 4 R_E. Fig. 1.30 shows the orbit superimposed on a cut of the magnetosphere. Cluster moves outbound over the northern polar cap, crosses the magnetopause and bow shock into the solar wind, before recrossing those boundaries in reverse order and moving over the southern polar cap back toward perigee. Fig. 1.30A applies to spring of each year. As the orbit is inertially fixed, it rotates around the Earth once a year. The apogee will be located in the geomagnetic tail half a year later, shown in Fig. 1.30B.

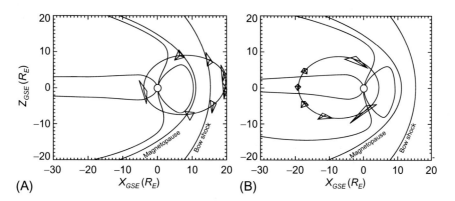

Fig. 1.30 Cluster orbit in spring (A) and in fall (B), as it cuts through the key regions of the magnetosphere and its boundaries. The spacecraft separations are exaggerated for clarity. From Paschmann, G., Escoubet, C.P., Schwartz, S.J., Haaland, S.E., 2005. Introduction. In: Paschmann, G., Schwartz, S., Escoubet, C.P., Haaland, S. (Eds.), Outer Magnetospheric Boundaries: Cluster Results. Springer, Netherlands, p. 2.

The orbits are tuned so that the four spacecraft are located at the vertices of a nearly regular tetrahedron when crossing the major boundaries. The separation distances between the spacecraft are adjusted from ~100 km to ~10,000 km during the mission. The four Cluster spacecraft carry identical sets of 11 scientific instruments, designed to measure the ambient electromagnetic fields and particle populations over a wide range of frequencies and energies. The instruments, the spacecraft, and mission are described in Escoubet et al. (1997). The truly international scale of the endeavor is evident from the fact that more than 200 investigators from ESA member states, the United States, Canada, China, the Czech Republic, Hungary, India, Israel, Japan, and Russia, are involved in analysis of the Cluster data.

1.16.8.2 THEMIS

The Time History of Events and Macroscale Interactions during Substorms (THEMIS) mission was designed to study magnetospheric substorms using a constellation of five satellites. All five were launched on a single Delta II launch vehicle on February 17, 2007. The name alludes to the Titan Themis, goddess of justice, wisdom, and good counsel. During the primary science mission, the spacecraft orbits were at apogee in the magnetotail, facilitating observations of substorm onset and reconnection events. This phase was performed during northern winter so that the ground-based instrumentation in the northern polar regions would be in darkness. The original THEMIS mission and instrumentation are detailed in the book The THEMIS Mission, edited by Burch and Angelopoulos (2009).

In 2010 THEMIS transitioned into an extended mission called ARTEMIS (Acceleration, Reconnection, Turbulence and Electrodynamics of the Moon's Interaction with the Sun). Two of the spacecraft (THEMIS B and THEMIS C) were redesignated as ARTEMIS P1 and ARTEMIS P2 and moved into lunar orbits. The remaining three spacecraft remained in Earth orbit. From lunar orbit, these two spacecraft are able to study the interaction of the unmagnetized Moon with the solar wind, including pick-up ions in the downstream flow created near the lunar surface.

1.16.8.3 MMS

Building off of the success of Cluster, MMS, the Magnetospheric Multiscale mission, also uses a formation of four spacecraft flying in formation to separate spatial and temporal variations. Launched on March 12, 2015, Burch and Torbert (2017) describe the goal of the mission "to study magnetic reconnection, a fundamental plasma physical process in which energy stored in a magnetic field is converted into the kinetic energy of charged particles and heat." This is accomplished through varying the distances between the spacecraft between 10 and 400 km, and by utilizing instruments capable of making extremely fast (down to 30 ms) measurements. MMS has the spatial and temporal resolution needed to resolve for the first time the microphysics of the electron diffusion region, where the magnetic field and the plasma become decoupled, allowing reconnection to occur. Phase one (September 2015 to February 2017) probed the dayside magnetopause, where the interplanetary and terrestrial magnetic fields reconnect. In the second phase (May 2017 to November 2017), MMS increased its apogee from 12 R_E to 25 R_E to investigate the magnetotail. MMS began an extended mission in January of 2018 to continue studying the magnetosphere and also investigate new phenomena such as turbulence in the solar wind and interplanetary shocks. Lower than expected fuel usage means that a total mission duration of more than 10 years is now likely (Williams et al., 2018).

1.16.8.4 RBSP

The Radiation Belt Storm Probes (RBSP), later renamed the Van Allen Probes, are a pair of spacecraft designed to study the Van Allen radiation belts. The two spacecraft carry identical instrument packages suited for the harsh environment of the radiation belts, including an Energetic Particle, Composition and Thermal Plasma (ECT) suite of instruments, an Electric and Magnetic Field Instrument Suite and Integrated Science (EMFISIS), an Ion Composition Experiment (RBSPICE), and a Relativistic Proton Spectrometer (RPS). The mission has led to fundamental changes in our understanding of the ring current during quiet and storm times, and how it is energized during geomagnetic storms. See Chapter 8 for more details. Launched on August 30, 2012, the probes are currently (as of late 2019) entering end-of-life activities after a successful scientific mission. Due to the need to combine the data from the two satellites in order to understand the dynamics of the radiation belts, the particle and field instruments were cross-calibrated before launch to a much greater degree than normal. This allows data from the two spacecraft to be intermixed as they probe various regions of the belts, as in Reeves (2015).

1.16.8.5 ICON

With a launch on October 10, 2019 during the final preparation of this edition, The Ionospheric Connection Explorer, or ICON mission is the newest of those listed here. Selected for development in April 2013 along with the Global-scale Observations of the Limb and Disk (GOLD), ICON's launch was repeatedly delayed between 2017 and 2019 due to issues with the Pegasus XL launch vehicle. Now that ICON has finally begun its 2-year primary mission, it will observe the equatorial ionosphere and thermosphere at an altitude of about 575 km.

The ICON mission has four primary science instruments, a Michelson Interferometer, FUV and EUV imagers, and an Ion Velocity Meter (IVM). The Michelson Interferometer for Global High-resolution Thermospheric Imaging (MIGHTI) instrument will observe the winds and temperature fluctuations in the thermosphere, which are driven by weather patterns closer to Earth's surface. In turn, the neutral winds drive the motions of the charged particles in the ionosphere. The two ultraviolet imagers will determine ionospheric and thermospheric density and composition by observing the airglow from high-energy particle collisions. The IVM is designed to measure the speed of local charged particle populations, which are pushed by high-altitude winds and associated electric fields. More information on the mission and instruments can be found in Rider et al. (2015).

1.17 Discussion

It is generally a good idea (even a necessity) to approach a model with the simplest physical assumptions, then to add complications as needed. This is especially true if the total amount of data available is very limited in either quantity or quality. Such is the case for the magnetosphere, where measurements were made on only a relatively few moving spacecraft. We owe a debt of gratitude to Dungey (1961), and to Axford and Hines (1961). Their simple proposals have guided us for over five decades. Of course, it goes without saying that any model must be falsifiable on the basis of observational evidence, should it prove to be contrary.

Dungey provided the idea, or process, of magnetic reconnection; later work has modified his process considerably. A key feature of this model (for southward IMF) is an electric field that always points from dawn to dusk in keeping with tailward flow. A magnetic X-line somewhere on the dayside as well as in the magnetotail seemed to be realistic in view of the changing topology. The convection velocity $\mathbf{v} = \mathbf{E} \times \mathbf{B}/B^2$ depends on this, or so it seemed according to observations as in Fig. 1.31A.

The process of magnetic reconnection was sketched (a) in the x,z noon-midnight meridian plane, while viscous interaction (b) uses the x,y equatorial plane, but both in 2D. The importance of three dimensions is beyond doubt; still, the difficulty in

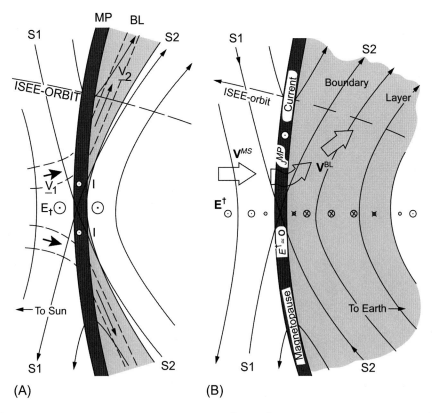

Fig. 1.31 (A) The reconnection geometry of the electric field. S1 and S2 represent different sheets of the separatrix. Standard Magnetic Reconnection depends on a tangential electric field. (B) A different topology of the electric field; here the plasma can move across closed field lines, since both **E** and **B** are reversed. Three-dimensional and time-dependent effects are important; this is General Magnetic Reconnection. From Heikkila, W.J., 1983. Comment on 'The causes of convection in the Earth's magnetosphere: A review of developments during the IMS' by S. W. H. Cowley. Rev. Geophys. 21(8), 1787.

conveying that idea on 2D paper seemed to be overwhelming. Birn et al. (2001) summarized the results of a coordinated study in the Geospace Environmental Modeling (GEM) program.

> *The conclusions of this study pertain explicitly to the 2D system. There is mounting evidence that the narrow layers which develop during reconnection in the 2D model are strongly unstable to a variety of modes in 3D system.*
>
> **Birn et al. (2001, p. 3718)**

However, rather taking the high road with 3D, the flat Earth system continues to be strong (e.g., see Fig. 1.16 or 1.20). Figs. 1.1 and 1.19 (and others) in this chapter are plainly in 3D. The physics of the LLBL just inside the magnetopause must imply

3D considerations (Where do the field lines go?) But the LLBL seems to have been overlooked in the analysis carried out thus far.

The trouble is in the mistaken concepts behind the so-called convection electric field: according to that, *there is just one electric field*:

$$\mathbf{E}_{conv} = -\mathbf{v} \times \mathbf{B} + \text{other terms} \tag{1.98}$$

It is forgotten that the electric field has two sources, induction (generally the larger) and charge separation (perhaps brought about by induction). The two types of field have different properties, one has a curl, the other does not (Heikkila and Pellinen, 1977).

$$\mathbf{E} = \mathbf{E}^{es} + \mathbf{E}^{ind} = -\nabla\phi - \frac{\partial \mathbf{A}}{\partial t} \tag{1.99}$$

By components:

$$\mathbf{E} = \mathbf{E}_{\perp}^{es} + \mathbf{E}_{\parallel}^{es} + \mathbf{E}_{\perp}^{ind} + \mathbf{E}_{\parallel}^{ind} \tag{1.100}$$

As already stated, this last equation is very important. The PTE depends on it, as does the LLBL.

The viscous interaction proposed by Axford and Hines was logical, in line with cause and effect according to Le Chatelier's principle. However, the exact mechanism has remained a mystery. The reality of the PTE phenomena and the LLBL is most likely the answer. Fig. 1.32 is based on the idea of Axford and Hines, but with the addition of a magnetospheric boundary layer discovered a decade later.

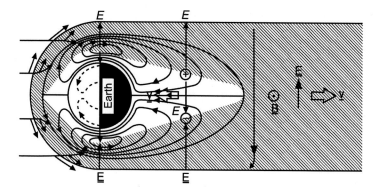

Fig. 1.32 The Axford and Hines (1961) model of the magnetotail, but with the addition of the LLBL. Since $\mathbf{E} \cdot \mathbf{J} < 0$ in the LLBL, plasma loses energy and momentum showing that it behaves as a viscous medium. From Heikkila, W.J., 1987. Neutral sheet crossings in the distant magnetotail. In: A.T.Y. Lui (Ed.), Magnetotail Physics. Johns Hopkins University Press, Baltimore, MD, p. 70.

Strangely enough, there are two articles that are highly relevant, published about the same time. Schmidt (1960) showed that a plasma cloud could travel through the plasma by creating a polarization electric field, with $\mathbf{E}\cdot\mathbf{J}<0$, a dynamo. Cole (1961) was also concerned with the subject of plasma flow through a boundary layer.

The great progress in auroral physics during the last few decades has created the awareness that auroral phenomena are various manifestations of dissipation processes associated with the solar wind/magnetosphere interaction. The Sun is continuously emitting enormous amounts of energy into space. This energy emission takes several forms (Akasofu, 1991; Gopalswamy et al., 2006), the first of which is the familiar blackbody radiation. The second mode of energy emission is the wind, which tends to confine the Earth and its magnetic field into a comet-shaped cavity called the magnetosphere. As the solar wind interacts with the magnetosphere, as much as 10^6 MW of electrical power is generated, discharged, and subsequently dissipated, partly through that portion of the upper atmosphere called the polar ionosphere. The third mode of energy emission, the solar X-ray and ultraviolet radiations, are responsible for producing the ionosphere.

Fig. 1.33 depicts three systems—the solar wind, the magnetosphere, and the ionosphere—which interact, transmitting and transforming the solar wind energy into energies of auroral phenomena, and eventually depositing most of it as heat energy in the ionosphere.

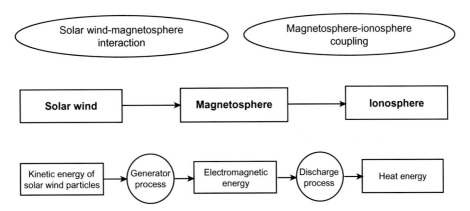

Fig. 1.33 Flow chart for solar wind/magnetosphere/ionosphere coupling. It shows the energy flow, energy conversion, and associated processes. Note that the first mechanism is a dynamo, necessary for the source of energy from the solar wind. This energy is dissipated in various ways to heat the magnetospheric and ionospheric plasmas and neutral gases. From Akasofu, S.-I., 1991. Auroral phenomena. In: Meng, C.-I., Rycroft, M.J., Frank, L.A. (Eds.), Auroral Physics. Cambridge University Press, New York, NY, p. 4.

Standard Magnetic Reconnection (SMR) occurs along an X-line, and has $\mathbf{E} \cdot \mathbf{J} > 0$, thus dissipation. It is not appropriate in the first block because it has the wrong sign.

> *We continue to struggle with age-old questions. This is the source of the angst ... At this ICS, we somehow managed to avoid debates that are fundamentally unresolvable, but still consider the dominant paradigms.*
>
> **Donovan, ICS8 (2006)**

> *We still have many long-standing, unsolved problems from the early days. [I] suggest that there is a possibility that the present paradigm may not be headed in the right direction. We have to recognize that several fundamental problems remain; ... what is needed is not just improvements to traditional theories or a mopping-up of residual problems. I am convinced that new thinking is needed to solve long-standing unsolved problems.*
>
> **Akasofu (2007, p. v)**

I propose that Fig. 1.32 offers a solution to the dilemma that these authors raise. It is indeed the model of Axford and Hines (1961) that is appropriate, but only with the addition of the LLBL. Since $\mathbf{E} \cdot \mathbf{J} < 0$ in the LLBL plasma loses energy and momentum, showing that it behaves as a viscous medium. Magnetic reconnection is involved, but it is in the general form in 3D proposed by Schindler et al. (1988).

> *More than a few of my friends of the same generation have a tendency, when they encounter one another at an AGU meeting, to grumble about the apparent lack of any significant progress in the field during the last 10 or 20 years.*
>
> **Axford (1994, p. 19199)**

Let us hope that we cannot say the same thing 20 years from now. The rest of this book is devoted to this objective.

1.18 Problems

1.1. What is the magnetic moment of an electron at the equator with a pitch angle of 30 degrees and kinetic energy of 14 keV at a magnetic L-shell of 3?

1.2. Show how Eq. (1.38) becomes Eq. (1.40) by using Eq. (1.39).

1.3. Derive Eq. (1.48). Start by crossing both sides of Eq. (1.47) with $\mathbf{B}_u$ and notice which term is zero in order to find that

$\mathbf{v}_{HT} \times \mathbf{B}_u = \mathbf{u}_u \times \mathbf{B}_u$. Cross each side of this equation with $\hat{\mathbf{n}}$ and use a vector identity on the left to expand.

1.4. What is the Alfvénic Mach number for a 0.50 eV electron for a typical value of v_A?

1.5. According to the image dipole model of Chapman and Ferraro, at what percentage of the distance from the origin to the magnetopause surface would you expect the image dipole contribution to be half as large as the Earth's dipole?

References

Aggson, T.L., Heppner, J.P., Maynard, N.C., 1983. Electric field measurements at the magnetopause, 1. Observation of large convective velocities at rotational magnetopause discontinuities. J. Geophys. Res. 88, 10000–10010.

Akasofu, S.-I., 1964. The development of the auroral substorm. Planet. Space Sci. 12 (4), 273–282. https://doi.org/10.1016/0032-0633(64)90151-5.

Akasofu, S.-I., 1968. Auroral observations by the constant local time flight. Planet. Space Sci. 16 (11), 1365–1368.

Akasofu, S.-I., 1979. Magnetospheric substorms and solar flares. Sol. Phys. 64 (2), 333–348.

Akasofu, S.-I., 1991. Auroral phenomena. In: Meng, C.-I., Rycroft, M.J., Frank, L.A. (Eds.), Auroral Physics. Cambridge University Press, New York, NY, pp. 223–239.

Akasofu, S.-I., 2007. Exploring the secrets of the aurora. Springer, New York, NY.

Alfvén, H., 1942. Existence of electromagnetic-hydrodynamic waves. Nature 150, 405–406.

Alfvén, H., 1950. Cosmical Electrodynamics. Oxford University Press, London and New York.

Anderson, K.A., Lin, R.P., Potter, D.W., Heetderks, H.D., 1978. An experiment to measure interplanetary and solar electrons. IEEE Trans. Geosci. Electron. 16 (3), 153–156.

Appleton, E.V., Barnett, M.A.F., 1925. On some direct evidence for downward atmospheric reflection of electric rays. Proc. Royal Soc. A 109, 621–641.

Arnoldy, R.L., 1974. Auroral particle precipitation and Birkeland currents. Rev. Geophys. 12 (2), 217–231.

Aubry, M.P., Russell, C.T., Kivelson, M.G., 1970. Inward motion of the magnetopause before a substorm. J. Geophys. Res. 75, 7018–7031.

Axford, W.I., 1962. The interaction between the solar wind and the Earth's magnetosphere. J. Geophys. Res. 67, 3791–3796.

Axford, W., 1994. The good old days. J. Geophys. Res. 99 (A10), 19199–19212.

Axford, W.I., Hines, C.O., 1961. A unifying theory of high-latitude geophysical phenomena and geomagnetic storms. Can. J. Phys. 39, 1433.

Benson, R.F., Bilitza, D., 2009. New satellite mission with old data: rescuing a unique data set. Radio Sci.. 44, RS0A04. https://doi.org/10.1029/2008RS004036.

Berchem, J., Okuda, H., 1990. A two-dimensional particle simulation of the magnetopause current layer. J. Geophys. Res. 95 (A6), 8133–8147.

Birkeland, K., 1908. The Norwegian Aurora Polaris Expedition 1902–1903, Vol. 1: On the Cause of Magnetic Storms and the Origin of Terrestrial Magnetism. H. Aschehoug and Co., Christiania

Birn, J., Drake, J.F., Shay, M.A., Rogers, B.N., Denton, R.E., Hesse, M., et al., 2001. Geospace environmental modeling (GEM) magnetic reconnection challenge. J. Geophys. Res. 106, 3715.

Boström, R., 1974. McCormac, B.M. (Ed.), Magnetospheric Physics. D. Reidel, Hingham, MA, pp. 45–59.

Breit, G., Tuve, M.A., 1925. A radio method of estimating the height of the conducting layer. Nature 116, 357.

Breit, G., Tuve, M.A., 1926. A test of the existence of the conducting layer. Phys. Rev. 28, 554.

Burch, J.L., Angelopoulos, V. (Eds.), 2009. The THEMIS Mission. Springer, New York, NY.

Burch, J.L., Torbert, R.B. (Eds.), 2017. Magnetospheric Multiscale: A Mission to Investigate the Physics of Magnetic Reconnection. New York, NY, Springer.

Burgess, D., 1995. Collisionless shocks. In: Kivelson, M.G., Russell, C.T. (Eds.), Introduction to Space Physics. Cambridge University Press, New York, NY, pp. 129–163.

Cahill, L.J., Amazeen, P.G., 1963. The boundary of the geomagnetic field. J. Geophys. Res. 68, 1835–1843.

Carlson, C.W., Torbert, R.B., 1980. Solar wind ion injections in the morning auroral oval. J. Geophys. Res. 85 (A6), 2903–2908.

Carovillano, R.L., Forbes, J.M. (Eds.), 1983. Solar-Terrestrial Physics. D. Reidel, Dordrecht, Netherlands.

Chapman, S., Ferraro, V.C.A., 1931. A theory of magnetic storms. Terr. Magn. Atmos. Electr. 36, 77–97.

Chen, F.F., 1984. Introduction to plasma physics and controlled fusion. In: Plasma Physics. second ed. vol. 1. Plenum Press, New York, NY.

Cole, K.D., 1961. On solar wind generation of polar geomagnetic disturbances. Geophys. J. R. Astron. Soc. 6, 103.

Davis, T.N., Sugiura, M., 1966. Auroral electrojet activity index AE and its universal time variations. J. Geophys. Res. 71, 785–801.

Dessler, A.J., 1984. The Vernov radiation belt (almost). Science 226 (4677), 915. https://doi.org/10.1126/science.226.4677.915.

Donovan, E., 2006. Preface. In: Syrjäsuo, M., Donovan, E. (Eds.), Proceedings of the Eighth International Conference on Substorms. University of Calgary, Alberta, CA, p. iii.

Dungey, J.W., 1961. Interplanetary magnetic field and the auroral zones. Phys. Rev. Lett. 6, 47–48.

Eastman, T.E., 2003. Historical review (pre-1980) of magnetospheric boundary layers and the low-latitude boundary layers. In: Newell, P.T., Onsager, T. (Eds.), Earth's Low-Latitude Boundary Layer. AGU Geophysical Monograph 133. American Geophysical Union, Washington, DC, pp. 1–11.

Egeland, A., Holter, Ø., Omholt, A., 1973. Cosmical Geophysics. Universitetsforlaget, Oslo, Norway.

Escoubet, C.P., Schmidt, R., Goldstein, M.L., 1997. CLUSTER—science and mission overview. Space Sci. Rev. 79 (1), 11–32.

Fairfield, D., Cahill Jr., L., 1966. Transition region magnetic field and polar magnetic disturbances. J. Geophys. Res. 71 (1), 155–169.

Fälthammar, C.-G., 1973. Motions of charged particles in the magnetosphere. In: Egeland, H., Omholt, (Eds.), Cosmical Geophysics. Universitetsforlaget, Oslo, Norway.

Faraday, M., 1832. Experimental researches in electricity. Philos. Trans. R. Soc. Lond. 122, 125–162.

Fennell, J., 1973. Access of solar protons to the Earth's polar caps. J. Geophys. Res. 78 (7), 1036–1046.

Frank, L.A., Craven, J.D., Gurnett, D.A., Shawhan, S.D., Weimer, D.R., Burch, J., 1986. The theta aurora. J. Geophys. Res. 91 (A3), 3177–3224.

Fukushima, N., 1969. Equivalence in ground geomagnetic effect of Chapman-Vestine's and Birkeland-Alfvén's electric current-systems for polar magnetic storms. Report Ionos. Space Res. Japan 23, 219–227.

Gold, T., 1959. Motions in the magnetosphere of the Earth. J. Geophys. Res. 64, 1219–1224. https://doi.org/10.1029/JZ064i009p01219.

Gopalswamy, N., Mewaldt, R., Torsti, J., 2006. Eruptions and energetic particles: an introduction. In: Gopalswamy, N., Mewaldt, R., Torsti, J. (Eds.), Solar Eruptions and Energetic Particles. AGU Geophysical Monograph 165. American Geophysical Union, Washington, DC, pp. 1–5.

Gosling, J.T., Thomsen, M.F., Bame, S.J., Elphic, R.C., Russell, C.T., 1990. Plasma flow reversals at the dayside magnetopause and the origin of asymmetric polar cap convection. J. Geophys. Res. 95 (A6), 8073–8084. https://doi.org/10.1029/JA095iA06p08073.

Greenwald, R.A., Baker, K.B., Dudeney, J.R., Pinnock, M., Jones, T.B., Thomas, E.C., et al., 1995. DARN/SuperDARN. Space Sci. Rev. 71 (1), 761–796.

Gringauz, K.I., Bezrukikh, V.V., Ozerov, V.D., Rybchin-sky, R.E., 1960. The study of the interplanetary ionized gas, high-energy electrons and corpuscular radiation of the sun, employing three-electrode charged particle traps on the second Soviet space rocket. Dokl. Akad. Nauk SSSR 131, 1302–1304 Translated in Soviet Physics Doklady, 5,361–364.

Gringauz, K.I., Bezrukikh, V.V., Ozerov, V.D., Rybchinskii, R.E., 1962. The study of interplanetary ionized gas, high-energy electrons and corpuscular radiation of the sun, employing three-electrode charged particle traps on the second Soviet space rocket. Planet. Space Sci. 9 (3), 103–107.

Hargreaves, J.K., 1992. The Solar-Terrestrial Environment. Cambridge Atmospheric and Space Science Series. Cambridge University Press, Cambridge, UK.

Heikkila, W.J., 1975. Is there an electrostatic field tangential to the dayside magnetopause and neutral line? Geophys. Res. Lett. 2, 154–157.

Heikkila, W.J., Pellinen, R.J., 1977. Localized induced electric field within the magnetotail. J. Geophys. Res. 82, 1610–1614.

Hines, C.O., Paghis, I., Hartz, T.R., Fejer, J.A. (Eds.), 1965. Physics of the Earth's Upper Atmosphere. Prentice-Hall, Inc., Englewood Cliffs, NJ, p. 434

Hudson, P.D., 1970. Discontinuities in an anisotropic plasma and their identification in the solar wind. Planet. Space Sci. 18, 161.

Hundhausen, A.J., 1995. The solar wind. In: Kivelson, M.G., Russell, C.T. (Eds.), Introduction to Space Physics. Cambridge University Press, Cambridge, UK, pp. 91–128.

Iijima, T., Potemra, T.A., 1976. The amplitude distribution of field-aligned currents at northern high latitudes observed by triad. J. Geophys. Res. 81, 2165–2174. https://doi.org/10.1029/JA081i013p02165.

Iijima, T., Potemra, T.A., 1978. The amplitude distribution of field-aligned currents associated with substorms. J. Geophys. Res. 83, 599–615.

Jørgensen, T., Rasmussen, 2006. Adam Paulson, a pioneer in auroral research. EOS Trans. AGU 87 (6), 61–66. https://doi.org/10.1029/2006EO060002.

Kamide, Y., Slavin, J.A. (Eds.), 1986. Solar Wind Magnetosphere Coupling, Astrophysics and Space Science Library. D. Reidel, Dordrecht, Netherlands.

Kellogg, P.J., 1962. The flow of plasma around Earth. J. Geophys. Res. 67, 3805.

Kivelson, M.G., Russell, C.T., 1995. Introduction to Space Physics. Cambridge University Press, New York, NY.

Korsmo, F.L., 2007. The genesis of the international geophysical year. Phys. Today 60 (7), 38.

Landmark, B., 1973. Formation of the ionosphere. In: Egeland, H., Omholt, (Eds.), Cosmical Geophysics. Universitetsforlaget, Oslo, Norway.

Lemaire, J., 1985. Plasmoid motion across a tangential discontinuity (with application to the magnetopause). J. Plasma Phys. 33 (3), 425.

Lemaire, J., Roth, M., 1978. Penetration of solar wind plasma elements into the magnetosphere. J. Atmos. Terr. Phys. 40 (3), 331–335.

Lemaire, J., Rycroft, M.J., Roth, M., 1979. Control of impulsive penetration of solar wind irregularities into the magnetosphere by the interplanetary magnetic field direction. Planet. Space Sci. 27 (1), 47–57.

Lincoln, J.V., 1967. Matsushita, S., Campbell, W.H. (Eds.), Geomagnetic Indices, Physics of Geomagnetic Phenomena. In: vol. I. Academic Press, New York, NY, p. 67.

Longmire, C.L., 1963. Plasma Physics and Thermonuclear Research. Pergamon Press, New York, NY.

Lui, A.T.Y. (Ed.), 1987. Magnetotail Physics. JHU Press, Baltimore and London, p. 433.

Mayaud, P.N., 1980. Derivation, Meaning, and Use of Geomagnetic Indices. AGU Geophysical Monograph 22. American Geophysical Union, Washington, DC.

McDiarmid, I., Burrows, J., Budzinski, E., 1976. Particle properties in the day side cleft. J. Geophys. Res. 81 (1), 221–226.

McPherron, R.L., 1979. Magnetospheric substorms. Rev. Geophys. 17 (4), 657–681. https://doi.org/10.1029/RG017i004p00657.

McPherron, R.L., 1995. Magnetospheric dynamics. In: Kivelson, M.G., Russell, C.T. (Eds.), Introduction to Space Physics. Cambridge University Press, New York, NY, p. 400.

Meng, C.-I., Rycroft, M.J., Frank, L.A., 1991. Auroral Physics, Proceedings. Cambridge University Press, Cambridge, UK, p. 494.

Menvielle, M., Berthelier, A., 1991. The K-derived planetary indices: description and availability. Rev. Geophys. 29 (3), 415–432. https://doi.org/10.1029/91RG00994.

Morfill, G., Scholer, M., 1973. Study of the magnetosphere using energetic particles. Space Sci. Rev. 15, 267.

Newell, P.T., Onsager T. (Eds.), 2003. Earth's Low-Latitude Boundary Layer. AGU Geophysical Monograph 133. American Geophysical Union, Washington, DC.

Nishida, A. (Ed.), 1982. Magnetospheric Plasma Physics. Center for Academic Publications, Tokyo, Japan, p. 348.

Odishaw, H. (Ed.), 1964. Research in Geophysics. MIT Press, Cambridge, MA, p. 574.

Parker, E.N., 1964. Odishaw, (Ed.), Research in Geophysics. MIT Press, Cambridge, MA, p. 108.

Parker, E.N., 1996. The alternative paradigm for magnetospheric physics. J. Geophys. Res. 10, 10.587.

Paschmann, G., Papamastorakis, I., Sckopke, N., Sonnerup, B., Bame, S., Russell, C., 1985. ISEE observations of the magnetopause: reconnection and the energy balance. J. Geophys. Res. 90 (A12), 12111–12120.

Phan, T.-D., Paschmann, G., Baumjohann, W., Sckopke, N., Lühr, H., 1994. The magnetosheath region adjacent to the dayside magnetopause: AMPTE/IRM observations. J. Geophys. Res. 99 (A1), 121–141. https://doi.org/10.1029/93JA02444.

Potemra, T.A. (Ed.), 1984. Magnetospheric Currents. AGU Geophysical Monograph 28. American Geophysical Union, Washington, DC.

Ratcliffe, J.A., 1960. Ratcliffe, J.A. (Ed.), Physics of the Upper Atmosphere. Academic Press, New York, NY.

Reeves, G.D., 2015. Radiation belt electron acceleration and role of magnetotail. In: Keilling, A., Jackman, C.M., Delamere, P.A. (Eds.), Magnetotails in the Solar

System. AGU Monograph 207. John Wiley & Sons, Inc., Washington, DC, pp. 345–359

Reiff, P.H., Burch, J.L., Hill, T.W., 1977. Solar wind plasma injection at the dayside magnetospheric cusp. J. Geophys. Res. 82, 479–491. https://doi.org/10.1029/JA082i004p00479.

Rider, K., Immel, T., Taylor, E., Craig, W., 2015. ICON: where's earth's weather meets space weather. In: 2015 IEEE Aerospace Conference, Big Sky, MT, pp. 1–10.

Roederer, J.G., 1967. On the adiabatic motion of energetic particles in a model magnetosphere. J. Geophys. Res. 72 (3), 981–992.

Rosenbauer, H., Grünwaldt, H., Montgomery, M.D., Paschmann, G., Sckopke, N., 1975. HEOS-2 plasma observations in the distant polar magnetosphere: the plasma mantle. J. Geophys. Res. 80, 2723.

Rostoker, G., 1972. Geomagnetic indices. Rev. Geophys. Space Phys. 10, 935–950.

Rostoker, G., 2007. Kamide, Y., Chian, A.C.-L. (Eds.), Substorms, Handbook of the Solar-Terrestrial Environment. Springer, Berlin, Germany, p. 376.

Rostoker, G., Akasofu, S.-I., Foster, J., Greenwald, R.A., Lui, A.T.Y., Kamide, Y., et al., 1980. Magnetospheric substorms—definition and signatures. J. Geophys. Res. 85 (A4), 1663–1668. https://doi.org/10.1029/JA085iA04p01663.

Schindler, K., Hesse, M., Birn, J., 1988. General magnetic reconnection, parallel electric fields, and helicity. J. Geophys. Res. 93, 5547–5557. https://doi.org/10.1029/JA093iA06p05547.

Schmidt, G., 1960. Plasma motion across magnetic fields. Phys. Fluids 3, 961.

Schmidt, G., 1979. Physics of High Temperature Plasmas, second ed. Academic Press, New York, NY.

Shen, C., Li, X., Dunlop, M., Liu, Z.X., Balogh, A., Baker, D.N., et al., 2003. Analyses on the geometrical structure of magnetic field in the current sheet based on cluster measurements. J. Geophys. Res. 108 (A5), 1168.

Sonnerup, B.U.Ö., Paschmann, G., Papamastorakis, I., Sckopke, N., Haerendel, G., Bame, S.J., et al., 1981. Evidence for magnetic field reconnection at the Earth's magnetopause. J. Geophys. Res. 86, 10049.

Sonnerup, B.U.Ö., Papamastorakis, I., Paschmann, G., Lühr, H., 1987. Magnetopause properties from AMPTE/IRM observations of the convection electric field: method development. J. Geophys. Res. 92 (A11), 12137–12159.

Sonnerup, B.U.Ö., Paschmann, G., Song, P., 1995. Fluid aspects of reconnection at the magnetopause: in situ observations. In: Song, P., Sonnerup, B.U.Ö., Thomsen, M.F. (Eds.), Physics of the Magnetopause. AGU Geophysical Monograph 90. American Geophysical Union, Washington, DC, pp. 167–180.

Spreiter, J.R., Alksne, A.Y., 1968. Comparison of theoretical predictions of the flow and magnetic field exterior to the magnetosphere with the observations of Pioneer 6. Planet. Space Sci. 16, 971–979.

Spreiter, J.R., Summers, A.L., Alksne, A.Y., 1966. Hydromagnetic flow around the magnetosphere. Planet. Space Sci. 14, 223–253.

Sullivan, W., 1961. Assault on the Unknown. McGraw-Hill, New York, NY.

Troshichev, O.A., 1988. The physics and meaning of the existing and proposed high-latitude geomagnetic indices. Ann. Geophys. 6, 601–609.

Van Allen, J.A., 1959. The geomagnetically trapped corpuscular radiation. J. Geophys. Res. 64, 1683–1689.

Walker, R.J., Russell, C.T., 1995. Solar-wind interactions with magnetized planets. In: Kivelson, M.G., Russell, C.T. (Eds.), Introduction to Space Physics. Cambridge University Press, New York, NY.

Williams, T., Godine, D., Palmer, E., Patel, I., Ottenstein, N., Winternitz, L., Petrinec, S., 2018. MMS extended mission design: evaluation of a lunar gravity assist option. In: 2018 AAS/AIAA Astrodynamics Specialist Conference, Snowbird, UT.

Willis, D.M., 1975. The microstructure of the magnetopause. Geophys. J. R. Astron. Soc. 41 (3), 355–389.

Winckler, J., Arnoldy, R., Hendrickson, R., 1975. Echo 2: a study of electron beams injected into the high-latitude ionosphere from a large sounding rocket. J. Geophys. Res. 80 (16), 2083–2088.

Wolf, R.A., 1983. The quasi-static (slow-flow) region of the magnetosphere. In: - Carovillano, R.L., Forbes, J.M. (Eds.), Solar-Terrestrial Physics. D. Reidel, Hingham, MA, pp. 303–368.

Wolfe, J.H., Intrilligator, D.S., 1970. The solar wind interaction with the geomagnetic field. Space Sci. Rev. 10, 511.

2

Kirchhoff's laws

Chapter outline

Mathematical rigor is important and cannot be neglected, but the theoretical physicist must first gain a thorough understanding of the physical implications of the symbolic tools he is using before formal rigor can be of help.

Morse and Feshbach (1953, p. 1)

Earth's Magnetosphere. https://doi.org/10.1016/B978-0-12-818160-7.00002-8

2.1 Introduction

The Boltzmann-Vlasov equation based on fundamental concepts provides the basis by kinetic theory for analytical treatment of plasma problems. For most purposes, the treatment of plasmas by this means is far too cumbersome and unnecessarily detailed. In hydrodynamics and aerodynamics, almost all practical problems can be accurately treated with fluid equations of motion and the thermodynamic state. For magnetized plasmas, there is a richer variety of phenomena, but it may still be possible to obtain equations of fluid-like behavior, or magnetohydrodynamics (MHD), from the Boltzmann equation by means of suitable approximations. This subject will be discussed in Chapter 4. Here we treat another approximation, circuit analysis.

2.2 Circuit analysis

Use of circuit theory for plasma problems in space research is not common. The magnetosphere is large and tenuous; lumped constant resistors R, capacitors C, and inductors L are unlikely for any meaningful analysis. Circuit analysis uses scalar equations useful for discussing energy, not vectors needed for analysis of the forces at work in a plasma. Circuit theory is essentially linear while equations of motion often exhibit nonlinear behavior. However, there is another side to the story.

> *The electric circuit laws of Ohm, Faraday, and Kirchhoff were based on experimental observations and antedated the electromagnetic theory of Maxwell and Lorentz. Indeed, the theory was developed as a generalization from these simpler and more restricted laws. It is interesting, but not surprising, then, to find that the circuit relations are just special cases of the more general field relations, and that they may be developed from the latter when suitable approximations are made.*
>
> **Jordan and Balmain (1968, p. 592)**

> *In a circuit problem there is often an applied voltage, and there are currents in the conductors of the circuit, charges on condensers in the circuit, ohmic losses, and power losses by radiation. These effects include almost everything that can happen when electric currents, charges, and conductors are let loose. The circuit problem is also one of the commonest problems illustrating the idea of cause and effect relationships. ... From the rigorous starting point of the fundamental laws [Maxwell's equations], it will be found that for circuits which are small compared with wavelength, this exact*

approach leads directly to the familiar circuit ideas based upon Kirchhoff's laws, and the concepts of lumped inductances and capacitances are sufficient for analysis.

Ramo and Whinnery (1953, p. 207)

Rigorous solutions to time-varying electromagnetic problems are obtainable only under very special circumstances, usually where the geometry is particularly simple. ... The reader may be familiar with the well-established techniques for discussing the properties of electrical networks. These are usually characterized by constant lumped-parameter elements such as resistors, capacitors, and inductors. Under steady-state conditions the properties of such networks may be established by setting up and solving a system of algebraic equations. The latter equations arise from an application of Kirchhoff's loop and node equations to the given network, with an assumed current-voltage relationship for each element.

Plonsey and Collin (1961, p. 326)

It is extremely helpful in physics to have graphical representations for mathematical quantities. Thus we represent a force or a velocity by a vector, which we draw as a directed line segment. Similarly, we find it useful to have a geometrical representation for differential equations, particularly for coupled equations. The representation that we introduce uses electrical circuit elements to represent physical quantities (which in themselves may have little or nothing to do with electricity). For example, mass is represented by inductance, restoring force by capacitance, and forces that dissipate energy by resistance. An external force is represented by a voltage generator. All this works, of course, because the differential equations that describe the motion of particles are analogous to the differential equations that describe networks. This applies to transient as well as periodic behavior. One can even, and again this is a consequence of the equations, make an analogy between the energy stored in the magnetic field of an inductor and the kinetic energy of motion of a mass. Similarly the potential energy stored in a spring is the analogue of the energy stored in the electric field of a capacitor.

Portis (1978, p. 679)

The use of fluid plasma concepts is one attempt to find solutions; the use of circuit analysis is definitely another. Electrical circuit theory is in a highly advanced state, and it may be the best way to understand some aspects of plasma physics, as a complementary approach. There are a vast number of theorems that can be used for the analysis, such as the superposition, Thévenin's and Tellegen's theorems, and various Reciprocity theorems (see Lorrain et al., 1988; Shen and Kong, 1995).

2.2.1 Lumped constants *R, C,* and *L*

There are three types of circuit elements to be considered: resistance *R*, capacitance *C*, and inductance *L*. These are usually small in size and are usually referred to as *lumped* as distinct from spatial. They are connected together by wires, called leads, to make a network. Fig. 2.1 is such a circuit, a battery connected to a resistor, perhaps the simplest circuit that can be imagined. From a field standpoint the leads should not dissipate or store any energy; they are treated as resistors with zero resistance. If energy is dissipated, for example using long leads with inadequate cross sectional areas, they are treated like true resistances.

It is also assumed that the maximum circuit dimensions are small compared with the wavelength; as a consequence, the fields will be quasistatic in nature. However, extension of the circuit idea to long transmission lines, and to transient conditions such as radiation, presents little difficulties in general.

The characteristics of resistor, capacitor, and inductor are specified on an energy basis. A resistor ideally dissipates energy; some energy storage is unavoidable but usually negligible. Resistance can be thought of as friction against the current flow. We consider the ideal capacitor as a lossless element which stores electric energy and the inductor the same for magnetic energy. These are analogous to inertia of the current flow. If the voltage and current vary with time, then the voltage $V(t)$ is considered to be the instantaneous difference in voltage between the terminals of an element. The current $I(t)$ is assumed to pass continuously through each element from one terminal to the other.

2.2.2 The resistor *R*

The most important relation in classical circuit theory is Ohm's law, published by the German physicist Georg Ohm in 1827:

$$V = IR \tag{2.1}$$

We consider it obvious today, but critics reacted to his treatment of the subject with hostility. Fortunately, Georg Ohm received recognition for his contributions to science well before he died.

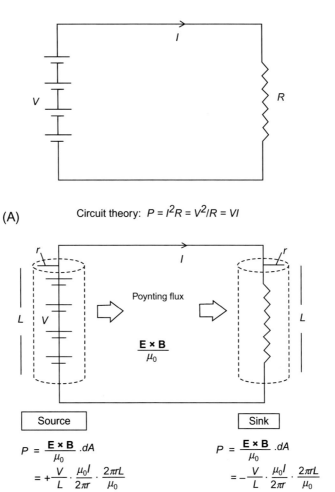

(A)   Circuit theory: $P = I^2R = V^2/R = VI$

Fig. 2.1 (A) The current is driven by a battery with a voltage V. The resistor is an *electrical load* for the battery. The energy imparted to the current carriers in falling through the voltage difference is VI. (B) The Poynting flux received from this external source and the power given by circuit analysis are *identical*. From Heikkila and Keith, after Heikkila, W.J., 1998. Cause and effect at the magnetopause. Space Sci. Rev. 83, 423.

The voltage V drives a current I, the value being determined by the resistance R. The resistances are usually rather constant, independent of frequency, but can vary over wide limits from well below $1\,\Omega$ to very high values, greater than $10^9\,\Omega$.

$$R = \frac{L}{\sigma A} \qquad (2.2)$$

Here L is the length, A is the cross sectional area of the resistor element, and the conductivity is σ. Ohm's Law may be written in general form as

$$\mathbf{J} = \sigma \mathbf{E} \qquad (2.3)$$

relating the current density $\mathbf{J}$ at a point to the total electric field $\mathbf{E}$ at that same location (assuming isotropic conductivity).

An important concept is the sign of $\mathbf{E} \cdot \mathbf{J}$. With dissipation this is positive:

$$\mathbf{E} \cdot \mathbf{J} > 0 \qquad (2.4)$$

The applied field results in an ohmic term with dissipation due to a current of the system. In a plasma, the current carriers receive energy from the electromagnetic field causing dissipation. This is true with direct voltages (independent of frequency as supplied by a battery) but also with alternating voltage, for example, with power from the electrical supply system. The current is always *in phase with the voltage* for a resistor.

On the other hand, this quantity can be negative:

$$\mathbf{E} \cdot \mathbf{J} < 0 \qquad (2.5)$$

This implies that the current carriers are losing energy: the medium is acting as a *dynamo*. Thus the sign of $\mathbf{E} \cdot \mathbf{J}$ is vital. The low-latitude boundary layer is a good example of a dynamo (illustrated in Chapter 9).

2.2.2.1 Circuit and wave approaches

Assume that a current I flows along a long straight resistive wire of length L and negligible thickness (see Fig. 2.1). The current is driven by a battery with a voltage V. The resistor is an *electrical load* for the battery. The energy imparted to the current carriers in falling through the voltage difference is VI; hence, the power dissipated in the resistor is very simply:

$$P = VI = I^2 R = V^2/R \qquad (2.6)$$

Now let us look at the problem from the field's point of view. The electric field is readily found by dividing V by the length L: $E = V/L$. The magnetic field produced by the current at a distance r is $B = \mu_0 I/2\pi r$. The total area about the wire is $A = 2\pi rL$ at the distance where we evaluated B. The *Poynting flux* intercepted by the wire:

$$\oint_A \frac{\mathbf{E} \times \mathbf{B}}{\mu_0} \cdot dA = \frac{1}{\mu_0} \cdot \frac{V}{L} \cdot \frac{\mu_0 I}{2\pi r} \cdot 2\pi rL = VI \qquad (2.7)$$

The Poynting flux received from this external source (the battery) and the power given by circuit analysis are *identical*. Whenever we have a Poynting flux, we need to follow the current path to find the dynamo (generator) to locate the source of energy. One need not be concerned with dimensional considerations. Being steady state, the travel time for energy transfer is irrelevant.

2.2.2.2 The steady state

The point shown earlier can be derived by a more general analysis. In the steady state, we can use the electrostatic scalar potential ϕ to describe the electric field:

$$
\begin{aligned}
-\oint_{\text{surf}} \mathbf{E} \times \mathbf{B} \cdot dS &= \oint_{\text{surf}} \nabla\phi \times \mathbf{B} \cdot dS \\
&= \oint_{\text{surc}} \nabla \times \phi\mathbf{B} \cdot dS - \oint_{\text{surc}} \phi \nabla \times \mathbf{B} \cdot dS \\
&= \iiint_{\text{vol}} \nabla \cdot \nabla \times \phi\mathbf{B} d\tau - \mu_0 \oiint_{\text{surf}} \phi\mathbf{J} \cdot dS
\end{aligned}
\tag{2.8}
$$

The volume integral vanishes due to a vector identity. Thus we have two *equivalent* results:

$$
\int_\tau \mathbf{E} \cdot \mathbf{J} d\tau = -\oint_{\text{surf}} \frac{\mathbf{E} \times \mathbf{B}}{\mu_0} \cdot dS
\tag{2.9}
$$

$$
= -\oint_{\text{surf}} \phi\mathbf{J} \cdot dS
\tag{2.10}
$$

agreeing with Fig. 2.1. Both are cause versus effect relationships; the cause is plasma yielding its energy to the electromagnetic field somewhere in the current circuit, while the effect is the opposite.

2.2.3 Electromotive force and sign of E·J

We shall now take into account the flow of charge, or current. The conservation of charge in a medium is expressed by the equation of continuity:

$$
\nabla \cdot \mathbf{J} = -\frac{\partial\sigma}{\partial t}
\tag{2.11}
$$

Stationary currents are possible *only if there are other sources of electric field present.* We should, in fact *must,* look for a process with $\mathbf{E} \cdot \mathbf{J} < 0$.

If we assume that such electromotive fields exist, and denote them by $\mathbf{E}'$, the conduction equation becomes (Panofsky and Phillips, 1962, p. 119)

$$
\mathbf{J} = \sigma(\mathbf{E} + \mathbf{E}')
\tag{2.12}
$$

We may note that an electromotive force (emf) ε exists as follows:

$$\varepsilon = \oint (\mathbf{E} + \mathbf{E}') \cdot d\mathbf{l} = \oint \mathbf{E}' \cdot d\mathbf{l} = \oint \frac{\mathbf{J} \cdot d\mathbf{l}}{\sigma} \qquad (2.13)$$

The conservative part of the field $\mathbf{E}$ drops out of the closed line integration; the current is entirely due to the nonconservative forces. This can be a battery, dynamo, or generator in which the current opposes the electric field. The best example of that is a change in the local magnetic flux Φ (discussed throughout this book) with an emf:

$$\varepsilon = -\frac{d\Phi}{dt} \qquad (2.14)$$

2.2.4 Capacitance C of an isolated conductor

Both capacitors and inductors involve time dependence; we refer to AC (alternating current) voltage and current. The frequency can be constant line frequency (either 50 or 60 Hz), but the concepts apply equally well for a general frequency as in plasma circuits.

Imagine a finite conductor like a sphere situated a long distance from any other body and carrying a charge Q. If Q changes, the conductor's potential also changes. The capacitance of the isolated conductor is by definition:

$$C = \frac{Q}{V} \text{ or } Q = CV \qquad (2.15)$$

The ratio Q/V is a constant and depends solely on the size and shape of the conductor. Thus the capacitance of an isolated element is equal to the charge required to increase its potential by 1 V. The unit of capacitance is the farad (F), or coulomb per volt; the earth is a capacitor of 4 farads. In practice, 1 farad is rather large, and microfarad or picofarad is more common.

There is work involved in charging a capacitor to a final charge Q. Now the current is not in phase with the voltage because of the time differentiation. In the usual case of a line voltage at 50 or 60 Hz, the current leads the voltage by 90 degrees as shown in Fig. 2.2. At some intermediate stage the incremental work is

$$dW = \left(\frac{q}{C}\right) dq \qquad (2.16)$$

The total work to go from $q = 0$ to $q = Q$ is

$$W = \int_0^Q \left(\frac{q}{C}\right) dq = \frac{1}{2} \frac{Q^2}{C} \qquad (2.17)$$

hence,

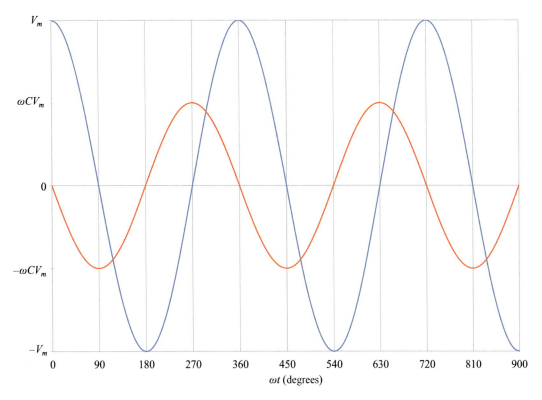

Fig. 2.2 For a capacitor connected to an alternating voltage source at a constant frequency, the instantaneous current (orange) leads the voltage (blue) by 90 degrees in the steady state $\omega C = 0.5$ in this example. From Keith.

$$W = \frac{1}{2}CV^2 \tag{2.18}$$

The energy stored in the field of an isolated conductor is

$$\varepsilon = \frac{QV}{2} = \frac{CV^2}{2} = \frac{Q^2}{2C} \tag{2.19}$$

Two or more capacitors may be joined together by connecting the conductors. In the parallel case, the same voltage appears across each capacitor and also across the combination. The equivalent capacitance is given by

$$C = \frac{Q_1}{V} + \frac{Q_2}{V} = C_1 + C_2 \tag{2.20}$$

If two uncharged capacitors are connected in series and subsequently charged, conservation of charge requires that each capacitor acquire the same charge. Thus the equivalent capacitance C of the combination is related to C_1 and C_2 by the expression

$$\frac{1}{C} = \frac{V_1 + V_2}{Q} = \frac{1}{C_1} + \frac{1}{C_2} \tag{2.21}$$

2.2.4.1 Alternating current in capacitors

Fig. 2.2 shows the voltage and current of a capacitor connected to an alternating voltage source at a constant frequency (usually 50 or 60 Hz). Now with an oscillating source with the (angular) frequency ω, the instantaneous current leads the voltage. The charge Q is also varying with the same frequency:

$$Q = CV_m \cos \omega t \tag{2.22}$$

The quantity V_m is the amplitude of the voltage, different from the root-mean-square, which is either 110 or 220 V in the usual line source. The current I is the rate of delivery of charge:

$$I = \frac{dQ}{dt} = -\omega CV_m \sin \omega t = \omega CV_m \cos \left(\omega t + \frac{\pi}{2} \right) \tag{2.23}$$

In the steady state, the current leads the voltage by $\pi/2$ (90 degrees).

In the commonly used phasor notation, engineers use j instead of i (since I is used for current):

$$I = j\omega Q = j\omega CV \tag{2.24}$$

2.2.4.2 Capacitance of a sphere

The field in the vacuum region can be obtained from Gauss' law. The charge Q acts as if it were a point charge; the field will be radial. We get

$$\phi_1 - \phi_2 = \int_1^2 \mathbf{E} \cdot d\mathbf{r} = \int_a^b E_r$$

$$= \int_a^b \frac{Qdr}{4\pi\varepsilon_0 r^2} = \frac{Q}{4\pi\varepsilon_0} \left(\frac{1}{a} - \frac{1}{b} \right) \tag{2.25}$$

The capacitance C simplifies to

$$C = \frac{4\pi\varepsilon_0 ab}{b-a} \tag{2.26}$$

With the outer sphere at infinity, $C = 4\pi\varepsilon_0 a$, proportional to its radius. Fig. 2.3 shows an experimental arrangement to accurately measure the capacitance of half sphere, the lower half serving as a guard.

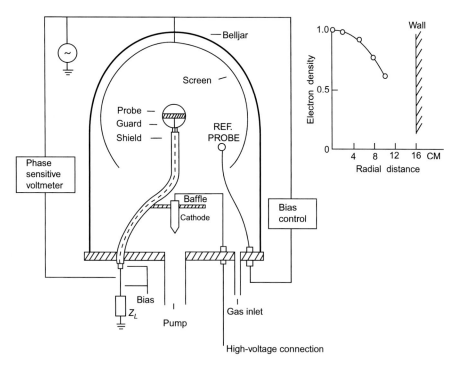

Fig. 2.3 An experimental arrangement to accurately measure the capacitance of a half sphere, the lower half serving as a guard. From Heikkila, W.J., 1969. Laboratory study of probe impedance. In: Thomas, J.O., Landmark, B.J. (Eds.), Plasma Waves in Space and in the Laboratory. Edinburgh University Press, Edinburgh, UK. p. 233.

2.2.4.3 The parallel-plate capacitor

A parallel-plate capacitor consists of two conducting plates of area A, separated by a distance s. The plates carry charges Q and $-Q$ (neglecting edge effects). From Gauss's law,

$$C = \frac{\varepsilon_0 A}{s} \tag{2.27}$$

The scalar potential ϕ arises essentially from charge stored on the capacitor plates, while the vector potential is due mainly to current in the leads. The LLBL is a capacitor due to charges on the two sides.

2.2.5 The inductor

The magnetic field energy is measured by an inductance L. This is often a small coil of wire, hopefully with negligible resistance. The self-inductance depends on the magnetic flux Φ and is defined by

$$L = \frac{d\Phi}{dI} \qquad (2.28)$$

We can expand the magnetic field in terms of the current I

$$\frac{d\Phi}{dt} = \frac{d\Phi}{dI}\frac{dI}{dt} \qquad (2.29)$$

Noting that the induction electric field opposes any change by Le Chatelier's principle and Lenz's law, and therefore the induced emf is

$$\varepsilon = -L\frac{dI}{dt} \qquad (2.30)$$

With the inductor, the current lags the voltage, contrary to a capacitor. From Ampere's circuital law, the magnetic induction inside the toroidal coil is

$$B = \frac{\mu_0 NI}{l} \qquad (2.31)$$

where N is the number of turns, l is the mean length of the winding, and I is the current in the winding (neglecting the variation of the magnetic induction over the cross sectional area). The flux linking each turn is

$$\Phi_1 = \frac{\mu_0 NIA}{I} \qquad (2.32)$$

and the total flux linking the N turns is

$$\Phi = \frac{\mu_0 N^2 A}{l} I \qquad (2.33)$$

The inductance is then simply

$$L = \frac{d\Phi}{dt} = \frac{\mu_0 N^2 A}{l} \qquad (2.34)$$

The SI unit of inductance is the henry (H), which is equal to one volt-second/ampere since the unit of emf is the volt. The dimensions of μ_0, which have been previously given as webers/ampere-meter or tesla-meters/ampere, can alternatively be given as henrys/meter.

When two or more inductors appear in the circuit, they behave in the same fashion as resistors (as distinct from capacitors). However, another quantity is essential (e.g., a transformer); the *mutual inductance M* between two circuits *a* and *b* is given by the Neumann equation

$$\Phi_{ab} = \frac{\mu_0 I_a}{4\pi} \oint_a \oint_b \frac{\mathbf{dl}_a \cdot \mathbf{dl}_b}{r} = M_{ab} I_a \qquad (2.35)$$

where

$$M_{ab} = \frac{\mu_0}{4\pi} \int_a \int_b \frac{\mathbf{dl}_a \cdot \mathbf{dl}_b}{r} \qquad (2.36)$$

The line integrals run around each circuit, and r is the distance between the elements $\mathbf{dl}_a$ and $\mathbf{dl}_b$. Because of the symmetry of the integral,

$$M_{ab} = M_{ba} = M \qquad (2.37)$$

The coupling coefficient between two circuits is

$$k = \frac{M}{(L_a L_b)^{1/2}} \qquad (2.38)$$

This coefficient takes the sign of M, and its magnitude is at most unity. The transformer is a good example of the mutual inductor.

2.2.5.1 Power and power factors

The power delivered to a resistor is determined by multiplying the voltage across the resistor by the current through the resistor. If $V(t)$ and $I(t)$ are the complex voltage and current, then the instantaneous power is

$$\overline{\mathrm{Re}(I_0 e^{j\omega t}) \, \mathrm{Re}(V_0 e^{j\omega t})} = \frac{1}{2} \mathrm{Re}\left(I_0^* V_0\right) \qquad (2.39)$$

where I_0^* is the complex conjugate of I_0.

The factor one-half in the previous equation represents the fact that the average of $\sin^2 \omega t = \cos^2 \omega t = 1/2$. An interesting factor is $\cos\phi$ that takes into account the fact that the current and voltage are not in phase. $\cos\phi$ is called the power factor of an AC circuit. The average power is a more important quantity, with the average being taken over either one full period or a very long time (many periods).

We should mention that the effective values of the voltage and current are often defined by

$$V_{\mathrm{eff}} = \frac{1}{\sqrt{2}} |V_0|, \quad I_{\mathrm{eff}} = \frac{1}{\sqrt{2}} |I_0| \qquad (2.40)$$

The virtue of these definitions is that a given V_{eff} applied to a resistance dissipates the same power as a constant voltage of the same magnitude.

2.2.5.2 Lumped electrodes in space

Plonsey and Collin (1961) show that lumped constant circuit elements C, L, and R can have a generalized meaning in space. In terms of the energy stored in the electric fields W_E, the magnetic fields W_M, and the energy dissipation P_E, they are

$$C = \frac{II^*}{4\omega^2 W_E} = \frac{4W_E}{VV^*} \tag{2.41}$$

$$L = \frac{4W_M}{II^*} = \frac{VV^*}{4\omega^2 W_M} \tag{2.42}$$

$$R = \frac{2P_E}{II^*} = \frac{VV^*}{2P_E} \tag{2.43}$$

where I and V denote current and voltage at a particular frequency ω and asterisk I^*, V^* denotes the complex conjugate.

Rostoker and Boström (1976), among others, have used these ideas showing that C, L, and R have values commensurate with the large stored energy in the vast magnetosphere. The circuit approach allows us to do the counting, oblivious to the dimensional concerns, related to cause and effect (Ramo and Whinnery, 1953, p. 207). The ability of using various network theorems, well known in the engineering community, such as superposition and Thévenin's theorems (Lorrain et al., 1988) can be of paramount importance, once recognized.

> The values of W_E, W_M, and P_1 are unique under a given set of terminal conditions. If the voltage V and current I can also be specified uniquely, then the circuit parameters—capacitance C, inductance L, and resistance R—are uniquely defined by the above formulas. Under static conditions it has been shown in earlier chapters that the above definitions for R, L, and C are equivalent to the geometrical definitions. However, the above definitions are more general in that they recognize the fact that ideal circuit elements do not exist physically. ... By defining capacitance in terms of electric energy storage, account is taken of all portions of the physical structure that contribute to the capacitance of the over-all device, and similarly for the inductance and resistance. ... Again we emphasize that it is necessary to be able to define unique terminal currents and voltages in order for these parameters to have unique values.

Plonsey and Collin (1961, p. 337)

2.3   Equations of circuit analysis

The properties of electrical networks are established by setting up and solving a system of algebraic equations; this is a distinct advantage as the alternative is the use of time-dependent integro-differential equations in three-dimensional space (also used in fluid plasma theory, see Chapter 4). These equations arise from an application of Kirchhoff's loop and node equations to the given network, with a current-voltage relationship for each element.

Fig. 2.4 shows a single mesh of an electric circuit. Points *A*, *B*, *C*, and *D* are called *nodes*; connections between nodes such as *AB* are *branches*. A closed circuit, such as *ABCD*, is a *mesh*. An active circuit comprises *sources*, while a passive circuit does not. A circuit is *linear* if the current through it is proportional to the voltage across it. The neglect of the displacement current is vital to this analysis. When this is not the case, there are other methods that can be used, for example, in transmission lines and antennas.

We assume the existence of four network parameters, the resistance *R*, the capacitance *C*, the inductance *L*, and mutual inductance *M*. The properties of these parameters are defined in terms of their voltage-current relationships as follows, where for simplicity the harmonic time variation $e^{j\omega t}$ is omitted:

$$V = RI = j\omega LI = \frac{I}{j\omega C} \tag{2.44}$$

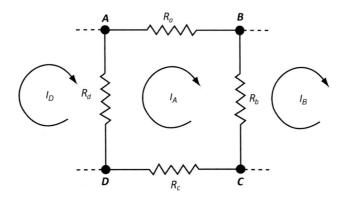

Fig. 2.4 The four nodes *A*, *B*, *C*, *D* are part of a larger circuit. The closed circuit ABCD is a mesh carrying the mesh current I_A. Other parts of the circuit (not shown) carry other mesh currents such as I_B and I_D. From Keith.

The previously mentioned may also be specified in terms of the inverse relations:

$$I = GV = j\omega CV = \frac{V}{j\omega L} \qquad (2.45)$$

where $G = 1/R$ is the conductance.

2.3.1 Kirchhoff's voltage (loop) law

The voltage $V(t)$ is considered to be the difference in voltage between the terminals of the element. For an arbitrary network, it is supposed that a unique voltage may be assigned for each circuit node. Kirchhoff's voltage law states that the sum of the voltages taken around any closed path is zero:

$$\oint_{circuit} \mathbf{E} \cdot d\mathbf{l} = 0 \qquad (2.46)$$

This is Kirchhoff's voltage law for the electric field.

2.3.2 Kirchhoff's current (node) law

Kirchhoff's current (or node) law states that at a node:

$$\sum_{n=1}^{N} I_n = 0 \qquad (2.47)$$

The idea behind this is that of charge conservation or its integral equivalent,

$$\oint_{surf} \mathbf{J} \cdot d\mathbf{S} = -\frac{\partial}{\partial t} \int_{vol} \rho d\tau \qquad (2.48)$$

If charge can accumulate somewhere, the term on the right denotes a capacitance current. Thus Kirchhoff's laws are summarized as follows:

- Algebraic sum of the voltage differences around any loop is zero:

$$\sum V_j = 0 \qquad (2.49)$$

- The algebraic sum of the currents toward a branch point is zero:

$$\sum I_j = 0 \qquad (2.50)$$

Circuits are linear, and so is the corresponding circuit analysis. The object of circuit analysis is to formulate a set of linear algebraic equations in n unknowns. Despite their apparent simplicity, Kirchhoff's laws are central to circuit analysis. Viewed as an electromagnetic boundary-value problem, it would be almost hopeless to find a field solution that satisfies the boundary conditions with field-aligned currents connecting the various shaped regions of the magnetosphere. To represent that with lumped constant resistors, capacitors, and inductors may be fruitful once the limitations are understood.

2.3.3   Wheatstone bridge

The Wheatstone bridge provides a good example of circuit analysis by the mesh and node methods (Hunten, 1964). It is easy to find the balance condition, but the off-balance sensitivity can be found only by a careful analysis.

The balance condition can be found by regarding R_1, R_2 and R_3, R_4 (Fig. 2.5A) as potential dividers; if they have an equal division ratio, then the potentials at the two ends of the galvanometer will be zero and no current will flow in it. Thus $R_1/R_2 = R_3/R_4$, or

$$R_1 R_4 = R_2 R_3 \qquad (2.51)$$

This "product" form of the balance condition is the most useful. In words, it is simply stated: "The products of opposite arms are equal."

In choosing the meshes, we note that the required answer is the galvanometer current, and that the other two currents need not be found. Therefore only one mesh current should traverse the galvanometer, as shown in Fig. 2.5. The internal resistance of the battery has been omitted for simplicity, but it can readily be added if necessary. The circuit equations are:

$$I_1(R_1 + R_2) + I_2(R_1 + R_2) + I_3 R_1 = V \qquad (2.52)$$

$$I_1(R_1 + R_2) + I_2(R_1 + R_2 + R_3 + R_4) + I_3(R_1 + R_3) = 0 \qquad (2.53)$$

$$I_1 R_1 + I_2(R_1 + R_3) + I_3(R_1 + R_3 + R_G) = 0 \qquad (2.54)$$

From this we can get the determinant D using standard techniques (see Arfken and Weber, 1995). It should be noted that the determinant is symmetrical about the main diagonal. This is a consequence of the fact that each branch is common to two meshes; it is always true that $R_{ij} = R_{ji}$.

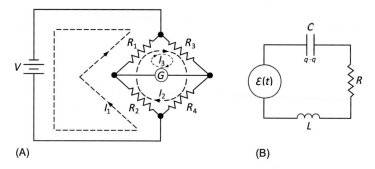

Fig. 2.5 (A) Wheatstone bridge. The balance condition can be found by regarding R_1, R_2 and R_3, R_4 as potential dividers; if they have an equal division ratio, then at balance $R_1 R_4 = R_2 R_3$. The products of opposite arms are equal. (B) An RLC series circuit. From Keith.

In the node method, complementary to the mesh method, the unknowns to be found are potentials rather than currents, the resistances are represented as conductances, and the power sources are represented as current generators. The required quantity is often a potential rather than a current, especially in circuits used with vacuum tubes and transistors. In the node method, we assume a potential for each node thus eliminating the need for Kirchhoff's first law. Because only potential differences have a physical significance, it is possible to set the potential of the reference node equal to zero and refer all the other potentials to it.

2.4 Series resonant circuit

Following the development in Wangsness (1986), we can analyze a circuit shown in Fig. 2.5B. A capacitance C and an inductance L have been added to a resistor R, and an applied emf $\varepsilon(t)$, as a function of time. We assume that I has the sense shown (if the opposite is correct the signs will be reversed). The charges on the capacitor plates are $\pm q$ as shown.

$$L\frac{dI}{dt} + RI + \frac{q}{C} = \varepsilon(t) \qquad (2.55)$$

We are primarily interested in the current and we can differentiate this with respect to the time,

$$L\frac{d^2I}{dt^2} + R\frac{dI}{dt} + \frac{I}{C} = \frac{d\varepsilon}{dt} \qquad (2.56)$$

This is a second-order differential equation with constant coefficients. If we suddenly apply a periodic force by an oscillator, the initial response (displacement) will be nonperiodic. After a long time, the displacement is found to be periodic in time with the same frequency as the applied force. The nonperiodic part of the solution that is eventually damped out is called a *transient*, while the periodic part that persists is called the *steady state*. It will be convenient to look at these two types of behavior separately.

A transient (damped) response can arise in general for any form of $\varepsilon(t)$, not only for a periodic one. The standard approach is to assume a solution of exponential form $I = ae^{\gamma t}$:

$$\left[L\gamma^2 + R\gamma + (1/C)\right]ae^{\gamma t} = 0 \tag{2.57}$$

The value of γ is found to be

$$\gamma = -\frac{R}{2L} \pm \delta \tag{2.58}$$

requiring two possible values of γ:

$$I(t) = \left(a_+ e^{\delta^+ t} + a_- e^{-\delta t}\right)e^{-Rt/2L} \tag{2.59}$$

The nature of the solutions depends on the values of R, C, and L. A common case corresponds to R being very small or even zero; then

$$I(t) = \frac{\varepsilon}{L\omega_n}e^{-Rt/2L}\sin\omega_n t \tag{2.60}$$

The current oscillates with the natural circular frequency ω_n, but with exponentially decreasing amplitude corresponding to the underdamped case. The steady-state solution

$$\varepsilon(t) = \varepsilon_0 e^{j\omega t} \tag{2.61}$$

where ε_0 is assumed to be real. Substituting this into the differential equation:

$$\left[-\omega^2 L + j\omega R + (1/C)I_0 e^{j\omega t} = i\omega\varepsilon_0 e^{j\omega t}\right] \tag{2.62}$$

so that

$$I_0 = \frac{\varepsilon_0}{Z} \tag{2.63}$$

where

$$Z = R + j\left(\omega L - \frac{1}{\omega C}\right) = R + jX \tag{2.64}$$

Z is the impedance and X the reactance. In general,

$$X = X_L + X_C \qquad (2.65)$$

where $X_L = \omega L$ and $X_C = -1/\omega C$ are the inductive and capacitive reactances. The physical significance of a complex impedance is that the steady-state current is not in phase with the applied emf. We can show this explicitly by writing

$$Z = R + jX = |Z|e^{j\vartheta} \qquad (2.66)$$

where

$$\tan\vartheta = \frac{X}{R} = \frac{\omega L - (1/\omega C)}{R} \qquad (2.67)$$

The real quantities are

$$\varepsilon(t) = \varepsilon_0 \cos\omega t \ \text{ and } \ I(t) = \frac{\varepsilon_0}{|Z|}\cos(\omega t - \vartheta) \qquad (2.68)$$

We see that impedance is a generalization of resistance. However, the impedance is a function of the applied frequency and hence is not completely characterized by the electromagnetic parameters of the system. This means that the steady-state current will have different values depending on the frequency. The frequency for which the current has its maximum value is the resonance frequency and corresponds to the minimum value of |Z|. We see ... that this corresponds to X = 0, which shows that the resonance frequency is exactly ω_0 that we previously found to correspond to the current being in phase with the applied emf. At resonance, the current has the simple value of ε/R.

Wangsness (1986, p. 456)

The resistance determines the sharpness of the resonance. It is convenient to describe this by introducing the dimensionless quantity Q defined as the ratio of the inductive reactance at resonance to the resistance:

$$Q = \frac{\omega_0 L}{R} = \frac{1}{R}\left(\frac{L}{C}\right)^{1/2} = \frac{1}{R}\sqrt{\frac{L}{C}} \qquad (2.69)$$

Fig. 2.6 shows the current amplitude as a function of ω/ω_0 for a few selected values of Q. The sharpness of the resonance is decreased as the resistance is increased in order to make Q smaller. The resistance of the inductor itself is often the deciding factor.

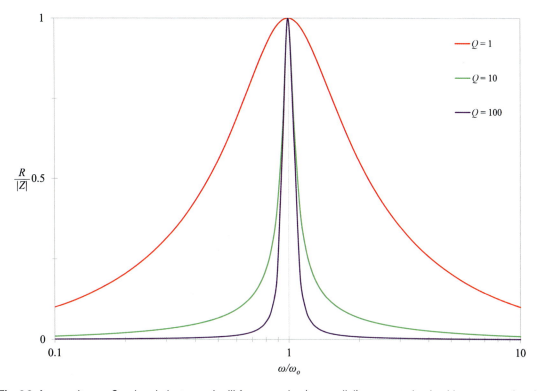

Fig. 2.6 A capacitance C and an inductance L will form a series (or parallel) resonant circuit with response sharply peaked. The resistance R determines the sharpness of the resonance with the dimensionless quantity Q defined as the ratio of the inductive reactance at resonance to the resistance. From Keith.

2.5   Other circuit theorems

Kirchhoff's laws are self-evident and appear trivial. This is a false impression as they are related to hundreds of circuit theorems.

> *There are two important theorems applying to all linear circuits, superposition theorem and Thévenin's theorem. They are of great use in simplifying analysis. It is far more important to understand them and be able to use them than to be able to derive them, but the derivations are not difficult.*

Hunten (1964, p. 8)

2.5.1   Superposition theorem

We have already come across the use of the superposition theorem in other situations. Here it may be stated as follows: in a *linear system*, each source of emf produces its own set of currents.

These do not depend on the presence or absence of other sources. When the current attributable to one source is to be found, all other sources can be omitted; however, it must be remembered that if a source is removed, its internal resistance must be left in the circuit. If there is more than one source in each mesh, by the superposition theorem the net effect is a sum of terms.

2.5.2 Thévenin's theorem

Thévenin's theorem states that *any* linear circuit having two terminals can be represented by the series combination of an emf in series with an impedance called the output impedance of the circuit. The values of these two parameters can be calculated from a knowledge of the circuit or by operations carried out at the terminals (see Hunten, 1964; Lorrain et al., 1988 for details).

Norton's theorem is the dual of Thévenin's: any active, linear, two-terminal circuit is equivalent to an ideal current source in parallel with an admittance, which is usually a conductance.

2.5.3 Tellegen's theorem

Tellegen's theorem is like Kirchhoff's laws: it is self-evident and appears trivial. It applies to direct voltages and currents in a circuit and also alternating currents in a circuit comprising only sources and linear passive components. Then the complex power flowing into branch b is $V_b I_b^*/2$, and the sum of the complex powers is zero,

$$\sum V_b I_b^* = 0 \tag{2.70}$$

This is just energy conservation. The sum of the complex powers in a circuit is zero (see Lorrain et al., 1988, p. 154).

2.6 Radiation from oscillating dipoles

It is usual to start the study of the generation of electromagnetic waves from the vector potentials satisfying the inhomogeneous wave equation with sources. We consider only the magnetic dipole (Fig. 2.7) as it is stored magnetic energy that is important for solar flares and substorms; a plasma transfer event (PTE) in Chapter 6, Figs. 6.19 and 6.21, is an example.

The approximations that are made limit the validity of the solutions to fields produced by slowly moving (nonrelativistic) charges, that is, those with velocity v small compared to the velocity of light, $v \ll c$. Nevertheless the results are applicable both to the emission of radio waves from antennas and of light from atoms. Relativistic considerations are treated in Chapter 3 with Liénard-Wiechert potentials.

2.6.1 Magnetic dipole radiation

A magnetic dipole antenna is a loop of wire carrying a current I, as in Fig. 2.7. By symmetry, the vector potential **A** is azimuthal: for any value of r' we have two symmetric **dl**s whose y components add and whose x components cancel; thus we need calculate only the y component of the **A** in the figure.

Let us calculate the radiation from a time-dependent charge/ current from an oscillating magnetic dipole. The dipole will be assumed to consist of a coil centered at the origin connected by a wire of negligible capacitance. The comparison of electric field from electric and magnetic dipole antennas is interesting (the electric dipole in Fig. 2.8A is well known). On the other hand, the magnetic dipole is somewhat surprising. A key difference is that the electric field for magnetic dipole must be time dependent, as Faraday found long ago. It is best if we assume a sinusoidal dependence, but in PTEs and substorms it can be more complex. At any one instant, the electric field is related to the variation of currents as shown in Fig. 2.8B. If the current is increasing, the

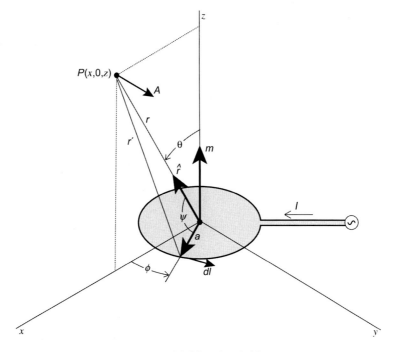

Fig. 2.7 Magnetic dipole antenna. The vector potential **A** is azimuthal by symmetry. From Keith.

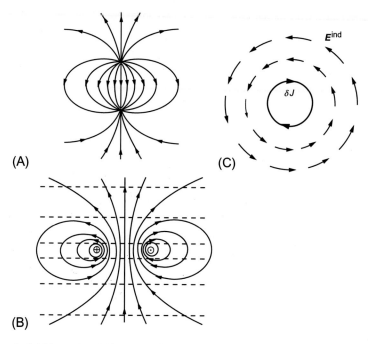

Fig. 2.8 (A) The electric field from electric (*solid lines*) and (B) magnetic (*dashed lines*) antennas. A key difference is that the electric field for magnetic dipole must be time dependent (Faraday, 1832), as indicated by (C) *dashed arrows* when (B) is viewed from above. The vector potential **A** is azimuthal and exists throughout space inversely proportional to *r*. From Heikkila and Brown.

induced electric field is felt throughout space, as shown by the dashed arrows.

Conservation of charge requires that the current I in the connecting wire be given by the motion of charge q. It must be noted that the condition of negligible capacitance of the wire and its concomitant uniform current can be satisfied only if the length of the dipole is small compared with the wavelength of the radiation. Suppose now that we have a wire loop of radius a around which we drive a sinusoidally varying current, at frequency ω:

$$I(t) = I_0 \cos \omega t \tag{2.71}$$

$$\mathbf{m}(t) = \pi a^2 I(t) \hat{\mathbf{k}} = m_0 \cos \omega t \hat{\mathbf{k}} \tag{2.72}$$

where

$$m_0 = \pi a^2 I_0 \tag{2.73}$$

is the maximum value of the magnetic dipole moment. The loop is uncharged, so the scalar potential is zero. The retarded vector potential is

$$\mathbf{A}(rt) = \frac{\mu_0}{4\pi} \int \frac{I_0 \cos \omega t(t - r/c)}{r} \mathbf{dl} \tag{2.74}$$

For a point P directly above the x-axis, $\mathbf{A}$ must aim in the y direction; x components from symmetrically placed points on either side of the x-axis will cancel. Thus,

$$\mathbf{A}(rt) = \frac{\mu_0}{4\pi} I_0 a \hat{\jmath} \int_0^{2\pi} \frac{I_0 \cos \omega t(t - r/c)}{r} \cos \phi \, d\phi \tag{2.75}$$

the $\cos\phi$ serving to pick out the y component of $\mathbf{dl}$. By the law of cosines,

$$r' = \sqrt{r^2 + a^2 - 2ra \cos \psi} \tag{2.76}$$

where ψ is the angle between the vectors $\mathbf{r}$ and $\mathbf{a}$: Thus,

$$r' = \sqrt{r^2 + a^2 - 2ra \sin \theta \cos \phi} \tag{2.77}$$

This allows several approximations to be made.

Approximation 1: $a \ll r$

This is the condition of a "perfect" dipole: we require that the loop be extremely small. To first order in a

$$r' = r\left(1 - 2\frac{a}{r} \sin \theta \cos \phi\right) \tag{2.78}$$

So that

$$\frac{1}{r'} = \frac{1}{r}\left(1 + \frac{a}{r} \sin \theta \cos \phi\right) \tag{2.79}$$

We next assume the radius is much less than the wavelength of the waves generated; this is the *near-zone* of the radiation pattern.

Approximation 2: $a \ll \frac{c}{\omega}$

The energy flux for magnetic dipole radiation is important for the radiation zone.

Approximation 3: $r \gg \frac{c}{\omega}$

In the radiation zone, these potential fields are in phase, mutually perpendicular, and transverse to the direction of propagation. The ratio of their amplitudes is likewise correct for an electromagnetic wave. They are, in fact, remarkably similar in structure to the fields of an oscillating electric dipole, only this time it is $\mathbf{B}$ that points in the θ direction and $\mathbf{E}$ in the ϕ, whereas for electric dipoles it is the other way around. The energy flux for magnetic dipole radiation is

$$\langle \mathbf{S} \rangle = \frac{\mu_o m_0^2 \omega^4}{32\pi^2 c^3} \left(\frac{\sin^2 \theta}{r^2}\right) \hat{r} \tag{2.80}$$

The total radiated power is therefore

$$P = \left(\frac{\mu_0 m_o^2}{12\pi c^3}\right)\omega^4 = \frac{1}{4\pi\varepsilon_0}\frac{m_0^2\omega^4}{3c^5} \qquad (2.81)$$

The intensity profile has the shape of a donut, and the power radiated goes as the fourth power of the frequency. There is, however, one important difference between electric and magnetic dipole radiation: for configurations with comparable dimensions, the power radiated electrically is enormously greater. Comparing the two:

$$\frac{P_{\text{magnetic}}}{P_{\text{electric}}} = \left(\frac{m_0}{p_0 c}\right)^2 \qquad (2.82)$$

For one particular example cited by Griffiths (1981), he gets:

$$\frac{P_{\text{magnetic}}}{P_{\text{electric}}} = \left(\frac{a\omega}{c}\right)^2 \qquad (2.83)$$

This quantity was assumed to be very small in Approximation 2, and here it appears squared. One should expect electric dipole radiation to dominate over the magnetic.

2.6.2 Radiation from a distribution of charges and currents

The same procedure is applicable to a configuration of charges and currents in substorms. We want to know the radiation pattern. It is best to take the pattern as being localized within some small region near the origin (Fig. 2.9).

For ease of visualization, it is convenient to consider the charge element subdivided, by planes perpendicular to the line from P to the charge, into thin slices which move with the charge.

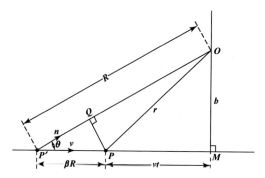

Fig. 2.9 Diagram for calculating the electric field of a moving charge. From Jackson, J.D.,1975. Classical Electrodynamics, second ed. Wiley, New York, NY, p. 658.

The retarded time is only specified implicitly, since the interval r/c by which it lags t is itself a function of time and has to be evaluated at the retarded time. This has been done by Reitz et al. (1993) and Griffiths (1981), and is discussed in Chapter 3.

2.7 Discussion

The necessary test for reconnection sites should make use of Kirchhoff's laws on circuit analysis; these are corollaries of Maxwell's equations in the low-frequency limit. By Kirchhoff's Current Law, the algebraic sum of the currents entering and leaving a node must be equal to zero. Since only one current, the dissipation current with $\mathbf{E} \cdot \mathbf{J} > 0$ is included in Standard Magnetic Reconnection (SMR) that cannot be accomplished. A closed circuit path must exist. We need another current, a dynamo current with $\mathbf{E} \cdot \mathbf{J} < 0$, to denote the source of energy. Schindler et al. (1988) have systems where the current is closed. They demonstrate that reconnection depends crucially on properties of the parallel electric field. General Magnetic Reconnection (GMR) must be used to discuss magnetic connectivity. None of the properties of 2D reconnection theory carry over into three dimensions (Pontin, 2011). That can be tested by Phan and Paschmann (1996), Phan et al. (1996), Zwickl et al. (1984), Pulkkinen et al. (1996), and Fujimoto et al. (1998). These are in conformity with a low latitude boundary layer on closed magnetic field by General Magnetic Reconnection, to be discussed in Chapter 5.

2.8 Summary

There are three types of circuit elements to be considered: resistance R, capacitance C, and inductance L. These are usually small in size and are often referred to as *lumped* as distinct from spatial.

- The characteristics of resistor, capacitor, and inductor are specified on an energy basis.
- If the voltage and current vary with time, then the voltage $V(t)$ is considered to be the instantaneous difference in voltage between the terminals of an element.
- The current $I(t)$ is assumed to pass continuously through each element from one terminal to the other.
- The equations arise from an application of Kirchhoff's loop and node equations to the given network, with an assumed current-voltage relationship for each element.

In a circuit problem there is often an applied voltage, there are currents in the conductors of the circuit, charges on condensers in the circuit, ohmic losses, and power losses by radiation. These effects include almost everything that can happen when electric currents, charges, and conductors are let loose. The circuit problem is also one of the most common problems illustrating cause and effect relationships. It is interesting, but not surprising, to find that the circuit relations are just special cases of the more general field relations, and that they may be developed from the latter when suitable approximations are made. Almost all practical problems can be accurately treated with fluid equations of motion and the thermodynamic state. For magnetized plasmas, there is a richer variety of phenomena, but it may still be possible to obtain equations of fluid-like behavior, or magnetohydrodynamics (MHD).

2.9 Problems

2.1. Apply Kirchhoff's current law to the four nodes of the Wheatstone bridge in Fig. 2.5. Label the total current in each branch separately this time. For example, label the current passing through the battery as I_B, and the current passing through R_3 as I_3. Using the same current labeling method, apply Kirchhoff's voltage law to all seven possible closed paths.

2.2. Note that not all of the equations from Problem 2.1 are linearly independent. How many independent equations are there? Assuming the source voltage and resistances are known, solve for the six currents.

2.3. Ampere's circuital law states that the integral of the magnetic field **B** along a closed path is equal to the product of current enclosed and permeability of the medium. Show that this leads to Eq. (2.31) for the magnetic induction inside a toroidal coil.

2.4. In Fig. 2.9, the β shown is the relativistic value $\beta = v/c$. Show that the distance traveled by the charge during the time it takes light to travel from P' to O is equal to βR. Using the various right triangles in the figure, find t as a function of R, b, v, and c.

2.5. In the circuit shown in Fig. 2.5B, the variable frequency source has an RMS output of 10V, $R = 4\,\Omega$, $L = 3\,\text{mH}$, and $C = 0.1\,\mu\text{F}$. Find the resonant frequency of the circuit and the resulting maximum current.

References

Arfken, G.B., Weber, H.J., 1995. Mathematical methods for physicists (4th ed.). Academic Press, New York, NY.

Faraday, M., 1832. Experimental researches in electricity. Philos. Trans. R. Soc. Lond. A 122, 125–162.

Fujimoto, M., Terasawa, T., Mukai, T., Saito, Y., Yamamoto, T., Kokubun, S., 1998. Plasma entry from the flanks of the near-earth magnetotail: geotail observations. J. Geophys. Res. 103, 4391.

Griffiths, D., 1981. Introduction to Electrodynamics, first ed. Prentice Hall, Englewood Cliffs, NJ.

Hunten, D.M., 1964. Introduction to Electronics. Holt, Rinehart, and Winston, New York, NY.

Jordan, E., Balmain, K., 1968. Electromagnetic Waves and Radiating Systems. Prentice Hall, London, UK.

Lorrain, P., Corson, D., Lorrain, F., 1988. Electromagnetic Fields and Waves, third ed. Freeman, San Francisco, CA.

Morse, P.M., Feshbach, H., 1953. Methods of Theoretical Physics. McGraw-Hill, New York, NY.

Panofsky, W.K.H., Phillips, M., 1962. Classical Electricity and Magnetism, second ed. Addison-Wesley, Reading, MA.

Phan, T.-D., Paschmann, G., 1996. The low-latitude dayside magnetopause and boundary layer for high magnetic shear: 1. Structure and motion. J. Geophys. Res. 101, 7801. https://doi.org/10.1029/95JA03752.

Phan, T.-D., Paschmann, G., Sonnerup, B.U.Ö., 1996. The low-latitude dayside magnetopause and boundary layer for high magnetic shear: 2. Occurrence of reconnection. J. Geophys. Res. 101, 7818–7828. https://doi.org/10.1029/95JA03751.

Plonsey, R., Collin, R.E., 1961. Principles and Applications of Electromagnetic Fields. McGraw-Hill, New York, NY.

Pontin, D.I., 2011. Three-dimensional magnetic reconnection regimes: a review. Adv. Space Res. 47, 1508–1522.

Portis, A.M., 1978. Electromagnetic Fields: Sources and Media. Wiley, New York, NY.

Pulkkinen, T.I., Baker, D.N., Owen, C.J., Slavin, J.A., 1996. A model for the distant tail field: ISEE-3 revisited. J. Geomagn. Geoelectr. 48, 455.

Ramo, S., Whinnery, J.R., 1953. Fields and Waves in Modern Radio, second ed. Wiley, New York, NY.

Reitz, J.R., Milford, F.J., Christy, R.W., 1993. Foundations of Electromagnetic Theory, fourth ed. Addison-Wesley, Reading, MA.

Rostoker, G., Boström, R., 1976. A mechanism for driving the gross Birkeland current configuration in the auroral oval. J. Geophys. Res. 81 (1), 235–244.

Schindler, K., Hesse, M., Birn, J., 1988. General magnetic reconnection, parallel electric fields, and helicity. J. Geophys. Res. 93, 5547–5557. https://doi.org/10.1029/JA093iA06p05547.

Shen, L.C., Kong, J.A., 1995. Applied Electromagnetism, third ed. PWS, Boston, MA.

Wangsness, R.K., 1986. Electromagnetic Fields, second ed. Wiley, New York, NY.

Zwickl, R.D., Baker, D.N., Bame, S.J., Feldman, W.C., Gosling, J.T., Hones Jr., E.W., et al., 1984. Evolution of the earth's distant magnetotail: ISEE 3 electron plasma results. J. Geophys. Res. 89 (A12), 11,007–11,012.

3

Helmholtz's theorem

Chapter outline

Earth's Magnetosphere. https://doi.org/10.1016/B978-0-12-818160-7.00003-X

Research is to see what everybody has seen and think what nobody has thought.

Szent-Györgyi (1957) (translated from Schopenhauer, 1851)

3.1 Introduction

The existence and variability of energetic particle populations in the Earth's magnetosphere has dominated investigations since early satellite observations. These are the key to understanding substorms and plasmoid structures. However, the mechanisms by which these charged particles can obtain such high energies are not well understood.

The acceleration of particles to high energies is a ubiquitous phenomenon at sites throughout the Universe. Impulsive solar flares offer one of the most impressive examples anywhere, releasing up to 10^{32} ergs of energy over timescales of several tens of seconds to several tens of minutes. ... It is generally accepted that the basic source of this released energy lies in stressed, current-carrying magnetic fields, if for no other reason than that this is the only energy reservoir of sufficient magnitude. ...The Sun [Earth's magnetosphere] is therefore a very efficient particle accelerator, although the mechanisms through which magnetic energy is transformed into accelerated suprathermal particles has yet to be satisfactorily elucidated.

Emslie and Miller (2003)

It is only the electric field that can energize charged particles by the force $\mathbf{F} = q\mathbf{E}$. There are several key points about the electric field that are sometimes overlooked especially with the **B**, **v** paradigm. Helmholtz's theorem (Arfken and Weber, 1995, p. 93) enunciates quite clearly the definition of any vector **V**, in particular the electric field **E**, in terms of the divergence and curl. Maxwell's equations are exactly such definitions.

3.1.1 Definition of a vector

Helmholtz's theorem states that a vector field is *uniquely defined* in three dimensions if the source density s (a scalar) and circulation density **c** (a vector) are given functions of the coordinates at all points in space, and if the totality of sources, as well as the source density, is zero at infinity. For a vector **V** this is stated by two equations,

$$\nabla \cdot \mathbf{V} = s \tag{3.1}$$

$$\nabla \cdot \mathbf{V} = \mathbf{c} \tag{3.2}$$

The right-hand sides of these equations are the inhomogeneous terms in the governing differential equations. By a vector identity, Eq. (3.2) is self-consistent only if the circulation density **c** is irrotational,

$$\nabla \cdot \mathbf{c} = 0 \tag{3.3}$$

Construct a scalar potential $\phi(\mathbf{r}_1)$

$$\phi(\mathbf{r}_1) = \frac{1}{4\pi} \int \frac{s(\mathbf{r}_2)}{r_{12}} d\tau_2 \tag{3.4}$$

and a vector potential $\mathbf{A}(\mathbf{r}_1)$

$$\mathbf{A}(\mathbf{r}_1) = \frac{1}{4\pi} \int \frac{\mathbf{c}(\mathbf{r}_2)}{r_{12}} d\tau_2 \tag{3.5}$$

Helmholtz's theorem will be satisfied if we can write **V** as the sum of two parts, one of which is irrotational, the other solenoidal:

$$\mathbf{V} = -\nabla \phi + \nabla \times \mathbf{A} \tag{3.6}$$

3.1.2 Source point versus field point

It is necessary to examine the notation of these equations before proceeding. The symbol $\mathbf{r}_1$ stands for x_1, y_1, z_1 at the *field point*, the location where we wish to evaluate the vector **V**. On the other hand, the symbol $\mathbf{r}_2$ stands for x_2, y_2, z_2 at the *source point*, the cause of the disturbance, the sources. The function $\mathbf{r}_{12}$ connects the two with $\mathbf{r}_{12}$ symmetric (see Fig. 3.1):

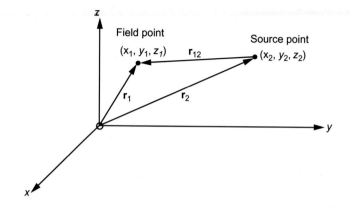

Fig. 3.1 Field point denoted by subscripts 1 and source point as 2. The complication arises when the two regions overlap. From Arfken, G.B., Weber, H.J., 1995. Mathematical Methods for Physicists, fourth ed. Academic Press, New York, NY, p. 94.

$$r_{12} = \left[(x_1 - x_2)^2 + (y_1 - y_2)^2 + (z_1 - z_2)^2 \right]^{1/2} \tag{3.7}$$

The positive direction is taken to be away from the source point toward the field. The sources s and $\mathbf{c}$ must vanish sufficiently rapidly at large distances so that the integrals exist. In integrals of this type these functional relationships will usually not be stated explicitly; rather, it is common to say

$$\phi = \frac{1}{4\pi} \int \frac{s}{r} d\tau \tag{3.8}$$

$$\mathbf{A} = \frac{1}{4\pi} \int \frac{\mathbf{c}}{r} d\tau \tag{3.9}$$

3.1.3 Dirac delta function

The primed and unprimed coordinates can treat the same volume of space. Sources of $\mathbf{E}$ and $\mathbf{B}$ are charges and currents within the plasma medium. The function $\mathbf{r} = \mathbf{r}(x_{i1} - x_{i2})$ vanishes whenever the source point and field point coincide, posing a problem for the integration because of the infinity at $\mathbf{r} = 0$. We can retrieve the scalar potential by taking the divergence of $\mathbf{V}$:

$$\nabla \cdot \mathbf{V} = -\nabla^2 \phi + \nabla \cdot \nabla \times \mathbf{A} \tag{3.10}$$

$$= -\nabla^2 \phi \tag{3.11}$$

$$= -\frac{1}{4\pi} \nabla^2 \left\{ \int \frac{s}{r} d\tau' \right\} \tag{3.12}$$

The Laplacian operator ∇^2 operates on the field coordinates; therefore we can interchange the order of integration and differentiation:

$$\nabla \cdot \mathbf{V} = -\frac{1}{4\pi} \int s \nabla^2 \left(\frac{1}{r}\right) d\tau'$$ (3.13)

$$= -\int s(x_i') \delta(\mathbf{r}) d\tau'$$ (3.14)

Here we have introduced the new function $\delta(\mathbf{r})$ where $\delta(\mathbf{r}) = 0, \mathbf{r} \neq 0$ (i.e., $x_i \neq x'_i$) otherwise $\int \delta(\mathbf{r}) d\tau' = 1$ ($\mathbf{r} = 0$ is included in volume of integration). This Dirac δ-function is not an analytical function; it is defined by its functional properties.

$$\delta(\mathbf{r}) = 0, \quad \mathbf{r} \neq 0$$ (3.15)

$$\int f(\mathbf{r}) \delta(\mathbf{r}) d\tau = f(0)$$ (3.16)

where $f(\mathbf{r})$ is any well-behaved function; the volume includes the origin.

Our source is at $\mathbf{r}_2$, not at the origin. This means that the 4π in Gauss's law appears if and only if the surface includes the point $\mathbf{r} = \mathbf{r}_2$. Also, noting that differentiating twice with respect to x_2, y_2, z_2 is the same as differentiating twice with respect to x_1, y_1, z_1, we have

$$\nabla_1^2 \left(\frac{1}{r_{12}}\right) = \nabla_2^2 \left(\frac{1}{r_{12}}\right) = -4\pi \delta(\mathbf{r}_1 - \mathbf{r}_2)$$ (3.17)

3.2 Helmholtz's theorem

This theorem states that a vector $\mathbf{V}$ satisfying Eqs. (3.1), (3.2) with both source and circulation densities vanishing at infinity may be written as the sum of two parts, one of which is irrotational, the other solenoidal:

$$\mathbf{V} = -\nabla \phi + \nabla \times \mathbf{A}$$ (3.18)

Rewriting Eq. (3.12) using the Delta function, we can integrate:

$$\nabla \cdot \mathbf{V} = -\frac{1}{4\pi} \int s(\mathbf{r}_2) \nabla_2^2 \left(\frac{1}{r_{12}}\right) d\tau_2$$ (3.19)

$$\nabla \cdot \mathbf{V} = -\frac{1}{4\pi} \int s(\mathbf{r}_2) \nabla_2^2 \left(\frac{1}{r_{12}}\right) d\tau_2$$ (3.20)

$$= s(\mathbf{r}_1)$$ (3.21)

This result shows that the assumed form of **V** and of the scalar potential are consistent with the source s. We now discuss the solenoidal part of Helmholtz's theorem by taking the curl of **V**.

$$\nabla \times \mathbf{V} = -\nabla \times \nabla\phi + \nabla \times (\nabla \times \mathbf{A}) \qquad (3.22)$$

$$= 0 + \nabla(\nabla \cdot \mathbf{A}) - \nabla^2 \mathbf{A} \qquad (3.23)$$

The first term vanishes by a vector identity. So does the second term if we use the Coulomb gauge where $\nabla \cdot \mathbf{A} = 0$; this vanishes because the circulation density is solenoidal. It can be shown in general that this term vanishes if the source falls off with distance sufficiently rapidly (Arfken and Weber, 1995, p. 96). With $\nabla(\nabla \cdot \mathbf{A}) = 0$, Eq. (3.23) now reduces to

$$\nabla \times \mathbf{V} = -\nabla^2 \mathbf{A} = -\frac{1}{4\pi}\int \mathbf{c}(\mathbf{r}_2)\nabla_1^2\left(\frac{1}{r_{12}}\right)d\tau_2 \qquad (3.24)$$

$$\nabla \times \mathbf{V} = -\frac{1}{4\pi}\int \mathbf{c}(\mathbf{r}_2)\nabla_2^2\left(\frac{1}{r_{12}}\right)d\tau_2 \qquad (3.25)$$

$$= -\int \mathbf{c}(\mathbf{r}_2)(-4\pi)\delta(\mathbf{r}_2 - \mathbf{r}_1)d\tau_2 \qquad (3.26)$$

$$= \mathbf{c}(\mathbf{r}_1) \qquad (3.27)$$

Note the permitted change from ∇_1^2 to ∇_2^2. This is the vector equivalent of Poisson's equation. Therefore Helmholtz's theorem is proved by Eqs. (3.21), (3.27).

3.2.1 Irrotational source: Conservative

An irrotational source is conservative and easy to comprehend because we have an excellent example with gravity. A force is conservative if the work done on a particle that moves between two points is the same for all paths connecting those points. A relevant example for plasma physics is the electrostatic field and equipotential surfaces.

3.2.2 Nonconservative source: Solenoidal

A nonconservative force requires a more complicated mathematical development. One example is friction. When a particle or object travels with a finite velocity; whatever its direction, it always leads to dissipation. Another example is the vortex motion of fluids and the electric field due to induction. This forces us to delve more deeply into the concept of the curl of a vector.

The curl is defined as a vector function whose component at a point in a particular direction is found by orienting an infinitesimal area normal to the direction at that point:

$$[\text{curl } \mathbf{V}]_i = \lim (\Delta S_i \to 0) \frac{\oint \mathbf{V} \cdot d\mathbf{l}}{\Delta S_i} \qquad (3.28)$$

where i denotes a particular direction, ΔS_i is normal to that direction, and the line integral is taken in the right-hand sense with respect to the positive i direction. The line integral around a closed path is called the net circulation of $\mathbf{V}$ around the chosen path. This integral is a measure of the tendency of the field's flow lines to "curl up." Laminar flow in a boundary layer with the boundary condition $V_\perp = 0$ at the surface has a finite curl, even in the steady state. The lift of an airfoil is directly proportional to the curl of the circulation (Landau and Lifshitz, 1959).

This definition has similarities to that of a divergence, defined as the integral taken about infinitesimal surface divided by the volume enclosed by that surface. Charge density is defined in precisely this way. However, the curl operation results in a vector instead of a scalar. This is the only additional complication in the curl over the divergence. Purcell (1985) has an excellent discussion of the curl, and distinguishing the physics from the mathematics.

3.3 Maxwell's equations

Maxwell's equations are an expression of Helmholtz's theorem for the two vector fields, $\mathbf{E}$ and $\mathbf{B}$. In terms of observables, we can define the two fields by Newton's laws and by the Lorentz force equation

$$\mathbf{F}_L = q(\mathbf{E} + \mathbf{v} \times \mathbf{B}) \qquad (3.29)$$

Maxwell's equations in their simplest forms for plasma in free space are:

$$\nabla \cdot \mathbf{E} = \rho/\varepsilon_0 \qquad (3.30)$$

$$\nabla \times \mathbf{E} = -\partial \mathbf{B}/\partial t \qquad (3.31)$$

$$\nabla \cdot \mathbf{B} = 0 \qquad (3.32)$$

$$\nabla \times \mathbf{B} = \mu_0 \mathbf{J} + \mu_0 \varepsilon_0 \partial \mathbf{E}/\partial t \qquad (3.33)$$

The two constants ε_0 (permittivity) and μ_0 (permeability) show that the International System of Units (SI) is used.

$$\mu_0 = 1.257 \times 10^{-6} \, \text{H/m} \qquad (3.34)$$

$$\varepsilon_0 = 8.854 \times 10^{-12} \, \text{F/m} \qquad (3.35)$$

The constant c, the velocity of light, is used instead with Maxwell's equations in Gaussian (CGS) units. The two systems are related by

$$c^2 \varepsilon_0 \mu_0 = 1 \qquad (3.36)$$

Maxwell's equations are a set of coupled equations, with both **E** and **B** occurring on the right as time derivatives; this involves the time rate of change of one vector as the solenoidal source for the other. As a result of this close interaction, it is common to say that Maxwell's equations are the equations for the electromagnetic field. The importance of Helmholtz's theorem for the two vector fields **E** and **B** is that s and **c** are identified as their *causes*, their *resources*, and the *starting place*. Divergence and curl are not just properties of **E** and **B**, they are much more.

3.3.1   Confusion about nonconservative forces

It is common to say "potential difference," for example, when we speak of auroral particle acceleration. That phrase implies a conservative force, without a curl. The electric field could be due to induction, rotational force with a finite curl, $\nabla \times \mathbf{E} = -\partial \mathbf{B}/\partial t$. A "voltage difference" is the better term to use, not implying anything about the force, conservative or not.

3.3.2   Two sources for the electric field

As Faraday (1832) first discovered, there is another source for the electric field. By Helmholtz's theorem and Maxwell's equations the electric field can also be produced by a time-varying magnetic field. The scalar s is the free charge density ρ producing the electrostatic field. The vector **c** causes a magnetic field; when **c** varies in time we get the induction electric field. The total electric field is **E** where

$$\mathbf{E} = \mathbf{E}^{es} + \mathbf{E}^{ind} = -\nabla \phi - \partial \mathbf{A}/\partial t \qquad (3.37)$$

The two types of sources are clearly independent, with no necessary connection between them. Each source yields an electric field that acts in the same way upon charged particles. Any connection between the two depends on the medium involved.

In a magnetized plasma, in particular the magnetosphere, we can split the electric field into components that are parallel and perpendicular to **B**:

$$\mathbf{E} = \mathbf{E}_{\parallel}^{es} + \mathbf{E}_{\parallel}^{ind} + \mathbf{E}_{\perp}^{es} + \mathbf{E}_{\perp}^{ind} \qquad (3.38)$$

As already stated in Chapter 1, we need to *understand each term* to appreciate the physics; the different terms imply different physics.

3.3.3 Charge separation can never extinguish induction fields

The secondary electric field due to polarization of charges can never extinguish the induction field. Polarization and induction electric fields are topologically different; an electrostatic field has a vanishing line integral around every closed contour, so it cannot affect the emf of the induced electric field. We can see this by forming the line integral around any closed contour enclosing a finite flux change:

$$\oint \mathbf{E} \cdot d\mathbf{l} = \oint (\mathbf{E}^{es} + \mathbf{E}^{\text{ind}}) \cdot d\mathbf{l} = 0 - \frac{d\Phi^M}{dt} \tag{3.39}$$

where Φ^M is the total magnetic flux enclosed by the contour. Whatever the distribution of the secondary field $\mathbf{E}^{es}$ is, the resultant field must remain finite and large enough to make the line integral finite and equal to $-d\Phi^M/dt$. A plasma is powerless to influence the emf due to changing magnetic fields by *any redistribution of charge*.

3.3.4 Principle of superposition

Maxwell's equations are linear differential equations in $\mathbf{E}$ and $\mathbf{B}$. This linearity is often taken for granted, for example, hundreds of different telephone calls on a single microwave link. The only way we have of detecting and measuring electric charges is by observing the interaction of charged bodies. We find that the force with which two charges interact is not changed by the presence of a third charge. This is the basis of the *principle of superposition*, which we shall invoke again and again in our study of space plasma physics.

The electrodynamics pertains to the force on a test charge q. It says that the total force on q is the vector sum of the forces attributable to the source charges individually:

$$\mathbf{F} = \mathbf{F}_1 + \mathbf{F}_2 + \cdots \tag{3.40}$$

Consider a static system at first with $\mathbf{v} = 0$ so we can neglect the magnetic force. Dividing through by q, we find that the electric field, too, obeys the superposition principle:

$$\mathbf{E} = \mathbf{E}_1 + \mathbf{E}_2 + \cdots \tag{3.41}$$

Integrating from the common reference point to P, it follows that:

$$\phi = \phi_1 + \phi_2 + \cdots \qquad (3.42)$$

That is, the potential at any point is the sum of the potentials due to all the source charges separately. This time it is an ordinary sum, not a vector sum, making it a lot easier to comprehend.

This principle must not be taken lightly; it is applicable even in a plasma. There may well be a domain of phenomena, involving very small distances or very intense forces, where superposition no longer holds. Indeed, we know of quantum phenomena in the electromagnetic field which do represent a failure of superposition, seen from the viewpoint of the classical theory (Jackson, 1975).

1. Groups of charges and currents produce electric and magnetic forces calculable by linear superposition.
2. If two or more electromagnetic fields are known, then the sum is also a solution.

But we must be careful, for example, for power calculations, with products of forces; also some instabilities may cause departure.

3.4 Gauss's law

Definition of the electrostatic field follows from Helmholtz's theorem; it involves just the one equation concerned with divergence. Any other definition, such as by the electron acceleration in a gyrotropic medium, or Ohm's law including the generalized form (see Sibeck et al., 1999, p. 211), would be contrived or artificial at best.

Often we want to consider charges that are distributed continuously over some region. The standard notation is as follows: charge spread out along a line is described by the letter λ (representing the charge per unit length), charge spread over a surface is described by σ (the charge per unit area), and charge spread throughout a volume is described by ρ (the charge per unit volume).

An electric field is the sum of the fields of its individual sources, by the principle of superposition. If we have a number of sources, $q_1, q_2, \ldots q_N$ the fields of which, if each were present alone, would be $\mathbf{E}_1, \mathbf{E}_2, \ldots \mathbf{E}_N$, the flux through some surface S in the actual field is in integral form

$$\oint_S \mathbf{E} \cdot \mathbf{n}\, da = \frac{q}{\varepsilon_0} \qquad (3.43)$$

Gauss's law and Coulomb's law are not two independent physical laws, but the same law expressed in different ways.

Symmetry is the key to the application of Gauss's law. Three kinds of symmetry are sufficient for analysis:
1. Spherical symmetry: make your Gaussian surface a concentric sphere.
2. Cylindrical symmetry: make your Gaussian surface a coaxial cylinder.
3. Plane symmetry: use a Gaussian "pillbox," which straddles the surface.

For the Gaussian charge layer in Fig. 3.2 (top-right), the electric field points away on both sides with a positive charge σ that is constant:

$$\mathbf{E} = \frac{\sigma}{2\varepsilon_0}\hat{\mathbf{n}} \qquad (3.44)$$

It may seem surprising, at first, that the field of an infinite plane is independent of how far away you are. The inverse square on distance in Coulomb's law is taken into account by the fact that as you move farther away from the plane, more charge comes into your "field of view."

Now take two infinite parallel planes that carry equal but opposite uniform charge densities as in Fig. 3.2. We can find the field in each of the three regions: (i) to the left of both, (ii) between them, and (iii) to the right of both. The left plate produces a field that points away from it to the left in region (i) and to the right in regions (ii) and (iii). The right plate, being negatively charged, produces a field that points toward it, to the right in regions (i) and (ii) and to the left in region (iii). The two fields cancel in regions (i) and (iii); they work together in region (ii). When there is an

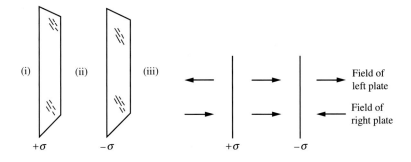

Fig. 3.2 Maxwell's equations are linear in **E** and **B**, so the principle of superposition applies. An electric field points away from positive charge (left plate) and toward negative charge (right plate). The net effect of a dipolar layer produces a strong field in between, and cancelation outside. From Griffiths, D., 2017. Introduction to Electrodynamics, 4th ed. Cambridge University Press, Cambridge. https://doi.org/10.1017/9781108333511, p. 75.

external field already present, an infinite dipole layer has no effect, transferring the field from one side to the other (see Chapter 9 on the low-latitude boundary layer).

3.5 Gauge conditions

The vector potential $\mathbf{A}$ is used to specify the magnetic field as $\mathbf{B} = \nabla \times \mathbf{A}$. Faraday's law in differential form can be stated using the vector potential as:

$$\nabla \times \mathbf{E} + \frac{\partial}{\partial t} \nabla \times \mathbf{A} = 0 \tag{3.45}$$

Assuming continuity of fields permit the interchange of temporal and spatial differentiations, this can be written as

$$\nabla \times \left\{ \mathbf{E} + \frac{\partial \mathbf{A}}{\partial t} \right\} = 0 \tag{3.46}$$

Therefore the vector $\{\mathbf{E} + \partial \mathbf{A} / \partial t\}$ has zero curl; it can be written as the gradient of a scalar function. After rearranging terms we have

$$\mathbf{E} = -\nabla \phi - \frac{\partial \mathbf{A}}{\partial t} \tag{3.47}$$

Now we have the electric and magnetic fields given in terms of a scalar potential ϕ and a vector potential $\mathbf{A}$. The curl of the vector $\mathbf{A}$ was used to define the magnetic field $\mathbf{B}$ but we are at liberty to specify its divergence, $\nabla \cdot \mathbf{A}$; this freedom forms a *gauge condition*.

3.5.1 Lorentz gauge

The Lorentz condition

$$\nabla \cdot \mathbf{A} + \varepsilon \mu \frac{\partial \phi}{\partial t} = 0 \tag{3.48}$$

has the property that both $\mathbf{A}$ and ϕ have the same form of wave equation. For the vector $\mathbf{A}$:

$$\nabla^2 \mathbf{A} - \varepsilon \mu \frac{\partial^2 \mathbf{A}}{\partial t^2} = -\mu \mathbf{J} \tag{3.49}$$

and for ϕ:

$$\nabla^2 \phi - \varepsilon \mu \frac{\partial^2 \phi}{\partial t^2} = -\frac{1}{\varepsilon} \rho \tag{3.50}$$

Energy in an electromagnetic wave alternates between electric and magnetic shown in Fig. 3.3A.

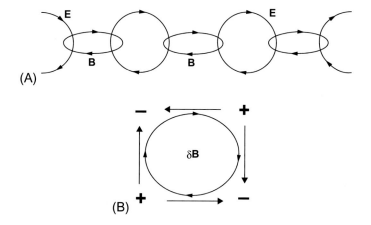

Fig. 3.3 (A) Energy in an electromagnetic wave alternates between electric and magnetic, making the Lorentz gauge condition appropriate. (B) Use of the Coulomb gauge achieves a more complete separation between longitudinal and transverse fields than any other choice of gauge would allow. From Konopinski, E.J., 1981. Electromagnetic Fields and Relativistic Particles. McGraw-Hill, New York, NY, p. 26.

3.5.2 Coulomb gauge

The Coulomb or transverse gauge is obtained using $\nabla \cdot \mathbf{A} = 0$. The source of $\mathbf{B}$ is an electrical current $\mathbf{J}$, so the vector potential is

$$\mathbf{A}(r) = \frac{\mu_0}{4\pi} \int_{V'} \frac{\mathbf{J}(r')d\tau'}{R'} \qquad (3.51)$$

where R' is the distance between the source point and the field point. Taking the divergence, one can see that the Coulomb gauge condition boils down to $\nabla \cdot \mathbf{J} = 0$. Within a closed volume, this means that all the currents are closed.

This is a good choice for plasma physics. Use of the Coulomb gauge achieves "a more complete separation between 'longitudinal' and 'transverse' fields than any other choice of gauge would allow" (Morse and Feshbach, 1953, p. 211). Fig. 3.3B illustrates this very well. A changing magnetic flux in some region will produce an induction electric field with a finite electromotive force (emf), shown by the circle. The plasma wants to respond, but that is limited by the tensor conductivity. In some area with a rotational discontinuity the plasma responds by charge separation as shown at the left and right using the direct conductivity; in that local region the electrostatic field will oppose the induction component, a matter of cause and effect. However, that same charge separation will enhance the induction component, top and also bottom, as shown. This complex pattern is well represented in the Coulomb or transverse gauge.

With both ϕ and $\mathbf{A}$ uniquely determined by this gauge condition, the electrostatic field is entirely due to the net charge distribution, described by the scalar potential, with zero curl. On the other hand, the inductive electric field describes the effect of changing currents, with zero divergence as opposed to the effects of charge separation. The solution for ϕ is the usual Coulomb integral (Cragin and Heikkila, 1981); in the Coulomb gauge there is no time retardation.

$$\nabla^2 \phi = -\frac{\rho}{\varepsilon} \qquad (3.52)$$

The solution for **A** (with retardation) is more complicated:

$$-\nabla^2 \mathbf{A} + \varepsilon\mu \frac{\partial^2 \mathbf{A}}{\partial t^2} + \varepsilon\mu \nabla \frac{\partial \phi}{\partial t} = \mu \mathbf{J} \qquad (3.53)$$

If there is free charge, then the scalar potential ϕ can be computed from it, but **A** *is not affected by the free charge* (Morse and Feshbach, 1953, p. 212).

The energization of particles is due to the force $q\mathbf{E}$. However, a particle does not gain energy at a single point; it must move in the electric field, and so the topology of the field is important. With an inductive electric field the particle gains energy (as in cyclotron) by going round and around. That is not possible with an electrostatic field.

3.6 Electrodynamics

Ampére, Faraday, and others had mapped out an essentially complete and exact description of the magnetic action of electric currents almost two centuries ago. Faraday's discovery of electromagnetic induction came less than 12 years after Oersted's experiment. Out of these experimental discoveries there grew the complete classical theory of electromagnetism, formulated mathematically by Maxwell (1873) and Lorentz (1899), and corroborated by Hertz's demonstration in 1887.

Relativity has its historical roots in electromagnetism. Lorentz, exploring the electrodynamics of moving charges, was led very close to the final formulation of Einstein's great paper of 1905, which was entitled "On the Electrodynamics of Moving Bodies," not "On the Theory of Relativity." We expect any complete physical theory to be relativistically invariant. As it turned out, physics already had one relativistically invariant theory, Maxwell's electromagnetic theory.

The central postulate of special relativity, which no observation has yet contradicted, is the equivalence of reference frames moving with constant velocity with respect to one another. The

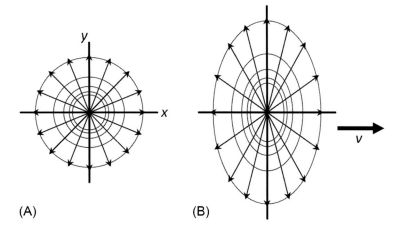

(A) (B)

Fig. 3.4 Electric field line distributions (arrows) and equipotentials for a point charge that is (A) a stationary and (B) moving along the x-axis at a velocity of 0.8c. The equipotentials near the origin are too close together to be shown. There is no radiation for a constant velocity. From Keith after Konopinski, E.J., 1981. Electromagnetic Fields and Relativistic Particles. McGraw-Hill, New York, NY, p. 319.

universal constant c appears in these formulas as a limiting velocity, approached by an energetic particle but never exceeded.

Purcell (1985) and Lorrain et al. (1988) have followed the historical path from Oersted to Einstein almost in reverse. They take special relativity as given, and ask how an electromagnetic system of charges and fields looks in another reference frame. We continue later in this chapter in a discussion of the emission of radiation by charged particles, and the Liénard-Wiechert potentials.

The elementary charge e remains invariant, while the distance and mass m vary with velocity. Electric field distributions (arrows) and equipotentials for a point charge moving to the right is shown in Fig. 3.4B for constant velocity at $\beta = v/c$. Fig. 3.4A shows Coulomb's law for a stationary charge. Say a macroscopic charge q moves at some arbitrary velocity v with respect to a reference frame S at time t. In the inertial reference frame S_0 occupied momentarily by q, the volume of the charge is V_0. In S, the charge is also q, but the volume is shorter in the direction of v. For electrons $\beta = 0.5$ for $v \sim 500\,\text{keV}$. There is no radiation.

3.6.1 The fields of moving charges

A plasma is an assembly of charged particles that interact with each other through polarization and induction effects. Our main objective here is to calculate the electric and magnetic fields of charges moving at velocities that may be large. Electromagnetic "news" travels at the speed of light. In the nonstatic case, it is not the present condition of the source that matters, but rather its condition at an earlier time when the "message" left t_r (called the retarded time, see Fig. 3.5). Since this message must travel a distance R to reach a distant field point, the delay is:

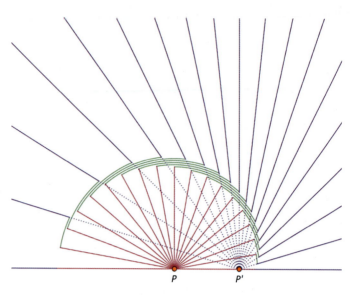

Fig. 3.5 The field of a point charge a short time after the charge has stopped at point *P* after previously traveling at a velocity of 0.8c. The kink travels outward at the speed of light. For a distant observer the field is that of a charge at *P'*, where the charge would have been if it had not stopped. In effect, this is an electric dipole antenna. From Frobenius (https://physics.stackexchange.com/users/110781/frobenius), Electric field associated with moving charge, URL (version: 2019-06-03): https://physics. stackexchange.com/q/426795.

$$t_r = t - \frac{R}{c} \tag{3.54}$$

The appropriate generalization for nonstatic sources is therefore the familiar solutions

$$\phi(\mathbf{r}) = \frac{1}{4\pi\varepsilon_o} \int \frac{\rho(\mathbf{r}', t_r)}{R} d\tau \tag{3.55}$$

$$\mathbf{A}(\mathbf{r}) = \frac{\mu_0}{4\pi} \int \frac{\mathbf{J}(\mathbf{r}', t_r)}{R} d\tau \tag{3.56}$$

Here $R = |\mathbf{r} - \mathbf{r}'|$ is the distance from the source to the field point at which the potentials are evaluated. The charge density $\rho(\mathbf{r}, t_r)$ is that which prevailed at point $\mathbf{r}$ at the retarded time t_r. Note that the retarded potentials reduce properly to the static case when ρ and $\mathbf{J}$ are independent of time.

3.6.2   Radiation from moving charges

Now we address the question of how moving charged particles radiate. A charge at rest does not generate electromagnetic waves, nor does a charge moving at a constant velocity (Griffiths, 1981;

Reitz et al., 1993). It takes an *accelerating charge* to radiate. Purcell (1985), in a fascinating textbook, shows the electric field of a charged particle moving with uniform velocity until $t=0$, when it was abruptly stopped at the origin as in Fig. 3.5. The news that it was stopped cannot reach any point farther than ct from the origin. The field outside the sphere of radius $R=ct$ must be that which would have prevailed if the electron had kept on moving at its original speed. The important new feature is the zigzag in the electric field lines near the middle of the interaction region.

It appears that J.J. Thomson was the first to calculate this **E**; he did so in 1903, 2 years before the publication of Einstein's historic paper on relativity. Imagine a charge that starts from rest, accelerates at the rate a for a short interval of time t, and then continues in a straight line at the constant velocity at. In effect, this is an electric dipole antenna. This will yield all the basic characteristics of radiation fields.

The apparent source of the outer field is displaced from the source of the inner field. ... We have a transverse electric field there, and one that, to judge by the crowding of the field lines [for an impulsive acceleration or deceleration] is relatively intense compared with the radial field. As time goes on, the zigzag in the field lines will move radially outward with speed c. But the thickness of the shell of transverse field will not increase, for that was determined by the duration of the deceleration process. The ever-expanding shell of transverse electric field would keep on going even if at some later time ... we suddenly accelerated the electron back to its original velocity. That would only launch a new outgoing shell... The field does have a life of its own! What has been created here before our eyes is an electromagnetic wave.

Purcell (1985, p.188)

3.6.3 The Liénard-Wiechert potentials

The Liénard-Wiechert potentials describe formally the electromagnetic effect of a moving charge. Built directly from Maxwell's equations, this potential describes the complete, relativistically correct, time-varying electromagnetic field for a point charge in arbitrary motion. These classical equations harmonize with the 20th century development of special relativity, but they are not corrected for quantum-mechanical effects. Electromagnetic radiation in the form of waves is a natural result of the solutions to these equations, a vector potential and a scalar potential. These equations were developed by Liénard (1898) and Wiechert (1901) and continued into the 1900s.

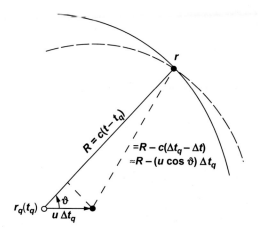

Fig. 3.6 $R = c(t - t_q)$ is the distance to a field point reached at time t. The signal after the particle has moved $u\Delta t_q$ reaches the same field point at a later time as shown. From Konopinski, E.J., 1981. Electromagnetic Fields and Relativistic Particles. McGraw-Hill, New York, NY, p. 287.

One plasma physics textbook, by Clemmow and Dougherty (1969), has treated the subject. It is an old subject dating back before the discovery of relativity. It was discussed by Landau and Lifshitz (1960), Panofsky and Phillips (1962), Konopinski (1981), Griffiths (1981), Reitz et al. (1993), Westgard (1997), and others.

The basic problem as illustrated in Fig. 3.6 is that there is a difference of scale between the time intervals Δt_q of the source traveling at velocity $\mathbf{v}$, and the time intervals Δt during which the corresponding changes occur at the field point shown at r. The length $R = c(t - t_q)$ is the distance traveled by the signal from the charge's position at the time t_q to the field point reached at time t. The signal that originates at the time $t_q + \Delta t_q$ reaches the field point at the later time $t + \Delta t$ after traveling for a distance $R - c(\Delta t_q - \Delta t)$.

There are various ways of doing the analysis. The oldest method, due to Liénard (1898) and Wiechert (1901), is first to derive expressions for ϕ and $\mathbf{A}$ for a moving point charge by a direct evaluation of the integrals, used by Clemmow and Dougherty (1969). Liénard-Wiechert potentials are:

$$\phi(\mathbf{r}, t) = \frac{1}{4\pi\varepsilon_o} \frac{q}{R\left(1 - \dfrac{\hat{R} \cdot \mathbf{v}}{c}\right)} \qquad (3.57)$$

$$\mathbf{A}(\mathbf{r}, t) = \frac{\mu_0}{4\pi} \frac{q\mathbf{v}}{R\left(1 - \frac{\hat{R} \cdot \mathbf{v}}{c}\right)} = \frac{\mathbf{v}}{c^2}\phi(\mathbf{r}, t) \tag{3.58}$$

According to Griffiths (1981), the total power radiated is

$$P = \frac{1}{4\pi\varepsilon_0} \frac{2}{3}\left(\frac{q^2 a^2}{c^3}\right)\gamma^6 \tag{3.59}$$

Notice that the distribution of radiation is the same whether the particle is accelerating or decelerating, as it only depends on the square of a. When a high-speed electron hits a metal target it rapidly decelerates, giving off what is called braking radiation, or *bremsstrahlung*. The dependence on the sixth power of γ

$$\gamma = \frac{1}{\sqrt{1 - v^2/c^2}} = \frac{1}{\sqrt{1 - \beta^2}} \tag{3.60}$$

is an indication that the radiated power increases enormously as particle velocity approaches c. The major difficulty here is that the potentials can no longer be expressed in terms of the present position of the charge; the retarded position and time appear explicitly.

3.6.4 Physical explanation of radiation

Radiation can be produced by two types of antennas, an oscillating electric dipole or a magnetic dipole antenna such as the one in Fig. 2.7. At small velocities, where $v \ll c$, the Liénard-Wiechert potential becomes identical to the steady-state potential, but retaining the time delay associated with retardation. At higher velocities, the radiation is sharply peaked in the forward direction (Fig. 3.5). The signal travels from the source to the field point at the speed of light c. By the time the potential is observed, the source may have moved from its initial location. Suppose that the source is moving at nearly the speed of light toward the observer, and consider a sequence of observed potentials due to a sequence of locations of the source. The source nearly catches up to the signal; consequently, the signals (potentials) received from the source are bunched up at the observer and received nearly simultaneously. The overlap of received signal increases the amount of the potential above what a static source would produce; in the extreme case, a charged particle traveling from infinity toward the observer would be perceived as a large electromagnetic pulse, followed immediately by the particle itself. This bunching of the signal is reminiscent of a "sonic boom."

3.7 Sporadic magnetopause beams

We now discuss an event which we interpret as a plasma transfer event (PTE) of solar wind plasma through the magnetopause into the low-latitude boundary layer (LLBL); more detail on the mechanism is in Chapter 6. On October 17, 1992, the GEOTAIL spacecraft skimmed the morningside magnetopause for several hours as shown in Fig. 3.7. Multiple crossings of the current sheet make this orbit excellent for a study of plasma processes relevant to solar wind-magnetospheric interaction, as reported by Heikkila (2003). Here we describe just the electron behavior, and these led to simulations that proved to be very revealing.

Data from the Comprehensive Plasma Instrumentation (CPI) of the University of Iowa provided density, energy, temperature, and velocity information for the entire 6-h period. Convection drift velocities are shown with 10-min averages. During the first hour, the spacecraft was in the magnetosheath and the velocities were smooth and regular, but during 1900 to 2100 UT they were quite variable.

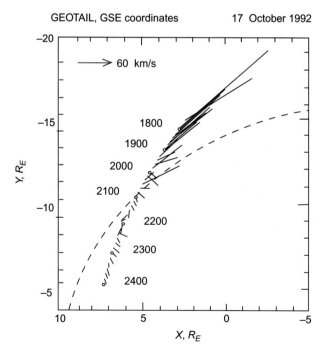

Fig. 3.7 GEOTAIL spacecraft made a prenoon magnetopause skimming orbit on October 17, 1992; the average location of the magnetopause is the *dashed line*. Convection drift velocities are shown with 10-min averages in the antisunward direction. From Heikkila and Paterson.

CPI executed three instrument cycles within the current layer between 19:23:28 and 19:24:16 UT depicted by Fig. 3.8A. The return fast crossing back into the magnetosphere is similar as can be seen in the magnetometer data at the top of Fig. 3.8. Special coordinates were used in the analysis of the CPI data for Fig. 3.9 in that z is always aligned with the magnetic field **B**. Here the ratio of the parallel and antiparallel fluxes is plotted on a logarithmic scale. Fig. 3.9 indicates that the ratio of parallel and antiparallel fluxes can be quite large (or low). At 1923:28 above 1×10^4 km/s on the x-axis, corresponding to an energy greater than 300 eV, the ratio is at about a factor of 3. This is true up to 4×10^4 km/s (4.6 keV). At 1924:16, the ratio is an order of magnitude at 2×10^4 km/s, corresponding to an energy of 1 keV, extending to 2 keV.

The new feature of great interest is the sporadic emissions of keV electrons exiting the magnetopause at certain times as in Fig. 3.9. These show a field-aligned component along B_n within the magnetopause current sheet. They appear to be energized

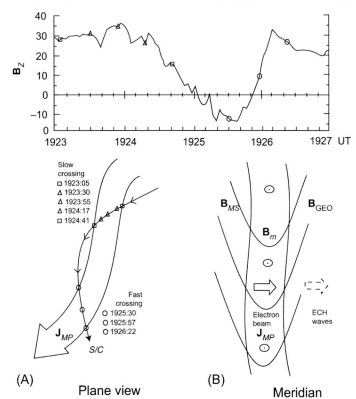

(A) Plane view   (B) Meridian

Fig. 3.8 The z component of the magnetic field in GSM coordinates covering the period 19:23 to 19:27. (A) A possible spacecraft trajectory through the magnetic record shown earlier. (B) A vertical cut showing the magnetic field topology, with a normal component of the magnetic field B_n associated with a rotational discontinuity. From Heikkila and Hairston.

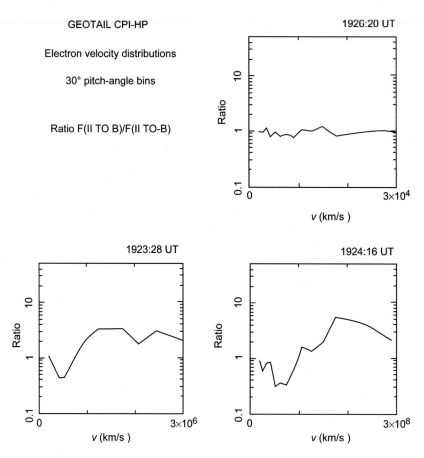

Fig. 3.9 *Top*: Data from the GEOTAIL spacecraft show counterstreaming electron fluxes interpreted as bouncing motion on closed field lines in the LLBL. *Bottom*: Two earlier data panels from exactly in the current sheet show that at high velocities (energies 1–10 keV) the parallel fluxes along B_n exceed the antiparallel fluxes by a factor that can be as high as 10. The electrons below $v = 10,000$ km/s (less than 200 eV) traveling with the ions antiparallel to B_n are involved in a shielding process. From Heikkila, W.J., 2003. Initial condition for plasma transfer events. In: Newell, P.T., Onsager, T. (Eds.), Earth's Low-Latitude Boundary Layer. AGU Geophysical Monograph 133. American Geophysical Union, Washington, DC, p. 164.

locally, quite often peaking at several keV in one direction along B_n sometimes by an order of magnitude. This tends to occur whenever the plasma velocity reverses.

The second feature that needs to be stressed is that at lower energies, below ~200 eV, the flux direction is reversed, with a ratio less than 1. These electrons are traveling in the same sense as the ions! This exciting observation led us to do the particle simulations given as follows.

3.8   Particle simulation in 1D

To truly understand magnetospheric plasma physics we must be able to follow the time development of various quantities in three dimensions. For example, we need at least two dimensions to describe an X-line in the magnetic field topology (often the x-z plane), but another set for an emf in the electric field (curl **E** in the x-y plane). That requires a formidable requirement for a global simulation study.

However, once a given model is understood (at least in principle) it may be possible to get some insight with fewer dimensions. Only two dimensions are needed for some aspects of plasma motion (such as convection with an elongated structure), while only one may be enough for a field-aligned study, with corresponding fewer simulation points.

Omura et al. (2003) used a 1D simulation to investigate the consequences of a parallel component of the electric field along magnetic field lines in a magnetoplasma (relevant to conditions in the plasma sheet). Such a situation may be hard to imagine in view of the high plasma conductivity along the field lines, but the earlier observations are difficult to explain in any other way. It is commonly believed that a parallel field with a voltage difference of several kilovolts (electric field strength $E_{\parallel} = 1$ V/km) is involved in auroral particle acceleration (Evans, 1974, 1975; Reiff et al., 1988). Satellite data on energetic particles during substorms to MeV energies reported by Kirsch et al. (1977) and Sarafopoulos and Sarris (1988) are consistent with a voltage drop (*not a potential drop!*) exceeding 1 MV (implying $E = \sim 10$ V/km over $\sim 10\ R_E$).

Fig. 3.10 illustrates the problem in the magnetotail during substorms in 3D. During the growth phase, the cross-tail current is enhanced locally between ~ 5 and $\sim 15\ R_E$ causing the tail-like feature. The natural tendency is for plasma to escape in the direction of the least magnetic pressure, outward between the high-pressure lobes (Heikkila and Pellinen, 1977). This implies a sudden local change in the current distribution in the field-reversal region. Creation of a southward component of the magnetic field is consistent with the induction electric field as depicted by the four arrows in Fig. 3.10 (with no z component).

Omura et al. (2003) assume a rather low value for the electric field corresponding to about 0.1 V/km. One-dimensional electrostatic particle simulations were performed with a constant external electric field applied to the Maxwellian thermal plasma with isotropic electrons and ions as in the magnetosheath. Periodic boundary conditions were used: the accelerated particles leaving

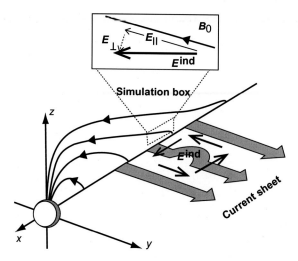

Fig. 3.10 An outward meander in current somewhere in the plasma sheet produces an inductive electric field with directions as shown. A parallel component $E_\parallel$ is produced, and we simulate the plasma response. From Omura, Y., Heikkila, W.J., Umeda, T., Ninomiya, K., Matsumoto, H., 2003. Particle simulation of plasma response to an applied electric field parallel to magnetic field lines. J. Geophys. Res. 108(A5), 2. http://doi.org/10.1029/2002JA009573.

one end of the boundary enter the system from the other side. By assuming the periodic boundary condition, the effect of excess charges that could appear at the boundaries of the acceleration region is neglected.

Table 1 of the article (Omura et al., 2003) lists all parameters for three different runs. A mass ratio of $m_i/m_e = 1600$, close to the real mass ratio assuming protons and electrons, was used. Space and time were normalized by the initial Debye length $\lambda_e = v_{e0}/\Pi_e$, and the plasma oscillation period Π_e^{-1}, respectively, where v_{e0} is the initial thermal velocity of electrons, and Π_e is the electron plasma frequency. All velocities are normalized by v_{e0}.

The magnitude of the induction electric field was based on the corresponding potential energy over the Debye length, where the potential energy is normalized by the initial thermal energy of electrons. A normalized value of the electric field C_E that satisfied

$$eE_\parallel^{\text{ind}}\lambda_e = C_E \frac{1}{2}m_e v_{e0}^2 \tag{3.61}$$

was used (e and m_e are electron charge and mass, respectively). $C_E = 0.01$ was assumed for all runs.

3.8.1   Simulation results

Three simulation runs were performed with different ratios of the ion temperature to the electron temperature; here we discuss only the first with $T_i \gg T_e$. In the presence of this constant and uniform electric field $E_\parallel^{\text{ind}} > 0$, the electrons and ions are accelerated in opposite directions. Initially, the drift energies of electrons and ions increase parabolically due to the acceleration by the constant electric field. When they reach a critical drift velocity of the counterstreaming electrons and ions, the plasma becomes unstable (a current driven instability called the Buneman instability).

Fig. 3.11 shows the time histories of the electric field energy and kinetic energies for electrons and ions. Kinetic energy of a

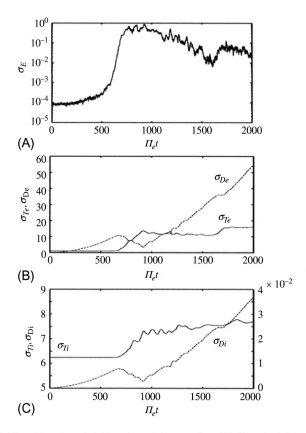

Fig. 3.11 Time histories of field and particle energies. (A) Electric field energy, (B) drift energy and thermal energy of electrons, and (C) drift energy and thermal energy of ions. At later times, the plasma has little effect on acceleration by the external electric field; counterstreaming beams (ion and electron) are produced at ever-increasing energies. From Omura, Y., Heikkila, W.J., Umeda, T., Ninomiya, K., Matsumoto, H., 2003. Particle simulation of plasma response to an applied electric field parallel to magnetic field lines. J. Geophys. Res. 108(A5), 3. http://doi.org/10.1029/2002JA009573.

group of particles is decomposed into a drift energy and a thermal energy. The growth rate increases turning from a negative value to a positive value as the drift velocity increases (not shown here). The electrostatic wave grows exponentially (Fig. 3.11A) from the thermal fluctuation level to a saturation level, at which some thermal electrons are trapped by the potential wells formed by the saturated coherent electrostatic wave. Because of the realistic mass ratio assumed the acceleration of the ions is negligibly small; the drift energy of ions varies almost proportionally to that of electrons (Fig. 3.11B and C).

The left panels of Fig. 3.12 show a series of phase space plots showing the distribution functions $f_{i,\,e}(x, v_x)$ for ions and electrons at different times of the simulation run. At $\Pi_e t = 400$, the relative drift velocity between ions and electrons is less than the critical velocity of the marginal stability, and at first there is no wave being excited. Once the drift velocity of the electrons exceeds the critical velocity, the electrostatic field begins to grow gradually due to the Buneman instability. As shown in Fig. 3.11, the electrostatic field energy increases exponentially during the period of $\Pi_e t = 600 \sim 750$. The left panel of Fig. 3.12B shows the distribution functions at saturation, which is due to trapping of some electrons. The drift energy of the electrons is converted to thermal energy during the period of $\Pi_e t = 700 \sim 900$. The trapped electrons oscillate in the potential wells to form vortices which coalesce with each other to form electrostatic solitary waves.

The continuous potentials of Fig. 3.12B coalesce with each other to form larger isolated potentials. Through repeated coalescences among the potentials in the simulation system, a large single potential is formed (Fig. 3.12C). The isolated potential is called an electron hole, which is very stable and close to the BGK equilibrium (Bernstein et al., 1957; Krasovsky et al., 1997), if there is no imposed accelerating electric field. Since the electrons are accelerated continuously under the constant electric field, there exists no stationary solution, such as a BGK mode, in the present simulation.

Fig. 3.13 from another simulation study by Chen (2002) shows the development of electron holes at an early stage. It seems that the ions become bunched like a rippling effect; the lower-energy electrons circle around the ion bunches, with a one-to-one correlation. This is likely Debye shielding of the ion bunches.

The right panel of Fig. 3.14 plots the effective total potential ϕ due to both induction and electrostatic electric fields in order to study the dynamics of electron holes formed by trapped and

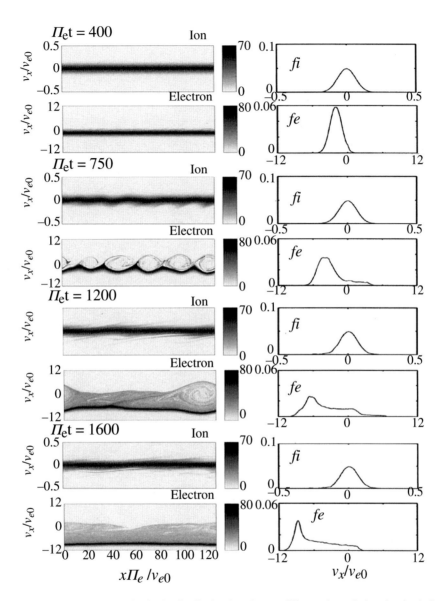

Fig. 3.12 Velocity phase space plots and velocity distribution functions at different times during the simulation. At first, all plasma particles are accelerated freely. Then a Buneman current instability sets in, and free acceleration is temporally hindered. As the particles gain more energy the instability diminishes, and almost free acceleration is maintained. From Omura, Y., Heikkila, W.J., Umeda, T., Ninomiya, K., Matsumoto, H., 2003. Particle simulation of plasma response to an applied electric field parallel to magnetic field lines. J. Geophys. Res. 108(A5), 4. http://doi.org/10.1029/2002JA009573.

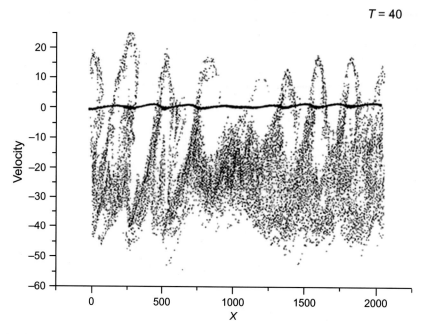

Fig. 3.13 The development of electron holes at an early stage during the simulation can be seen. Low-energy electrons follow the ion bunches to shield their charge. From Tao Chen.

untrapped resonant electrons. The spatial structure at different times is also plotted in the left panels. The initial slope is due to the induction electric field $E_{\parallel}^{\mathrm{ind}}$. When the electrostatic field grows to exceed the magnitude of $E_{\parallel}^{\mathrm{ind}}$, the trapping of electrons sets in to form the isolated potentials through coalescence. When they coalesce with each other, there arises an attraction force between two potentials as we can find at $x\Pi_e/v_{e0} = 20 \sim 80$ during the period of $\Pi_e t = 1100 \sim 1150$. The merged potential becomes larger, and it is further accelerated in the negative direction of E_n, because the potential is formed by the electrons (note the change of scale in the left plots).

Untrapped electrons, on the other hand, move freely with a fairly large drift velocity that is continuously increasing. On the right panels of Fig. 3.12 are also plotted the distribution functions $f_{i,v}(v)$ of ions and electrons averaged over the position x. While the trapped electrons oscillate around the ion drift velocity to form the symmetric thermal component, the untrapped electrons contribute to the asymmetry, forming a bump on the high-energy tail of the electron velocity distribution function. This is an electron beam energized directly by the parallel electric field.

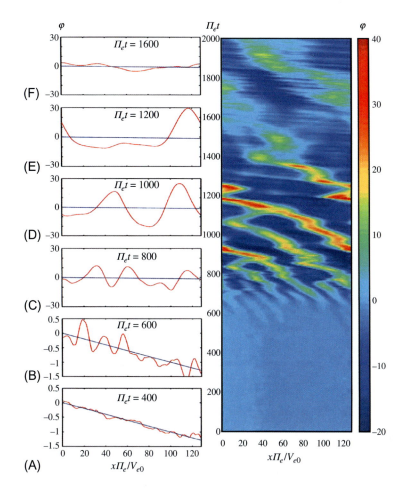

Fig. 3.14 Left panel: spatial structure of total potential at different times (note the change in scale). Small disturbances coalesce with each other producing very large electrostatic fields. They then expire, as shown dramatically on the right. From Omura, Y., Heikkila, W.J., Umeda, T., Ninomiya, K., Matsumoto, H., 2003. Particle simulation of plasma response to an applied electric field parallel to magnetic field lines. J. Geophys. Res. 108(A5), 5. http://doi.org/10.1029/2002JA009573.

3.8.2 Confirmation

In order to confirm this amazing result, Büchner and Elkina (2005) used their newly developed highly accurate conservative Vlasov simulation scheme (Fig. 3.15). Like Omura et al. (2003), they applied a constant external electric field to an initially Maxwellian plasma. In such setup, both electrons and ions are accelerated in the external electric field in different directions, the more inertial ions at a slower path. For better comparison, they have used, in particular, the same parameters, like the ratio of ion to electron thermal velocities, similar to the external electric field.

The outcome of these simulations shows the evolution of the average over the whole simulation box length. Indeed, as one can see in Fig. 3.15A for an almost isothermal plasma, the electron distribution function stays stable until reaching an electron drift

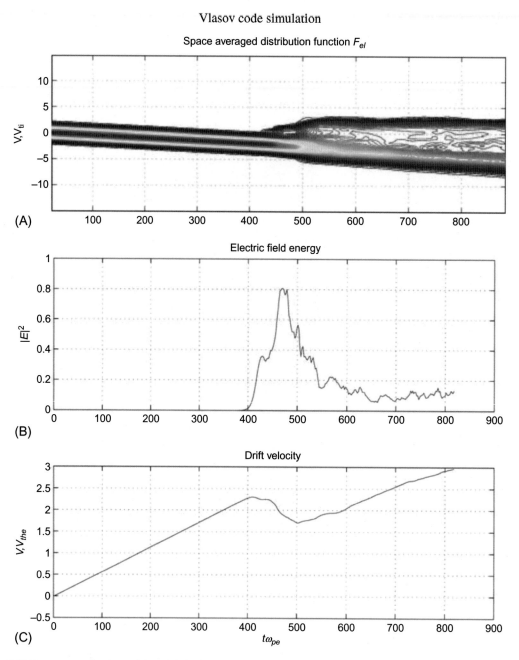

Fig. 3.15 The common horizontal time axis is time normalized to ω_{pe}. (A) Evolution of the spatially averaged electron distribution function after the application of an electric field accelerating the electrons in the negative velocity direction until the threshold of an ion instability is reached. (B) Time evolution of the energy of the electric field fluctuations and (C) of the average particle drift (current) velocity. From Büchner, J., Elkina, N., 2005. Vlasov code simulation of anomalous resistivity. Space Sci. Rev. 121, 247. http://doi.org/10.1007/s11214-006-6542-6.

speed larger than the electron thermal velocity. Then the accelerated drifting Maxwellian distribution function becomes Buneman unstable. Fig. 3.15B shows that the energy of the electric field fluctuations grows strongly, reaching a maximum, then drops; the energy of the electrostatic fluctuations drops by an order of magnitude to an almost stationary level. The evolution of the electron drift velocity is depicted in Fig. 3.15C. There one sees that after reaching a certain level, the fluctuations cease to grow. The drift velocity keeps on increasing for some electrons, but at a slightly reduced rate as compared to free space.

3.9 Exceptional electron beam observation

During a passage of the plasma sheet, the Cluster 3 spacecraft made an exceptional high-resolution measurement of a beam of electrons with energies up to 400 keV (Taylor et al., 2006). The beam was only fully resolved by combining the energy range coverage of the Plasma Electron and Current Experiment (PEACE) and Research with Adaptive Particle Imaging Detector (RAPID) electron detectors. Its pitch angle distribution evolved from antiparallel, through counterstreaming, to parallel over a period of 20 s. Although energetic electron fluxes (a few hundred keV) are frequently observed in the plasma sheet, electron beams of this nature have been rarely, if ever, reported. The global conditions of the magnetosphere at this time were analyzed using multiple spacecraft, ground, and auroral observations. The beam event is clearly associated with a substorm.

Fig. 3.16 presents an overview of the electron and magnetic field observations for August 11, 2002 from Cluster 3. As time progresses (from 1340 to 1429 UT), the magnetic field magnitude and B_x component begin to increase along with a gradual decrease in the peak plasma sheet electron temperature (as shown by the overall reduction of electron flux observed by RAPID). At around 1422 UT, Cluster observes a local maximum in the magnetic field, accompanied by a burst of energetic electrons up to 400 keV observed only by Cluster 3. Immediately after the burst, the magnetic field becomes much more disturbed and continues to be highly variable, as Cluster exits the plasma sheet and enters the lobe. At around 1445 UT Cluster reenters the plasma sheet, the entry characterized by the sudden return of >keV electrons (again with elevated RAPID fluxes) and a significant drop in the magnetic field magnitude pertaining to the proximity of the neutral sheet. Cluster remains close to the neutral sheet until just after 1447 UT, when the field magnitude increases again, but remains well

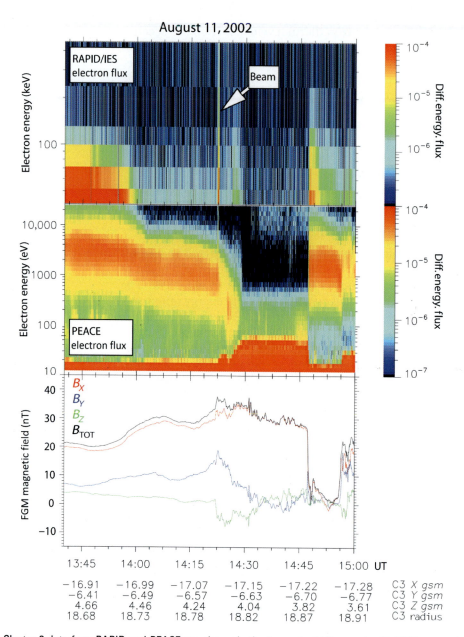

Fig. 3.16 Cluster 3 data from RAPID and PEACE are shown in the top and middle panel, with FGM magnetic field data (GSM) in the lower panel, at 1340–1500 UT, August 11, 2002. The beam appears just after 1422 UT.

From Taylor, M.G.G.T., Reeves, G.D., Friedel, R.H.W., Thomsen, M.F., Elphic, R.C., Davies, J.A., et al., 2006. Cluster encounter with an energetic electron beam during a substorm. J. Geophys. Res. 111, 3. http://doi.org/10.1029/2006JA011666.

below its earlier lobe and plasma sheet value. The population at 1422 UT is much more than a simple brief encounter with a similar energetic population, with highly dynamic pitch angle and energy evolution.

3.9.1 Details of the electron beam

In Fig. 3.17 we focus on the electron beam itself using high-resolution data from the FGM, PEACE HEEA, and RAPID IES instruments onboard Cluster 3, over a time period from 1422:03.221 to 1422:55.224 UT. Each column represents data from a specific energy channel, with the value noted at the top of each column. Energy increases from left to right, so that the first four columns are from the PEACE instrument and the final three from the RAPID instrument. Each row represents data from a single spin, as denoted by the labeling on the left, which orders the data in increasing time from top to bottom after 1422 UT. Each individual panel represents an angle-angle during one spin, with contour lines marking the magnetic field direction. PEACE data are sampled in 32 azimuths and 6 polar bins, whereas RAPID data are divided into 3 heads in the polar direction, with each head comprised of 3 polar bins, giving a total of 9 polar directions. The high fluxes during the appearance of the beam in RAPID are at least an order of magnitude higher than they are in the lobe.

The data at 03.221 (row 1) show the condition of the plasma sheet before the appearance of the beam. The high energies ($\geq 9112\,eV$) have very low fluxes, in contrast with high fluxes in the bins centered at 541 and 2009 eV. We also note the contrast in pitch angle distribution, with a comparatively isotropic electron population between 541 and 2009 eV, compared to the distinctly field-aligned population within the lowest energy bin (73 eV). We note that all four spacecraft observe a similar low energy field-aligned population in the plasma sheet at this time (not shown here). However, unique to Cluster 3, beginning at ~ 19.180 (row 5), there is a notable decrease in the fluxes anti-parallel to the field at energies at and below 541 eV, with a corresponding increase in the same direction at higher energies. At the highest energies ($\geq 47.2\,keV$), RAPID observes an evolution of the pitch angle, moving from an initial anti-parallel through to bi-directional and finally being more field-aligned. Overall, this feature related to the beam persists for ~ 15–$20\,s$, recovering first at the lowest energies, then the higher energies, until after 39.149 (row 10) when the plasma returns to prebeam-like appearance, albeit with a lower overall peak energy.

Taylor et al. (2006)

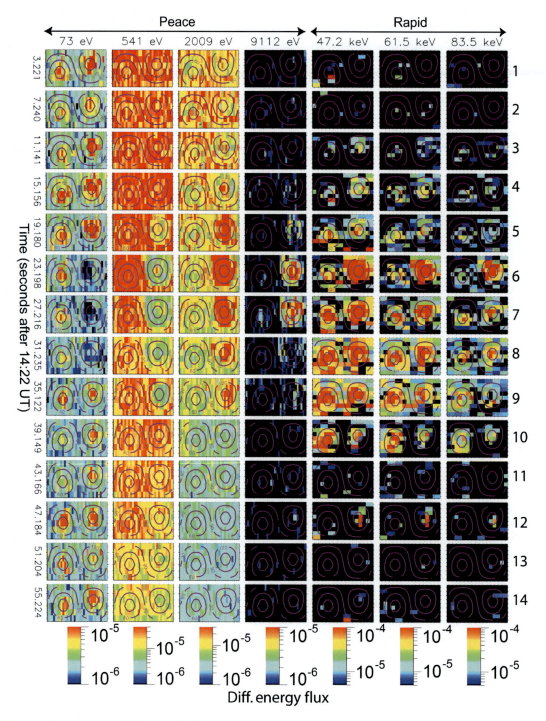

Fig. 3.17 High-resolution PEACE and RAPID electron data presented in the common azimuthal angle-polar angle format, 32 × 6 for PEACE and 16 × 9 for RAPID, for each spin, at different times (rows) and different energies (columns). The first four columns are PEACE data and the final three columns RAPID. Flux units are PEACE (ergs/(cm² s eV str)) and RAPID (ergs/(cm² s keV str)). Row numbers are marked for easy reference. From Taylor, M.G.G.T., Reeves, G.D., Friedel, R.H.W., Thomsen, M.F., Elphic, R.C., Davies, J.A., et al., 2006. Cluster encounter with an energetic electron beam during a substorm. J. Geophys. Res. 111, 5. http://doi.org/10.1029/2006JA011666.

3.9.2   Evolution of the beam

Fig. 3.18 highlights the evolution of the beam in the form of a distribution function (s^3/km^6) from four single spins of data corresponding to marked periods. Fig. 3.18A shows the situation at 1422:03.221 UT and is representative of the prebeam period where all four spacecraft observe a field-aligned low-energy electron population, indicated by the light gray block arrows in the schematic. This is shown in the spectra by the dominance of parallel

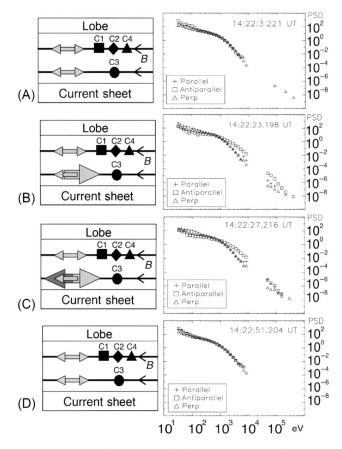

Fig. 3.18 Schematic representations of the four spacecraft observations, along with corresponding PSD spectra from the data shown in Fig. 3.17. In the schematic, the *black lines and arrows* indicate the *B* field direction, along with the small, *light gray arrows* representing the low-energy (<400 eV) electrons, the large *light gray arrows* representing the >500 eV electrons, and the *darker gray arrows* representing the higher-energy >40 keV electrons. The spectra show PSD (s^3/km^6) versus energy (eV) spectra for the parallel *(cross)*, perpendicular *(triangle)*, and parallel *(square)* pitch angle directions for the times (A) 14:22:03.221, (B) 14:22:23.198, (C) 14:22.27.216, and (D) 14:22:51.204. From Taylor, M.G.G.T., Reeves, G.D., Friedel, R.H.W., Thomsen, M.F., Elphic, R.C., Davies, J.A., et al., 2006. Cluster encounter with an energetic electron beam during a substorm. J. Geophys. Res. 111, 6. http://doi.org/10.1029/2006JA011666.

and antiparallel fluxes up to about 200 eV. Fig. 3.18B represents the situation at 1422:23.198 UT, where Cluster 3 observes the onset of the beam. It is characterized in the antiparallel direction by enhanced phase space density (PSD) at energies $\sim >500$ eV (larger light gray block arrow in the schematic) and the reduction of <400 eV electrons. Fig. 3.18C shows the situation at 1422:27.216 UT, where the parallel component of the high-energy beam is observed, only in the RAPID energy range (larger dark gray arrow). After the beam, the situation returns to that Fig. 3.18D, 1422:51.204 UT with the return of the field-aligned electrons and low-energy electrons. Cluster 3 was located -0.45 R_E lower in the Z direction closer to the neutral sheet. This event was fortuitous in that the burst mode facilitated the examination of the fine structure of the high-energy beam.

3.9.3 Comparison to simulations

We note the similarities to Fig. 3.9 in the sense of the low-energy electrons at the magnetopause observed with GEOTAIL being the same as the ions. This is seen in Fig. 3.17 for three spins beginning at 14:22:23.198 UT, for 73 and 541 eV. This is the same sense as in the simulations, Figs. 3.12 and 3.14. We will comment further in Section 3.11, "Discussion."

3.10 Other observations of energization

There are a vast number of observations of energization. Here will just list three.

3.10.1 Counterstreaming particles

Observations of energetic proton ($E_p \sim 0.29$ MeV) and electron ($E_e \sim 0.22$ MeV) bursts were reported by Kirsch et al. (1977); they were encountered with the IMP-8 satellite in the magnetotail on November 26, 1973. Large fluxes of both electrons and protons appeared at the location of IMP-8. Protons and electrons exhibit large anisotropies (30:1 and 5:1 for protons and electrons, respectively), with the electrons propagating tailward while the protons are propagating earthward. These oppositely directed anisotropies last as long as 1 min and are not necessarily aligned with the local magnetic field. While these enhanced fluxes are observed, there exist significant depressions in the local magnetic field accompanied with the appearance of both positive and negative components of B_x and B_z, that is, a signature of the presence of the neutral sheet. "We interpret these observations as suggesting the presence of an electric field in association with the acceleration region in the magnetotail."

3.10.2 Inverse velocity dispersion bursts

Fine-time resolution ($\sim$10 s) observations of bursts of energetic particles inside the Earth's plasma sheet by the IMP-8 spacecraft were reported by Sarafopoulos and Sarris (1988). This has revealed that the detailed structure of the time-intensity profiles of the ion bursts often display an "inverse velocity dispersion" effect (Fig. 3.19), whereby the lower-energy ($\sim$300 keV) ion intensity

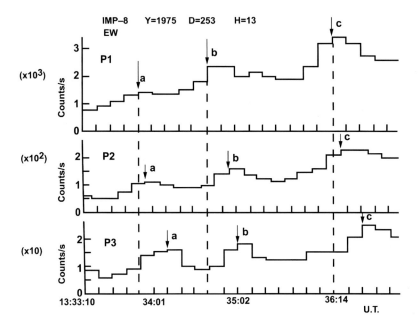

Fig. 3.19 Observations of bursts of energetic particles inside the Earth's plasma sheet by the IMP-8 spacecraft. Ion bursts often display the "inverse velocity dispersion" where the low-energy (P1-300 keV) ion intensity enhancements are detected before the mid (P2-500 keV) and high (P3-1 MeV) energy channels. From Sarafopoulos, D.V., Sarris, E.T., 1988. Inverse velocity dispersion of energetic particle bursts inside the plasma sheet. Planet. Space Sci. 36, 1187.

enhancements are detected before the higher energy (1 MeV) ones. In particular, it was found that long duration (several minutes or more) bursts of energetic particles are composed of a series of highly anisotropic, short duration impulsive bursts, which last a few tens of seconds. Furthermore, the inverse velocity dispersion effect is observed: (a) both inside the plasma sheet and at its high-latitude boundary, (b) during periods of thinning as well as expansion of the plasma sheet, and (c) for both tailward and earthward streaming of the energetic ions.

3.10.3 Global substorm onset

The magnetospheric substorm that occurred shortly after 04:00 UT on August 27, 2001 has been a topic of detailed study (see Chapter 11). These studies included analysis of data from energetic electron detectors aboard the Cluster and Polar spacecraft that were located in the magnetotail during the substorm. Serendipitously, Chandra, the X-ray observatory, also carrying an energetic electron detector, was located in the magnetotail several hours earlier in local time (Blake et al., 2005). Examination of data from the three magnetotail positions revealed that energetic electrons ($>35 \, keV$) appeared simultaneously at all three locations (Fig. 3.20).

The locations of the spacecraft are shown in Fig. 3.21. Energetic electrons appeared simultaneously (within the uncertainties) at Cluster, Chandra, and Polar, whose spacecraft had large separations in local time, latitude, and radial distance. Time-correlated electron bursts were seen for more than an hour prior to the substorm. Whatever the entire physical description of substorms may turn out to be, it must be congruent with these energetic electron observations.

3.11 Discussion

It must be noted that only the electric field can energize charged particles by the force $\mathbf{F} = q\mathbf{E}$. Helmholtz's theorem enunciates quite clearly the definition of the electric field $\mathbf{E}$ in terms of the divergence and curl; Maxwell's equations are exactly such definitions. As a direct result, the electric field has two components, $\mathbf{E} = \mathbf{E}^{es} + \mathbf{E}^{ind}$, one conservative $\mathbf{E}^{es} = -\nabla\phi$ the other solenoidal $\mathbf{E}^{ind} = -\partial\mathbf{A}/\partial t$. Both components have parallel and transverse components in realistic magnetic fields everywhere.

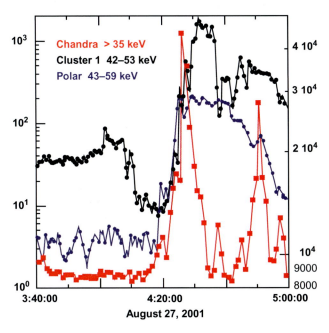

Fig. 3.20 The magnetospheric substorm at 0400 UT on August 27, 2001 included energetic electron detectors aboard the Cluster and Polar spacecraft. Chandra, the X-ray observatory was located in the magnetotail several hours earlier in local time. Examination of data from the three magnetotail positions revealed that energetic electrons (>35 keV) appeared simultaneously at all three locations. The physical description of substorms must be congruent with these energetic electron observations. From Blake, J.B., Mueller-Mellin, R., Davies, J.A., Li, X., Baker, D.N., 2005. Global observations of energetic electrons around the time of a substorm on 27 August 2001. J. Geophys. Res. 110, 4. http://doi.org/10.1029/2004JA010971.

Fluid theory considers only one convection electric field $\mathbf{E}^{conv} = -\mathbf{v} \times \mathbf{B}$ (plus other terms), used in the generalized form of Ohm's law (Sibeck et al., 1999, p. 211; Vasyliunas, 2001). As a definition of the electric field it is contrived or artificial at best. Convection electric field is transverse to $\mathbf{B}$; the reality of a parallel component is problematic.

3.11.1 Simulations of plasma response

Previous simulations used ad hoc particle distributions in order to get the desired instabilities (e.g., Omura et al., 1994, 1996). With more recent simulations (Omura et al., 2003), these instabilities grow out of a well-behaved stable plasma. Now, their origin is easier to understand. The Buneman instability in a plasma is due to a parallel electric field, easier to justify than a specific type of plasma distribution, such as a beam.

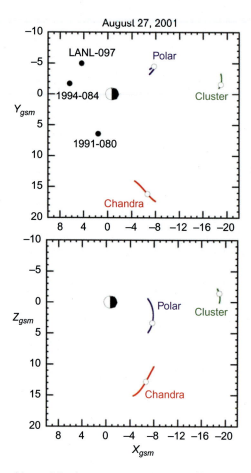

Fig. 3.21 The positions of the three magnetotail spacecraft plotted as a function of time for the first 6 h of August 27, 2001; the centroid of the Cluster constellation is plotted. The locations at substorm onset, which was about 04:01 are shown as *open circles*. The locations of three GEO spacecraft are shown in the X_{gsm}, Y_{gsm} plot at the same time as *solid circles*. From Blake, J.B., Mueller-Mellin, R., Davies, J.A., Li, X., Baker, D.N., 2005. Global observations of energetic electrons around the time of a substorm on 27 August 2001. J. Geophys. Res. 110, 2. http://doi.org/10.1029/2004JA010971.

3.11.2 Electron holes and Debye shielding

In the process leading to the Buneman instability there arise electron holes, low-energy electrons orbiting a series of positive potentials, some incredibly large (see Fig. 3.14). They modify the electric field from its initial value locally, as shown in Figs. 3.12 and 3.14. The low-energy electrons form a cloud around the ion bunches, a form of Debye shielding. It is likely that these low-energy electrons, the hole distributions, are the ones responsible for anomalous resistivity. They are not effective in providing resistivity for a fluid treatment of plasma.

3.11.3 Untrapped electrons

A substantial part of the accelerated electrons remain untrapped during, and after, the nonlinear evolution. These untrapped electrons form an electron beam which leaves the acceleration region; they continue to gain energy. *This is a major result of the simulation.* It is now apparent that the presence of a plasma has only a marginal effect on the result of applying an electric field in the long term.

3.11.4 Broadband electrostatic noise

Satellite observations over the past 30 years have indicated the presence of broadband electrostatic noise (BEN) in different regions of the Earth's magnetosphere (Matsumoto et al., 1994). BEN is found to have waveforms of solitary bipolar electric field pulses, which were called Electrostatic Solitary Waves (ESWs). Kojima et al. (1994) investigated the variation of the ESW pulse width in time (i.e., pulse duration, w) and their repetition period (i.e., duration between the pulses, T). They found that even though the pulse width (w) and repetition period (T) change very quickly, their ratio w/T is more or less constant. From this observation, they concluded that the spatial size of the solitary potentials is almost constant.

> *The nonlinear BEN waveforms do not show continuous broadband noise [but] are composed of bipolar pulses in most cases. The bipolar signature is characterized by solitary spikes composed of a half sinusoid-like cycle followed by a similar half cycle having the opposite sign.*
>
> **Moolla et al. (2007)**

The GEOTAIL spacecraft has measured ESWs. It is likely that is the reason for broadband activity as shown in Figs. 3.12 and 3.14.

3.11.5 Another source of high energies

There is still another source of high-energy particles. Evidence presented by Chen (2008) in the high-altitude cusp reveals that the charged particles can be energized locally. The power spectral density of the cusp magnetic fluctuations shows increases by up to four orders of magnitude in comparison to an adjacent region. Large fluctuations of the cusp electric fields have been observed with amplitudes of up to 350 mV/m. The measured left-hand polarization of the cusp electric field at ion gyrofrequencies indicates that the cyclotron resonant acceleration mechanism (CRAM) is working in this region. The CRAM can energize electrons and ions from keV to MeV in seconds. This is the topic of Chapter 7.

3.12 Summary

By Helmholtz's theorem any vector field **V** has two parts, one irrotational with divergence **V** = s, the other solenoidal with curl **V** = **c**. The sources s and **c** are identified as the *causes* of the fields. Maxwell's equations are just these types, with the sources for **E** as $s = \rho/\varepsilon_0$, **c** = $-\partial \mathbf{B}/\partial t$.

3.12.1 Parallel component

Both parallel and transverse components of the electric field will exist whatever the source in a realistic magnetic field (not in 2D). The problem for the plasma to handle is to cancel the parallel component due to induction by charge polarization to maintain **E·B** = 0 everywhere. This problem is real, and it is omnipresent; it may be the cause of broadband noise.

3.12.2 Transverse component

The actual electric field **E** is extremely difficult to calculate as the potential depends in an exceedingly sensitive way on the actual differential displacements of positive and negative particles. However, regardless of the complexities involved in calculating the actual electric field we can identify several important features.

1. If the polarization of charges leads to suppression of the magnetic-field-aligned electric field component $E_{\parallel}^{\text{ind}}$, the requirement on the integral around an arbitrary contour implies that the value of the transverse component $E_{\perp}^{\text{ind}}$ *is enhanced* instead (see Fig. 3.2).

2. Discharge would energize particles impulsively, nonadiabatically (Fig. 3.20 further discussed in Chapter 11).

3. Further energization is caused by betatron acceleration, with the result that a few particles can reach MeV energies in a few seconds.

4. The total power radiated by an energetic particle is

$$P = \frac{1}{4\pi\varepsilon_0}\frac{2}{3}\left(\frac{q^2 a^2}{c^3}\right)\gamma^6 \qquad (3.62)$$

The distribution of radiation is the same whether the particle is accelerating or decelerating, as it only depends on the square of a. Dependence on the sixth power of $= 1/\sqrt{1-\beta^2}$ is awesome.

3.12.3 Global onset

Energetic particles are very common in all space observations; we cite just three instances. Kirsch et al. (1977) saw counterstreaming electrons and protons with IMP-8 in the magnetotail. The anisotropies lasted as long as a minute and were not necessarily aligned with the local magnetic field. Sarafopoulos and Sarris (1988) observed bursts of energetic particles inside the plasma sheet, also using IMP-8 (Fig. 3.19). Lower-energy enhancements were often seen before the higher energies in an "inverse velocity dispersion" effect. Blake et al. (2005) reported that energetic electrons appeared simultaneously (within the uncertainties) at Cluster, Chandra, and Polar, whose spacecraft had large separations in local time, latitude, and radial distance. Time-correlated electron bursts were seen for more than an hour prior to the substorm (Fig. 3.20). Whatever the entire physical description of substorms (Fig. 3.22) may turn out to be, it must be congruent with these energetic electron observations. The answer may lie with the Liénard-Wiechert potentials.

3.12.4 Lightning strike!

Something remarkable is happening! The pulse is evidence of a brief yet strong radiation field, a *discharge, a lightning strike*! The conductivity along the X-line goes from essentially zero (Pedersen) to high (direct). The conductivity along the X-line reaches a critical level and the current increases rapidly, as will be apparent in Chapter 11.

3.12.5 Simulations agree with observations

The outcome of the simulations shows that the electron distribution function stays stable until reaching an electron drift speed larger than the electron thermal velocity. Then the accelerated drifting Maxwellian distribution function becomes Buneman unstable. The energy of the electric field fluctuations grows

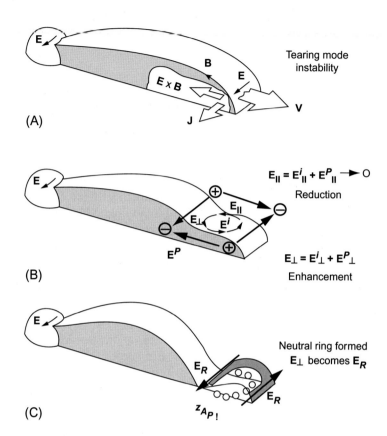

Fig. 3.22 A likely sequence of events during a substorm. Everything is stable and adiabatic at the beginning (A), but evolves during the growth phase (B) toward a breakdown (C) that can be called a discharge. From Heikkila, W.J., 1983. The reason for magnetospheric substorms and solar flares. Sol. Phys. 88, 332.

strongly, reaching a maximum, then drops; the energy of the electrostatic fluctuations drops by an order of magnitude to an almost stationary level. The evolution of the electron drift velocity is that only after reaching a certain level, the fluctuations cease to grow. The drift velocity keeps on increasing for some electrons, but at a slightly reduced rate as compared to free space.

3.12.6 Energization to very high energies

The acceleration along the neutral line (the discharge) provides adequate particle energies so that, when the particles leave the neutral line with its small value of B and enters into regions of higher B in the surrounding magnetic field, it has sufficient magnetic moment to benefit from the betatron acceleration. As shown by Bulanov and Sasorov (1975), Pellinen and Heikkila (1978), and Heikkila (1984), for reasonable values of the magnetic field changes during the substorm, initially low-energy particles can be energized to MeV energies in a few seconds. It has been shown that the acceleration process consists of a direct acceleration by the electric field in the nonadiabatic region near the zero line and a betatron acceleration in the drift region. The characteristic energy of charged particles is determined in the nonrelativistic and ultrarelativistic limits. It was found that the energetic spectra have an exponential form in the high-energy range. The use of Liénard-Wiechert potentials is a necessity in the treatment of very high energies; but thus far it has been ignored.

The dependence of the mechanism on the neutral line region (Fig. 3.22) also illustrates how to achieve selective acceleration, rather than an equitable distribution of available energy among the whole particle population, which is a firm requirement.

There is still another source of high-energy particles. The measured left-hand polarization of the cusp electric field at ion gyrofrequencies indicates that the cyclotron resonant acceleration mechanism (CRAM) is working. The CRAM can energize ions from keV to MeV in seconds.

Emslie and Miller (2003) in *Dynamic Sun* noted that "Despite decades of observations in X-rays and gamma-rays, the mechanism for particle acceleration remains an enigma." We should, and can, learn from our research with the Earth's magnetosphere, for the Earth is our cosmic laboratory.

3.13 Problems

3.1. Assume that Fig. 3.5 occurs at time $t_1 = 0$, and the stationary charge at point P is 5 cm from the point P'. (a) At what negative time t_0 did the charge stop? (b) How far from the stationary charge would a point P'' be if $t_2 = 2.5 \times 10^{-10}$ s?

3.2. What is the power radiated by an electron, initially traveling at $0.7c$, that strikes a barrier and stops in a distance of 1 mm?

3.3. Demonstrate that letting the source density $s=0$ leads to a **V** that is purely solenoidal, and letting the circulation density **c**$=0$ gives a **V** that is purely irrotational.

3.4. Show that Eq. (3.44) is true by explicit application of Eq. (3.43). Also, use Eq. (3.43) to find the electric field inside and outside a spherical shell of radius R carrying a uniform charge density σ.

3.5. The 1D particle simulation discussed in this chapter used an electric field of 0.1 V/km. If a single electron were accelerated in such a field, how far would it travel before reaching relativistic (0.1c) speeds? How does this distance compare with the size of the magnetosphere?

References

Arfken, G.B., Weber, H.J., 1995. Mathematical Methods for Physicists, fourth ed. Academic Press, New York, NY.

Bernstein, I.B., Greene, J.M., Kruskal, M.D., 1957. Exact nonlinear plasma oscillations. Phys. Rev. 108, 546.

Blake, J.B., Mueller-Mellin, R., Davies, J.A., Li, X., Baker, D.N., 2005. Global observations of energetic electrons around the time of a substorm on 27 August 2001. J. Geophys. Res. 110, A06214. https://doi.org/10.1029/2004JA010971.

Büchner, J., Elkina, N., 2005. Vlasov code simulation of anomalous resistivity. Space Sci. Rev. 121, 237–252. https://doi.org/10.1007/s11214-006-6542-6.

Bulanov, S.V., Sasorov, P.V., 1975. Energetic spectrum of particles accelerated in the neighborhood of zero line of the magnetic field. (in Russian) Astron. J. 52, 763.

Chen, J., 2008. Evidence for particle acceleration in the magnetospheric cusp. Ann. Geophys. 26, 1993–1997.

Chen, T., Liu, Z.-X., Heikkila, W., 2002. Special effect of parallel inductive electric field. Chin. Phys. Lett. 19, 881–884.

Clemmow, P.C., Dougherty, J.P., 1969. Electrodynamics of Particles and Plasmas. Addison-Wesley, Reading, MA.

Cragin, B.L., Heikkila, W.J., 1981. Alternative formulations of magnetospheric plasma electrodynamics. Rev. Geophys. Space Phys. 19 (223), 1981.

Emslie, A.G., Miller, J.A., 2003. Particle acceleration. In: Dwivedi, B.N. (Ed.), Dynamic Sun. Cambridge University Press, Cambridge, UK, pp. 262–287.

Evans, D.S., 1974. Precipitating electron fluxes formed by a magnetic field aligned potential difference. J. Geophys. Res. 79, 2853. https://doi.org/10.1029/JA079i019p02853.

Evans, D.S., 1975. Evidence for the low altitude acceleration of auroral particles. In: Physics of the Hot Plasma in the Magnetosphere. Proceedings of the Thirtieth Nobel Symposium, Kiruna, Sweden, April 2–4. Plenum Press, New York, NY, pp. 319–340.

Faraday, M., 1832. Experimental researches in electricity. Philos. Trans. R. Soc. Lond. A 122, 125–162.

Griffiths, D., 1981. Introduction to Electrodynamics, first ed. Prentice Hall, Englewood Cliffs, NJ.

Heikkila, W.J., 1984. Magnetospheric topology of fields and currents. In: Potemra, T.A. (Ed.), Magnetospheric Currents. AGU Geophysical Monograph 28. American Geophysical Union, Washington, DC, pp. 208–222.

Heikkila, W.J., 2003. Initial condition for plasma transfer events. In: Newell, P.T., Onsager, T. (Eds.), Earth's Low-Latitude Boundary Layer. AGU Geophysical Monograph 133. American Geophysical Union, Washington, DC, pp. 157–168.

Heikkila, W.J., Pellinen, R.J., 1977. Localized induced electric field within the magnetotail. J. Geophys. Res. 82, 1610–1614.

Jackson, J.D., 1975. Classical Electrodynamics, second ed. Wiley, New York, NY.

Kirsch, E., Krimigis, S.M., Sarris, E.T., Lepping, R.P., Armstrong, T.P., 1977. Evidence for an electric field in the magnetotail from observations of oppositely directed anisotropics of energetic ions and elections. Geophys. Res. Lett. 4, 137.

Kojima, H., Matsumoto, H., Miyatake, T., Nagano, I., Fujita, A., Frank, L.A., et al., 1994. Relation between electrostatic solitary waves and hot plasma flow in the plasma sheet boundary layer: GEOTAIL observations. Geophys. Res. Lett. 21 (25), 2919–2922.

Konopinski, E.J., 1981. Electromagnetic Fields and Relativistic Particles. McGraw-Hill, New York, NY.

Krasovsky, V., Matsumoto, H., Omura, Y., 1997. Bernstein-Greene-Kruskal analysis of electrostatic solitary waves observed with Geotail. J. Geophys. Res. 102 (A10), 22131–22139.

Landau, L.D., Lifshitz, E.M., 1959. Fluid Mechanics. (translated from the Russian by J. B. Sykes and W. H. Reid) Pergamon Press/Addison-Wesley, London, UK/ Reading, MA.

Landau, L.D., Lifshitz, E.M., 1960. The classical theory of fields. (translated from the Russian by Morton Hamermesh) Pergamon Press, Oxford, New York, NY.

Liénard, A., 1898. Champ electrique et magnetique produit par une charge concentree en un point et animee d'un mouvement quelconque. L'Eclairage Elec. 16, 5.

Lorentz, H.A., 1899. Simplified theory of electrical and optical phenomena in moving systems. In: Proceedings Academy of Science, Amsterdam, vol. 1, pp. 427–442.

Lorrain, P., Corson, D., Lorrain, F., 1988. Electromagnetic Fields and Waves, third ed. Freeman, San Francisco, CA.

Matsumoto, H., Kojima, H., Miyatake, T., Omura, Y., Tsutsui, M., 1994. Electrostatic solitary waves (ESW) in the magnetotail: BEN wave forms observed by GEO-TAIL. Geophys. Res. Lett. 21, 2915.

Maxwell, J.C., 1873. A Treatise on Electricity and Magnetism. vol. I. Clarendon, Oxford, UK.

Moolla, S., Bharuthram, R., Singh, S.V., Lakhina, G.S., Reddy, R.V., 2007. An explanation for high-frequency broadband electrostatic noise in the earth's magnetosphere. J. Geophys. Res. 112, A07214. https://doi.org/10.1029/2006JA011947.

Morse, P.M., Feshbach, H., 1953. Methods of Theoretical Physics. McGraw-Hill, New York, NY.

Omura, Y., Kojima, H., Matsumoto, H., 1994. Computer simulation of electrostatic solitary waves: a nonlinear model of broadband electrostatic noise. Geophys. Res. Lett. 21, 2923.

Omura, Y., Matsumoto, H., Miyake, T., Kojima, H., 1996. Electron beam instabilities as generation mechanism of electrostatic solitary waves in the magnetotail. J. Geophys. Res. 101, 2685.

Omura, Y., Heikkila, W.J., Umeda, T., Ninomiya, K., Matsumoto, H., 2003. Particle simulation of plasma response to an applied electric field parallel to magnetic field lines. J. Geophys. Res. 108 (A5), 1197. https://doi.org/10.1029/2002JA009573.

Panofsky, W.K.H., Phillips, M., 1962. Classical Electricity and Magnetism, second ed. Addison-Wesley, Reading, MA.

Pellinen, R.J., Heikkila, W.J., 1978. Energization of charged particles to high energies by an induced substorm electric field within the magnetotail. J. Geophys. Res. 83, 1544.

Purcell, E.M., 1985. Electricity and Magnetism, Berkeley Physics Course. vol. 2. McGraw-Hill, New York, NY.

Reiff, P.H., Collin, H.L., Craven, J.D., Burch, J.L., Winningham, J.D., Shelley, E.G., 1988. Determination of auroral electrostatic potentials using high- and low-altitude particle distributions. J. Geophys. Res. 93, 7441.

Reitz, J.R., Milford, F.J., Christy, R.W., 1993. Foundations of Electromagnetic Theory, fourth ed. Addison-Wesley, Reading, MA.

Sarafopoulos, D.V., Sarris, E.T., 1988. Inverse velocity dispersion of energetic particle bursts inside the plasma sheet. Planet. Space Sci. 36, 1181.

Schopenhauer, A., 1851. Parerga und Paralipomena: Kleine Philosophische Schriften. vol. 2. A. W. Hayn, Berlin, Germany, p. 93 Section: 76.

Sibeck, D.G., Paschmann, G., Treumann, R.A., Fuselier, S.A., Lennartsson, W., Lockwood, M., et al., 1999. Plasma transfer processes at the magnetopause. Space Sci. Rev. 88, 207–283.

Szent-Györgyi, A., 1957. Bioenergetics. Academic Press, New York, NY, p. 57.

Taylor, M.G.G.T., Reeves, G.D., Friedel, R.H.W., Thomsen, M.F., Elphic, R.C., Davies, J.A., et al., 2006. Cluster encounter with an energetic electron beam during a substorm. J. Geophys. Res. 111, A11203. https://doi.org/10.1029/2006JA011666.

Vasyliunas, V.M., 2001. Electric field and plasma flow: what drives what. Geophys. Res. Lett. 28 (2177), 2001.

Westgard, J.B., 1997. Electrodynamics: A Concise Introduction. Springer-Verlag, New York, NY.

Wiechert, E., 1901. Elektrodynamische Elementargesetze. Ann. Phys. 309, 667–689.

4

Magnetohydrodynamic equations

I am not urging that [plausible hypotheses] be no longer used in real hydrodynamics—even in pure mathematics [their use is] very important. In hydrodynamics, progress would hardly be possible without [their] free use—and complete rigor is seldom possible. I am only insisting that plausible arguments must be checked, either by the discipline of rigorous proof … or by experiment, before they can be accepted as scientifically established.

Birkhoff (1960, p. 5)

Earth's Magnetosphere. https://doi.org/10.1016/B978-0-12-818160-7.00004-1

4.1 Introduction

In Chapter 2, one approximate method for understanding plasma dynamics was introduced, circuit analysis. A different and highly developed approximation is to treat the plasma as a fluid. As described in Chapter 1, the ability to treat a collisionless plasma in terms of bulk fluid flow depends on the presence of a magnetic field to transmit information between particles, and this type of hydrodynamics is called magnetohydrodynamics (MHD).

4.2 Basic magnetohydrodynamic equations

This will be brief, as the subject is well known and treated in several textbooks (e.g., Baumjohann and Treumann, 1996; Kivelson and Russell, 1995; Davidson, 2016; Galtier, 2016). We first discuss the nature of MHD waves that occur in plasmas.

> *When physical systems experience perturbations, it is common for them to respond by emitting waves. For example, a sound wave in a gas like the atmosphere is produced by a change in pressure at the source of the wave, whether it is a hi-fi speaker system or a dynamite blast. The pressure perturbation then travels through the atmosphere. By knowing the properties of the atmosphere, one can predict the speed at which the signal will propagate. For a closed system, the oscillations normally are combinations of standing waves whose frequencies are governed by the size of the system, as well as by its material properties.*
>
> *In a plasma, as in a gas, we might expect to find waves that are similar to sound waves; however, a plasma is composed principally of charged particles that carry currents. Thus, its electromagnetic properties are of paramount importance, but plasma density and pressure are also relevant. As a consequence, plasma waves differ from both sound waves and electromagnetic waves.*
>
> **Kivelson (1995, p. 330)**

MHD waves are found as solutions to the equations to express the conservation laws and Maxwell's equations. They are presented by Kivelson (1995, p. 332) and repeated here for convenience. Eq. (4.1) guarantees that mass is conserved as the fluid moves:

$$\frac{\partial \rho}{\partial t} + \nabla \cdot \rho \mathbf{u} = 0 \tag{4.1}$$

Momentum conservation is assured by this equation in which we assume neither sources nor losses:

$$\rho\left(\frac{\partial \mathbf{u}}{\partial t} + \mathbf{u} \cdot \nabla \mathbf{u}\right) = -\nabla p + \mathbf{j} \times \mathbf{B} \qquad (4.2)$$

Maxwell's equations in the low-frequency limit will be needed. These equations are

$$\frac{\partial \mathbf{B}}{\partial t} = -\nabla \times \mathbf{E} \text{(Faraday's law)} \qquad (4.3)$$

$$\nabla \times \mathbf{B} = \mu_0 \mathbf{J} \text{(Ampere's law)} \qquad (4.4)$$

$$\nabla \cdot \mathbf{B} = 0 \qquad (4.5)$$

and the ideal Ohm's law

$$\mathbf{E} + \mathbf{u} \times \mathbf{B} = 0 \qquad (4.6)$$

These do not form a complete set; we must find another equation in order to have the number of equations equal to the number of unknowns. An equation of state is usually chosen. The specific entropy (entropy per unit volume) is conserved in the convecting magnetized plasma:

$$\left(\frac{\partial}{\partial t} + \mathbf{u} \cdot \nabla\right)\left(\frac{p}{\rho^\gamma}\right) = 0 \qquad (4.7)$$

Here, as before, p is the pressure, ρ is the mass density, $\mathbf{u}$ is the flow velocity, $\mathbf{J}$ is the electric-current density, $\mathbf{B}$ is the magnetic field (magnetic induction), μ_0 is the magnetic permeability of free space, and $\mathbf{E}$ is the electric field. The quantity gamma, γ, is the ratio of the specific heat at constant pressure to the specific heat at constant volume where N is degrees of freedom:

$$\gamma = (2 + N)/N \qquad (4.8)$$

γ is frequently referred to as the polytropic index.

As the derivative acting on the expression in parentheses on the right in Eq. (4.7) is just the time rate of change in a frame that follows the plasma as it flows through the system, the equation requires that the plasma obey an adiabatic equation of state. In most space plasmas, best agreement between observations and theory is achieved with a γ of 5/3, appropriate for three dimensions (Chen, 1984).

We can express the current in terms of the magnetic field from Ampere's law, and Eq. (4.2) becomes

$$\rho\left(\frac{\partial \mathbf{u}}{\partial t} + \mathbf{u} \cdot \nabla \mathbf{u}\right) = -\nabla p + (\nabla \times \mathbf{B}) \times \mathbf{B}/\mu_0 \qquad (4.9)$$

4.2.1 Equations for linear waves

For simplicity, we assume that the perturbations carried by the waves are small. Let us assume that the plasma is initially at rest, which means that there are neither flows nor electric fields, and also assume that no currents are flowing. The wave perturbations introduce finite but small $\mathbf{E}$, $\mathbf{u}$, and $\mathbf{J}$. The magnetic field, mass density, and pressure also change, so that

$$\mathbf{B} \rightarrow \mathbf{B} + \mathbf{b}, \quad \rho \rightarrow \rho + \delta\rho, \quad p \rightarrow p + \delta p \qquad (4.10)$$

All of the perturbed quantities, $\mathbf{b}$, $\delta\rho$, δp, $\mathbf{u}$, $\mathbf{E} = -\mathbf{u} \times \mathbf{B}$, and $\mathbf{j} = \nabla \times \mathbf{b}/\mu_0$, are assumed to be small enough that only terms linear in any of them need be retained. This means that squares or high powers and cross products will be dropped. The perturbed quantities then must satisfy the equations

$$\frac{\partial \delta p}{\partial t} + \rho \nabla \cdot \mathbf{u} = 0 \qquad (4.11)$$

$$\rho \frac{\partial \mathbf{u}}{\partial t} = -\nabla \delta p + (\nabla \times \mathbf{b}) \times \mathbf{B}/\mu_0 \qquad (4.12)$$

$$\frac{\partial \mathbf{b}}{\partial t} = \nabla \times (\mathbf{u} \times \mathbf{B}) \qquad (4.13)$$

We must have $\nabla \cdot \mathbf{b} = 0$ as usual. The adiabatic requirement also becomes an initial condition, because

$$\frac{\partial \delta p}{\partial t} = \frac{\gamma p}{\rho} \frac{\partial \delta p}{\partial t} = c_s^2 \frac{\partial \delta p}{\partial t} \qquad (4.14)$$

$$\frac{\partial}{\partial t} \left(\frac{\delta p}{c_s^2 \delta \rho} \right) = 0 \qquad (4.15)$$

and the constant value of the ratio of δp to $\delta\rho$ is set by the initial conditions. Substitution of δp in terms of $\delta\rho$ leaves us with seven unknowns that describe the wave perturbations: $\delta\rho$, u, and b, and seven equations, once again counting each component of a vector equation separately.

4.2.2 Equation of state

This linear approach is ideal for situations where the instability grows out of a stable plasma. It not useful in the case of the sudden invasion of a new phenomenon, a blast wave from outside, or inside the local plasma.

Space physics has progressed by making approximations, quite understandably (see quote from Birkhoff (1960) at the beginning of the chapter). Instead of solving the fundamental

Boltzmann tranport equation and equations of electrodynamics, the simpler MHD equations have been used. MHD theory relies on the conservation equations of mass, momentum, and energy obtained from the first three velocity moments of the Boltzmann equation; consequently, the results need to be taken seriously. Great success has been claimed, at least for some models.

However, an equation of state is needed to achieve a complete set.

At this point, let us count the number of unknowns and the number of equations. Assuming, as noted, that we need not solve for the negligibly small charge density there are 14 unknowns: E, B, j, u, ρ, and p. A vector has three components, and so it corresponds to three unknowns: … we have only 10 independent equations, not enough equations to determine all the unknowns. Energy conservation has not yet been invoked … . The heat flux [q], on the other hand, adds to the set of equations a vector that is not a function of the original set of unknowns. This means that we must either use an approximation to express it in terms of the original set or add further equations governing q. Many treatments avoid introducing energy conservation explicitly. Instead, they obtain an additional equation from the assumption that there is no change in the entropy of a fluid element as it moves through the system. This means that the pressure and the density are related by [an equation of state].

Kivelson (1995, p. 47)

An equation of state is essentially a point function, while admittedly the plasma can communicate along magnetic field lines bringing information from distant regions.

The physical content of the required approximation is that there should be sufficient "localization" of the particles in physical space. For if we hope to obtain differential equations in space and time, the rate of change of hydrodynamic quantities at a point must depend on their present values at that point or in its neighborhood. This is not possible if particles are able to move substantial distances in a short time, as the evolution of local properties is then influenced by the arrival from various directions of fresh particles from quite separate parts of the flow field. The only distant influence which can be included in the hydrodynamic scheme is a self-consistent field, which appears as a body force.

Clemmow and Dougherty (1969, p. 344)

The guiding center theory (first-order orbit theory) of single particles that was developed in the first chapter helps us to understand how particles move around in complex electromagnetic fields. While this theory has given us important insights, it is not adequate for a system that contains many particles. The single-particle picture must be augmented with information on the dynamics of a group of particles. MHD theory is one approximate approach, but we must be cautious.

4.3 Example of MHD for magnetospheric research

As an example of MHD simulations, there is the Grand Unified Magnetosphere Ionosphere Coupling Simulation (GUMICS-4) (Janhunen, 1996; Palmroth et al., 2001; Laitinen et al., 2006).

Laitinen et al. (2006) used GUMICS-4 to simulate the location and motion of the magnetospheric cusp and the subsolar magnetopause under various interplanetary magnetic field (IMF) directions and solar wind dynamic pressures. It is a global MHD simulation code, with the simulation box covering the solar wind out to 32 R_E upwind, and the entire magnetosphere, out to 224 R_E in tail. The MHD solar wind and magnetosphere are coupled to a planar ionosphere, where electrostatic current continuity holds.

The ionospheric electron density is solved from the recombination and ionization production rates in a three-dimensional ionospheric grid with 20 nonuniform levels of height. The ionization production rate, caused by magnetospheric processes, needs the source plasma density, temperature, and the local magnetosphere-ionosphere parallel potential drop, which, however, is assumed zero for simplicity. The local Pedersen and Hall conductivities are computed from the electron density and used in the calculation of the height-integrated current continuity equation.

GUMICS-4 uses an adaptive Cartesian grid where the grid refinement is semiautomatic; whenever the code detects large gradients, the grid is refined. To save memory and time, the refinement threshold depends on location, and smaller gradients are required to refine the grid in the near Earth region than, for example, in the distant tail. The basic grid spacing is 8 R_E, and the refinement is done up to a maximum adaptation level, which is given as a model input. Increasing the maximum adaptation level by one decreases the smallest grid spacing into half, that is, adaptation level 1 would correspond to 4 R_E. The code also uses subcycling, which means that the time step is not constant throughout the simulation box but is smaller in the near Earth region and larger in the distant tail. The time step must be everywhere smaller than the travel time at which information is transferred, the travel time of an Alfvén wave across a grid cell. Elliptic cleaning is used to keep div **B** small.

While there are no particles in MHD, the diamagnetic depression in the cusp caused by the incoming magnetosheath plasma is still clearly discernible. In MHD simulation the cusp is a region of enhanced pressure, which indicates that there is a surrounding pressure gradient and thus current to maintain pressure balance according to the equation $\mathbf{J} \times \mathbf{B} = \nabla P$. The current results from a disturbance magnetic field, and as the field is in the opposite direction, the diamagnetic effect is observed.

The authors use a model of the solar wind and the magnetosphere shown in Fig. 4.1 in the noon-midnight meridian plane. The important methodological advance in their work is the use of topological properties of magnetic field lines to locate the reconnection line as the separator line at the junction of domains of different field lines.

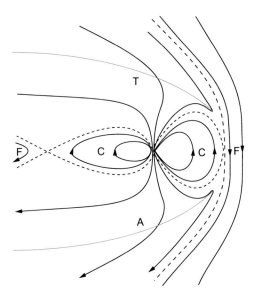

Fig. 4.1 The four topologically different magnetic field regions in the magnetosphere during southward IMF. T stands for toward the Earth, A for away, C for closed, and F for free field lines. The figure is drawn for southward IMF, and under these conditions the four regions are seen to meet at the magnetopause and magnetotail reconnection sites. Fig. 4.2 will show that the separator drapes around the Earth along the flanks of the magnetopause. From Laitinen, T.V., Janhunen, P., Pulkkinen, T.I., Palmroth, M., Koskinen, H.E.J., 2006. On the characterization of magnetic reconnection in global MHD simulations. Ann. Geophys. 24, 3062.

The conventional definition of reconnection rate as the electric field parallel to an X-line is problematic in global MHD simulations for several reasons: the X-line itself may be hard to find in a non-trivial geometry such as at the magnetopause, and the lack of realistic resistivity modeling leaves us without reliable non-convective electric field. In this article we describe reconnection characterization methods that avoid those problems and are practical to apply in global MHD simulations. We propose that the reconnection separator line can be identified as the region where magnetic field lines of different topological properties meet, rather than by local considerations. The global convection associated with reconnection is then quantified by calculating the transfer of mass, energy, or magnetic field across the boundary of closed and open field line regions. The extent of the diffusion region is determined from the destruction of electromagnetic energy, given by the divergence of the Poynting vector. Integrals of this energy conversion provide a way to estimate the total reconnection efficiency.

Laitinen et al. (2006, p. 3062)

This method, called four-field junction, is especially useful on the magnetopause; the complicated geometry makes it difficult to locate the reconnection line by examining only the local behavior.

4.3.1 The four-field junction

Quantitative study of reconnection in the GUMICS simulation is based on energy conversion; this is computed using the divergence of the Poynting vector (the Poynting flux through a closed surface). The distribution of reconnection-related energy conversion on the magnetopause is analyzed using energy conversion surface density, and the total efficiency of reconnection is quantified using reconnection power. The most important parameters regulating the magnetopause reconnection power are the solar wind speed and the direction of the IMF. The separator is the line where four separate regions of magnetic field meet. The field lines can be classified in four classes:

- **Free**, or disconnected. These are solar wind field lines not attached to the Earth.
- **Away**. These are open field lines whose "back end" is attached to the Earth. They emanate from a region of northern magnetic polarity. In the magnetosphere, the southern tail lobe is formed by away-type field lines.
- **Toward**. Open field lines whose "front end" is attached to the Earth. They emanate from a region of southern polarity and form the northern tail lobe.
- **Closed**. Attached to the Earth from both ends.

In a global MHD simulation, the junction of these four regions of magnetic field can be found in the following manner. First, set up a Cartesian grid in a region where a separator line is expected to exist. Starting from each grid point, trace the magnetic field both forwards and backwards, that is, integrate the field line. In practice, the field line integration is the computationally heavy part of this procedure, so it is wise to save the classification of grid points as an intermediate result.

The final result is, strictly speaking, not a line but a set of points. However, these points tend to form a line or ribbon, as illustrated in Fig. 4.2. The figure was produced from a test run on GUMICS-4 with slowly (10 degrees every 10 min) rotating IMF. The separator line crosses the subsolar magnetopause in a tilted orientation, which depends on the IMF clock angle. Along the flanks of the magnetopause the line reaches tailward, it crosses the tail between the tail lobes, and thus forms a loop around the Earth.

Note that the "field line velocity" $\mathbf{E} \times \mathbf{B}/B^2$ would twist the magnetic fields as depicted by Fig. 4.3, depending on the electric field. The concept of moving field lines dates to Hannes Alfvén who, in an early paper (Alfvén, 1942), noted that in an infinitely conducting medium "the matter of the liquid is fastened to the lines of force." This phenomenon became known as "frozen-in magnetic field lines." Since then, an overwhelming amount of empirical data have proven that magnetic-field-aligned electric fields exist and are of key importance in the physics of auroras. Alfvén, who had introduced the concept, became a strong critic of "moving" magnetic field lines (Alfvén, 1976). Thus the concept of Fig. 4.3 must be taken with due caution.

The two gaps in the cross-tail part are caused by the coarseness of the grid used: the x-configuration in the tail is so thin that no nearby grid points happened to touch the closed and open segments in those parts of the tail. Very small IMF clock angles are also problematic for the four-field junction technique. Otherwise, the four-field junction is a stable and well-behaved representation of the separator line, and it also evolves smoothly in time as solar wind conditions change.

Fig. 4.4 illustrates the usability of energy conversion surface density by showing it on the magnetopause under four different IMF orientations. Solar wind dynamic pressure was 2 nPa and IMF magnitude 10 nT. (This is the same run that was used for Fig. 4.2.) The integration range was 1.5 R_E to both directions from the magnetopause surface, which was determined using plasma flow lines as defined by Palmroth et al. (2003).

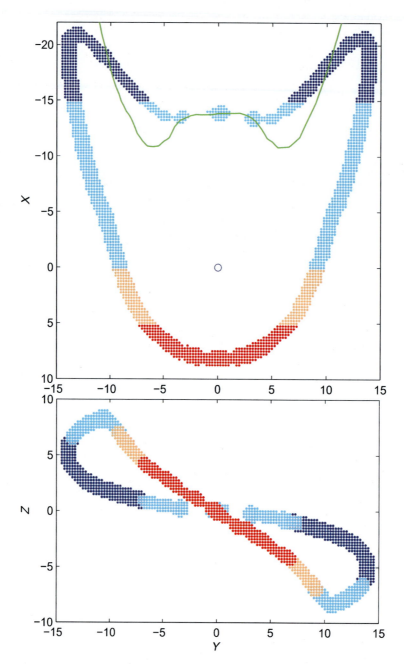

Fig. 4.2 The four-field junction in Gumics-4, seen from north (upper panel) and from the Sun (lower panel). The points are colored by their *X* coordinate to aid in identifying different parts of the junction in the lower panel. The *green line* is the magnetotail *x*-line given by an independent search procedure based on the local field geometry. From Laitinen, T.V., Janhunen, P., Pulkkinen, T.I., Palmroth, M., Koskinen, H.E.J., 2006. On the characterization of magnetic reconnection in global MHD simulations. Ann. Geophys. 24, 3063.

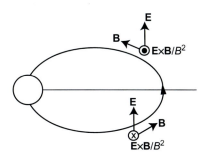

Fig. 4.3 Dipole magnetic field in the presence of a uniform electric field parallel to the dipole axis. Note that the "field line velocity" $\mathbf{E} \times \mathbf{B}/B^2$ would twist the magnetic fields. From Fälthammar, C.-G., Mozer, E.S., 2007. On the concept of moving magnetic field lines. EOS *Trans. AGU* 88(15), 169.

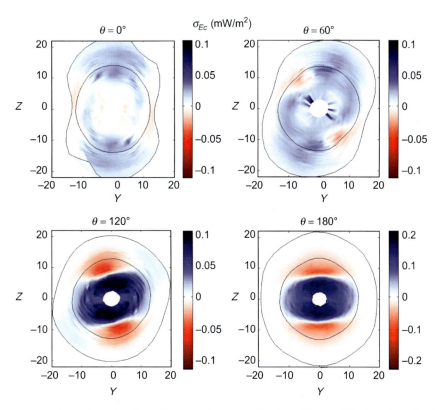

Fig. 4.4 Energy conversion surface density on the magnetopause under four different IMF orientations. The IMF direction is shown by an *arrow* and also given in degrees above each panel. The lines are intersections of the magnetopause with the planes $X=0$ and $X=-30$. The white hole at the subsolar point is a technical feature caused by the plotting procedure. Note the different color scale in the lower-right panel! From Laitinen, T.V., Janhunen, P., Pulkkinen, T.I., Palmroth, M., Koskinen, H.E.J., 2006. On the characterization of magnetic reconnection in global MHD simulations. Ann. Geophys. 24, 3067.

Reconnection power is the amount of magnetic energy converted to other forms per unit time (shown by color):

$$P_{MR} = -\int_V \nabla \cdot \mathbf{S} dV = -\oint_{\text{surf}} \mathbf{S} \cdot d\mathbf{A} \qquad (4.16)$$

where V is an integration volume containing the entire reconnection region and $\mathbf{S}$ is the Poynting vector. As an integral quantity it measures the total effectiveness of reconnection taking into account both the intensity and the size of the process. The magnetopause reconnection region appears as a slab-like region of negative Poynting vector divergence with blue color conforming to the shape of the current sheet.

4.3.2 Changes in the energy conversion

However, on the magnetopause there are places of strong positive Poynting vector divergence shown in red very close to the sinks of magnetic energy in blue. The intensity of both strengthens and weakens approximately simultaneously, and therefore these regions would cancel out large portions of each other in a volume integration. This would hide the changes in the energy conversion process. Therefore the definition of reconnection power was changed into the form wherein it picks only the negative values of Poynting vector divergence.

$$P_{MR} = -\int_V \nabla \cdot \mathbf{S} \Theta(-\nabla \cdot \mathbf{S}) dV \qquad (4.17)$$

Here Θ is a step function. The volume V in which the magnetopause reconnection power is calculated was chosen to be the set of those points whose x coordinate is positive and whose distance from the magnetopause is smaller than 1.5 R_E.

The figure reveals that during southward IMF magnetic energy is consumed in a wide region around the subsolar point. All this supports the view that reconnection is a large-scale process that cannot be described exhaustively by inspecting only the x-line or its surroundings.

Laitinen et al. (2006)

Given that the observational techniques are now firmly in place we ought to be able to resolve the consequences for this intricate system. That we cannot mean that some errors are present, either numerical or physical.

There is no general rule for telling apart numerical and physical phenomena. In fact the division is artificial, as all phenomena in a numerical simulation are products of numerical calculations and in that sense of numerical origin. ... Thus classifying something as physical is not a statement about correspondence between the simulation and nature—it only says that within the reference frame of the physical model the phenomenon is real. Comparing the model to observations is then another question.

Laitinen (2007, p. 114)

4.4 Recent advances in MHD simulations

MHD simulation models are constantly improving as increasing computer processing power allows for ever greater fidelity with increased temporal and spatial resolution. This often results in a more complicated and realistic magnetosphere, rather than the "smooth and boring" models of earlier years. Following are just two examples of the types of improvements that are now possible.

4.4.1 Turbulent flow

As with any other type of fluid dynamics, the most difficult type of flow to model is turbulent flow. It is not surprising, then, that much recent work in MHD has been focused on various regimes of turbulence. For example, weak Alfvén wave turbulence occurs in a plasma subjected to a strong magnetic field. This corresponds to a regime where a large ensemble of waves interact in a nonlinear fashion. One reason that it is currently the subject of intense research is that modern experiments and simulations can reach this regime, confirming theoretical predictions or even finding new features. Galtier (2016) describes a number of recent works on the topic, including Aubourg and Mordant (2015) and Campagne et al. (2015).

From Meyrand et al. (2015), we learn that one of the most striking features of strong hydrodynamic turbulence is the presence of both a complex chaotic spatial/temporal behavior and a remarkable degree of coherence. The small-scale correlations of turbulent motion show significant deviations from the Gaussian statistics that are usually expected in systems with many degrees of freedom. This phenomenon, known as intermittency, has been the subject of much research and controversy since Batchelor et al.'s (1949) original observations. Intermittency can be measured with the probability density function (PDF) of the velocity differences between two points separated by some distance. In the presence of intermittency, PDFs develop more and more stretched and fatter tails when the separation distance decreases, increasing the probability of extreme events.

This nonself-similarity of PDFs in hydrodynamics reflects the fact that the energy dissipation of turbulent fluctuations is not evenly distributed in space, but concentrated in very intense vorticity filaments. Interest has been increasing in the study of intermittency in the regime of weak turbulence, the study of the long-time statistical behavior of a sea of weakly nonlinear dispersive waves for which a natural asymptotic closure can be found. The energy transfer between waves occurs mostly among resonant sets of waves and the resulting energy distribution, far from a thermodynamic equilibrium, is characterized by a wide power law spectrum. Weak turbulence is a very common natural phenomenon studied in space plasmas and elsewhere. Intermittency has been observed both experimentally and numerically, and is attributed to the presence of coherent structures. It is linked to the breakdown of the weak nonlinearity assumption induced by the dynamics of weak turbulence and is not an intrinsic property of this regime. Intermittency is actually at odds with classical weak turbulence theory because of the random phase approximation which allows the asymptotic closure and resultant derivation of the weak turbulence equations.

Weak magnetohydrodynamic (MHD) turbulence differs significantly from other cases because of the singular role played by slow modes for which $k_\parallel = 0$, where $\mathbf{k}$ is a wave vector in Fourier space, and the subscript $\parallel$ indicates the component of $\mathbf{k}$ parallel to the guide field $\mathbf{b}_0$. Since Alfvén waves have frequencies $\omega_k^{\pm} = \pm k_\parallel v_A$ (where v_A is the Alfvén speed) and only counter-propagating waves can interact, the resonance condition, $\omega_{k1}^{+} + \omega_{k2}^{-} = \omega_{k3}^{\pm}$ and $\mathbf{k}_1 + \mathbf{k}_2 = \mathbf{k}_3$, implies that at least one mode must have $k_\parallel = 0$. This mode, which acts as a catalyst for the nonlinear interaction, is not a wave but rather a kind of two-dimensional condensate with a characteristic time $\tau_A \sim 1/(k_\parallel v_A) = +\infty$ and cannot be treated by weak turbulence. The standard way to overcome this complication has been to assume that the $k_\parallel$ spectrum of Alfvén waves is continuous across $k_\parallel = 0$. Under this assumption, the weak MHD theory was established by Galtier et al. (2000) and a $k_\perp^{-2}$ energy spectrum was predicted in the simplest case of zero cross-helicity with a direct cascade toward small scales. This prediction has been confirmed both observationally (Saur et al., 2002) and numerically (Perez and Boldyrev, 2008). Work in this area is ongoing, and more advances are expected in the coming years.

4.4.2 Buoyancy waves

The Rice Convection Model is a well-established simulation for modeling the magnetosphere using MHD equations. Toffoletto et al. (2020) have recently added a new element to the model quantifying buoyancy waves in the magnetotail caused by turbulence on the dayside magnetosphere. The oscillations are similar to those of the Earth's neutral atmosphere; a small parcel of air is displaced downward, and gravity causes the parcel to bob up and down. The oscillatory motion observed in the plasma sheet is very similar to those buoyancy oscillations in the neutral atmosphere, except that the buoyant force is caused by the curvature of magnetic field lines rather than by gravity. By constraining the field lines of the tail to the xz plane, they were able to derive a coupled pair of differential equations that describe the linear oscillations of a thin magnetic filament sliding through a stationary medium. The inclusion of realistic magnetospheric fields that transition from tail-like to quasi-dipolar resulted in the simulation showing mixtures of wave modes. The authors combatted this by limiting their investigation to linear normal modes, specifically, finding the frequency of the lowest even mode, considered to be a fundamental parameter for the Earth's magnetosphere.

The normal mode calculation uses a variable grid spacing, especially where the field lines are sharply curved in the equatorial tail region. The coupled equations were solved using second-order finite differences accounting for the variable grid, resulting in an eigenvalue problem that could identify all possible wave modes. Only the lowest frequency modes symmetric about the equatorial plane were retained for study. It was found that for tail field lines that extend out in the plasma sheet, the wave mode could be characterized as a buoyancy wave where the maximum perpendicular displacement on a field line was larger or comparable to the maximum parallel displacement on the same field line. Nightside field lines that were more dipolar (with equatorial crossings of less than about 10 R_E) showed waves that were best characterized as long-wavelength slow mode, where the perturbation velocity is almost parallel to the magnetic field.

Well out into the tail (18+ R_E), the buoyancy mode resembles the simple theory of magnetohydrodynamic interchange (Bernstein et al., 1958; Chandrasekhar, 1960). Interchange represents a clear, pure picture of buoyancy, but is idealized because it assumes that the system is in force equilibrium, so that the inertial effects are neglected. The work of Toffoletto et al. (2020) does not neglect inertia and shows that deviations from pure interchange modes get larger along field lines closer to the Earth.

4.5 Discussion

We must look critically at the whole question of solar wind interaction with the magnetosphere. Three systems—the solar wind, the magnetosphere, and the ionosphere—interact, transmitting and transforming the solar wind energy, depositing some of it as heat energy in the ionosphere.

4.5.1 Magnetospheric convection

The plasma flow in the magnetosphere is not caused by temperature gradients, so the term *convection* is misleading. The phenomenon would be better described by the word *advection*. However, convection is the term used in space physics and will therefore also be used in this book.

Our present-day understanding of the magnetospheric convection to which Laitinen et al. (2006) subscribes is based on the model of two neutral lines, *X*-lines, or reconnection lines, presented by Dungey (1961) in Fig. 1.15 of Chapter 1. In this chapter's Figs. 4.1 and 4.5 it is the magnetopause and magnetotail *X*-lines (shown in projection). The southern field lines (A) and the northern field lines (T) meet at the neutral line where they *reconnect*. Behind the reconnection line in the magnetotail, plasma becomes released from its magnetic connection to the Earth, flows into the interplanetary space (F), and is mixed with the solar wind. The plasma that remains on closed field lines (C) is transported earthward and flows past the Earth back to the dayside.

Actually, the plasma flow is more complicated than that. The electric field the plasma can produce by charge separation $-\nabla \phi$

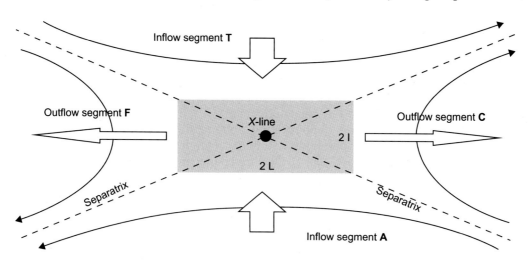

Fig. 4.5 Simple reconnection at an *X*-line, or a separator. The four separatrices that meet at the separator divide the space into four topologically distinct segments, which are labeled as F(ree), T(owards), A(way), and C(losed) for comparison with Fig. 4.1. The *shaded area* represents the diffusion region (DR). From Laitinen, T.V., Janhunen, P., Pulkkinen, T.I., Palmroth, M., Koskinen, H.E.J., 2006. On the characterization of magnetic reconnection in global MHD simulations. Ann. Geophys. 24, 3060.

has zero curl. It has no effect on the induction electric field $-\partial \mathbf{B}/\partial t$ which is solenoidal.

4.5.2 The convection electric field

The use of MHD fluid plasma physics theory for three-dimensional activity may be (partly) the answer to magnetospheric dynamics, but that leaves open one obvious question: what is the convection electric field?

Assuming that it is the usual one that obeys the ideal MHD law $\mathbf{E}^{conv} + \mathbf{v} \times \mathbf{B} = 0$, its value is given by:

$$\mathbf{E}_{\parallel}^{conv} = \left[-\frac{\partial \mathbf{A}}{\partial t} - \nabla \phi \right]_{\parallel} = 0 \qquad (4.18)$$

That equation has to be satisfied everywhere in a realistic space; otherwise we cannot use ideal MHD. With changing conditions, the magnetic fields are produced by currents whose properties depend on the tensor conductivities. They vary in space in a complicated way. The plasma has to fight induction at all locations, simultaneously.

4.6 Summary

- In MHD approximation, a key assumption is that perturbations are small enough that only linear terms need be retained.
- This leaves open the important question of the arrival of fresh particles from quite separate parts of the flow field.

Reconnection is a large-scale process that cannot be described exhaustively by inspecting only the x-line or its surroundings. A method, called four-field junction, is especially useful on the magnetopause where the complicated geometry makes it difficult to locate the reconnection line.

4.7 Problems

4.1. What "field line velocity" is implied by and electric field $\mathbf{E} = 0.3$ V/km in the $+x$ direction and a magnetic field $\mathbf{B} = 4$ nT at an angle of 30 degrees with respect to the $+x$ direction?

4.2. Use Ampere's law to eliminate $\mathbf{J}$ from the momentum Eq. (4.2), then use a vector identity to show that the $\mathbf{J} \times \mathbf{B}$ term can be written as $\frac{1}{\mu_o}(\mathbf{B} \cdot \nabla)\mathbf{B} - \nabla \left(\frac{B^2}{2\mu_o} \right)$.

4.3. Show that if the solar wind is "frozen" to the magnetic field, the shape of the field in the solar equatorial plane is an Archimedes spiral $\phi = \phi_o - \omega r/u$, where ϕ is the azimuthal angle, ω

is the angular velocity, r is the distance from the Sun, and u is the solar wind speed.

4.4. In Fig. 4.4, the solar wind dynamic pressure is said to be 2 nPa. What strength of magnetic field would give an equivalent value for the magnetic pressure (see Chapter 1)?

4.5. Section 4.3.1 describes a class of field lines as "free or disconnected." Is this really possible? What would these "free" field lines likely be connected to in reality?

References

Alfvén, H., 1942. Existence of electromagnetic-hydrodynamic waves. Nature 150, 405–406.

Alfvén, H., 1976. On frozen-in field lines and field-line reconnection. J. Geophys. Res. 81, 4019–4021.

Aubourg, Q., Mordant, N., 2015. Nonlocal resonances in weak turbulence of gravity-capillary waves. Phys. Rev. Lett. 114, 144501.

Batchelor, G.K., Townsend, A.A., Jeffreys, H., 1949. The nature of turbulent motion at large wave-numbers. Proc. R. Soc. Lond. 199, 238–255.

Baumjohann, W., Treumann, R.A., 1996. Basic Plasma Space Physics. Imperial College Press, London, UK.

Bernstein, I.B., Frieman, E.A., Kruskal, M.D., Kulsrud, R.M., 1958. An energy principle for hydromagnetic stability problems. Proc. R. Soc. Lond. A244 (1236), 17–40. https://doi.org/10.1098/rspa.1958.0023.

Birkhoff, G., 1960. Hydrodynamics: A Study in Logic Fact and Similitude, third ed. Princeton University Press, Princeton, NJ.

Campagne, A., Gallet, B., Moisy, F., Cortet, P.-P., 2015. Disentangling inertial waves from eddy turbulence in a forced rotating-turbulence experiment. Phys. Rev. E. 91(4), 043016.

Chandrasekhar, S., 1960. Plasma Physics. The University of Chicago Press, Chicago, IL.

Chen, F.F., 1984. Introduction to plasma physics and controlled fusion. In: Plasma Physics, second ed. vol. 1. Plenum Press, New York, NY.

Clemmow, P.C., Dougherty, J.P., 1969. Electrodynamics of Particles and Plasmas. Addison-Wesley, Reading, MA.

Davidson, P., 2016. Introduction to Magnetohydrodynamics. Cambridge Texts in Applied Mathematics, Cambridge University Press, Cambridge, UK. https://doi.org/10.1017/9781316672853.

Dungey, J.W., 1961. Interplanetary magnetic field and the auroral zones. Phys. Rev. Lett. 6, 47–48.

Galtier, S., 2016. Introduction to Modern Magnetohydrodynamics. Cambridge University Press, New York, NY.

Galtier, S., Nazarenko, S.V., Newell, A.C., Pouquet, J., 2000. Weak turbulence theory for incompressible magnetohydrodynamics. J. Plasma Phys. 63 (5), 447–488.

Janhunen, P., 1996. GUMICS-3: a global ionosphere-magnetosphere coupling simulation with high ionospheric resolution. In: Proceedings ESA 1996 Symposium on Environment Modelling for Space-Based Applications ESA SP-392.

Kivelson, M.G., 1995. Physics of space plasmas (p. 27–55), and pulsations and magnetohydrodynamic waves (p. 330–353). In: Kivelson, M.G., Russell, C.T. (Eds.), Introduction to Space Physics. Cambridge University Press, New York, NY.

Kivelson, M.G., Russell, C.T., 1995. Introduction to Space Physics. Cambridge University Press, New York, NY.

Laitinen, T.V., Janhunen, P., Pulkkinen, T.I., Palmroth, M., Koskinen, H.E.J., 2006. On the characterization of magnetic reconnection in global MHD simulations. Ann. Geophys. 24, 3059.

Laitinen, T.V., 2007. Rekonnektio Maan Magnetosfäärissä - Reconnection in Earth's Magnetosphere (Ph.D. thesis). Finnish Meteorological Institute.

Meyrand, R., Kiyani, K.H., Galtier, S., 2015. Weak magnetohydrodynamic turbulence and intermittency. J. Fluid Mech. 770, R1.

Palmroth, M., Laakso, H., Pulkkinen, T., 2001. Location of high-altitude cusp during steady solar wind conditions. J. Geophys. Res. 106, 21109.

Palmroth, M., Pulkkinen, T.I., Janhunen, P., Wu, C.-C., 2003. Stormtime energy transfer in global MHD simulation. J. Geophys. Res. 108, 1048. https://doi.org/10.1029/2002JA009446.

Perez, J.C., Boldyrev, S., 2008. On weak and strong magnetohydrodynamic turbulence. Astrophys. J. 672, L61–L64.

Saur, J., Politano, H., Pouquet, A., Matthaeus, W.H., 2002. Evidence for weak MHD turbulence in the middle magnetosphere of Jupiter. Astron. Astrophys. 386, 699–708. https://doi.org/10.1051/0004-6361:20020305.

Toffoletto, F.R., Wolf, R.A., Schutza, A.M., 2020. Buoyancy waves in Earth's nightside magnetosphere: normal-mode oscillations of thin filaments. J. Geophys. Res. 125. https://doi.org/10.1029/2019JA027516.

5

Poynting's energy conservation theorem

Chapter outline

> *Once it has achieved the status of paradigm, a scientific theory is declared invalid only if an alternate candidate is available to take its place.*
>
> **Kuhn (1970)**

5.1   Introduction

Many problems in mechanics are simplified by means of energy considerations; this is equally true in plasma physics. Under static conditions energy exists as potential energy; in

Earth's Magnetosphere. https://doi.org/10.1016/B978-0-12-818160-7.00005-3

plasma, it is the electrostatic energy defined by the conservative scalar potential. Kinetic situations are more complex with solenoidal vector potentials being important; these are nonconservative and can be counter-intuitive.

Poynting (1852–1914) formulated his conservation of energy theorem in 1884, based directly on Maxwell's equations. These equations in free space (vacuum) are:

$$\varepsilon_0 \nabla \cdot \mathbf{E} = \sigma \tag{5.1}$$

$$\nabla \times \mathbf{E} = -\partial \mathbf{B}/\partial t \tag{5.2}$$

$$\nabla \cdot \mathbf{B} = 0 \tag{5.3}$$

$$\nabla \times \mathbf{B} = \mu_0 \mathbf{J} + \varepsilon_0 \mu_0 \partial \mathbf{E}/\partial t \tag{5.4}$$

When a material medium is considered, it is conventional to divide charge and current into two broad classes, "bound" charge and "free" charge. Bound charge and current densities arise from polarization **P** and magnetization **M** of a medium (Panofsky and Phillips, 1962; Jackson, 1975; Wangsness, 1986; Reitz et al., 1993). These are included in the definition of new quantities **D** and **H** by constitutive relations; for a linear, isotropic, homogenous medium these are:

$$\nabla \cdot \mathbf{D} = \sigma \tag{5.5}$$

$$\nabla \times \mathbf{E} = -\dot{\mathbf{B}} \tag{5.6}$$

$$\nabla \cdot \mathbf{B} = 0 \tag{5.7}$$

$$\nabla \times \mathbf{H} = \mathbf{J} + \dot{\mathbf{D}} \tag{5.8}$$

$$\mathbf{D} = \mathbf{D}[\mathbf{E}, \mathbf{B}] = \varepsilon_0 \mathbf{E} + \mathbf{P} = \varepsilon \mathbf{E} \tag{5.9}$$

$$\mathbf{H} = \mathbf{H}[\mathbf{E}, \mathbf{B}] = \mathbf{B}/\mu_0 - \mathbf{M} = \mathbf{B}/\mu \tag{5.10}$$

In addition, for conducting media there is the generalized Ohm's law:

$$\mathbf{J} = \mathbf{J}[\mathbf{E}, \mathbf{B}] \tag{5.11}$$

The square brackets are used to signify that the connections are not necessarily simple; they may be nonlinear, depend on past history, etc. (Jackson, 1975, p. 14). For example, the cross-tail current in the plasma sheet is driven by gradients in the magnetic field, not by an electric field. Thus the current is in the dawn-dusk sense on the nightside no matter what the sense of the electric field may be. With a dawn-dusk electric field across the plasma sheet also in the same direction, $\mathbf{E} \cdot \mathbf{J}$ is positive to denote an electrical load. With the opposite sense (e.g., for a northward IMF, seeChapter 9), $\mathbf{E} \cdot \mathbf{J}$ would be negative to designate a dynamo (the current carriers would be losing energy). On the dayside for a southward IMF the current is in the dusk-dawn sense, being driven by the same gradients in the magnetic field despite the opposing electric field. Now $\mathbf{E} \cdot \mathbf{J}$ is negative, with the ring current acting as a dynamo. History comes into the picture because the effects of the currents (cross-tail, field aligned, partial ring, and ring currents) affect the structure of the magnetic field, modifying the current.

5.2   The electric displacement: D field

Matter is made up of atoms and molecules that have equal amounts of positive and negative charges. Bound charges ρ_b arise from the constituents of matter, and we generally have no control over their distribution. Free charges ρ_f are essentially the rest of the charges; we can control their distribution to a large extent by physically moving them about. An electric field will move these charges; the molecule now has an induced dipole moment and it has become polarized.

We define the polarization $\mathbf{P}$ as the electric dipole moment per unit volume, so that the total dipole moment $d\mathbf{p}$ in a small volume $d\tau$ at $\mathbf{r}$ is

$$d\mathbf{p} = \mathbf{P}(\mathbf{r})d\tau \tag{5.12}$$

Thus the total dipole moment of a volume V of material will be

$$\mathbf{P}_{\text{total}} = \int_V \mathbf{P}(\mathbf{r})d\tau \tag{5.13}$$

From its definition, $\mathbf{P}$ will be measured in C/m^2. $\mathbf{P}$ is generally expected to be a function of position within the material. We are assuming $d\tau$ to be small on a macroscopic scale but large on a microscopic atomic scale so that $d\tau$ actually contains many molecules. If there are n molecules per unit volume each of which has a dipole moment $\mathbf{p}$, then $\mathbf{P} = n\mathbf{p}$ provided that all of the dipoles are in the same direction. We expect there will be a functional relation between $\mathbf{P}$ and $\mathbf{E}$. The dipole moment of the volume element $d\tau$ contributing to the potential at $\mathbf{r}$ is

$$d\phi = \frac{d\mathbf{p} \cdot \hat{\mathbf{R}}}{4\pi\varepsilon_0 R^2} = \frac{\mathbf{P}(\mathbf{r}) \cdot \hat{\mathbf{R}}d\tau}{4\pi\varepsilon_0 R^2} \tag{5.14}$$

where, as usual, $\mathbf{R} = \mathbf{r} - \mathbf{r}'$. In order to find the total potential, we integrate this over the volume V of the material and we eventually get:

$$\phi(\mathbf{r}) = \frac{1}{4\pi\varepsilon_0} \int_V \frac{-\nabla\cdot\mathbf{P}d\tau}{R} + \frac{1}{4\pi\varepsilon_0} \int_S \frac{\mathbf{P}\cdot\hat{\mathbf{n}}da}{R} \tag{5.15}$$

where S is the surface bounding V and $\hat{\mathbf{n}}$ is the outer normal to the surface. We see that this is exactly the potential that would be produced by a volume charge density ρ_b distributed throughout the volume, and a surface charge density σ_b on the bounding surface where

$$\rho_b = -\nabla\cdot\mathbf{P} \tag{5.16}$$

The dielectric can be replaced by a distribution of volume and surface charge densities that are related to the polarization $\mathbf{P}$.

$$\nabla\cdot(\varepsilon_0\mathbf{E}+\mathbf{P}) = \rho_f \tag{5.17}$$

The form of this equation where only the free charge density appears suggests that it may be useful to define a vector field $\mathbf{D}$ as follows:

$$\mathbf{D} = \varepsilon_0\mathbf{E}+\mathbf{P} \tag{5.18}$$

Replace Eq. (5.17) with an equation depending only on the free charge:

$$\nabla\cdot\mathbf{D} = \rho_f \tag{5.19}$$

The vector $\mathbf{D}$ is often called the electric displacement, or the $\mathbf{D}$ field. The principal characteristic of $\mathbf{D}$ is the property that its divergence depends only on the free charge density. The dimensions of $\mathbf{D}$ are the same as those of $\mathbf{P}$, and thus $\mathbf{D}$ will be measured in C/m^2.

Finally, it turns out that the electric energy density of a system of charges is

$$u_E = \frac{1}{2}\mathbf{D}\cdot\mathbf{E} = \frac{1}{2}\varepsilon\mathbf{E}\cdot\mathbf{E} = \frac{\mathbf{D}^2}{2\varepsilon} \tag{5.20}$$

And therefore

$$U_E = \int_{\text{allspace}} \frac{1}{2}\mathbf{E}\cdot\mathbf{D}d\tau \tag{5.21}$$

5.3 The magnetic field H

We have already met the Lorentz force on a charged particle defining **B**. The magnetic force is more complicated; there are considerably more twists and turns than with the electrostatic force.

We have to think 3D, not 1D or 2D. We should be prepared for some surprises! The most important action of the magnetic field on particles is that they spiral around the magnetic field lines. Being charged, this constitutes a very small current I, and the formation of a magnetic dipole moment $\mathbf{m}$:

$$\mathbf{m} = I\mathbf{A} \tag{5.22}$$

with the surface $\mathbf{A}$ defined by the area swept out by the particle. The expression

$$\mathbf{m} = \frac{1}{2}I\oint_C \mathbf{r} \times d\mathbf{l} \tag{5.23}$$

is the magnetic dipole moment.

The "curl equation" is the differential form of Ampere's circuital law. Here we must be careful to include all types of currents that can produce a magnetic field. Hence, in the general case, this equation is properly written as

$$\nabla \times \mathbf{B} = \mu_0(\mathbf{J} + \mathbf{J}_M) \tag{5.24}$$

where $\mathbf{J}$ is the transport current density and $\mathbf{J}_M$ is the magnetization current density. This equation may be combined to yield

$$\nabla \times \left(\frac{1}{\mu_0}\mathbf{B} - \mathbf{M}\right) = \mathbf{J} \tag{5.25}$$

which is equivalent to the following expression:

$$\mathbf{H} = \mathbf{B}/\mu_0 - \mathbf{M} \tag{5.26}$$

That is, the auxiliary magnetic vector $\mathbf{H}$ is related to the transport current density through its curl. This equation is the fundamental magnetic field equation in the presence of matter. It may be preferable to use an integral formulation of the theory. With the aid of Stokes's theorem, it may be converted to

$$\int \nabla \times \mathbf{H} \cdot d\mathbf{a} = \oint_C \mathbf{H} \cdot d\mathbf{l} = \int_S \mathbf{J} \cdot \mathbf{n}\, da \tag{5.27}$$

$$\oint_C \mathbf{H} \cdot d\mathbf{l} = I \tag{5.28}$$

In other words, the line integral of the tangential component of the magnetic field $\mathbf{H}$ around a closed path C is equal to the entire transport current through the area bounded by the curve C.

We should assume the existence of a functional relation between $\mathbf{M}$ and $\mathbf{B}$, that is, we would write $\mathbf{M} = \mathbf{M}(\mathbf{B})$. As logical as this might seem, it is not what is usually done; instead, one begins by writing a relation between $\mathbf{M}$ and $\mathbf{H}$, that is, $\mathbf{M} = \mathbf{M}(\mathbf{H})$.

The simplest possible case is that of a linear isotropic homogeneous magnetic material for which the magnetization is proportional to and parallel to the magnetic field. Then we can write

$$\mathbf{M} = \chi_m \mathbf{H} \tag{5.29}$$

Here χ_m is called the *magnetic susceptibility* and is a constant characteristic of the material. The previous relation is another example of a constitutive equation and is not a fundamental equation of electromagnetism.

Again, it turns out that the magnetic energy density of a system of charges is

$$u_M = \frac{1}{2}\mathbf{H} \cdot \mathbf{B} = \frac{\mathbf{B}^2}{2\mu} = \frac{1}{2}\mu \mathbf{H}^2 \tag{5.30}$$

$$U_M = \int_{\text{allspace}} \frac{1}{2}\mathbf{H} \cdot \mathbf{B} d\tau \tag{5.31}$$

5.4 Poynting's theorem

We have seen that $\mathbf{J}_f \cdot \mathbf{E}$ is associated with dissipation.

$$W = \int_{\text{vol}} w d\tau = \int_{\text{vol}} \mathbf{J}_f \cdot \mathbf{E} d\tau \tag{5.32}$$

Taking $\mathbf{J}_f$ from Eq. (5.8):

$$W = \int_{\text{vol}} \mathbf{E} \cdot (\nabla \times \mathbf{H}) d\tau - \int_{\text{vol}} \mathbf{E} \cdot \frac{\partial \mathbf{D}}{\partial t} d\tau \tag{5.33}$$

Taking note of Faraday's equation together with a vector identity, we can write the first integrand as

$$\mathbf{E} \cdot (\nabla \times \mathbf{H}) = \mathbf{H} \cdot (\nabla \times \mathbf{E}) - \nabla \cdot (\mathbf{E} \times \mathbf{H}) = -\mathbf{H} \cdot \frac{\partial \mathbf{B}}{\partial t} - \nabla \cdot (\mathbf{E} \times \mathbf{H})$$

$$\tag{5.34}$$

Using this result, the divergence identity, we obtain *Poynting's theorem*:

$$-\int_{\text{vol}} \left(\mathbf{E} \cdot \frac{\partial \mathbf{D}}{\partial t} + \mathbf{H} \cdot \frac{\partial \mathbf{B}}{\partial t} \right) d\tau = \int_{\text{vol}} \mathbf{J} \cdot \mathbf{E} d\tau + \oint_{\text{surf}} (\mathbf{E} \times \mathbf{H}) \cdot d\mathbf{s} \tag{5.35}$$

The left-hand side represents the rate of change of the stored electric and magnetic energies. Where does the energy go?

Fig. 5.1 shows a possible location on the magnetopause containing an X-line where we might apply this action. The first term

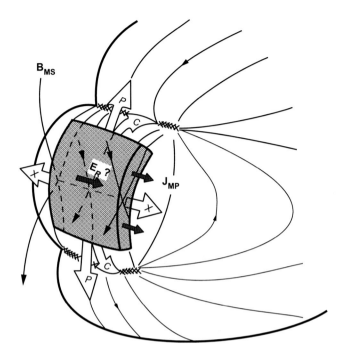

Fig. 5.1 Standard magnetic reconnection (SMR) at the magnetopause for southward IMF. The electric field is dawn-dusk, parallel to the X-line. It is parallel to magnetopause current, thus $\mathbf{E} \cdot \mathbf{J}$ is positive, indicating dissipation. The possible source of the energy is not mentioned. From Heikkila, W.J., 1978. Criticism of reconnection models of the magnetosphere. Planet. Space Sci. 26, 122.

on the right shows that there can be energization of the current carriers within the volume. The surface term denotes propagation through the bounding surface. This analysis makes use of the Poynting vector

$$\mathbf{S} = \mathbf{E} \times \mathbf{B} / \mu_0 = \mathbf{E} \times \mathbf{H} \qquad (5.36)$$

involved with power flow of the electromagnetic field. We write this theorem with the order of the terms changed, in vacuum:

$$\iiint_{\text{vol}} \mathbf{E} \cdot \mathbf{J} d\tau = -\frac{1}{\mu_0} \oiint_{\text{surf}} \mathbf{E} \times \mathbf{B} \cdot d\mathbf{s} - \frac{d}{dt} \iiint_{\text{vol}} \frac{1}{2} \left(\varepsilon_0 E^2 + \frac{1}{\mu_0} B^2 \right) d\tau \quad (5.37)$$

The dissipation term is on the left side, an electrical load as a consequence of the assumption $\mathbf{E} \cdot \mathbf{J} > 0$ as in Fig. 5.1. The possible sources for this energy integral are described by the terms on the right. Double and triple integrals are used to accentuate the dimensions.

5.4.1 The steady state

Steady-state theories ($\partial/\partial t = 0$) use only the first source term in Poynting's theorem. In the steady state we can use the electrostatic scalar potential ϕ to describe the electric field (neglecting the last two right-hand terms of Eq. (5.37)). As shown in Chapter 4, we have two *equivalent* results:

$$\int_{\text{vol}} \mathbf{E} \cdot \mathbf{J} d\tau = -\oint_{\text{surf}} \mathbf{E} \times \mathbf{H} \cdot d\mathbf{s} \tag{5.38}$$

$$= -\oint_{\text{surf}} \phi \mathbf{J} \cdot d\mathbf{s} \tag{5.39}$$

These are cause and effect relationships. The first of these describes a Poynting vector as being the cause, yielding its energy to the electromagnetic field. The second uses the circuit analysis technique to discover a current which is an electrical load; the current carriers are energized because of the voltage (potential) drop (see Fig. 5.1). In both cases a source of energy is required, in the same current circuit, to replace the energy that is dissipated. The Poynting flux originates from this external source, a dynamo somewhere in the current circuit. With the circuit approach it is clear that a dynamo (such as a battery) with $\mathbf{E} \cdot \mathbf{J} < 0$ is the natural explanation.

5.4.2 Electric energy

All observations from the various regions of the magnetosphere explored with spacecraft indicate great variability in both space and time; therefore, it is necessary to consider local plasma structures with explicit time dependence. Since the dielectric constant is a tensor, we must use the auxiliary vector $\mathbf{D}$ and $\mathbf{H}$, as doing so brings in the currents. The change in electric energy is

$$\delta U_E = \int \mathbf{E} \cdot \delta \mathbf{D} \tag{5.40}$$

Taking only the second source term in Poynting's theorem,

$$\iiint_{\text{vol}} \mathbf{E} \cdot \mathbf{J} d\tau = -\iiint_{\text{vol}} \mathbf{E} \cdot \frac{\partial \mathbf{D}}{\partial t} d\tau = -\iiint_{\text{vol}} \mathbf{E} \cdot \frac{\varepsilon}{2} \frac{\partial \mathbf{E}}{\partial t} d\tau \tag{5.41}$$

Note that the polarization current density is $\mathbf{J}_P = \varepsilon \partial \mathbf{E} / \partial t$, the term is a source of electromagnetic energy enabled by charge polarization.

$$\iiint_{\text{vol}} \mathbf{E} \cdot \mathbf{J} d\tau = -\iiint_{\text{vol}} \mathbf{E} \cdot \mathbf{J}^{\text{POL}} d\tau \tag{5.42}$$

Lemaire and Roth (1978, 1991) have advanced a theory for the penetration of solar wind plasma through the magnetopause, as shown in Fig. 5.2. They note that the solar wind plasma must

Uniform steady solar wind

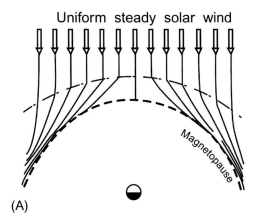

(A)

Nonuniform unsteady solar wind

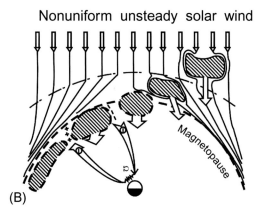

(B)

Fig. 5.2 We repeat the important Fig. 1.18 from Chapter 1. (A) When the solar wind is steady and uniform, the magnetopause is a smooth surface. The solar wind does not penetrate except by SMR supposedly due to anomalous resistivity along the X-line in the magnetic field topology, whose detailed properties are still not known. (B) When the solar wind is nonuniform and unsteady, plasma density irregularities (plasmoids) carried in the solar wind will be able to penetrate deeper into the geomagnetic field if they have an excess momentum density, as explained by Schmidt (1960, 1979). From Lemaire, J., Roth, M., 1978. Penetration of solar wind plasma elements into the magnetosphere. J. Atmos. Terr. Phys. 40(3), 332.

be patchy, and blobs with more momentum than the adjacent plasma can penetrate through the magnetopause current sheet.

Let us consider what occurs when there is a local density clump in the plasma. Let the density gradient be toward the center, and suppose KT is constant. There is then a pressure gradient toward the center. Since the plasma is quasineutral, the gradient exists for both the electron and ion fluids. Only when the forces are equal and opposite can a steady state be achieved.

However, another electric field may be present due to the momentum of the entire clump. It is this that is considered by Poynting's theorem. It represents the mass motion of the plasma. Schmidt (1960, 1979) has demonstrated the mechanism, which depends on a polarization current. A confirmation is found by the two-dimensional (three velocity components) electrostatic simulations of Koga et al. (1989) and Livesey and Pritchett (1989). This will be treated in Chapter 6.

5.4.3 Magnetic energy

Another local source is that described by the rate of change of stored magnetic energy. Taking only the last source term in Poynting's theorem, the energization becomes

$$\int_{\text{vol}} \mathbf{E} \cdot \mathbf{J} d\tau = -\int_{\text{vol}} \mathbf{H} \cdot \frac{\partial \mathbf{B}}{\partial t} \tag{5.43}$$

$$= -\frac{d}{dt} \int_{\text{vol}} \frac{B^2}{2\mu_0} d\tau \tag{5.44}$$

The right-hand term can also be expressed in another form (Wangsness, 1986, p. 357):

$$-\int_{\text{vol}} \frac{\mathbf{B}}{\mu_0} \cdot \frac{\partial \mathbf{B}}{\partial t} = \int_{\text{vol}} \frac{\mathbf{B}}{\mu_0} \cdot \nabla \times \mathbf{E} d\tau \tag{5.45}$$

$$= \frac{1}{\mu_0} \oint_{\text{vol}} \nabla \cdot (\mathbf{E} \times \mathbf{B}) d\tau + \frac{1}{\mu_0} \oint_{\text{vol}} \mathbf{E} \cdot \nabla \times \mathbf{B} d\tau \tag{5.46}$$

$$= \frac{1}{\mu_0} \oint_{\text{vol}} \nabla \cdot (\mathbf{E} \times \mathbf{B}) d\tau + \oint_{\text{vol}} \mathbf{E} \cdot \mathbf{J} d\tau \tag{5.47}$$

The interpretation of this result is dependent on the nature of the magnetic field at the magnetopause, the difference between tangential or rotational discontinuities, even the physics of the magnetopause itself. What is the direct connection between electromagnetic fields on the two sides of the magnetopause?

5.4.3.1 Tangential discontinuity

If $B_n = 0$ then we have a tangential discontinuity. The meaning of this first term in Eq. (5.46) can be clarified by using the divergence theorem:

$$\oint_{\text{vol}} \nabla \cdot (\mathbf{E} \times \mathbf{B}) d\tau = \oint_{\text{surf}} (\mathbf{E} \times \mathbf{B}) \cdot \hat{\mathbf{n}} ds \tag{5.48}$$

The plasma particles have no access across the magnetopause. The remaining term has a positive sign, so it would be an electrical load. However, in the moving reference frame the electric field vanishes; the induction electric field is sufficient to move the magnetopause in step with the plasma.

5.4.3.2   Rotational discontinuity

Alternatively, if B_n is finite then we must have a normal component of the magnetic field at the magnetopause, through the current sheet as in Fig. 5.3. With the finite B_n, the reality of this corresponds to a rotational discontinuity at the magnetopause. The plasma particles can flow field aligned along B_n. Now the plasma can respond to the induction field at the edges of the perturbation in response to the induction electric field. We have to look at both terms in Eq. (5.46) again. The first term is no longer zero as the plasma particles can react to the induction electric field. That creates charge separation that produces an opposing electrostatic field; the total electric field is therefore modified. This is discussed in the next section and is a major topic for a plasma transfer event (PTE) in Chapter 6.

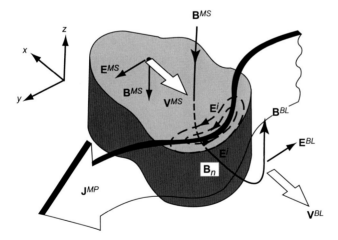

Fig. 5.3 Model for impulsive plasma transport across an *open* magnetopause. A plasma cloud (a flux tube) with excess momentum is assumed to distort the magnetopause current locally. This creates an inductive electric field, everywhere opposing the current perturbations by Lenz's law (the plasma response by charge separation is not included). Note that the induction electric field cuts though the magnetopause at both sides forcing general magnetic reconnection (GMR). From Heikkila, W.J., 1984. Magnetospheric topology of fields and currents. In: Potemra, T.A. (Ed.), Magnetospheric Currents. AGU Geophysical Monograph 28. American Geophysical Union, Washington, DC, p. 211.

5.5 Discussion

As discussed in Chapter 1, two theories for the interaction of the solar wind plasma with the magnetosphere were proposed in the same year: viscous-like interaction by Axford and Hines (1961) and (standard) magnetic reconnection by Dungey (1961). Poynting's theorem has a lot to say about this interaction; unfortunately that theorem is rarely used in space plasma physics (but see Longmire, 1963).

Theoretical work on Standard Magnetic Reconnection (SMR) concentrates on the X-line with zero-B, and the electrical load with $\mathbf{E} \cdot \mathbf{J} > 0$. The source of energy $\mathbf{E} \cdot \mathbf{J} < 0$ is not included. In like fashion, most observational papers report the results in two dimensions (2D), in the noon-midnight plane, like Fig. 5.4. We need to find out how the inclusion of the source of energy, a dynamo, affects magnetic reconnection.

Magnetic reconnection is a phenomenon of great importance in solar system plasmas and, presumably, in astrophysical plasmas,

Fig. 5.4 Magnetic reconnection in Earth's magnetosphere. (A) Noon-midnight cut, showing magnetopause and magnetotail reconnection sites, the former for southward IMF. (B) Zoom-in on the region around the X-line, with the ion and electron diffusion regions indicated by the *shading and the rectangular box*, respectively. The quadrupolar Hall magnetic field is pointing in and out of the plane of the figure. The Hall electric field is shown by the *(red) arrows*, while the *(blue) arrows* mark the oppositely directed jets in the outflow regions. Note that entry and acceleration occur all the way along the current sheet. From Paschmann, G., 2008. Recent in-situ observations of magnetic reconnection in near-earth space. Geophys. Res. Lett. 35, 2 (L19109). doi:10.1029/2008GL035297.

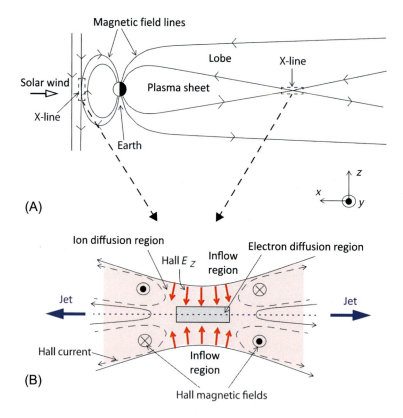

*because it converts energy stored in magnetic fields into particle
kinetic energy and changes the magnetic field topology, allowing
effective exchanges of mass, momentum, and energy between
differently magnetized plasma regions. For reconnection to occur,
the "frozen-in field" condition of magneto-hydrodynamics (MHD)
must break down in a localized region, commonly called the
"diffusion region", of the current sheet separating the plasmas.
There the magnetic field lines that are flowing in from both sides
become "cut" and "reconnected" [as shown in Figure 5.4], forming
an X-type configuration. Associated with the inflow is an
electric field pointing out of the plane of the diagram parallel to
the X-line. In component reconnection, the magnetic field also has
a component (the "guide field") into or out of the plane [in
Figure 5.4].*

Paschmann (2008)

That statement, and also the often quoted report by Vasyliunas
(1975), does not recognize that SMR is based on 2D. The comment
applies to Fig. 5.4; the same identical process is shown for the dayside and for the magnetotail. This is at variance with Fig. 1.33 of
Chapter 1, which shows a dynamo for the dayside and an electrical load for the nightside according to Akasofu (1991). It forgets
that SMR is based on dissipation, not recognizing the source of
this dissipated energy. One must use the correct physics, the basis
to get the correct mathematics, to solve any physical problem.

Schindler et al. (1988) have advanced the idea of general
magnetic reconnection (GMR); they use the real electric field
$E = -\partial A/\partial t - \nabla \phi$, not the so-called convection electric field
$E = -v \times B$. GMR may occur with a finite value of a magnetic field
B, a guide field. GMR was advocated as a process affecting the
topology of the magnetic field in three dimensions allowing B
to be finite (not an X-line).

GMR is based on the *breakdown of magnetic connection*, with a
changing magnetic helicity.

$$E = -v \times B + R \qquad (5.49)$$

where R is a vector function that may be complicated. Magnetic
connection means the following: two plasma elements A and B
are connected by a magnetic field line at a given time t_1, then
as they both move they could be connected by a field line at
any other time t_2, as illustrated by Fig. 5.5. This *line conservation*
is a well-known property called the *frozen-in-field* of ideal MHD.
GMR depends on a finite electric field in some diffusion region,
producing a breakdown of magnetic connection. It is involved
in changes in magnetic helicity (Schindler et al., 1988).

Fig. 5.5 GMR occurs at finite values of the magnetic field, along a guide field. Two plasma elements **A** and **B** are connected by a magnetic field line at a given time t_1, then as they both move they may be disconnected at any other time t_2 depending on a parallel electric field from the diffusion region. From Fälthammar, C.-G., 2004. Magnetic-field aligned electric fields in collisionless space plasmas—a brief review. Geofis. Int. 43, 227.

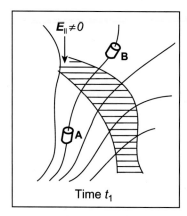

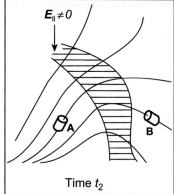

Inspect Fig. 5.3 again: due to its assumed localization with zero divergence $\nabla \cdot \delta \mathbf{J} = 0$, there must be an electric field parallel to B_n with a rotational discontinuity at the two sides. This satisfies the condition imposed by Schindler et al. (1988): we have GMR. An electric field parallel to the magnetic field was investigated by Omura et al. (2003), as reported in Chapter 3. We found several aspects that were surprising. Electron hole distributions developed out of stable plasma; particles continued to gain energy from the electric field.

In Fig. 5.6 it is assumed that the frame of reference is fixed to the magnetopause in the center of the perturbation, as in reconnection models. It is for an outward motion of the magnetopause (inward motion of solar wind plasma), no doubt caused by a sunward motion due to surface waves. In this frame the magnetopause, top and bottom, is moving sunward as shown by large arrows; it is here that induction is felt. With a rotational discontinuity the plasma polarizes; the plasma is trying to reduce the parallel component of the induction field. In so doing, it enhances the transverse component, reversing as indicated. We see that this charge distribution will drive the magnetosheath plasma into the current sheet and out on the other side as impulsive injection explained by Lemaire and Roth (1978).

5.5.1 Concept of a plasma transfer event

There is no question about the reality of a PTE through the magnetopause; observations come from a variety of sources beginning with the rocket results of Carlson and Torbert (1980). What is not known is the openness or closure of **B** (see the reviews

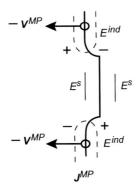

Fig. 5.6 Here it is assumed that the frame of reference is fixed to the magnetopause in the center of the perturbation, as in reconnection models. Given a rotational discontinuity the plasma polarizes; the plasma is trying to reduce the parallel component of the induction field. In so doing, it enhances the transverse component, reversing as indicated. We see that this charge distribution will drive the magnetosheath plasma into the current sheet and out on the other side as impulsive injection, explained by Lemaire and Roth (1978). From Heikkila, W.J., 1997. Interpretation of recent AMPTE data at the magnetopause. J. Geophys. Res. 102(A5), 2121.

by Lundin, 1988; Lundin et al., 2003; Lemaire and Roth, 1978, 1991; Heikkila, 1998).

Impulsive penetration (IP) after Lemaire and Roth (1978) uses electric energy, that is, plasma in motion, while PTE uses magnetic energy, organized motion. *They are not the same process* since they depend on different terms of Poynting's theorem.

The electric field has two sources, charge separation and induction. The total field **E** is

$$\mathbf{E} = \mathbf{E}^{\mathrm{ind}} + \mathbf{E}^{es} = -\partial \mathbf{A}/\partial t - \nabla \phi \qquad (5.50)$$

which has induction and electrostatic components. We write it in this order to emphasize the relative importance of the two (the induction field is generally larger).

The plasma response to the imposition of the induced electric field leads to the creation of an electrostatic field, at least when conductivities are not all zero. A PTE is a three-dimensional object, that is, 2D to show the magnetic topology (e.g., x-z plane in GSM coordinates), and another set to show localization involving curl **E** (x-y plane). These two types of field have different topological characteristics, one being solenoidal using the Coulomb gauge, discussed extensively by Morse and Feshbach (1953), the other being irrotational (conservative) with zero curl. Consequently, they can never cancel each other; the most the plasma can do is to redistribute the field while maintaining the curl.

It is instructive to express the induction electric field in integral form:

$$\text{emf} = \varepsilon = \oint_{circuit} \mathbf{E} \cdot d\mathbf{l} = -\frac{d\Phi^{MAG}}{dt} \tag{5.51}$$

where Φ^{MAG} is the magnetic flux through the circuit used for the integration. It is only by this emf that we can *tap stored magnetic energy*.

The electric field experiment on the Polar satellite (Mozer, 2005) collected bursts of data on occasion at high rates of 1600 and 8000 samples/s to measure the parallel and perpendicular components of the vector electric field. One event (Fig. 5.7) was discussed by Mozer and Pritchett (2010) with perpendicular electric field components as large as 200 mV/m and parallel electric field components as large as 80 mV/m. These measurements, made in the low-latitude boundary layer (LLBL) just earthward of the magnetopause, will be discussed in Chapter 6.

A more pertinent choice toward understanding magnetic reconnection is to concentrate on the magnetopause itself. The magnetopause is determined by the magnetic field, specifically the B_z component. This is done in Fig. 5.7 where the B_z reverses sign. Although the magnitudes are somewhat lower than in the LLBL they are still quite hefty, up to 50 mV/m. An interesting aspect is that the fluctuation rate is larger at 09:37:49.4 and again at 09:37:50 UT as marked by * signs.

5.6 Plasma transfer event seen by Cluster

This process was seen by C3 as shown by Fig. 5.8 (more details in Chapter 6). The plasma density is increased, quadrupled in the middle, for over 1 min, in a burst of plasma entering through the magnetopause. PEACE electron data show several aspects in agreement with a PTE.

- The pitch angle distribution for the higher-energy channels has a maximum near 90 degrees, indicating field lines are closed on either side.
- In the heart of the PTE, robust fluxes of lower-energy electrons are evident with maximum field-aligned pitch angles, 180 degrees at 07:30 UT and 0 degree at 07:40 UT, in agreement with the PTE concept.
- This result may indicate open magnetic field lines, a firm requirement for the PTE process.

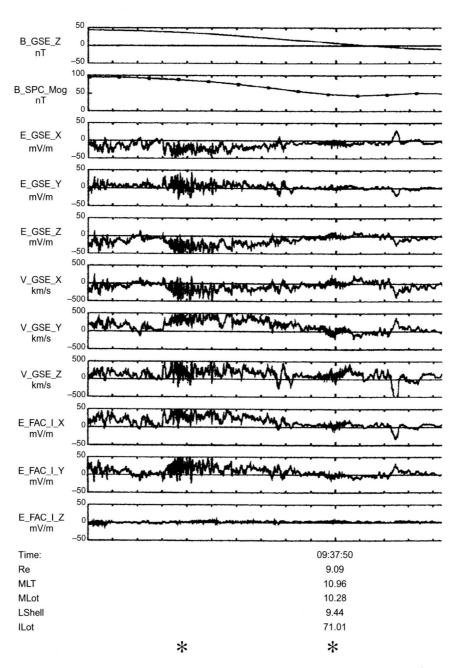

Fig. 5.7 A short sample of Polar data at the magnetopause on April 13, 2001, where B_z reverses sign (good crossing of about 1 second). The magnetopause is determined by the magnetic field, specifically the B_z component. An interesting aspect is that the repetition rate is larger at 09:37:49.4 and again at 09:37:50 UT, at the magnetopause as marked by * signs. This is an indication that GMR is taking place. 1.4 s of data after Mozer, F.S., Pritchett, P.L., 2010. Spatial, temporal, and amplitude characteristics of parallel electric fields associated with subsolar magnetic field reconnection. J. Geophys. Res. 115, 5 (A04220). https://doi.org/10.1029/2009JA014718, p. 2.

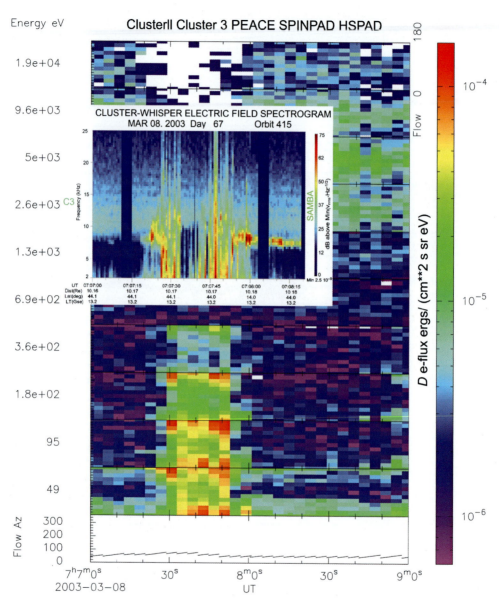

Fig. 5.8 Data from Cluster 3 on March 8, 2003. PEACE shows the localized nature of the solar wind pressure pulse. Each energy level is organized by pitch angle with zero at the top. WHISPER (inset) shows the plasma wave emissions associated with GMR where **E** is parallel to **B** at the beginning and end of the perturbation. From Heikkila, W.J., Canu, P., Dandouras, I., Keith, W., Khotyaintsev, Y., 2006. Plasma transfer event seen by cluster. In: Cluster and Double Star Symposium—5th Anniversary of Cluster in Space. ESA SP-598, p. 3.

- WHISPER instrument showed two bursts of plasma emissions at the beginning and end of the PTE.

It is likely that these bursts coincided with a parallel component of the induction electric field, as shown in Fig. 5.3. These bursts probably produced the emissions, evidence of GMR.

The two observations by PEACE and WHISPER are shown together to mark a GMR event. A similar claim cannot be made for SMR; researchers are still looking for the X-line with its massive energy output.

- **Plasma transfer event**. The PTE process was recorded by C3 on March 8, 2003. CIS/CODIF data shows penetration of solar wind plasma lasting for over 1 min at the time of the event. PEACE data showed soft electron fluxes on two sides of the perturbation in Fig. 5.8. These were likely the cause of the emissions by WHISPER (inset). All these effects were caused by the large electric field measured by EFW. These are the marks of GMR.

5.7   Three systems

We now look critically at the whole question of solar wind interaction with the magnetosphere, and especially at the paradigm of magnetic reconnection (MR). MR is based on a 2D model, as illustrated in Fig. 5.1; it is assumed that MR is caused by an electric field $\mathbf{E}_R$ along an X-line. The model assumes that it is spatially constant (being in two dimensions), thus curl $\mathbf{E}_R = 0$. Faraday (1832) discovered that curl $\mathbf{E} = -\partial \mathbf{B}/\partial t$; whenever the magnetic field is changing (due to the assumed reconnection), curl $\mathbf{E}$ is finite. Therefore the 2D model that is commonly used to *define magnetic reconnection* is flawed.

Three systems—the solar wind, the magnetosphere, and the ionosphere—interact, transmitting and transforming the solar wind energy into energies of auroral phenomena, and eventually depositing most of it as heat energy in the ionosphere. Fig. 1.33 of Chapter 1 is a flow chart for solar wind/magnetosphere/ionosphere coupling. It shows the energy flow, energy conversion, and associated processes. In his book Akasofu reiterates:

We have to recognize that several fundamental problems remain; what is needed is not just improvements to traditional theories or a mopping-up of residual problems. I am convinced that new thinking is needed to solve long-standing unsolved problems.

Akasofu (2007)

I learned that we had a system that needed to be replaced.

Greer (1994), in an interview with Mikhail Gorbachev

Standard magnetic reconnection (SMR) does not qualify as the generator process (on the left in Fig. 1.33) because it has the wrong sign, positive (a load) instead of negative (a dynamo). SMR does not use a closed current system, nor recognize the concept of a dynamo.

5.8 3D reconnection

The existence of the third dimension can radically alter a physical situation, particularly in regard to stability. A two-dimensional tight-rope walker confined to the vertical plane containing the rope could never fall off.

Dungey (1958)

Pontin (2011) has reviewed the possible regimes of three-dimensional magnetic reconnection, observing that reconnection models that are even weakly three-dimensional are crucially different from two-dimensional (or even 2.5-dimensional) models.

5.8.1 Magnetic topology

Magnetic reconnection always requires the presence of a current sheet (that is, an intense, localized current layer), whether the plasma is collisional or collisionless. As such, the determination of possible reconnection sites starts with understanding where current sheets form. In two dimensions, reconnection occurs at magnetic "X-points," but in the absence of two-dimensional symmetry, the number of possible sites for current sheet formation and reconnection increases greatly. These sites can be either topological (preserved by an arbitrary smooth deformation of the field) or geometrical (not preserved under deformations).

Two primary sites for potential current sheet formation are at 3D null points, an extension of the 2D X-point concept, or at a separator line, a field line that runs from one null point to another. Examples of the two field line structures are shown in Fig. 5.9. The magnetic nulls must be hyperbolic in order to preserve $\nabla \cdot \mathbf{B} = 0$. The field lines surrounding the null point asymptotically approach (or recede from) it in opposite directions, forming the spine of the null, while the field lines receding from (or approaching) the null form a surface known as the fan plane, separating distinct volumes of magnetic flux as a *separatrix*. The separator line (Fig. 5.9B) is defined by the transverse intersection of the fan

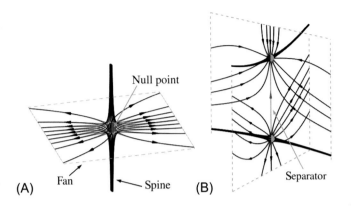

Fig. 5.9 Potential magnetic field line structure in the vicinity of (A) an isolated 3D null point, and (B) a generic fan-fan separator. From Pontin, D.I., 2011. Three-dimensional magnetic reconnection regimes: a review. Adv. Space Res. 47, 2.

planes of the two nulls. This configuration is topologically stable, and observations by the Cluster mission show both single nulls and collections of nulls in the Earth's magnetotail current sheet (Deng et al., 2009). Dorelli et al. (2007) have also found such structures in magnetospheric simulations.

5.8.2 Properties of 3D reconnection

Following the framework of general magnetic reconnection (GMR) put forward by Schindler et al. (1988), reconnection in 3D is defined by a breakdown of magnetic field line and flux conservation. That is, a breakdown in the magnetic connectivity between different plasma elements. This occurs when a component of the electric field parallel to the magnetic field ($E_\parallel$) is spatially localized in all three dimensions. The change of connectivity (reconnection rate) is the maximum value over all field lines of

$$\Phi = \int E_\parallel ds \tag{5.52}$$

where the integral is performed along magnetic field lines from one side of the diffusion region to the other. The new properties of reconnection in 3D can be understood by reviewing the frozen-in field conditions developed in Section 1.5 and noticing that if we assume that $\mathbf{B}$ and $\mathbf{v}$ are 2D, then we must have $\mathbf{E} \cdot \mathbf{B} = 0$ and $\mathbf{R} \cdot \mathbf{B} = 0$, so we can write

$$\mathbf{R} = d\mathbf{v} \times \mathbf{B} \tag{5.53}$$

which leads to

$$\mathbf{E} + (\mathbf{v} - d\mathbf{v}) \times \mathbf{B} = 0 \tag{5.54}$$

So, a flux transporting flow exists everywhere in 2D given by

$$\mathbf{v} = d\mathbf{v} + \mathbf{E} \times \mathbf{B}/B^2 \qquad (5.55)$$

In 3D reconnection, however, by definition $\mathbf{E} \cdot \mathbf{B} \neq 0$, and the conditions under which the conservation of magnetic topology, field lines, and magnetic flux hold are much more subtle (Schindler et al., 1988; Pontin, 2011). In the case of 2D reconnection, the singularity of the flux transport velocity at an X-point is due to the fact that the reconnection process involves magnetic field lines being cut and rejoined at that point. Changes in field line connectivity are thus discontinuous at the null. Since the flux transport velocity is smooth and continuous everywhere outside the X-point, the field lines evolve as if they are reconnected only at this one point and occur in a one-to-one pairwise fashion.

According to Pontin (2011), none of the previous properties of 2D reconnection carry over to the three-dimensional case. It can be proven generally (Priest et al., 2003) that in a localized region within which $\mathbf{E} \cdot \mathbf{B} \neq 0$, a simple flux transport velocity *does not exist*.

> The result is that, if one follows magnetic field lines from footpoints commoving in the ideal flow, they appear to split as soon *as they enter the non-ideal region*, and the connectivity changes *continually and continuously* as they pass through the non-ideal region.
>
> **Pontin (2011, p. 5)**

This means that over any small increment of time, every field line within the nonideal region experiences a change in magnetic connectivity, so the field lines are not reconnected in a one-to-one fashion as they are in two dimensions. As seen in Fig. 5.10, consider two field lines about to enter the diffusion region, one connects plasma elements A and B, and the other connects elements C and D. If the field lines are chosen such that after passing through the diffusion region A is connected to C, then B will NOT be connected to D. Thus in three dimensions, we cannot think of reconnection as a simple cut and paste of field line pairs. As the flux tubes enter the diffusion region, they immediately begin to split, with previously connected plasma elements no longer being coincident. Frames 2–5 of Fig. 5.10 show the apparent "slip-running" of the field lines, with the solid sections of the flux tubes traces from ideal commoving footpoints (black bands) moving at the local plasma velocity outside the diffusion region. The transparent sections correspond to field lines traced into and beyond the diffusion region, and appear to slip past each other at a different velocity while inside the nonideal region. Although frame 1 shows two flux tubes, frame 6 of Fig. 5.10 shows that the

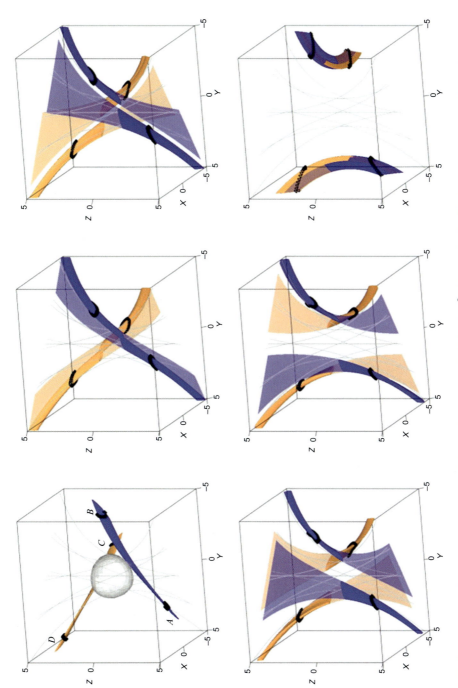

Fig. 5.10 Reconnection of two representative flux tubes in the magnetic field $\mathbf{B} = (y, k^2x, 1)$, with $k = 1.2$. The flux tubes are traced from ideal comoving footpoints *(marked black)*, and the solid sections move at the local plasma velocity (outside D), while the transparent sections correspond to field lines that pass through the diffusion region. A localized diffusion region *(shaded surface in the first frame)* is present around the origin. From Pontin, D.I., 2011. Three-dimensional magnetic reconnection regimes: a review. Adv. Space Res. 47, 6.

four cross sections do not line up to form two unique flux tubes. These 3D properties hold regardless of the magnetic topology at the reconnection region.

5.9 Scientific paradigms

Historian of science Thomas Kuhn gave this word *paradigm* its contemporary meaning. It refers to the set of practices that define a scientific discipline during a particular period of time. A scientific paradigm is composed of many aspects:

- What is to be observed and scrutinized.
- The kind of questions that are supposed to be asked.
- How these questions are to be structured.
- How results of scientific investigations should be interpreted.

The prevailing paradigm often represents a more specific way of viewing reality, or limitations on acceptable programs for future research, rather than the much more general scientific method. Grant funding, and publication of the results, would be more difficult to obtain for such experiments in proportion to the amount of departure from accepted standard models. A simplified analogy for paradigm is a habit of reasoning or the box in the commonly used phrase: "thinking outside the box" (see, for example, Parks, 2004 on the MHD box). Thinking inside the box is analogous with normal accepted science. The box encompasses the thinking of normal science, and thus the box is analogous with paradigm. "Thinking outside the box" would be what Kuhn calls revolutionary science. Revolutionary science is usually unsuccessful, and only rarely leads to a new paradigm. When they are successful they lead to large-scale changes in the scientific worldview.

- **Paradigm shifts** tend to be most dramatic in sciences that appear to be stable and mature.
- **Paradigm paralysis** is the inability to see beyond the accepted model of thinking.

The revolution with continental drift took only some 40 years. With SMR, involving an electric field along the X-line has already persisted longer than that. The most of what can be claimed for SMR is that it is only a half-theory, at best. At worst, it has been badly proposed.

A consequence of this reconnection was noted by Paschmann (2008): it should be able to explain that SMR "converts energy stored in magnetic fields into particle kinetic energy and changes the magnetic field topology." In his summary he notes:

Reconnection is commonly considered to be a means for accelerating particles to high energies. It is thus interesting that no such particles have been reported for the solar wind reconnection events, perhaps because they might be confined to very thin sheets and thus not resolved. ... The unfreezing of the ions from the magnetic field has only been established in a few fortuitous cases, the problem being the general inability of present plasma measurements to resolve the narrow ion diffusion region structures.

Paschmann (2008, p. 5)

Alternatively, Schindler et al. (1988) advanced the idea of Generalized Magnetic Reconnection (GMR); this may occur with a finite magnetic field **B** as in Figs. 5.7 and 5.8. GMR uses the complete electromagnetic field $\mathbf{E} = -\partial \mathbf{A}/\partial t - \nabla \phi$.

5.10 Summary

Poynting's theorem deals with energy, a scalar; it can be expressed by a single equation. It is the equivalent of all of Maxwell's equations which deal with forces, that is, vectors. That feature is worth exploiting.

- Poynting's theorem divides all electromagnetic phenomena into two groups, with and without explicit time dependence.
- It also divides all by electric versus magnetic energy.
- It divides by cause versus effect according to the sign of the terms.
- **Physics is time dependent in three dimensions.** The development of the physics of this interaction was begun in 2D, Dungey in the meridian plane, Axford and Hines in the equatorial plane. While these authors probably had a three-dimensional picture in mind, our research has largely been guided by these two-dimensional drawings (see Fig. 5.4). Even the data analysis often used is in the noon-midnight meridian, thus neglecting local time variations.
- **All currents must be closed** in any physical device in accordance with Kirchhoff's principles; it must be so to analyze cause and effect by the sign of $\mathbf{E} \cdot \mathbf{J}$. The product $\mathbf{E} \cdot \mathbf{J} > 0$ is an electrical load, while the source of the dissipated energy is governed by $\mathbf{E} \cdot \mathbf{J} < 0$.
- **The dynamo region is generally neglected** in space research. The source of the dissipated energy with $\mathbf{E} \cdot \mathbf{J} < 0$ is hardly ever discussed in space plasma physics.

- **Generalized (finite-B) magnetic reconnection.** Schindler et al. (1988) advanced the idea of GMR. GMR uses $\mathbf{E} = -\partial \mathbf{A}/\partial t - \nabla \phi$, the real electric field.
- **3D reconnection.** Based on the ideas of GMR, reconnection in three dimensions has little in common with reconnection in 2D.

5.11 Problems

5.1. In cylindrical coordinates, the magnetic field of a long straight wire is $\mathbf{B} = \mu_o I/2\pi r$ in the phi direction, and the electric field inside the wire is $\mathbf{E} = I/\pi R^2 \sigma$ $\mathbf{E}$ in the direction of the wire, where $R =$ the radius of the wire. What is the Poynting vector at the surface of the wire?

5.2. An electromagnetic wave is made up of a sinusoidal electric field oscillating in the y dimension with magnitude E_o, and a magnetic field oscillating in the x dimension with magnitude E_o/c. What is the magnitude and direction of the Poynting vector?

5.3. Show that the power delivered by the Lorentz force is equal to the left-hand side of Eq. (5.37) by computing $\mathbf{P} = \int \mathbf{F} \cdot \mathbf{v} d\tau$.

5.4. This chapter has a lot to say about the interactions of electric and magnetic fields and particles in a plasma, but fundamentally, what is the only thing that can change the kinetic energy of a charged particle? Explain in terms of $\mathbf{E} \cdot \mathbf{J}$.

5.5. What would be the stored magnetic energy of aluminum in a magnetic field of 100 nT?

References

Akasofu, S.-I., 1991. Auroral phenomena. In: Meng, C.-I., Rycroft, M.J., Frank, L.A. (Eds.), Auroral Physics. Cambridge University Press, New York, NY, pp. 223–239.

Akasofu, S.-I., 2007. Exploring the Secrets of the Aurora. Springer, New York, NY.

Axford, W.I., Hines, C.O., 1961. A unifying theory of high-latitude geophysical phenomena and geomagnetic storms. Can. J. Phys. 39, 1433.

Carlson, C.W., Torbert, R.B., 1980. Solar wind ion injections in the morning auroral oval. J. Geophys. Res. 85 (A6), 2903–2908.

Deng, X.H., Zhou, M., Li, S.Y., et al., 2009. Dynamics and waves near multiple magnetic null points in reconnection diffusion region. J. Geophys. Res. 114, A07216. https://doi.org/10.1029/2008JA013197.

Dorelli, J.C., Bhattacharjee, A., Raeder, J., 2007. Separator reconnection at earth's dayside magnetopause under generic northward interplanetary magnetic field conditions. J. Geophys. Res. 112, A02202. https://doi.org/10.1029/2006JA011877.

Dungey, J.W., 1958. Cosmical Electrodynamics. Cambridge University Press, New York, NY.

Dungey, J.W., 1961. Interplanetary magnetic field and the auroral zones. Phys. Rev. Lett. 6, 47–48.

Faraday, M., 1832. Experimental researches in electricity. Philos. Trans. R. Soc. Lond. A 122, 125–162.

Greer, C., 1994. The well-being of the world is at stake, with a quote from Mikhail Gorbachev. Parade Mag. January 23.

Heikkila, W.J., 1998. Cause and effect at the magnetopause. Space Sci. Rev. 83, 373–434.

Jackson, J.D., 1975. Classical Electrodynamics, second ed. Wiley, New York, NY.

Koga, J., Geary, J.L., Fujinami, T., Newberger, B.S., Tajima, T., Rostoker, N., 1989. Numerical investigation of a plasma beam entering transverse magnetic fields. J. Plasma Phys. 42 (Part 1), 91–110.

Kuhn, T.S., 1970. The Structure of Scientific Reductions. The University of Chicago Press, Chicago, IL.

Lemaire, J., Roth, M., 1978. Penetration of solar wind plasma elements into the magnetosphere. J. Atmos. Terr. Phys. 40 (3), 331–335.

Lemaire, J., Roth, M., 1991. Non-steady-state solar wind-magnetosphere interaction. Space Sci. Rev. 57, 59–108.

Livesey, W.A., Pritchett, P.L., 1989. Two dimensional simulations of a charge-neutral plasma beam injected into a transverse magnetic field. Phys. Fluids B 1, 914.

Longmire, C.L., 1963. Plasma Physics and Thermonuclear Research. Pergamon Press, New York, NY.

Lundin, R., 1988. On the magnetospheric boundary layer and solar wind energy transfer into the magnetosphere. Space Sci. Rev. 48, 263–320.

Lundin, R., Nilsson, H., Yamauchi, M., 2003. Critical aspects of magnetic reconnection. In: Lundin, R., McGregor, R. (Eds.), Proceedings of the Magnetic Reconnection Meeting, IRF Scientific Report 280. Swedish Institute of Space Physics, Kiruna, Sweden, pp. 2–5.

Morse, P.M., Feshbach, H., 1953. Methods of Theoretical Physics. McGraw-Hill, New York, NY.

Mozer, F.S., 2005. Criteria for and statistics of electron diffusion regions associated with subsolar magnetic field reconnection. J. Geophys. Res. 110, A12222. https://doi.org/10.1029/2005JA011258.

Mozer, F.S., Pritchett, P.L., 2010. Spatial, temporal, and amplitude characteristics of parallel electric fields associated with subsolar magnetic field reconnection. J. Geophys. Res. 115, A04220. https://doi.org/10.1029/2009JA014718.

Omura, Y., Heikkila, W.J., Umeda, T., Ninomiya, K., Matsumoto, H., 2003. Particle simulation of plasma response to an applied electric field parallel to magnetic fieldlines. J. Geophys. Res. 108 (A5), 1197. https://doi.org/10.1029/2002JA009573.

Panofsky, W.K.H., Phillips, M., 1962. Classical Electricity and Magnetism, second ed. Addison-Wesley, Reading, MA.

Parks, G.K., 2004. Why space physics needs to go beyond the MHD box. Space Sci. Rev. 113, 97–125.

Paschmann, G., 2008. Recent in-situ observations of magnetic reconnection in near-earth space. Geophys. Res. Lett. 35, L19109. https://doi.org/10.1029/2008GL035297.

Pontin, D.I., 2011. Three-dimensional magnetic reconnection regimes: a review. Adv. Space Res. 47, 1508–1522.

Poynting, J.H., 1884. XV. On the transfer of energy in the electromagnetic field. Philos. Trans. R. Soc. 175, 343–361.

Priest, E.R., Hornig, G., Pontin, D.I., 2003. On the nature of three-dimensional magnetic reconnection. J. Geophys. Res. 108, 1285.

Reitz, J.R., Milford, F.J., Christy, R.W., 1993. Foundations of Electromagnetic Theory, fourth ed. Addison-Wesley, Reading, MA.

Schindler, K., Hesse, M., Birn, J., 1988. General magnetic reconnection, parallel electric fields, and helicity. J. Geophys. Res. 93, 5547–5557. https://doi.org/10.1029/JA093iA06p05547.

Schmidt, G., 1960. Plasma motion across magnetic fields. Phys. Fluids 3, 961.

Schmidt, G., 1979. Physics of High Temperature Plasmas, second ed. Academic Press, New York, NY.

Vasyliunas, V.M., 1975. Theoretical models of magnetic field line merging, 1. Rev. Geophys. Space Phys. 13, 303–336.

Wangsness, R.K., 1986. Electromagnetic Fields, second ed. Wiley, New York, NY.

6

Magnetopause

Chapter outline

Earth's Magnetosphere. https://doi.org/10.1016/B978-0-12-818160-7.00006-5

215

There is always a strong inclination for a body of professionals to oppose an unorthodox view. Such a group has a considerable investment in orthodoxy ... To think the whole subject through again when one is no longer young is not easy and involves admitting a partially misspent youth. Clearly it is more prudent to keep quiet, to be a moderate defender of orthodoxy, or to maintain that all is doubtful, sit on the fence, and wait in statesmanlike ambiguity for more data.

Bullard (1975)

6.1   Introduction

Quite often, the course of some human endeavor is greatly advanced by one person with imagination and foresight. Such has been the case with solar-terrestrial physics: Dungey (1961) recognized that the interplanetary magnetic field (IMF), although weak, might influence and perhaps determine the nature of solar wind coupling with the magnetosphere. His foresight has inspired much of the research in this field for nearly a half a century and has even affected thinking in related fields from laboratory plasma experiments to astrophysics.

Unfortunately, even at this late stage, it is still not clear what are the actual processes involved. As summarized by Sonnerup:

Observational information concerning the location of reconnection sites on the magnetopause is incomplete but suggests a more complicated picture than originally envisaged, possibly with multiple sites and moving reconnection lines; the controversy over component merging versus antiparallel merging remains unresolved.

Sonnerup et al. (1995)

In short, the simple Dungey model requires modification, even serious reconsideration; this in no way diminishes the significance of his foresight referred to earlier. The opportunity now exists to evolve a distributed observatory to meet the goals of space research. The power of simultaneous observations at multiple vantage points has been clearly demonstrated, for example, with the Cluster, Themis, and MMS missions. What is offered are excellent ground-based instruments and a fleet of widely deployed geospace, planetary, solar, and heliospheric spacecraft working together to help understand solar activity, its interaction with geospace, and other planetary systems throughout the solar system.

In this chapter we shall explore, from first principles, some of these questions as applied to the magnetopause. It is at the magnetopause that these questions are met head on, directly and most simply, with plenty of observations. If we cannot understand the magnetopause, what hope do we have in explaining magnetospheric physics in general with more sparse data? We first note some key observations. We will then use Poynting's theorem, reviewed in Chapter 5, as the starting point.

- Only three processes need to be considered, the three terms on the right-hand side in Poynting's theorem.
- The term with the Poynting vector, without explicit time dependence, is central to magnetic reconnection, as presently defined; it was thought that it brings in the required energy from somewhere.
- The second term, time dependent and three-dimensional, draws upon electric energy; that is, kinetic energy of the plasma's bulk motion.
- The third term, also time dependent in three dimensions, invokes magnetic energy. It is the magnetic field that holds everything together; intense events happen when this field loses control.

The standard reconnection model (SMR) with an X-line has serious flaws, even more serious than expressed by Sonnerup. We must use time-dependent terms, in three dimensions, right from the start.

6.2 Solar wind-magnetopause interaction

From an experimental point of view, the past 60 years of space exploration have permitted the accumulation of enough measurements to calculate typical plasma parameters at the dayside magnetopause (see review by Haerendel and Paschmann (1982), and the conference on the Physics of the Magnetopause with Song et al. (1995) serving as editors). Magnetosheath plasma flow speeds vary from one hundred to a few hundred kilometers per second. These velocities are comparable with the sound speed, $v_s \sim 200\text{--}400\,\text{km/s}$, but remain larger than the Alfvén velocity (v_A) which is of the order of $100\,\text{km/s}$ for average density and magnetic field $(n_{MS} = 10\,\text{cm}^{-3},\ B_{MS} = 15\,\text{nT})$.

Theoretical treatments are almost exclusively based on the assumption that the solar wind can be adequately described by the continuum equations for a perfect gas. The fundamental MHD equations based on conservation of mass, momentum, and entropy, together with the modified form of Maxwell's equations, were adopted. An equation of state is also needed because the number of MHD equations is less than the number of unknowns (see Chapter 4). The bow shock and magnetopause were regarded as ideal discontinuities, treated by solving jump relations matching the continuous fluid solutions on either side. Modeling the flow around the magnetopause with an adiabatic assumption of $\gamma = 5/3$ appears to be the best possible one within the limitations of the one-fluid approach. However, this pertains to motions transverse to the magnetic field; parallel flows are difficult to treat, at best.

Spreiter and Alksne (1968, 1969) have simplified, considerably, the task of calculating the flow around the magnetosphere by dropping all terms containing the magnetic field **B** from the fluid equations, solving Maxwell's equations separately (see Chapter 1). This approach is justified by the observation that the flow velocities, except near the stagnation point, are largely super-Alfvénic with little influence by the weak magnetic field. Once **v** has been found, the frozen-in magnetic field (which is thought to be passively transported by the fluid) can be calculated.

6.2.1 Superposed epoch analysis

The obvious thing to do is to map various parameters within the magnetopause current sheet, but detailed examinations of the magnetopause itself are difficult. The magnetopause is usually thin, and it moves in or out like a surface wave with substantial speed past a satellite. Conditions are constantly changing owing to its nonstationary nature. This is a serious obstacle because of the limited time resolution of the data. The typical duration of a magnetopause crossing is usually no more than a few seconds; in many cases, this is less than the time required for a plasma measurement.

In order to study the variations of key plasma parameters and the magnetic field in the magnetopause and the adjacent magnetosheath in a systematic way, a superposed epoch analysis was performed by Phan et al. (1994) using AMPTE/IRM data for 69 crossings of the magnetopause. The averages hopefully would be more meaningful than the individual samples. Ideally, one would prefer to examine the spatial profiles. However, this would imply translating time scales into spatial scales; this requires knowledge of the motion of the region relative to the spacecraft which is not possible with a single spacecraft. Instead, an analysis of temporal profiles can be done with the expectation that the resulting profiles will approximate the spatial variations in the region.

For a quantitative analysis of the reconnection process at the magnetopause we require, in addition to the criteria already stated earlier, that all the plasma parameters and the magnetic field in the magnetosheath near the magnetopause be rather stable. This condition is necessary in order to have a good baseline against which the variations across the magnetopause can be measured. This restriction reduced the total number of suitable events to 69. Because of the nature of the AMPTE/IRM orbit, all 69 crossings occur within 30 degrees of the equatorial plane, with most northern (southern) hemisphere crossings before (after) local noon. This effect tends to convolve local time effects with latitude effects.

First, a key time is defined, the magnetopause crossing time. For high-shear cases the key time is defined by the maximum magnetic field rotation rate encountered by the satellite as it goes from the low-latitude boundary layer (LLBL) to the magnetosheath or vice versa. For the low-shear case, the magnetopause is identified by the net changes in proton and electron temperatures and their anisotropies. Phan et al. (1994) included 20 min of data on the magnetosheath side and 2 min on the boundary layer/ magnetosphere side of the key time. The choice of 20 min was made because the magnetosheath proper is generally reached within 20 min after (or prior to) the magnetopause encounter. For outbound crossings, the order of the time series is reversed in the analysis so that the magnetosheath is on the left-hand side of our key time in all cases. The following discussions are given in terms of inward motion of the spacecraft from the magnetosheath toward the magnetosphere.

Since the absolute magnitude of some parameters may vary considerably from one case to the next, and since the focus of this study is on the magnetosheath, they normalized the averages of these parameters over the entire magnetosheath interval to one. Without such a renormalization, cases with larger-than-average magnitude would dominate. They superimposed all crossings by adding them logarithmically, using 10-s time bins. Logarithmic (or geometric) averaging reduces the dominance of cases with larger-than-average dynamic range. A total of 13 low-shear cases and 25 high-shear cases satisfied the earlier stated selection criteria for use in the superposed epoch analysis shown in Fig. 6.1. Note that the magnetic shear angle across the 13 low-shear magnetopauses ranges from about 0 to 30 degrees, while across the 25 high-shear magnetopauses it ranges over a much broader range from 60 to 180 degrees. The final plasma and magnetic field profiles corresponding to low and high magnetic shears are shown side by side to facilitate comparison. Numerical values for the final profiles averaged over two 10-min intervals in the magnetosheath are shown together with the extrema of the individual profiles averaged over the same time intervals. The first 10-min interval, from 20 to 10 min before the encounter of the magnetopause, is somewhat representative of the condition of the magnetosheath proper; the second interval comprises the 10 min immediately preceding the magnetopause crossing. Substantial differences in the properties of plasma and magnetic field in these two intervals would indicate the presence of a magnetosheath transition layer adjacent to the magnetopause.

The key times of the low-shear and high-shear crossings are defined by the times of the proton temperature change and maximum magnetic field rotation, respectively. The magnetosheath is to the left and the magnetosphere is to the right of the key time. The low-shear cases are displayed in the left panels, and the high-shear cases are displayed in the right panels. From top to bottom in Fig. 6.1, the figure displays the magnetic field rotation angle φ_B, the magnetic pressure P_B, the perpendicular thermal pressure $P_{p\perp}$, the total pressure P_{tot}, the plasma β, and the field elevation angle λ_B. The error of the mean is indicated by vertical bars.

6.2.1.1 Magnetic field rotation on magnetopause

The field rotation from the magnetospheric direction to the magnetosheath direction starts earthward of the key time. In the low-shear case (left panels), the gradual rotation continues well into the magnetosheath region, generally extending several

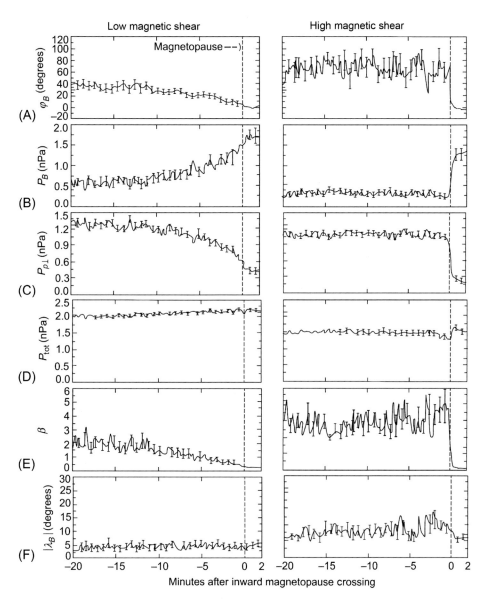

Fig. 6.1 Superposed epoch analysis of 13 low-shear and 25 high-shear magnetopause crossings. The key times of the low-shear and high-shear crossings are defined by the times of the proton temperature change and maximum magnetic field rotation, respectively. The magnetosheath is to the left and the magnetosphere is to the right of the key time. The low-shear cases are displayed in the left panels, and the high-shear cases are displayed in the right panels. From top to bottom, the figure displays (A) the magnetic field rotation angle φ_B, (B) the magnetic pressure P_B, (C) the perpendicular thermal pressure $P_{p\perp}$, (D) the total pressure P_{tot}, (E) the plasma β, and (F) the field elevation angle λ_B. From Phan, T.-D., Paschmann, G., Baumjohann, W., Sckopke, N., Lühr, H., 1994. The magnetosheath region adjacent to the dayside magnetopause: AMPTE/IRM observations. J. Geophys. Res. 99(A1), 130. doi:10.1029/93JA02444.

minutes into that region. Since this rotation occurs in the magnetic field pileup region, a significant amount of current is present there. In the high-shear case (right panels), on the other hand, 90% of the field rotation occurs abruptly, on average within one sampling interval of the plasma instrument, 4.35 s. The abrupt rotation generally takes place where the field is weak; a more gradual rotation occurs in the LLBL, where the field is stronger. The duration of a satellite crossing of a high-shear current layer is therefore much shorter than that of a low-shear one. This implies that even after taking into account the fact that on average the low-shear magnetopause moves three times more slowly than the high-shear one (Paschmann et al., 1993), the region of current flow in the low-shear case is still at least an order of magnitude thicker than for high shear. However, they do not believe that the extended region of field rotation for the low-shear case should be considered part of the magnetopause.

6.2.1.2 Magnetic pressure

The magnetosheath magnetic pressure P_B increases toward the low-shear magnetopause, indicating a pileup of magnetic field against the magnetopause. The pileup is most significant close to the magnetopause, where an associated decrease in plasma density and pressure occurs. The situation is fundamentally different adjacent to a high-shear magnetopause, where a constant value or gradual decrease of the magnetic field toward the magnetopause is generally observed. Immediately earthward of the key time, the magnetic pressure increases drastically. It should also be noted that as one moves from the bow shock toward the magnetopause, the magnetic pressure always increases.

6.2.1.3 Perpendicular plasma pressure

A magnetosheath transition layer is observed (Fig. 6.1) adjacent to the low-shear magnetopause, where the average plasma pressure and density decrease gradually as one moves toward the magnetopause. The decrease is more important in the 10-min interval immediately preceding the magnetopause encounter, where the magnetic pressure is enhanced. On average, the plasma pressure at the magnetopause is half of its value in the magnetosheath proper. The transition layer is not observed adjacent to the high-shear magnetopause: in this case, the plasma pressure and density remain rather constant until the magnetopause key time is reached.

6.2.1.4 Total pressure

In the low-shear case, both perpendicular plasma pressure and magnetic pressure change significantly across the magnetopause and magnetosheath regions but their sum remains rather constant throughout these regions except for a slight trend of decreasing total pressure as one moves away from the magneto-pause. This trend can be explained by the fact that the majority (11 of 13) of the low-shear crossings included in this study are inbound; these crossings usually occur as a result of the magne-topause moving outward in response to a temporal decrease in the magnetosheath total pressure. In the high-shear case, the magnetosheath magnetic and plasma pressures both remain rather uniform in the entire region within the 20 min preceding the magnetopause crossing, so that P_p is also constant, except for a distinct decrease just outside the magnetopause. Across the magnetopause, the plasma and magnetic pressures vary sig-nificantly, and their sum generally has a small jump across this boundary: a deficiency of P_{tot} on the magnetosheath side and an excess on the LLBL side of the magnetopause are often observed.

6.2.1.5 Plasma β

As the magnetic pressure increases and the plasma pressure drops on approach to the low-shear magnetopause, β ($=P_p/P_B$) decreases appreciably in the magnetosheath transition layer. The level of fluctuation, shown by the sizes of the error bars, is much lower in the magnetosheath transition layer than in the magnetosheath proper. In the high-shear case, the plasma β is generally high and fluctuates throughout the magnetosheath region all the way to the magnetopause. The average value of β is about 0.4 immediately outside the low-shear magnetopause and about four adjacent to the high-shear (rising just before the magnetopause).

6.2.2 Magnetic field normal to magnetopause

The angle λ_B measures the angle between the magnetic field vector and the plane of the magnetopause. The magnetic field is parallel and perpendicular to the magnetopause when $\lambda_B = 0$ and 90 degrees, respectively. Since λ_B is as likely to be positive as negative, $<\lambda_B> = 0$. The more relevant quantity to examine is its absolute value $|\lambda_B|$. In the low-shear cases, $|\lambda_B| = 5°$ throughout the near-magnetopause magnetosheath region; that is, the mag-netic field lies mainly in a plane parallel to the low-shear magnetopause. In fact, $|\lambda_B|$ is close to zero all the way to the bow shock in all cases, indicating that it is the bow shock that rotates the magnetic field into a plane parallel to both the bow shock and the dayside magnetopause. In the high-shear case, $|\lambda_B|$ is again very small in the magnetosheath proper. However, closer to the high-shear magnetopause, $|\lambda_B|$ is larger and generally more variable. Since the normal directions used for each event in this study are derived from a model rather than determined by minimum-variance analysis of the actual data, this larger value may be an indication that high-shear magnetopauses are generally more curvy than low-shear ones, which is consistent with other results. The curvature of the magnetopause may be the result of surface waves and/or a time-variable reconnection rate.

6.2.2.1 *Plasma velocity tangential to magnetopause*

The component of the magnetosheath plasma velocity tangential to the magnetopause V_t increases gradually on approach to the low-shear magnetopause (Fig. 6.2A). Phan et al. (1994) compute the parallel and perpendicular components of the flow relative to a fixed reference magnetic field direction, choosing the magnetosheath field immediately outside the magnetopause as the reference. The results are shown in panels (B) and (C) for $V_{t\perp}$ and $V_{t\parallel}$, respectively, and also in terms of the angle $(V_{t\perp}/V_{t\parallel})$ (panel D), which measures the relative angle between the flow and the magnetic field. This angle is 0 degrees (90 degrees) when the flow is parallel (perpendicular) to the reference magnetic field.

The result shows that initially, the average flow is directed at 45 degrees to the field. As one approaches the magnetopause, the flow rotates to become more perpendicular to the reference field. Close to the magnetopause, it is about 67.5 degrees. Across the magnetopause, the flow direction changes abruptly. This change seems to indicate that the definition of the low-shear magnetopause, based on changes in thermal properties of the plasma, also corresponds to a topological boundary. Inspections of the individual events show that when the tangential flow is already perpendicular to the magnetic field in the magnetosheath proper, it remains so going into the magnetosheath transition layer.

This is in strong contrast to the high-shear magnetopause; the magnetosheath flow direction is more variable and changes strongly from case to case. A sudden enhancement of the tangential flow is sometimes observed in, and earthward of, the magnetopause current layer; this has been observed on numerous occasions notably by Mozer and Pritchett (2010) concerned with auroral motions in the low-latitude boundary layer (Chapter 9).

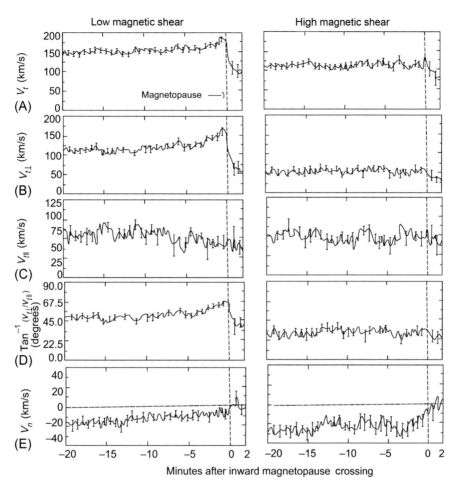

Fig. 6.2 Superposed epoch plots of the tangential-to-magnetopause component of (A) the bulk plasma flow v_t, (B) the components of the tangential flow perpendicular and (C) parallel to the reference field $v_{t\perp}$ and $v_{t\parallel}$, (D) the relative angle between the flow and the reference field $\tan^{-1}(v_{t\perp}/v_{t\parallel})$, and (E) the bulk velocity normal to the magnetopause v_n. Note the flow enhancement and rotation of the tangential flow on approach to the low-shear magnetopause. Also, note the nearly linear decrease of $|v_n|$ on approach to the low-shear magnetopause and higher $|v_n|$ adjacent to the high-shear magnetopause. From Phan, T.-D., Paschmann, G., Baumjohann, W., Sckopke, N., Lühr, H., 1994. The magnetosheath region adjacent to the dayside magnetopause: AMPTE/IRM observations. J. Geophys. Res. 99(A1), 134. doi:10.1029/93JA02444.

6.2.2.2 Plasma depletion in the transition layer

The plasma depletion phenomenon in the magnetosheath transition layer, whose signatures are a decrease of plasma density and a simultaneous increase of magnetic pressure toward the magnetopause, is shown in this systematic survey to be present only for the low-shear magnetopause. Generally, no changes in the plasma parameters are observed adjacent to the high-shear magnetopause. Of the 13 low-shear cases examined in this study, all show the magnetosheath region adjacent to the magnetopause to be of lower density and higher magnetic pressure than in the magnetosheath proper, although the degree of plasma depletion varies from case to case.

> *Remarkably, none of the 25 high-shear cases examined shows evidence for lower density and higher magnetic pressure immediately outside the magnetopause, even in cases in which accelerated flows, which are normally associated with magnetic reconnection, are not observed at the magnetopause. It should be noted that further into the magnetosheath, a quasi-monotonic increase of the average magnetic pressure as one moves toward the magnetopause is always present. However, this increase is balanced not by a simultaneous decrease in the plasma pressure but mainly by a decrease of the dynamic pressure.*
>
> **Phan et al. (1994, p. 135)**

6.2.2.3 *Plasma bulk velocity normal to magnetopause*

The normal velocity v_n is negative throughout most of the magnetosheath region, indicating simply that the plasma flows toward the magnetopause. v_n decreases almost linearly on approach to the low-shear magnetopause, as would be expected in a flow around a rigid obstacle. It should be noted that the decrease in v_n is less prominent than the increase of the tangential velocity v_t, in the immediate vicinity of the magnetopause. As a result, the bulk speed rises on approach to the magnetopause. Right at the magnetopause, v_n vanishes.

Adjacent to the high-shear magnetopause, v_n is generally very variable, as shown by the large error bars, and more negative. At the magnetopause, v_n has a finite value, $v_n \sim -10\,\text{km/s}$, which is consistent with the magnetopause being more open.

An astounding result is shown by Fig. 6.3: there is a significant difference between inbound and outbound crossings of the normal plasma velocity v_n for high shear. This topic of the velocity normal to the magnetopause was again addressed by Phan and Paschmann (1996) within the current layer itself. They found the superposed epoch analysis of the bulk velocity normal to the magnetopause v_n differed for (a) inbound and (b) outbound crossings of the magnetopause. For an inbound crossing the normal plasma velocity v_n is consistent with the idea that the plasma just follows the moving magnetopause. In contrast, for an outbound pass (i.e., where the magnetopause is moving inward), it

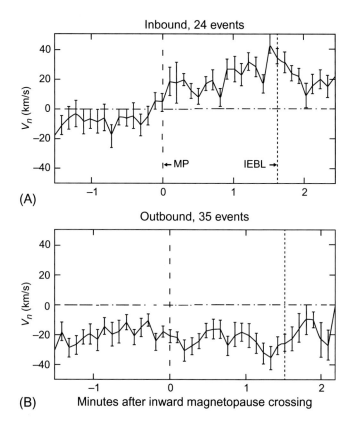

(A)

(B) Minutes after inward magnetopause crossing

Fig. 6.3 Superposed epoch analysis of the bulk velocity normal to the magnetopause v_n for (A) inbound and (B) outbound crossings of the magnetopause/LLBL for high shear. Note the difference in the sign of v_n measured in the LLBL and the outer magnetosphere for the two subsets. From Phan, T.-D., Paschmann, G., 1996. The low-latitude dayside magnetopause and boundary layer for high magnetic shear: 1. Structure and motion. J. Geophys. Res. 101, 7808. doi:10.1029/95JA03752.

is consistently antisunward, $v_n \sim$ 20–30 km/s. The value of 20 km/s for the normal component of plasma velocity is highly significant; it is large compared with the motion of the magnetopause itself. The LLBL is assumed here to be the region between the magnetopause and the inner edge of the boundary layer (IEBL), defined as the location where the density has dropped to 5% of its magnetosheath value.

6.2.2.4 Dawn/dusk asymmetries

The orbit of the AMPTE/IRM spacecraft was such that local time effects on the magnetopause flanks could not be established. More recently, data from the Cluster (Haaland et al., 2014) and THEMIS (Haaland et al., 2019) multisatellite missions have been able to show statistically that a dawn/dusk asymmetry also exists.

Haaland et al. (2014) identified about 5800 times from 2001 to 2010 when one or more of the Cluster satellites crossed the magnetopause. Utilizing both single- and multispacecraft analysis techniques, they were able to calculate velocity, orientation, and thickness of the magnetopause. Between 2001 and 2006, they were also able to use the four-spacecraft curlometer method to derive detailed current profiles for a number of crossings. After that time, the spacecraft configuration at the magnetopause was not suitable for such calculations. The results showed that the dawn-side magnetopause was consistently thicker (1699 km or 18.9 ion gyroradii on average), with a lower current density (34.7 nA/m^2), while the dusk side was thinner (1496 km or 14.6 ion gyroradii) with a higher current density (41.8 nA/m^2). These two effects tended to cancel out such that the total current carried on the two flanks was approximately equal.

THEMIS multisatellite observations between 2007 and 2009 (Haaland et al. 2019) are consistent with the Cluster data. Although the total number of identified magnetopause crossings was much smaller than with Cluster (538 crossings), THEMIS also saw a thicker average magnetopause on the dawn side (1407 km or 14.9 ion gyroradii), and a thinner average magnetopause on the dusk side (1149 km or 8.0 ion gyroradii). In this case, however, the current densities were very similar (16.7 nA/m^2 dawn, 16.3 nA/m^2 dusk). Both studies also saw an asymmetry in the plasma velocity, with the dawn-side plasma being 30% (THEMIS) to 60% (Cluster) faster.

The main underlying cause of this local time asymmetry is not yet understood. The solar wind dynamic pressure or the orientation of the IMF cannot alone explain it. Other external factors, in the form of different plasma properties in the dusk and dawn magnetosheath most likely play a significant role, while properties and processes inside the magnetosphere probably also contribute. An intrinsic asymmetric dependence of magnetopause structure on the sense of the flow relative to the magnetic field may also play a role.

6.3 ISEE observations

An electric field variation in the boundary layer was deduced by Heikkila (1982b) using Mozer's instrument on ISEE. The electric field instrument consists of two spherical sensors separated by 73.5 m in the spin plane, nearly coinciding with the ecliptic. In the preferred mode, the sensors are biased positive relative to the satellite. The voltage difference is measured as shown in Figs. 6.4 and 6.5. A more complete description, and an overview of the pass, were reported by Mozer et al. (1978). Magnetic field measurements are used to indicate the magnetopause current

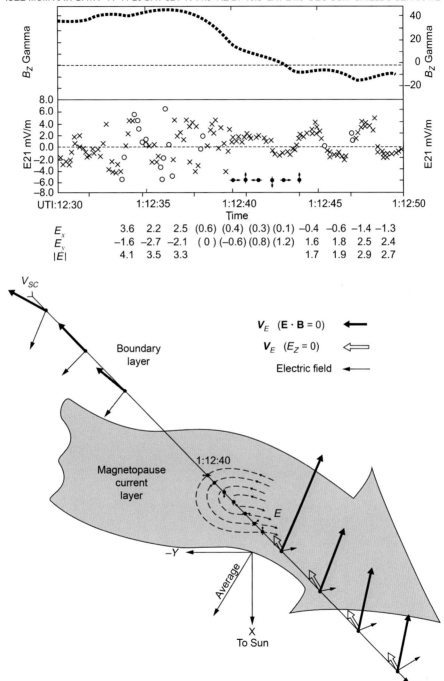

Fig. 6.4 The electric field data, and $\mathbf{E} \times \mathbf{B}/B^2$ assuming that $\mathbf{E} \cdot \mathbf{B} = 0$ or $E_z = 0$, for the exit crossing. The velocity component of the flow is everywhere in the earthward direction, with the plasma flow faster than the motion of the magnetopause. The plasma convection is inward at all times. From Heikkila, W.J., 1982. Inductive electric field at the magnetopause. Geophys. Res. Lett. 9(8), 879.

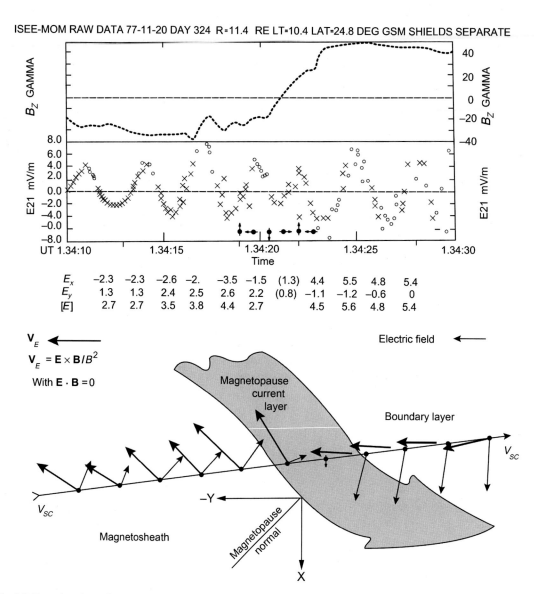

Fig. 6.5 Boundary layer flow pushes the magnetopause to the left. The magnetosheath plasma has to flow tangentially to the magnetopause in front because of continued solar wind flow. From Heikkila, W.J., 1982. Inductive electric field at the magnetopause. Geophys. Res. Lett. 9(8), 879.

sheet: the magnetometer is described by Russell (1978). Magnetopause crossings were chosen to be free of interference. The data have been fitted to a sinusoidal function for a single spin period of 3.04 s advancing by a half-spin to obtain reliable data points for the electric field vector separated by 1.5 s.

The chosen exit from the magnetosphere occurred at 1:12:42 UT (see Fig. 6.4). The fitting procedure yielded three consistent electric field measurements in the boundary layer. The deduced plasma flow was mainly earthward and toward the dawn side in the boundary layer. At the beginning of the interval when the satellite entered the current layer at 38–40 s the data points are somewhat scattered. After 01:12:40 UT in the middle of the current layer the instrument showed an almost constant reading of 1.5 mV/m for about 2 s, during which time the satellite spin carried the booms about two-thirds of a turn. This measurement is well above the possible errors (a few tenths of a mV/m); there was very little scatter in these data points, indicating very little turbulence. The boom orientation is shown for this period, with an upward arrow being in the x direction. From a single instrument it is impossible to tell whether the booms were pointed at a constant direction to an electric field of constant magnitude, or at some changing angle to a changing field; however, there is no question that the field vector is rotating since the data points do not cross the zero line for at least two-thirds of a spin. Thus this unique observation indicates that the electric field vector is turning, roughly at the satellite spin rate. Thereafter, the electric field was pointed mainly in the $+y$ and $-x$ direction in the magnetosheath, as appropriate to antisunward flow with a southward IMF; this is true even in the outer part of the current layer.

The electric field vectors, and the drift velocity, are plotted with the assumption that the spacecraft velocity through the magnetopause perturbation was an inward movement of 20 km/s and a duskward movement of 150 km/s. Determination of the drift velocity would require the simultaneous measurement of all three components of the electric field vector. As such data is not available some assumptions must be made, and we assume that $\mathbf{E}\cdot\mathbf{B}=0$; an alternative is $E_z=0$, also shown. This flow vector is shown as a solid arrow every half-spin (1.5 s) whenever these are reliable.

An example of entrance into the magnetosphere is shown in Fig. 6.5 at 1:34:21 UT. The exact times when the spin plane containing sensor No. 1 was in the solar direction are indicated by tick marks at the bottom of the abscissa axis. Our interpretation is that in Fig. 6.5 there is a strong flow across the magnetopause, in agreement with Fig. 6.3.

6.4 Profile of magnetopause electron temperature

Fig. 6.6 shows the electron density and temperature profile across the magnetopause. The partial energetic (1.8–30 keV) electron density (Fig. 6.6A) is much higher in the magnetosphere than

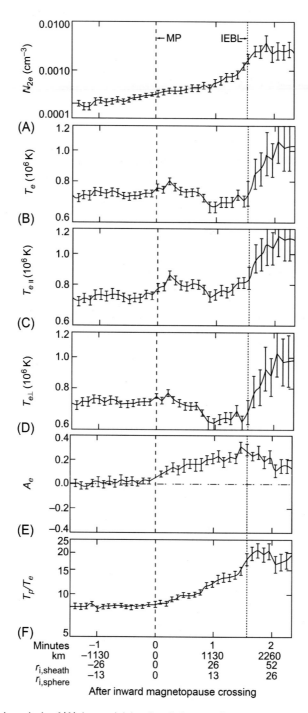

Fig. 6.6 Superposed epoch analysis of (A) the partial density of electrons between 1.8 and 30 keV, N_{2e}, (B) the electron temperature T_e, (C) the parallel $T_{e\parallel}$ and (D) perpendicular, $T_{e\perp}$ temperatures, (E) temperature anisotropy $A_e = T_{e\parallel}/T_{e\perp} - 1$, and (F) the proton-to-electron ratio T_p/T_e. Note the depletion of magnetospheric energetic electrons at the IEBL. From Phan, T.-D., Paschmann, G., 1996. The low-latitude dayside magnetopause and boundary layer for high magnetic shear: 1. Structure and motion. J. Geophys. Res. 101, 7807. doi:10.1029/95JA03752.

in the LLBL and the magnetosheath, which is indicative of their magnetospheric origin. As one moves from the magnetosheath toward the magnetosphere, the steepest increase in this plasma population does not occur at the magnetopause key time but near the IEBL key time, where cold magnetosheath plasma first appears. In Fig. 6.6E the temperature anisotropy parameter, which is around zero in the magnetosheath, becomes positive near the magnetopause and remains so throughout the LLBL.

6.4.1 Dynamo versus electrical load

If dissipation is implied by an increase in T_e, an electrical load with $\mathbf{E}\cdot\mathbf{J}>0$, then a reduction in T_e implies a dynamo with $\mathbf{E}\cdot\mathbf{J}<0$. Since the sense of the current is known by Ampere's law (dawn to dusk), the electric field must reverse within the current sheet. Phan et al. (1996) present velocity changes across the magnetopause on the crossing on November 1, 1984. The spacecraft moved across the magnetopause/LLBL during the crossing.

6.4.2 One detailed crossing

To illustrate their method they present details on one crossing shown in Fig. 6.7. The flow speed increases as the spacecraft moves earthward across the magnetopause and roughly doubles at $t=22$ s (vertical dashed line) before it reaches the magnetospheric level at the outer edge of the LLBL at the right edge of the figure.

What really matters is the change in flow velocity vector, not just the change in speed. To illustrate this, Fig. 6.7B shows two curves labeled $|\Delta\mathbf{V}_{\mathrm{obs}}|$ and $|\Delta\mathbf{V}_{\mathrm{th}}|$. The magnitude of the measured change in flow velocity $|\Delta\mathbf{V}_{\mathrm{obs}}|$ relative to the velocity at the magnetosheath reference point is seen to increase from below 50 km/s to about 245 km/s before it drops to low values while still inside the current layer (as seen in the bottom panel, F). The much larger dynamic range is due to a substantial rotation of the flow velocity. The second curve in Fig. 6.7B, labeled $|\Delta\mathbf{V}_{\mathrm{th}}|$ is the change in velocity one would expect from tangential stress balance across a one-dimensional rotational discontinuity (RD), as given by the Walén relation (Hudson, 1970):

$$\Delta\mathbf{V}_{th} = \pm[(1-\alpha_1)/\mu_0\rho_1]^{1/2}[\mathbf{B}_{2t}(1-\alpha_2)/(1-\alpha_1)-\mathbf{B}_{1t}] \qquad (6.1)$$

where $\alpha=(P_\parallel-P_\perp)\mu_0/B^2$ is the anisotropy factor, computed from the measured parallel and perpendicular pressures and the magnetic field $\mathbf{B}$. The mass density ρ is based on the measured number density, assuming a mixture of 95% protons and 5% alpha

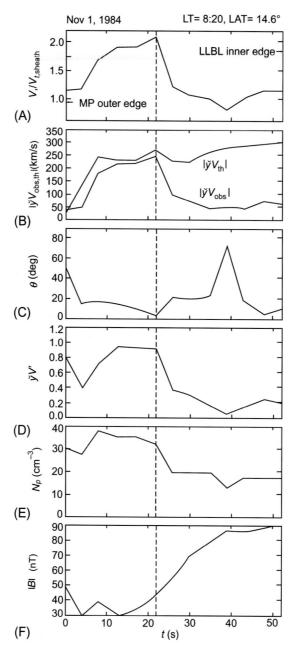

Fig. 6.7 Example of magnetopause/LLBL crossing. The (relative) time runs from the outer edge of the magnetopause to the inner edge of the LLBL. (A) Measured tangential flow speed normalized to the average speed in the magnetosheath reference interval; (B) magnitude of the observed change in flow velocity ΔV_{obs} (bottom curve) and the change $|\Delta V_{th}|$ (top curve) predicted by the Walén relation; (C) angle θ between ΔV_{obs} and ΔV_{th}; (D) scalar measure of the agreement between the two velocities, ΔV^*. Note that the maximum in $|\Delta V_{obs}|$ *(dashed line)* coincides with a local minimum in θ where ΔV^* is also close to its maximum value, (E) density and (F) magnetic field profiles. From Phan, T.-D., Paschmann, G., Sonnerup, B.U.Ö., 1996. The low-latitude dayside magnetopause and boundary layer for high magnetic shear: 2. Occurrence of magnetic reconnection. J. Geophys. Res. 101(A4), 7818. doi:10.1029/95JA03751.

particles. In Eq. (6.1), the subscript t denotes components along the magnetopause and, assuming earthward plasma flow across the magnetopause, the plus and minus signs distinguish crossings north and south of the reconnection line, respectively. The indices 1 and 2 refer to the reference interval in the magnetosheath and to a point within the magnetopause/LLBL, respectively.

Eq. (6.1) was derived under the assumption that the quantity $\rho(1-\alpha)$ is constant across the magnetopause. Verification of this condition requires plasma composition information which is not available on AMPTE/IRM.

As Fig. 6.7(B) illustrates, $|\Delta \mathbf{V}_{th}|$, computed according to Eq. (6.1) from the measured magnetic field, plasma density, and pressure anisotropy α, is seen to rise much like $|\Delta \mathbf{V}_{obs}|$, reaching 270 km/s at the time it peaks. The authors excuse the fact that $|\Delta \mathbf{V}_{th}|$ continues to stay high after this time as due to an artifact of the use of Eq. (6.1) beyond its region of relevance. That is, its use where there is significant mixing of magnetospheric and magnetosheath plasma, when it is valid only for plasma that has actually penetrated the magnetopause and experienced the associated magnetic stresses.

Fig. 6.7C demonstrates that the angle between $|\Delta \mathbf{V}_{obs}|$ and $|\Delta \mathbf{V}_{th}|$ is small, $<20°$, when $|\Delta \mathbf{V}_{obs}|$ is large, reaching a minimum ($<3°$) at the time when $|\Delta \mathbf{V}_{obs}|$ reaches its maximum ($t=22\,\mathrm{s}$). One concludes that there is good agreement between measured and predicted velocity changes, particularly at the time of maximum measured flow speed.

To get a single, scalar, and dimensionless measure of agreement between observed and predicted flows, they compute a reconnection quality factor, $\Delta V^* = \Delta V_{obs} \cdot \Delta V_{th}/|\Delta V_{th}|^2$, which represents the projection of $\Delta \mathbf{V}_{obs}$ onto $\Delta \mathbf{V}_{th}$ normalized to $|\Delta \mathbf{V}_{th}|$. The sign is chosen such that ΔV^* is nonnegative. With this definition, $\Delta V^*=0$ occurs whenever $\Delta \mathbf{V}_{obs}$ is either 0 or is perpendicular to $\Delta \mathbf{V}_{th}$. On the other hand, $\Delta V^*=1$ indicates either perfect agreement in magnitude as well as direction, or in case the two vectors are not aligned, agreement between their projection. Fig. 6.7(D) demonstrates that for the example chosen ΔV^* exceeds 0.9 in the interval $13\,\mathrm{s}<t<22\,\mathrm{s}$ and then drops off rapidly for $t>22\,\mathrm{s}$, that is, beyond the time of the maximum.

A feature that is seen in this example and is typical of many others is the rapid drop in ΔV^* immediately earthward of the location of maximum $|\Delta \mathbf{V}_{obs}|$, which is marked by the dashed line in Fig. 6.7. The actual situation is more complicated.

While recent observational efforts have revealed many new features and consequences of various magnetopause processes,

"... they leave crucial questions about the interplay between the upstream magnetosheath conditions and the structure and dynamics of the magnetopause boundary unanswered. ... Various speculative scenarios can be imagined, but a firm understanding of these interrelations is not presently at hand."

At present, no analytical or numerical reconnection model exists which incorporates [all of the possible effects]. However, it appears certain that this point is located within a region of open field lines which probably extends substantially earthward of the point to incorporate essentially the entire time interval in which intermixed plasma populations are observed.

Phan et al. (1996)

Finally, the relationship between local magnetopause processes and signatures on the one hand and the global interaction between the shocked solar wind and the magnetosphere on the other remains largely unexplored. We will continue this debate in Section 6.12, at the end of this chapter.

6.5 Impulsive penetration

The idea that solar wind plasma-field irregularities with an excess momentum density penetrate deeper into the geomagnetic field was introduced by Lemaire (1977) and Lemaire and Roth (1978). This can occur even with tangential discontinuities, as rationalized by Lemaire (1985). Plasma blobs or filaments of magnetosheath plasma inside the dayside magnetopause have been seen by a number of investigators using a variety of spacecraft and simulations (Heikkila, 1982a; Lundin and Dubinin, 1984; Lundin, 1997; Echim and Lemaire, 2000; Lundin et al., 2003; Gunell et al., 2012, 2014; Lyatsky et al., 2016a,b). The concept of impulsive penetration is based on the observation that the solar wind is generally patchy over distances smaller than the diameter of the magnetosphere. It is indeed unusual to find magnetograms of periods of more than 30 s when the IMF does not change at least by a few percent. The presence of these plasma irregularities indicates that the solar wind momentum density is not uniform, and that the dynamic pressure the solar wind inflicts upon the magnetosphere must be patchy, nonuniform, rapidly changing in time. The magnetopause itself might be in motion, perhaps as a surface wave. It is not proper to assume that these effects are not important.

6.5.1 Plasma entry across the magnetopause

Plasma entry requires that the electric field reverse across the magnetopause in step with the reversal of the magnetic field for continued tailward flow in the LLBL.

The existence of a low-latitude boundary layer demonstrates that magnetosheath plasma can penetrate into the magnetosphere. How this actually happens has remained an unresolved question in magnetospheric physics, despite numerous studies based on single-spacecraft observations.

De Keyser (2005)

One of the key factors is indicated by the question mark at the magnetopause in Fig. 6.8: it is necessary to change the polarity of the electric field. The mechanism was based on a theory first proposed by Schmidt (1979); he suggested that the plasma could maintain its motion by electrically polarizing, with regions of positive and negative charge maintaining an electrostatic field, thus continuing the drift.

Our finding supports the long-held idea that reconnection has a role in dissipating the energy associated with plasma turbulence in space and astrophysical systems, although the scale for dissipation

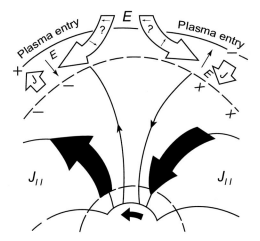

Fig. 6.8 Plasma entry requires that the electric field reverse across the magnetopause in step with the reversal of the magnetic field for continued tailward flow in the LLBL. The perturbation current with localized plasma penetration provides the answer to this dilemma (due to curl **E** explained in Figs. 6.20 and 6.21). From Heikkila, W.J., 1998. Cause and effect at the magnetopause. Space Sci. Rev. 83, 415.

by reconnection would be at the electron scale instead of the ion scale. To assess the importance of reconnection in dissipating turbulence energy in small systems quantitatively, the basic properties of electron-only reconnection (such as the rate, duration and onset conditions of reconnection) will need to be investigated theoretically and observationally. These properties could differ substantially from those known from the standard model of reconnection.

Phan et al. (2018)

One of the key factors is the reversal of the magnetic field in the current sheet for continued tailward flow. This fact has now been discovered in a recent article by Phan et al. (2018). The large-scale plasma and magnetic flux convection patterns within the magnetosphere are set up by the interaction of the shocked solar wind with the internal geomagnetic field. The magnetic field lines of the planet's magnetic field are not stationary. They are continuously joining or merging with magnetic field lines of the interplanetary magnetic field. The physics of this process was first explained by Dungey (1961) using magnetic reconnection.

6.5.2 Relevant plasma experiments

Relevant laboratory plasma experiments were carried out by Bostick (1957), Baker and Hammel (1962, 1965), and Demidenko et al. (1966). These experiments have demonstrated that diamagnetic plasma streams injected impulsively across uniform or nonuniform magnetic fields are able to penetrate the magnetic field. The mechanism was based on a theory first proposed by Schmidt (1960, 1979), who suggested that the plasma could maintain a motion by electrically polarizing, with regions of positive and negative charge maintaining an electrostatic field thus continuing the drift.

These plasma elements are now called plasmoids, a word which means plasma-magnetic field entities, according to the definition by Bostick (1957). Such a generic name can also include many other **B**-field signatures seen in the vicinity of the magnetopause and magnetotail. Collisionless plasmoids are thrown into the geomagnetic field, just like rain droplets penetrate impulsively through the surface of a lake. This idea is illustrated in Fig. 6.9; there is little doubt about considering this (Schmidt, 1960, 1979).

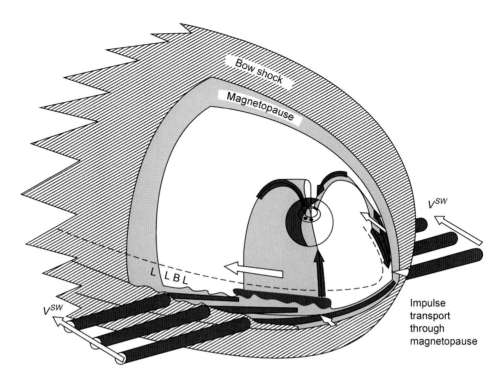

Fig. 6.9 Draping of magnetosheath flux in the equatorial plane with continued tailward flow. The magnetosphere becomes engulfed with magnetosheath plasma if it can get through the magnetopause. Here the LLBL is indicated (not to scale), and a short latitudinal section on the inside edge of the LLBL is also shown, carrying field-aligned currents to the ionosphere. From Heikkila, W.J., 1998. Cause and effect at the magnetopause. Space Sci. Rev. 83, 416.

6.5.3   Plasma weakly diamagnetic ($\beta \ll 1$)

Let us consider one of these many plasma density enhancements with an increase of the density ($dn > 0$) moving toward the magnetopause with the background speed v_{sw}. If this plasmoid corresponds to an excess density but has the same bulk velocity as the surrounding solar wind background, its momentum density $(n + dn)mv_{sw}$ is then necessarily larger than the average (nmv_{sw}).

This plasma element will conserve its excess momentum. Therefore it can plow its way through the magnetosheath toward the magnetopause, with less deflection than the average flow. Unlike the other plasmoids, which have a lower momentum density than average, the former one will reach the position of the mean magnetopause with an excess momentum and an excess kinetic energy. At the mean magnetopause position, where the normal component of background magnetosheath plasma velocity becomes equal to zero, the plasmoid has a residual velocity (GC) given by

$$v_e = v_{sw} dn / (n + dn) \qquad (6.2)$$

The penetration mechanism of the plasmoid into the region of the magnetospheric field lines lies between the following two extreme cases: weakly diamagnetic, low-β plasmoids on one hand, and strongly diamagnetic, high-β plasmoids on the other hand.

In the case of low-β plasmoids, the plasmoid enters the geomagnetic field by means of an electric polarization and its associated $\mathbf{E} \times \mathbf{B}$ drift. Collective plasma effects lead to the accumulation of polarization charges which generates a local (internal) electric field inside the moving plasma element such that its initial momentum can be preserved. Lemaire (1985) has deduced the rate of change of the electric field is:

$$\frac{\partial \mathbf{E}}{\partial t} = -\frac{nm}{\varepsilon_0 B^2} \mathbf{B} \times \left[\frac{\mu^+ + \mu^-}{m} \nabla B + \frac{d}{dt} \left(\frac{\mathbf{E} \times \mathbf{B}}{B^2} \right) \right] \qquad (6.3)$$

The plasma dielectric constant is very large so that the expression in the square brackets must be very small in order to prevent unreasonably large values of the rate of change of the electric field intensity.

$$\frac{\mu^+ + \mu^-}{m} \nabla B + \frac{d}{dt} \left(\frac{\mathbf{E} \times \mathbf{B}}{B^2} \right) \cong 0 \qquad (6.4)$$

This means that

$$\frac{dv}{dt} = v \frac{dv}{dx} = \frac{\overline{\mu}^+ + \overline{\mu}^-}{m} \frac{dB}{dx} \qquad (6.5)$$

Lemaire shows:

$$\frac{1}{2} mv^2 + kT_\perp^+ + kT_\perp^- = \text{constant} \qquad (6.6)$$

The bulk velocity of the plasmoid is nearly constant and equal to the original velocity.

6.5.4   Plasma strongly diamagnetic ($\beta \gg 1$)

In this case, the diamagnetic currents flowing at the surface of the plasmoid produce additional magnetic fields which are comparable to, or larger than, the background magnetic field. This also means that the total kinetic energy of the particles is at least of the order of the magnetic energy. From the point of view of ideal

MHD, the plasmoid excludes the geomagnetic field from its interior as it penetrates the magnetosphere; the plasmoid "pushes" the field lines aside and "passes" between them. The existence of finite parallel electric fields invalidates the MHD approximation near the surface of real plasmoids. It is clear that induced electric fields cannot be ignored when β is large. Simulations by Echim et al. (2005) found that the forward motion of a plasma slab drives backward motion of the adjacent layers. The two boundary layers are not symmetrical. This has been confirmed by Nakamura et al. (2004) from Cluster observations; velocity gradient at the duskward edge tends to be sharper than at the dawnward edge.

6.5.5 Simulations of plasma beams

Two-dimensional (three velocity components) electrostatic simulations were performed by Livesey and Pritchett (1989) to study charge-neutral beam injection across a uniform vacuum magnetic field in Fig. 6.10. Parameters are chosen that allow the beam to penetrate across the magnetic field by the polarization drift mechanism.

Upon injection, the beam polarizes by virtue of the Lorentz force, forming space-charge boundary layers, and continues to propagate across the field at the injection velocity. The beam path

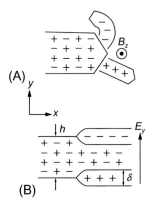

Fig. 6.10 Schematic diagram of the injected beam. The uniform magnetic field **B** exists throughout the simulation system. In (A), the beam particles are shown responding to the Lorentz force upon injection. In (B), space-charge layers form resulting in the polarization field. The beam continues to **E** × **B** drift into the magnetic field. From Livesey, W.A., Pritchett, P.L., 1989. Two dimensional simulations of a charge-neutral plasma beam injected into a transverse magnetic field. Phys. Fluids B 1, 914.

curves in the direction opposite to that of ion gyration. Ions gyrating out of the positive space-charge layer allow a net electron current to flow from the head of the beam to its source. The resulting $\mathbf{J} \times \mathbf{B}$ force is of the correct direction and magnitude to account for the observed beam deflection. The presence of a tenuous ($n-/n+=1/100$) ambient plasma enhances the shielding of the ions in the positive space-charge layer and permits their escape in greater quantities. The $\mathbf{J} \times \mathbf{B}$ force exceeds that of the vacuum case, and a more pronounced beam curvature is observed. In the presence of a marginally dense ($n-/n+=1/10$) ambient plasma, the beam deflects sharply and partially separates into ion and electron streams. The streams then recombine, and the reconstituted beam deflects in the opposite direction.

> *In summary, two-dimensional electrostatic and electromagnetic PIC codes have been used to investigate plasma-beam propagation across magnetic fields. The electrostatic simulations show that a plasma beam can propagate across magnetic fields via $\mathbf{E} \times \mathbf{B}$ drifts. This is in agreement with theory (Peter and Rostoker, 1982). There is little distortion of the beam. However, losses of electrons and ions from charge layers at the edges of the beam appear and undulations in the potential are observed, which reflect small oscillations in the beam drift velocity. ... From a series of high-drift-kinetic-β runs we determine that the magnetic fields diffuse into the beam on time scales much shorter than classical diffusion times ...*
>
> **Koga et al. (1989)**

6.5.6 Draping of magnetosheath plasma

There is draping of magnetosheath flux in the equatorial plane with continued tailward magnetosheath flow, as illustrated in Fig. 6.9.

Fig. 6.11 indicates this significant process is a sequence of four steps, meant to be a temporal sequence. The speed of the magnetopause is less than the normal speed of the plasma v_n, or (to put it the other way around) $v_n > v_{MP}$ The solar wind plasma is moving right through the eroding magnetopause, as is supported by the data shown in Fig. 6.6. It will reverse on the other side, since the geomagnetic field is opposite to the IMF, as elaborated by Lemaire and Roth (1991).

On the other hand, the sense of the induction electric field will be reversed for an inbound pass (outward motion of the magnetopause) compared to an outbound pass (inward motion of the magnetopause), as a close inspection of Fig. 6.3 will reveal.

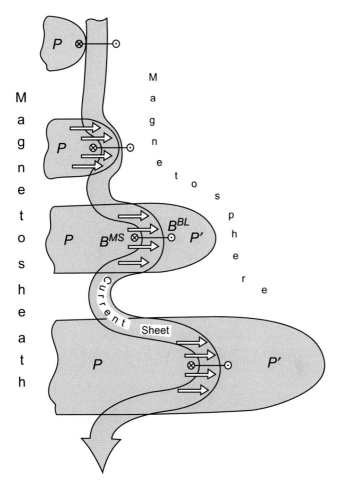

Fig. 6.11 A sequence of four views for an inward meander of the magnetopause current sheet, depicting that the normal plasma velocity is greater than the velocity of the magnetopause. By that motion the plasma is doing work, as is now evident by the behavior of T_e in Fig. 6.6. From Heikkila, W.J., 1998. Cause and effect at the magnetopause. Space Sci. Rev. 83, 401.

6.5.7 Transient auroral event

On December 5, 1986, high-latitude magnetometer stations in Greenland, as well as Iqaluit and the South Pole, showed a strong perturbation lasting for about 10 min beginning at 0930 UT in an otherwise quiet period. A pair of field-aligned currents separated in the east-west sense and moving westward (tailward) at 4–5 km/s is consistent with the data, producing a twin vortex pattern of Hall currents.

Similar perturbations, but with reduced intensity, were also recorded on the afternoon side at Svalbard, Heiss Island, and several locations in northern Siberia. This morning and afternoon occurrence is expected, according to Fig. 6.9. The perturbation was also observed with the incoherent scatter radar at Sondrestrom, these data agreeing with the twin vortex pattern.

The perturbation was accompanied by auroral forms overhead at Sondrestrom that also traveled westward. Meridian scanning photometer recordings at the radar site showed the cleft, located about 3–5 degrees poleward in latitude; the cleft did not move from the far northern sky for several hours, even while the disturbance was observed. Viking and Polar Bear satellites passed just before the disturbance over Greenland, and DMSP encountered the disturbance near Baffin Island a few minutes later. These spacecraft observations increased our confidence in the interpretation of the data. ISEE 1/2 and IMP 6 recorded a magnetic disturbance in the solar wind, the likely cause of this event. Similar observations by others have been associated with transfer events.

However since the observed event occurred on closed field lines, our interpretation is quite different. It is that an impulsive penetration (IP) of solar wind plasma on an interplanetary magnetic flux tube took place through the magnetopause, ending up in the LLBL (Fig. 6.12). Some efficient mechanism is required to feed the boundary layer with the total amount observed; we

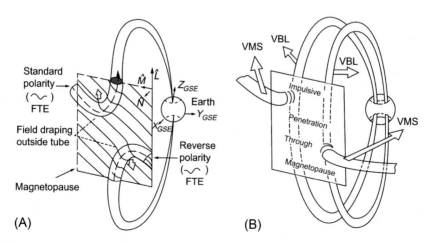

(A) (B)

Fig. 6.12 (A) Sketch illustrating the effects of a localized burst of reconnection FTE at the magnetopause. (B) With IP of solar wind plasma through the magnetopause onto closed field lines, two separate disturbances would result on closed field lines, on the morning side as well as on the afternoon side. From Heikkila, W.J., Jorgensen, T.S., Lanzerotti, L.J., Maclennan, C.G., 1989. A transient auroral event on the dayside. J. Geophys. Res. 94(A11), 15303.

identify it as IP with energy supplied from stored electric energy due to the bulk motion of the plasmoid.

6.6 Flux transfer event

The fundamental signature of the passage of a flux transfer event (FTE) (Russell and Elphic, 1978) located near the magneto-pause is a bipolar pulse in the magnetic field normal to the unperturbed magnetopause surface. This signature is interpreted as the passage of a flux tube, embedded in the magnetopause and created by Sonnerup et al. (2004) of a patchy and impulsive reconnection. Data from Cluster (Fig. 6.13) have been used to study the structure of an FTE seen near the northern cusp. They employ Grad-Shafranov reconstruction (Fig. 6.14) using measured fields from all four spacecraft to produce a map of the FTE cross section.

The FTE consists of a flux rope of approximate size 1 R_E and irregular shape, embedded in the magnetopause. Its axis z is tangential to the magnetopause. Since no reconnection signatures are seen, the map provides a fossil record of the prior reconnection process that created the flux rope: the strong core field indicates that it was generated by component merging. An average reconnection electric field >0.18 mV/m must have occurred in the burst of reconnection that created the FTE. The total axial (z) current and magnetic flux in the FTE were − 0.66 MAmp and +2.07 MWeber, respectively.

Sonnerup et al. (2004)

Several important items emerge from this study (but see Section 6.12):

- "The high correlation coefficient for the reconstructed map indicates that many of its features are reliable." In particular, the orientation of the invariant axis has been accurately determined in the optimization process.
- It is clear that this *flux rope had a strong core field*.
- This analysis [suggests] "an average reconnection electric field [*is small*], ~0.18 mV/m."
- "The absence of reconnection signatures implies that, by the time the FTE reaches Cluster, it is nearly a fossil structure. It is in approximate force balance but is far from force free."
- While the plasma flow speeds in the HT frame, seen by C1 as it traverses the FTE, are very small, "this is not the case for C3 where the flow speeds are in fact supersonic."

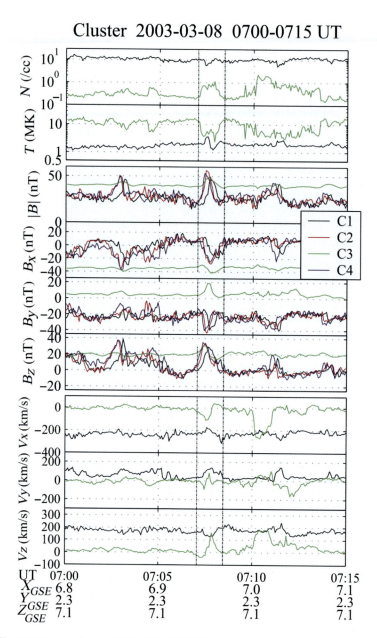

Fig. 6.13 Cluster data on 2003, 0700–0715 UT. From top, the panels show: number density; temperature; field magnitude and GSE components; GSE velocity components. Time interval between the two vertical lines was used for the reconstruction. From Sonnerup, B.U.Ö., Hasegawa, H., Paschmann, G., 2004. Anatomy of a flux transfer event seen by cluster. Geophys. Res. Lett. 31, 2, L11803. doi:10.1029/2004GL020134.

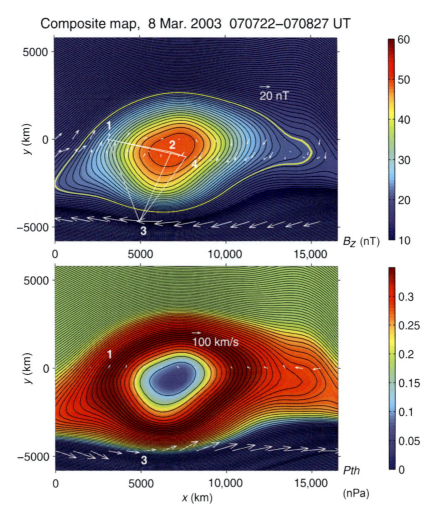

Composite map, 8 Mar. 2003 070722–070827 UT

Fig. 6.14 Reconstructed field lines projected onto the transverse plane, with axial magnetic field (top panel) or plasma pressure (bottom panel). In the top panel, Cluster tetrahedron and measured transverse field, $\boldsymbol{B}_T = (B_x, B_y)$ are shown in *white*. In the bottom panel, *white arrows* show measured transverse velocity in the HT frame. Equatorward edge of the map is to the right with magnetosphere on bottom. From Sonnerup, B.U.Ö., Hasegawa, H., Paschmann, G., 2004. Anatomy of a flux transfer event seen by cluster. Geophys. Res. Lett. 31, 3, L11803. doi:10.1029/2004GL020134.

6.7 Cluster observations of plasma transfer

The Cluster spacecraft had a near optimum tetrahedron configuration on March 8, 2003, with C1, C2, C4 lying in a plane close to parallel to that of the magnetopause, while C3 was located about 0.7 R_E closer to Earth. This is the same data set that was used by Sonnerup et al. (2004) in Figs. 6.13 and 6.14. The spacecraft all made simultaneous observations of magnetospheric and magnetosheath parameters, Plasma Electron and Current Experiment (PEACE) measurements of the differential energy flux of 10 eV to 26 keV electrons from each spacecraft, FGM measurements of the magnetic field components, and strength in boundary normal coordinates. WHISPER illustrates the emissions observed up to 40 kHz on the four spacecraft during the FTE2 event identified, and EFW made electric field measurements. The three spacecraft C1, C2, C4 crossed the magnetopause and moved into the sheath at ~0653 UT. This is associated with a significant change in the orientation of the magnetic field B_L, B_M, the disappearance of the higher-energy electron population, and the appearance of colder, denser electrons. Prior to magnetopause crossing and after, these spacecraft observe similar signatures which were interpreted as FTEs at C3, as displayed in Fig. 6.13. C3 was in the magnetosphere until 07:00 or later.

6.7.1   CIS ion data

A PTE was observed by Cluster 3 on March 8, 2003, beginning at 07:07 UT. Fig. 6.15 displays the ion data obtained by the CIS experiment (Rème et al., 2001) obtained onboard spacecraft C3 between 07:07 and 07:09 UT. The top five panels give the energy-time ion spectrograms from the HIA sensor (no mass discrimination), for ions arriving in the 90° × 180° sector with a field-of-view pointing in the Sun, dusk, tail, and dawn directions, respectively, and then the omnidirectional ion flux. The following four panels show the omnidirectional ion flux measured by the CODIF sensor, separately for H^+, He^{++}, He^+, and O^+ ions. All spectrogram units are in particle differential energy flux (keV/cm^2 s sr keV). The density values are given in the bottom two panels, for the HIA sensor (no mass discrimination) and the CODIF sensor (separately for H^+, He^{++}, He^+, and O^+ ions). The PTE associated burst of plasma is clearly seen in the data. The density, during the event, increases by a factor of four, and the presence of He^{++} ions confirms its solar wind origin.

6.7.2   WHISPER plasma emissions

Fig. 6.16 illustrates the emissions observed up to 40 kHz by WHISPER (Pickett et al., 2003) on the four spacecraft during the FTE2 event identified in Fig. 6.13. The differences in the signatures observed between Cl, C2, C4, in the magnetosheath and C3 in the magnetosphere are evident. The faint emissions close to 30–35 kHz

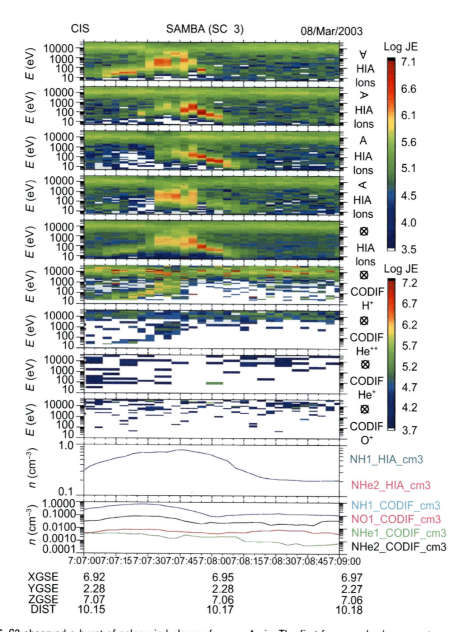

Fig. 6.15 C3 observed a burst of solar wind plasma for over 1 min. The first four panels show spectrograms in different directions. Observation of He^{++} confirms the identity as the solar wind. The increase in density is up to a factor of 4. From Heikkila, W.J., Canu, P., Dandouras, I., Keith, W., Khotyaintsev, Y., 2006. Plasma transfer event seen by cluster. In: Cluster and Double Star Symposium—5th Anniversary of Cluster in Space. ESA SP-598, p. 2.

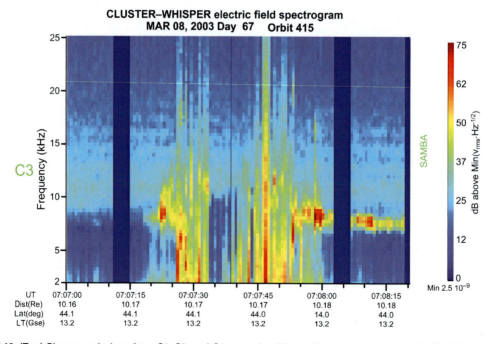

Fig. 6.16 (Top) Plasma emissions from C1, C2, and C4 are quite different from those observed by C3. C3 shows intense bands on both sides of the PTE; there are no emissions in the middle. This can be readily understood as a result of the induction electric field (see Fig. 6.17) as a result of General Magnetic Reconnection (GMR). (Bottom) A double noise burst was noted by C3 centered on the PTE. This is included in Fig. 5.8 in Chapter 5. From Heikkila, W.J., Canu, P., Dandouras, I., Keith, W., Khotyaintsev, Y., 2006. Plasma transfer event seen by cluster. In: Cluster and Double Star Symposium—5th Anniversary of Cluster in Space. ESA SP-598, p. 3.

are the local plasma frequency, corresponding to a local density of ~10–15 cm^3 for Cl, C2, and C4. The bursty broadband emissions observed at low frequencies are also common in this region and due to solitary potential structures. In the magnetosphere, C3 detected a lower-density plasma, identified here by the low frequency cutoff of the continuum radiation at 8 kHz (N_e~0.8 cm^3). The signatures associated with the boundary of the PTE are very strong bursts, up to ~1 mV/m, of upper hybrid emissions, which are probably triggered by the low-energy field-aligned beams observed by the PEACE instruments. Intense broadband emissions, more than two orders of magnitude above background, possibly triggered by the counterstreaming electron beams reported from PEACE data, are observed when C3 penetrates in the PTE.

6.7.3 PEACE electron data

Electron measurements were obtained by PEACE. PEACE consists of two sensors with hemispherical electrostatic analyzers, each with a 180 degrees field of view radially outwards and perpendicular to the spin plane. Together, the sensors cover an energy range from 0.6 eV up to 26 keV over 12 polar sectors (Johnstone et al., 1997). The data shown were taken by the High Energy Electron Analyzer (HEEA) sensor on Cluster 3, covering the energy range from 34 eV to 22 keV. Pitch angle distributions were determined onboard at one spin resolution. Thirteen pitch angle bins and 30 energy steps are telemetered in this mode, which were reduced to the 10 energy bins shown in Fig. 6.17. Full-resolution pitch angle data is retained within each energy bin, going from 0° at the bottom of each panel to 180° at the top. The bottom panel labeled "Flow AZ" shows the spin angle between the direction of incoming particles at the start of the spin and the magnetic field direction.

6.7.4 EFW data

Fig. 6.18 presents electric field measurements during the interval 07:07–07:09 UT after Gustafsson et al. (2001). The upper panel shows negative of the spacecraft potential, which indicates a density/temperature change around 07:07:10–07:08:00. The middle panel shows a simultaneous enhancement of wave activity in a frequency range between 0.25 and 10 Hz. One should note that the electric field presented is coming from only one probe pair, and thus represents an incomplete measurement of the electric field. The last panel shows spin resolution electric field measured

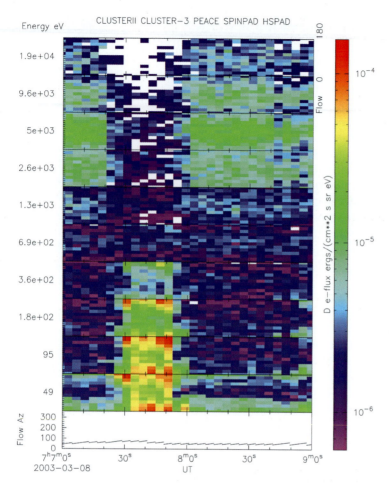

Fig. 6.17 Data from Cluster 3 on March 8, 2003. PEACE electron data divided into 10 channels with the center energy indicted on the left. Each channel is divided into pitch angle with 0° at the bottom, 180° at the top. The electron counting rate at the highest energies maximizes near 90°, indicating trapping on closed field lines surrounding the event. At lower energies, the data show intense electron bursts in the PTE. From Heikkila, W.J., Canu, P., Dandouras, I., Keith, W., Khotyaintsev, Y., 2006. Plasma transfer event seen by cluster. In: Cluster and Double Star Symposium–5th Anniversary of Cluster in Space. ESA SP-598, p. 3.

by the EFW and CIS HIA. One can see a clear bipolar signature in E_y between 07:07:10 and 07:08:00, where E_y is changing from negative to positive in the middle of the structure at 07:07:40 UT. Gustafsson et al. (2001) shows a possible path the spacecraft took through the event. The first response was in the negative y direction, then in the positive y, as predicted by the model. This constitutes a dynamo followed by a load.

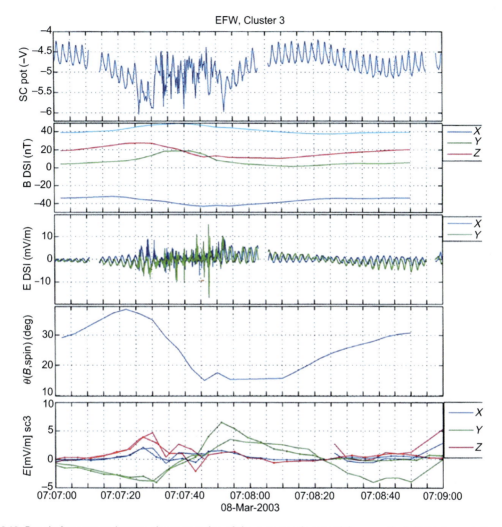

Fig. 6.18 Panels from top to bottom show: negative of the spacecraft potential, magnetic field, high-resolution electric field from one boom pair, angle between the spacecraft spin plane and the magnetic field, spin resolution electric field (the z component electric field is calculated assuming **E** · **B** = 0, the *dash-dot lines* are components of **v** × **B** from CIS HIA). From Heikkila, W.J., Canu, P., Dandouras, I., Keith, W., Khotyaintsev, Y., 2006. Plasma transfer event seen by cluster. In: Cluster and Double Star Symposium—5th Anniversary of Cluster in Space. ESA SP-598, p. 3.

6.8   Plasma transfer event

The second time-dependent term in Poynting's theorem involves magnetic energy. One would expect that it is of the utmost importance for a theory of solar wind interaction since it represents, by far, the energy source that is dominant in space plasmas (Emslie and Miller, 2003).

Both types of fields can, in general, have components parallel and transverse to the magnetic field in any realistic circumstance:

$$\mathbf{E} = \mathbf{E}_{\parallel}^{es} + \mathbf{E}_{\parallel}^{ind} + \mathbf{E}_{\perp}^{es} + \mathbf{E}_{\perp}^{ind} \qquad (6.7)$$

We must evaluate all four terms before we know how the plasma will react to a given situation. The Coulomb (transverse) gauge spells out the differences (see Chapter 3, Fig. 3.3). A changing current will create an induction electric field, by Faraday's and Lenz's laws. The induction component can be quite strong in space plasmas. The induction electric field in integral form is:

$$emf = \varepsilon = \oint \mathbf{E} \cdot d\mathbf{l} = -d\Phi^M / dt \qquad (6.8)$$

where Φ^M is the magnetic flux through the circuit used for the integration. It is only by this *emf* that we can tap stored magnetic energy.

6.8.1 Fundamentals of a plasma transfer event

To begin at the beginning, we note that the electric field has two sources, charge separation and induction. For this reason alone it is better to use the *E,J* paradigm rather than the *B,V* discussed by Parker (1996); with the *B,V* there is only one electric field, the so-called convection electric field.

Fig. 6.19 shows a model for impulsive plasma transport across an *open* magnetopause with the geomagnetic field lines reaching out into the magnetosheath. A plasma cloud (a flux tube) with excess momentum is assumed to distort the magnetopause current locally. Le Chatelier's principle and Lenz's law state that this is an inductive electric field, everywhere opposing the current perturbations.

Note that *the induction electric field cuts though the magnetopause at both sides*, in opposite directions, with a parallel electric field along B_n. This can happen anywhere on the magnetopause surface.

The only source of a magnetic field is a current **J** by Ampere's law; therefore, to study changes in the magnetic field we should consider perturbation electric currents $\delta\mathbf{J}$, the source of $\delta\mathbf{B}$. A changing current will create an induction electric field. We focus on the electric field directly, noting that the total field **E** is

$$\mathbf{E} = \mathbf{E}^{es} + \mathbf{E}^{ind} = -\nabla\phi - \frac{\partial \mathbf{A}}{\partial t} \qquad (6.9)$$

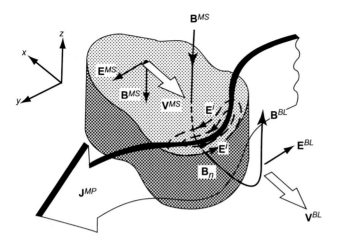

Fig. 6.19 Model for impulsive plasma transport across an *open* magnetopause. A plasma cloud (a flux tube) with excess momentum is assumed to distort the magnetopause current locally. This creates an inductive electric field; everywhere opposing the current perturbations by Lenz's law (the plasma response by charge separation is included later). Note that the induction electric field cuts though the magnetopause at both sides forcing GMR. From Heikkila, W.J., 1984. Magnetospheric topology of fields and currents. In: Potemra, T.A. (Ed.), Magnetospheric Currents. AGU Geophysical Monograph 28. American Geophysical Union, Washington, DC, p. 211.

and has induction and electrostatic components. We write it in this order to emphasize the relative importance of the two. The plasma response to the imposition of the induced electric field leads to the creation of an electrostatic field, at least when conductivities are not all zero.

A PTE is a three-dimensional object: 2D to show the magnetic topology (e.g., x–z plane in GSM coordinates), and another set to show localization involving curl $\mathbf{E}$ (x–y plane). These two types of fields have different topological characteristics, one being solenoidal using the Coulomb gauge as discussed extensively by Morse and Feshbach (1953), the other being irrotational (conservative) with zero curl. Consequently, they can never cancel each other; the most the plasma can do is to redistribute the field while maintaining the curl.

6.8.2 Localized pressure pulse

The inferred immediate cause of a plasma transfer event is a localized pressure pulse from the magnetosheath, an inward push by solar wind plasma associated with erosion first found by Aubry et al. (1970). The pressure pulse is likely to be in some small region, not extending to infinity in the y direction. Only the induction electric field is shown in Fig. 6.19; the plasma response through charge separation (creating an electrostatic field) is treated in the next three subsections. On the left is the undisturbed magnetopause current $\mathbf{J}(0)$ before the pressure pulse; on the right is the condition after the first strike (the tangential velocity is assumed to vanish here; its effect will be discussed in Section 8.4). The total current perturbation $\delta\mathbf{J}$ is this *change in current* which induces a voltage:

$$\mathbf{E}^{\text{ind}} = -\frac{\partial \mathbf{A}}{\partial t} = -\mu_o \iiint_\tau \frac{\partial \mathbf{J}}{\partial t} d\tau \qquad (6.10)$$

where $\mathbf{A}$ shows the dependence on the time rate of change of the current. The field is in the reference frame of the undisturbed current, the laboratory frame, everywhere opposing the perturbation (note the negative sign).

6.8.3 Response of the plasma: $B_n = 0$

Any plasma response is hindered by the magnetic field if B_n vanishes. Because B_z is the dominant component of the magnetic field on either side of the magnetopause (at least for high as well as low shears), the very low (effectively zero) Pedersen conductivity $\sigma_1 \sim 0$ for a collisionless plasma in the tangential y direction limits polarization of charge in that direction. The induction electric field alone is the field that determines the motion of the plasma over the bumpy surface, a velocity that is everywhere tangential to the local magnetopause within the perturbation.

6.8.4 Response of the plasma: B_n finite

The plasma response changes dramatically with an open magnetosphere (Fig. 6.20). In this case, a rotational discontinuity will be present, with a finite B_n. Electron and ion mobilities are high along the magnetic field. Now we can use the very large direct conductivity σ_o; the plasma can polarize along the magnetic field lines as shown in Fig. 6.21, top and bottom, in different senses, causing an electrostatic field tangential to the magnetopause, reversing as indicated. Thus we see that this E^{es} will drive the solar wind plasma into the current sheet. On the other side, since both $\mathbf{B}$ and $\mathbf{E}$ reverse, the electric drift $\mathbf{E} \times \mathbf{B}$ will be also earthward. This is the answer to the question marks in Fig. 6.8: PTE is produced.

Comparing this result with that of the $B_n = 0$ given earlier is very important: it establishes a dependence on the IMF.

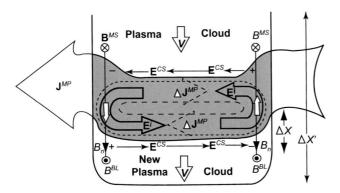

Fig. 6.20 Impulsive penetration is produced by an electrostatic field due to a charge distribution created by an induction electric field. Charged particles from the old plasma cloud go through the current sheet along B_n and form a new plasma cloud on closed magnetic field lines. From Heikkila, W.J., 1984. Magnetospheric topology of fields and currents. In: Potemra, T.A. (Ed.), Magnetospheric Currents. AGU Geophysical Monograph 28. American Geophysical Union, Washington, DC, p. 211.

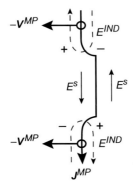

Fig. 6.21 Here it is assumed that the frame of reference is fixed to the magnetopause in the center of the perturbation, as in reconnection models. The magnetopause, top and bottom, is moving sunward as shown by *large arrows*; now it is here that induction field is felt. With a rotational discontinuity the plasma polarizes, top and bottom, causing an electrostatic field tangential to the magnetopause, reversing as indicated. We see that this charge distribution will drive the magnetosheath plasma into the current sheet and out on the other side. From Heikkila, W.J., 1997. Interpretation of recent AMPTE data at the magnetopause. J. Geophys. Res. 102(A5), 2121.

6.8.5 Tangential motion

It is essential to include the tangential motion (to the magnetopause) to understand the effects of a PTE upon the physics of the magnetosphere. In the magnetosheath all the solar wind

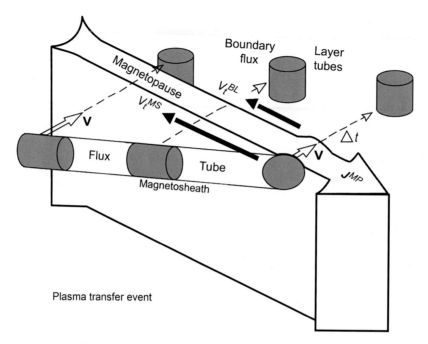

Fig. 6.22 In the magnetosheath, all the solar wind plasma is moving antisunward, even super-Alfvénically toward the flanks. Whatever plasma penetrates through the magnetopause must face conditions is a new medium. Two *black arrows* are meant to denote the tailward motion of the plasma, higher in the magnetosheath, lower in the LLBL. From Heikkila, W.J., Canu, P., Dandouras, I., Keith, W., Khotyaintsev, Y., 2006. Plasma transfer event seen by cluster. In: Cluster and Double Star Symposium—5th Anniversary of Cluster in Space. ESA SP-598, p. 7.

plasma is moving antisunward, even super-Alfvénically toward the flanks. Whatever plasma penetrates through the magnetopause must face conditions in a new medium. The two black arrows in Fig. 6.22 are meant to denote the tangential motion of the plasma, higher in the magnetosheath, lower in the LLBL. For example, this difference makes it possible to have multiple injection events in the LLBL due to repeated swipes from the magnetosheath as observed by Carlson and Torbert (1980) and Woch and Lundin (1992).

6.9 Skimming orbit of GEOTAIL

On October 17, 1992, the GEOTAIL spacecraft skimmed the morningside magnetopause for several hours; multiple crossings of the current sheet make this orbit excellent for a study of plasma processes relevant to solar wind-magnetospheric interaction.

Data from the Comprehensive Plasma Instrumentation (CPI) of the University of Iowa provided density, energy, temperature, velocity, and Alfvén velocity information for the entire 6-h period.

We have considered the CPI data in great detail for 4 min 19:23 to 19:27 as a typical sample for magnetopause crossings; the results are summarized in Chapter 3, as seen in Fig. 3.7. The spacecraft is spinning, 3 s per spin. CPI data was analyzed for the exit from the magnetosphere (often called an outbound crossing) and entry (inbound crossing). The spacecraft was nearly stationary, and the crossings were accomplished by the relatively fast inward/outward motions of the magnetopause. Kawano et al. (1994) covered the wave-like behavior uncovered on this orbit of GEOTAIL. The slow crossing takes place first (with three instrument cycles), followed by the fast crossing (with one cycle).

6.9.1 Overview of Comprehensive Plasma Instrumentation data

The plasma data are from the HP analyzer that is one element of the CPI on the GEOTAIL spacecraft (Frank et al., 1994). The HP analyzer provides three-dimensional observations of the velocity distributions of electrons and ions with good resolution of energies and angles. Charged particles with velocities directed into a fan-shaped region along the spin axis of the spacecraft are intercepted by one of the electron or ion sensors that divide the field of view into nine approximately equal polar sections. A programmed sequence of logarithmically spaced voltage steps applied to the electrostatic deflection plates selects particles with energies per charge in the range of approximately 1 eV to 48 keV. Azimuth angles are determined by sectioning the spin into equal azimuth sectors.

The HP analyzer was operating in a mode that acquired samples in eight spin sectors at 28 energy steps in the range 6 eV to 48 keV. A full set of samples was acquired in seven spins or approximately 21 s. An eighth spin was used to acquire samples at energies per charge as low as 1 V. Plasma moments are computed as weighted sums of the measured phase-space densities. Improved statistical accuracy is achieved by summing three consecutive instrument cycles prior to computation of the moments.

6.9.2 Transition diagram

The data allow us to plot the transition parameter in Fig. 6.23 (Hapgood and Bryant, 1990). This plot is based upon the idea that the LLBL is a mixture of magnetosheath and magnetospheric

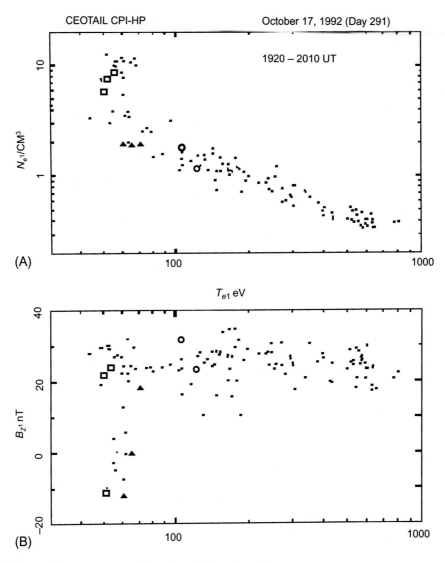

Fig. 6.23 (A) The transition parameter for 50 min from 19:20 to 20:10 UT, a scatter plot of plasma density against electron temperature (energy). Higher plasma density implies a location close to the magnetopause, or the magnetosheath, while a higher average energy implies a location on inner edge of the LLBL. The special symbols refer to detailed plasma analysis discussed in the text. Each point is for an instrument cycle with no averaging. (B) The magnetic field B_z at the times when the data in (A) were taken. From Heikkila, W.J., 2003. Initial condition for plasma transfer events. In: Newell, P. T., Onsager, T. (Eds.), Earth's Low-Latitude Boundary Layer. AGU Geophysical Monograph 133. American Geophysical Union, Washington, DC, p. 161.

plasma. A scatter plot of plasma density against energy is made for a chosen interval; higher plasma density implies a location close to the magnetopause, or even the magnetosheath, while a higher average energy (temperature) implies a location on inner edge of the boundary layer (IEBL) or the magnetosphere proper (Lockwood and Hapgood, 1997). It is possible to fit a curve to the plotted points, and then to divide this curve into equal lengths to get a "transition parameter." Instead, we show the transition plot in Fig. 6.23 for the longer period from 19:20 to 20:10 UT. We plot plasma density as a function of energy, with no averaging (each point corresponds to an instrument cycle). The special symbols are squares for the end points, triangles for the intermediate samples, and open circles for the others, for the shorter period 19:23 to 19:27. At the left is the value appropriate to the magnetosheath, while the right part is the geomagnetic field in the LLBL or the outer magnetosphere.

At these moderate to high energies, loss cones will be developed within a few seconds, half a bounce period, with the loss of particles. This observation is typical for bouncing motion between northern and southern hemispheres, on closed field lines. The technique has been used quite frequently, for example, by McDiarmid et al. (1976) and Phan et al. (2005); they too found that solar wind plasma penetrates to closed field lines.

6.9.3 Inward and outward exists

We can use this approach for an outward motion of plasma as well; the roles of dynamo and load will be reversed. For an inward motion in Fig. 6.24A with a decreasing current density, the value of $\mathbf{E} \cdot \mathbf{J} > 0$ corresponds to an electrical load. The view for an outward meander of the magnetopause current sheet is shown in Fig. 6.24B. The sense of the induction field is reversed, with an increasing current density, $\mathbf{E} \cdot \mathbf{J} < 0$ corresponds to a dynamo as indicated by the battery symbol.

6.10 Electric field at high sampling rates

The electric field experiment on the Polar satellite collected bursts of data at rates of 1600 and 8000 samples/s (Mozer and Pritchett, 2010). These high rates are required for rapid events to measure the parallel and perpendicular components of the vector electric field at much better resolution. Some 150 such bursts from 2000 through 2003 were examined during 5-month intervals when the spacecraft was on the dayside of the Earth and the apogee was 9.5 R_E. Transverse components of 200 mV/m and parallel electric field components as large as 80 mV/m were found.

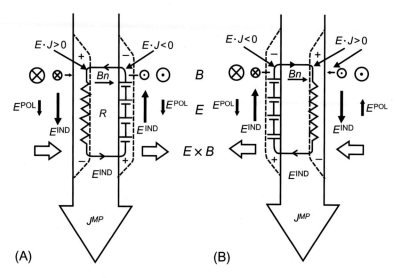

Fig. 6.24 By Lenz's law, the induction electric field is everywhere opposed to the perturbation current. (A) For an inward motion of a localized perturbation, with the decreasing current density, **E·J**>0 corresponding to an electrical load. (B) The view of an outward meander of the magnetopause current sheet, showing that the sense of the induction field is reversed, with an increasing current density **E·J**<0 corresponding to a dynamo as indicated by the battery symbol. From Heikkila, W.J., 1998. Cause and effect at the magnetopause. Space Sci. Rev. 83, 398.

The authors plotted the data for 15 s on April 13, 2001 in Fig. 6.25. The high rates all appeared on the magnetospheric side of the current sheet in the outer boundary layer (OBL, see Chapter 9). Parallel electric fields were associated with significant plasma density depletions, and their anticorrelations with plasma density are found in the simulations of Chapter 3.

Two seconds in which B_z reverses sign are replotted in Fig. 6.26. Plasma density goes very quickly from its magnetospheric value of less than $1\,\mathrm{cm}^{-3}$ to greater than $2\,\mathrm{cm}^{-3}$ (somewhat less than typical for the magnetosheath). The plasma data goes from the magnetospheric side of the current sheet to the magnetosheath (top panel) in a monotonic manor, smoothly and rapidly in 1 s; magnetopause motion has to be involved. The magnetopause itself is located from 09:36:49.2 UT (this includes a precursor of 0.1 s) to 09:36:50.3 UT (including an addendum of 0.2 s discussed later).

There are two major bursts of rapid fluctuations, one beginning at 49.3 s, the other at 49.9 s, separated by 0.6 s in the central part of the perturbation. This suggests that something important is happening here. I believe it is GMR with a changing magnetic helicity, connected with a PTE. That is evidenced in the values for the plasma velocity, v_GSE_x being negative (antisunward).

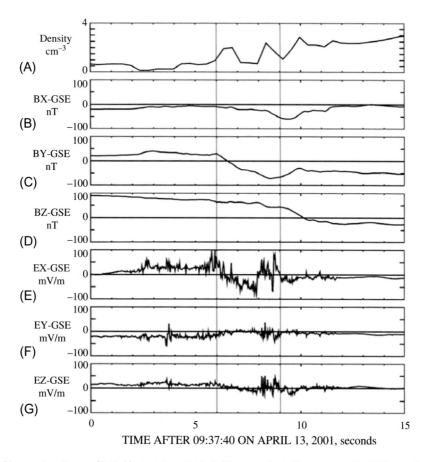

Fig. 6.25 (A) Plasma density and (E, F, G) electric and (B, C, D) magnetic fields measured in GSE coordinates during a 15 s crossing of the subsolar magnetopause on the Polar satellite. From Mozer, F.S., Pritchett, P.L., 2010. Spatial, temporal, and amplitude characteristics of parallel electric fields associated with subsolar magnetic field reconnection. J. Geophys. Res. 115, 2, A04220. doi:10.1029/2009JA014718.

A large value of the x and y components of the IMF tend to obscure this; the results are most striking in field-aligned values in Fig. 6.26. The field-aligned coordinate system (FAC) is magnetic field aligned with the Z_FAC axis parallel to the magnetic field, B, while the X_FAC axis is perpendicular to B in the plane containing the magnetic field line and it is positive inward. The Y_FAC axis defines the third component of this right-hand coordinate system by being perpendicular to B and pointing generally in the westward direction.

The sharp spike in E after 50.2 s may be associated with the changing B, the subject of Phan and Paschmann (1996). They define

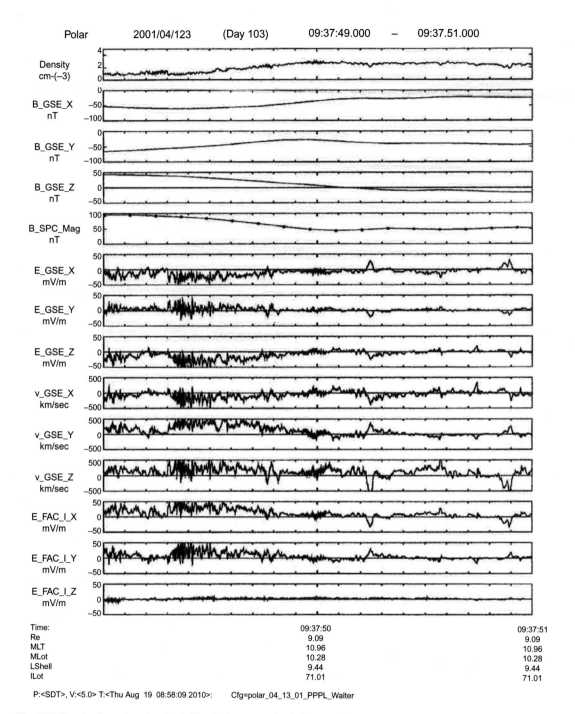

Fig. 6.26 Results from the electric field experiment on the Polar satellite for 2 s. The plasma density change takes place smoothly and rapidly in 1.1 s at a rapid crossing of the magnetopause, a PTE. It is most striking in field-aligned values (especially the *y* component) 09:36:49.3 until 09:37:50.3 in a dynamo/load pair. There are two bursts at 49.4 and 50.0 s associated with GMR and changing magnetic helicity, as illustrated by Fig. 6.24. From Mozer and Heikkila, after Mozer, F.S., Pritchett, P.L., 2010. Spatial, temporal, and amplitude characteristics of parallel electric fields associated with subsolar magnetic field reconnection. J. Geophys. Res. 115, 2, A04220. doi:10.1029/2009JA014718.

two key times: the crossing time of the magnetopause and the crossing time of the IEBL. The magnetopause key time is defined to be the outer edge of the magnetopause current layer, which is readily identified in the data (as one moves from the magnetosheath to the magnetosphere) by the abrupt onset of rotation of the magnetic field in the plane tangential to the magnetopause.

Being transient in nature, the question of energy transfer becomes important. That is likely to depend on the Alfvén velocity, only a few hundred kilometers per second. The distant neutral line of SMR becomes a problem; the actual magnetic field is tens of nT in this instance. We interpret this as indicating that GMR is occurring.

Schindler et al. (1988) have considered a highly relevant article on GMR in nonvanishing magnetic fields (finite-B reconnection) in Fig. 6.27. Fälthammar (2004) has reviewed it in Fig. 5.5 in Chapter 5.

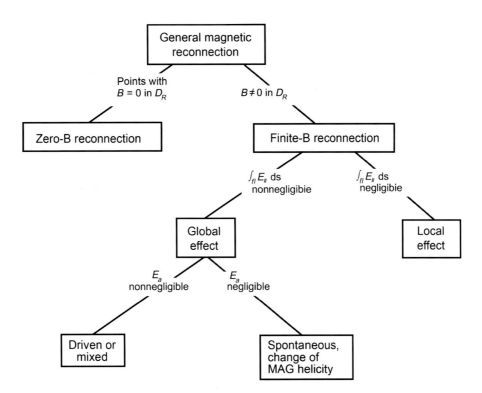

Fig. 6.27 Regimes of magnetic reconnection. SMR and GMR occur in separate branches. E_a denotes the asymptotic electric field magnitude at large distances from the diffusion region. From Schindler, K., Hesse, M., Birn, J., 1988. General magnetic reconnection, parallel electric fields, and helicity. J. Geophys. Res. 93(A6), 5551.

We show that $E_{\parallel}$ *(the electric field component parallel to the magnetic field) plays a crucial physical role in finite-B reconnection, and we present two theorems involving* $E_{\parallel}$. *The first states a necessary and sufficient condition on* $E_{\parallel}$ *for global reconnection to occur. Here the term "global" means the generic case where the breakdown of magnetic connection occurs for plasma elements that stay outside the nonideal region. The second theorem relates the change of magnetic helicity to* $E_{\parallel}$ *for cases where the electric field vanishes at large distances.*

Schindler et al. (1988)

These results provide new insight into three-dimensional reconnection processes; this will be explored in Chapters 10 and 11 in terms of plasmoid formation.

6.11 MMS observations at the magnetopause

6.11.1 Mission details

The Magnetospheric MultiScale (MMS) mission, launched in 2015, is designed to investigate magnetic reconnection in the boundary regions of the Earth's magnetosphere, particularly along its dayside boundary with the solar wind and the neutral sheet in the magnetic tail (Burch et al., 2016a). Unlike previous missions, which used the spacecraft spin period to sweep out 3D particle distributions, the MMS Fast Plasma Instrument uses multiple sensors with large fields of view. Four sensors each for electrons and ions with $45° \times 180°$ fields of view, accomplished with the use of electrostatic scanning. This allows for extremely high time resolution data (30 ms for electrons and 150 ms for ions) to be collected while in burst mode, enabling accurate measurements of currents and electron drift velocities. Also, the four spacecraft constellations have closer spacing (about 10 km) than previous missions such as Cluster in order to distinguish features of the electron diffusion region. This region within the magnetopause may be only a few kilometers thick and can move at tens of kilometers per second, so the high time resolution is necessary in order to be able to collect multiple electron distribution functions. For example, a 5 km thick region traveling at 50 km/s will contain a spacecraft for only 0.1 s, or about three samples of the 3D electron distribution. Ion diffusion regions are much larger, on the order of 200 km, so even with the lower time resolution of the ion detectors, a 50 km/s passage would take 4 s, or about 25 samples. In order to maximize the spacecraft's orbital "dwell time" at the magnetopause, for that phase of the mission the orbits were configured to have an apogee in the sunward direction at about $12\,R_E$. The enormous data rate implied by the speed of the samples means that only select time periods can be returned to Earth. This is accomplished with a large (96 GB) capacity on board storage and autonomous evaluation of data quality in near real time to select samples for downlink. The large amount of storage also allows time (up to 3 days) for scientists to review the lower resolution survey data and, if necessary, override the selections of burst mode data to be returned. Both mechanisms have been reported to be effective (Burch et al., 2016b).

6.11.2 Observations

Details of the electron distributions near the electron diffusion region (EDR) were first reported by Burch et al. (2016b). The electron distributions have crescent-shaped features (Fig. 6.28) that indicate the electron currents near the X-line. The crescents shown on the left side of Fig. 6.28 are said by Burch et al. (2016b) to be the result of demagnetization of solar wind electrons as they flow into the reconnection site, and their acceleration and deflection by an outward-pointing electric field that is set up at the magnetopause boundary by plasma density gradients. As they

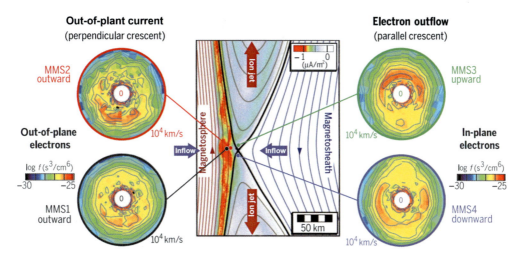

Fig. 6.28 The data in the *circles* show electrons with velocities from 0 to 10^4 km/s carrying current out of the page on the left side of the X-line and then flowing upward and downward along the reconnected magnetic field on the right side. The most intense fluxes are *red* and the least intense are *blue*. The plot in the center shows magnetic field lines and out-of-plane currents derived from a numerical plasma simulation using the parameters observed by MMS. From Burch, J.L., et al., 2016. Electron-scale measurements of magnetic reconnection in space. Science 352, 1189, aaf2939. doi:10.1126/science.aaf2939.

are deflected, the solar wind electrons mix in with magneto-spheric electrons and are accelerated along a meandering path that straddles the boundary, picking up the energy released in annihilating the magnetic field. The crescent-shaped velocity distributions are diverted along the newly connected magnetic field lines in a narrow layer just at the boundary (right side of Fig. 6.28). Similar crescent features were reported by Berkowitz (2019) in the magnetotail, where there persistence during rapid fluctuations of the electromagnetic fields was interpreted to mean that turbulent effects were not dominant during reconnection, but rather that the field was able to release magnetic energy efficiently and continuously accelerate the electrons for extended periods of time.

The magnetospheric conditions under which the EDR was encountered by MMS was investigated by Fuselier et al. (2017) using the 12 encounters that had been identified out of the 4500 total magnetopause crossings. They found that EDR encounters occurred over a wide variety of solar wind and magnetospheric conditions, including a large range of magnetic shear, plasma beta gradients, and magnetosheath flow velocities. The 12 encounters also occurred over a large range of local times (6:00 to 18:00 LT) and dynamic pressures (0.9–4.6 nPa). All but one of the events occurred during high magnetic shear of greater than 120 degrees with negative B_z. This is not surprising given that MMS always crosses the boundary at low latitudes.

Previous studies (Phan et al., 2000; Dunlop et al., 2011) have shown that the reconnection line can extend as a stable feature across the entire dayside magnetopause, implying that reconnection is determined by large-scale interactions between the solar wind and the magnetosphere, rather than by local conditions at the magnetopause. Whistler waves that can produce anomalous resistivity by affecting electrons' motion have been observed in the EDR by Cao et al. (2017). They used MMS data to find large-amplitude whistler waves propagating away from the X line with a very small wave-normal angle. These waves were most likely generated by the perpendicular temperature anisotropy of the ~ 300 eV electrons inside the EDR, significantly affecting the electron-scale dynamics of magnetic reconnection. Zhang et al. (2020) found that near the magnetopause there are ultra-low frequency (ULF) perturbations that can modulate whistler-mode waves. While close to the magnetopause, whistler wave bursts were seen by the THEMIS mission to have the same periodicity as the ULF perturbations. This study showed that much of the outer magnetosphere is significantly influenced by ULF perturbations excited by magnetopause dynamic responses to the solar wind.

Webster et al. (2018) have reported a total of 32 encounters with the EDR while near the dayside magnetopause, representing instances of energy flow from the magnetic field to the particles (i.e., a load with $E \cdot J > 0$). As expected, these events all show phase space distributions with a thermalized core population colocated with the energized crescent contribution described by Burch et al. (2016b) responsible for producing a bulk flow perpendicular to B. In the 3D data, it can be seen that the "crescents" are actually more like a partial toroid.

6.12 Discussion

The central problem of magnetospheric research concerns mass, momentum, and energy transfer from the solar wind into the magnetosphere. There is no doubt that this process is influenced by conditions in the solar wind, especially the IMF, including its direction. One of the most dramatic examples of this dependence is the work by Phan and his colleagues. Two aspects stand out loud and clear:

- The bulk velocity normal to the magnetopause v_n for inbound and outbound crossings in Fig. 6.3 of the magnetopause/LLBL for high shear is not the same, as assumed in SMR. "There is a significant difference between inbound and outbound crossings of the normal plasma velocity v_n for high shear" (Phan and Paschmann, 1996).
- The temperature shows an increase on the left in Fig. 6.29 as expected for dissipation, but a surprising decrease on the right (Heikkila, 1998). Since the plasma is losing energy, that fact calls for a dynamo.

It can no longer be said that the plasma "shares the inward/outward motion of the magnetopause" (e.g., Sonnerup et al., 2004). This failure is a direct result of the use of a two-dimensional model for SMR. Recall what was said in Chapter 1 when summarizing the results of a coordinated study in the Geospace Environmental Modeling (GEM) program:

> *The conclusions of this study pertain explicitly to the 2D system.*
> *There is mounting evidence that the narrow layers which develop*
> *during reconnection in the 2D model are strongly unstable to a*
> *variety of modes in 3D system.*
>
> **Birn et al. (2001, p. 3718)**

Fig. 6.19 shows the view of the *localized* interaction when using 3D (but viewed in the equatorial plane). A plasma cloud (a flux tube) with excess momentum distorts the magnetopause current

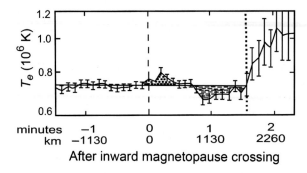

Fig. 6.29 The pivotal result of Phan and Paschmann (1996) for the superposed epoch analysis of total electron temperature T_e with high magnetic shear. Both inbound and outbound crossings are shown, it being assumed that the one is the image of the other in SMR. The region on the left within the magnetopause indicates the dissipation region with a modest increase in T_e. On the right, the analyses show a region where significant cooling has taken place in GMR. It is here that the solar wind plasma is giving up energy to the electromagnetic field. From Heikkila, W.J., 1998. Cause and effect at the magnetopause. Space Sci. Rev. 83, 429.

locally. The plasma response by charge separation is not included in SMR; the idea is that the first push is what sets everything off in GMR. This creates an *inductive electric field*, everywhere opposing current perturbations, by Lenz's law.

6.12.1 Electrostatic and induction fields

Two important matters distinguish GMR from SMR. For one, the electric field is the real electromagnetic field including both induction and charge separation. The other is the diffusion region; something happens here that is not ideal MHD as everybody knows.

Everyone should know that the real electric field has induction and electrostatic components:

$$\mathbf{E} = \mathbf{E}^{es} + \mathbf{E}^{ind} = -\nabla\phi - \frac{\partial \mathbf{A}}{\partial t} \qquad (6.11)$$

The plasma response to an induced electric field $\mathbf{E}^{ind} = -\partial\mathbf{A}/\partial t$ (the *cause* that can be robust) leads to the creation of an electrostatic field by charge separation $\mathbf{E}^{es} = -\nabla\phi$ (the *effect*). Of course, that response is affected by the conductivities of the medium.

This point is especially clear if considered in the frame of reference of the moving or meandering magnetopause, within the localized perturbation. Let us consider a very simple meander, that of a straight line current bounded at each end with an abrupt

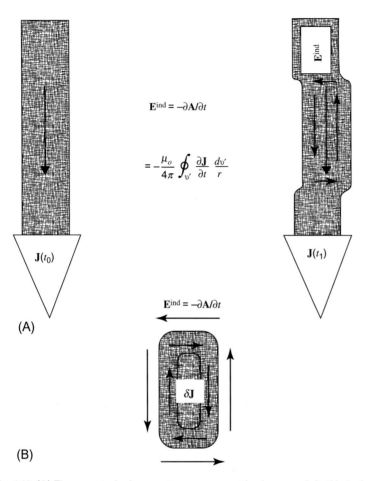

$$\mathbf{E}^{\text{ind}} = -\partial\mathbf{A}/\partial t$$

$$= -\frac{\mu_o}{4\pi} \oint_{v'} \frac{\partial\mathbf{J}}{\partial t} \frac{dv'}{r}$$

Fig. 6.30 (A) The unperturbed magnetopause current is shown on left; this is the frame used in reconnection theories where the reconnection electric field is shown embedded (supposedly due to anomalous resistivity along the X-line). PTE assumes a different approach, that of a localized meander of the magnetopause current on the right. This is associated with an induction electric field in the frame of the unperturbed magnetopause current. (B) The perturbation current by itself and the induction. From Heikkila, W.J., Canu, P., Dandouras, I., Keith, W., Khotyaintsev, Y., 2006. Plasma transfer event seen by cluster. In: Cluster and Double Star Symposium—5th Anniversary of Cluster in Space. ESA SP-598, p. 5.

shear as in Fig. 6.30. Locally, there would be no change in the current in this frame and hence no inductive electric field (again, by Faraday's and Lenz's laws). Now, the magnetopause current away from the perturbation is receding in the opposite direction. An inductive electric field is present there in the dusk-dawn sense on the sunward side and opposite on the earthward side for the sunward recession of the magnetopause current. In the deforming part (at the edges of the localized perturbation), there again will be a normal component of the inductive electric field, exactly as before. This must be so, since the parallel component is not affected by the Lorentz transformation. The inductive electric field will drive charges as before, provided that the conductivities are not zero; these charges will create an electrostatic field within the perturbation, reversing on the two sides of the magnetopause current because of the quadrupole nature of the charge distribution.

6.12.2   The dynamo

With the 2D used to describe SMR, it is impossible to show both an electrical load $\mathbf{E} \cdot \mathbf{J} > 0$ and a dynamo $\mathbf{E} \cdot \mathbf{J} < 0$. Consequently, the dynamo is omitted in discussions of standard reconnection theory.

Returning to this chapter's Fig. 6.7, the vertical line denotes the spot where the analysis could be explained by reconnection theory. The authors excuse the fact that "$|\Delta V_{\text{th}}|$ continues to stay high after this time is *an artifact* of the use ... beyond its region of relevance ..." It would be better to say that the sense of the electric field is the cause: it should be in the sense of a dynamo (opposed to the current) rather than a load. The PTE concept has the correct sense.

Let us look at the difference between inbound and outbound crossings of the normal plasma velocity v_n for high shear. A localized pressure change in the solar wind (say an increase as in Fig. 6.3A) is accompanied by a localized displacement or meander (toward the Earth) of the magnetopause current. An inductive electric field is created, enhanced by the plasma as discussed earlier, in the sense specified by Lenz's law opposing the current perturbation everywhere. It is an attractive feature of the PTE concept that the energy source is located where solar wind plasma first contacts the obstacle, the magnetopause. It should be noted that the load and the dynamo must exist in the same electric circuit to describe cause and effect.

Due to the speed of the magnetopause (at least 10 km/s), it follows that the surface is wave-like; it must come back, as in Fig. 6.24B. The feature illustrated in Fig. 6.5 shows a marked difference to that of Fig. 6.4. Now the sense of the induction field is reversed and so are the relative positions of the dynamo and the load. The pool in Fig. 6.29 comes ahead of the rise with an outbound spacecraft. To summarize, Figs. 6.5 and 6.7 verify this enhancement concept. In the region where the current is decreasing (by assumption for outbound crossings) the enhanced induction electric field has the property that $\mathbf{E}\cdot\mathbf{J} > 0$, and thus energy is going from the electromagnetic field into the plasma forming an electrical load. On the other side, where the current is increasing (again by assumption) we have $\mathbf{E}\cdot\mathbf{J} < 0$: energy is going from incoming plasma to the electromagnetic field. The observed behavior of T_e is in excellent agreement with the dynamo concept.

6.12.3 Agreement with Walén relation

AMPTE/IRM plasma data, with full three-dimensional coverage and 4 s time resolution, permitted a detailed study of plasma acceleration events at the magnetopause. Fig. 6.31 from Sonnerup (1985) and Sonnerup et al. (1995) shows correlations between the plasma velocity in the HT frame and the Alfvén speed as the spacecraft travels through the magnetopause. The event shown in Fig. 6.31A is representative of the quality of agreement with the Walén relation one finds: the flow speed in the HT frame is about 80% of the Alfvén speed.

> *Fig. 6.31B shows an instance where no agreement is present, even though an excellent HT frame existed, and even though the magnetic shear angle across the magnetopause was large. In this second case, the plasma flow in the HT frame is seen to be small, indicating that the magnetic structures within this magnetopause layer were simply convected past the spacecraft with the ambient plasma flow.*
>
> **Sonnerup et al. (1995)**

This latter result is in agreement with the PTE concept. In the middle of the perturbation in Fig. 6.31 the solar wind plasma simply goes by $\mathbf{E}\times\mathbf{B}$ drift (see Figs. 6.20 and 6.21); both $\mathbf{E}$ and $\mathbf{B}$ reverse, so $\mathbf{E}\times\mathbf{B}$ does not. Only at the edges are the particles accelerated by the induction electric field $E_{\parallel}^{\text{ind}}$, with energization.

6.12.4 Response of the plasma: B_n finite

The plasma response changes dramatically with an open magnetosphere. In this case, a rotational discontinuity will be present, with a finite B_n. Electron and ion mobilities are high along the magnetic field. Now we can use the very large direct conductivity σ_0; the plasma can polarize along the magnetic field lines as shown in Fig. 6.21, top and bottom, in *different senses*, causing an electrostatic field tangential to the magnetopause, reversing

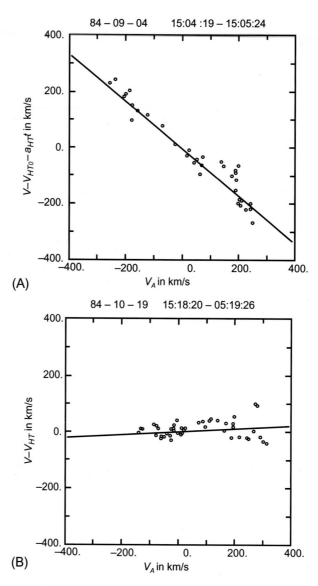

Fig. 6.31 Walén test. (A) Scatter plot of the GSE X, Y, Z components of the plasma velocity in the de Hoffmann-Teller frame, V versus the corresponding components of the modified Alfvén velocity V_A for magnetopause crossing by AMPTE/IRM on September 4, 1984. Good agreement with the Walén relation is seen. (B) Same AMPTE/IRM crossing on October 19, 1984, showing no agreement with the Walén relation. From Sonnerup, B.U.Ö., Paschmann, G., Phan, T.-D., 1995. Fluid aspects of reconnection at the magnetopause: in situ observations. In: Song, P., Sonnerup, B.U.Ö., Thomsen, M.F. (Eds.), Physics of the Magnetopause. AGU Geophysical Monograph 90. American Geophysical Union, Washington, DC, p. 172.

as indicated, *enhancing the field due to induction.* Thus we see that this $\mathbf{E}^{\mathrm{es}}$ will drive the solar wind plasma into the current sheet, as in the reconnection frame. On the other side, since both $\mathbf{B}$ and $\mathbf{E}$ reverse, the electric drift $\mathbf{E} \times \mathbf{B}$ will be also earthward. A PTE is created.

A finite B_n is crucial to the reality of a PTE. The tangential component of the induction field will be enhanced by the plasma, thus a dependence on the IMF.

6.12.5   Bohm diffusion

The equations for diffusion in fully ionized plasmas in a steady-state plasma (Chen, 1984, p. 186) should be:

$$\mathbf{J} \times \mathbf{B} = \nabla p \tag{6.12}$$

$$\mathbf{E} + \mathbf{v} \times \mathbf{B} = \eta \mathbf{J} \tag{6.13}$$

The parallel component of the latter equation is simply

$$E_{\parallel} = \eta_{\parallel} J_{\parallel} \tag{6.14}$$

which is Ohm's law (provided we know the resistivity). The perpendicular component is found by taking the cross-product with $\mathbf{B}$: Eq. (6.15)

$$\mathbf{E} \times \mathbf{B} + (\mathbf{v} \times \mathbf{B}) \times \mathbf{B} = \eta_{\perp} \mathbf{J} \times \mathbf{B} = \eta_{\perp} \nabla p \tag{6.15}$$

$$\mathbf{E} \times \mathbf{B} - v_{\perp} B^2 = \eta_{\perp} \nabla p \tag{6.16}$$

$$v_{\perp} = \frac{\mathbf{E} \times \mathbf{B}}{B^2} - \frac{\eta_{\perp}}{B^2} \nabla p \tag{6.17}$$

The first term is just the $\mathbf{E} \times \mathbf{B}$ drift of both species together. The second term is the diffusion velocity in the direction of $-\nabla p$. The flux associated with diffusion is

$$\Gamma_{\perp} = n v_{\perp} = \frac{\eta_{\perp} n (kT_i + kT_e)}{B^2} \nabla n \tag{6.18}$$

With diffusion coefficient

$$D_{\perp} = \frac{\eta_{\perp} n \sum kT}{B^2} \tag{6.19}$$

This is the so-called classical diffusion coefficient. Note that
- $D_{\perp}$ is proportional to $1/B^2$.
- It is not a constant but it is proportional to n.
- It decreases with increasing temperature in a fully ionized gas due to the resistivity term.

- Diffusion is automatically ambipolar.
- If a transverse **E** field is applied to a uniform plasma, both species drift together with the $\mathbf{E} \times \mathbf{B}$ velocity.

Although the theory of diffusion via Coulomb collisions had been known for a long time, laboratory verification of the $1/B^2$ dependence of $D_\perp$ in a fully ionized plasma eluded all experimenters until the 1960s. Diffusion scaled as B^{-1} rather than B^{-2}, and the decay of plasmas with time was found to be exponential rather than reciprocal. Furthermore, the absolute value was far larger than that given by $D_\perp$.

Chen (1984, p. 190)

This anomalously poor magnetic confinement was first noted by Bohm et al. (1949). The semiempirical formula follows the Bohm diffusion law:

$$D_\perp = \frac{1}{16} \frac{kT_e}{eB} \equiv D_B \qquad (6.20)$$

where D_B is independent of density, giving a decay exponential with time. It increases, rather than decreases, with temperature, and though it decreases with B, it decreases more slowly than expected.

The scaling of D_B with KT_e can easily be shown to be the natural one whenever the losses are caused by $\mathbf{E} \times \mathbf{B}$ drifts, either stationary or oscillating. Let the escape flux be proportional to the $\mathbf{E} \times \mathbf{B}$ drift velocity:

$$\Gamma_\perp = n v_\perp \propto nE/B \qquad (6.21)$$

Because of Debye shielding, the maximum potential in the plasma is given by

$$e\phi_{\max} \approx kT_e \qquad (6.22)$$

If R is a characteristic scale length of the plasma (of the order of its radius), the maximum electric field is then

$$E_{\max} \approx \frac{\phi_{\max}}{R} \approx \frac{kT}{eR} \qquad (6.23)$$

$$\Gamma_\perp \approx \gamma \frac{nkT_e}{R\,eB} \approx -\gamma \frac{kT_e}{eB} \nabla n = -D_B \nabla n \qquad (6.24)$$

where γ is some fraction less than unity. Thus that D_B is proportional to kT_e/eB is no surprise. The value $\gamma = 1/16$ has no theoretical justification but is an empirical number agreeing with most experiments. Thus the PTE process with its $\mathbf{E} \times \mathbf{B}$ dependence is likely to be linked to Bohm diffusion.

6.12.6   Diversion of the magnetosheath flow

The diversion of the magnetosheath flow around the magneto-spheric obstacle (the magnetopause) involves more than 90% of the solar wind intercepted by the bow shock, about 10^{28} ions/s. Sonnerup et al. (2004) have handled it in an amazing way, as we saw in Fig. 6.14. However, they should not have included C3 in their analysis. With C3 left out they would be released from at least three difficulties.

(1) The fit would be better; as they said: "This relatively low value [of the correlation coefficient] is a consequence of the fact that C1 and C3 separately gave somewhat different $\mathbf{v}_{HT}$ vectors."

(2) The velocities for C3 were supersonic, violating the conditions of their analysis.

(3) The temperature minimum on C3 at FTE2 is quite visible in Fig. 6.13. The reconnection process requires dissipation with $\mathbf{E}\cdot\mathbf{J} > 0$, implying a temperature maximum.

Instead of a "fossil record of the prior reconnection process that created the flux rope" (Sonnerup et al., 2004), a PTE offers the real solution of a transfer into the LLBL (Heikkila et al., 2006). This explains the problems raised with Figs. 6.7 and 6.8.

• It is essential to include the tangential motion of the perturbation to the magnetopause to understand the effects of a PTE upon the physics of the magnetosphere. In the magnetosheath all the solar wind plasma is moving antisunward, even super-Alfvénically toward the flanks. Whatever plasma penetrates through the magnetopause must face conditions as a new medium. The two black arrows in Fig. 6.32 denote the tailward motion of the plasma, higher in the magnetosheath, lower in the LLBL. This difference permits multiple injection events in the LLBL due to successive blasts from the magnetosheath.

• There is an example from hydrodynamics: the vertical lift force F_y resulting from a flow (Landau and Lifshitz, 1959, p. 139) is directly proportional to the curl of the circulation by Zhukovskii's theorem (Zhukovskii, 1906). It seems possible that the same is true for magnetic reconnection following Faraday's law.

6.12.7   Change of magnetic interconnection

There is no doubt that the interconnection of the geomagnetic field with the IMF is continually changing, and it is appropriate to call the process that affects that change *magnetic reconnection* or merging. Unfortunately, there appear to be at least two very different connotations of this concept (Heikkila, 1983) but the

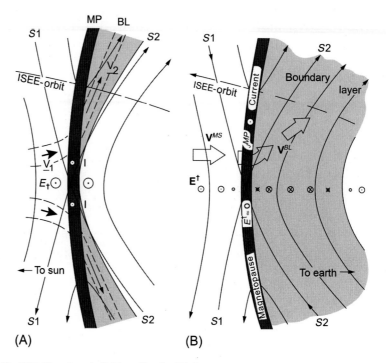

Fig. 6.32 The electric field profiles for (A) the reconnection model and (B) for a PTE. The first is a 2D model, while the second is 3D (a cut through the middle of the localized perturbation); the latter has a finite curl implying time dependence by Faraday's law. Here the plasma can move across closed field lines, since both **E** and **B** are reversed, in contrast to the situation shown in (A). In the PTE model, three-dimensional and time-dependent effects are important. From Heikkila, W.J., 1983. Comment on 'The causes of convection in the Earth's magnetosphere: a review of developments during the IMS' by S. W. H. Cowley. Rev. Geophys. 21(8), 1787.

difference is never openly expressed, nor even appreciated. One is colloquial, the other mathematical. It is likely that true reconnection is accomplished by the electromotive force (and only by the electromotive force) $emf = -d\Phi^M/dt$ which energy can be interchanged with stored magnetic energy.

6.12.8 Plasma flow

Plasma flow is nearly always tailward in the distant magnetotail beyond 180 R_E. Heikkila (1987a, 1988) and Pulkkinen et al. (1996) have found that the magnetic field is northward in the field reversal region for low to moderate plasma velocities (higher velocities are related to transient effects). The latter finding was true even with high levels of geomagnetic activity. The implication

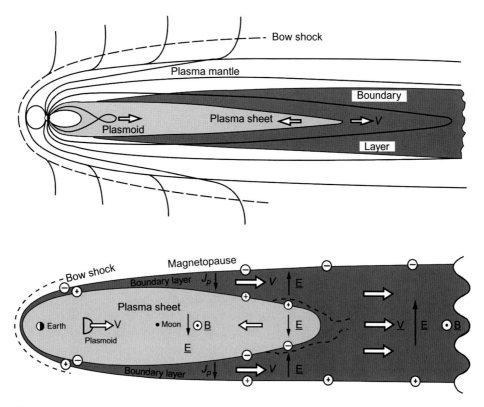

Fig. 6.33 The northward component of the magnetic field in the distant magnetotail shown by the ISEE-3 observations (in the absence of plasmoids) implies that the electric field is from dusk to dawn, the reverse of its sense in the plasma sheet. These two cuts of the entire magnetosphere show that the plasma sheet is a small cavity of low-density plasma within a sea of boundary layer and mantle plasma; the LLBL should be visualized as being wrapped around the plasma sheet, at least partly on closed field lines. From Heikkila, W.J., 1990. Magnetic reconnection, merging, and viscous interaction in the magnetosphere. Space Sci. Rev. 53, 4.

is that the magnetic field lines are closed in the steady state. This is inconsistent with the Dungey model in Fig. 1.15A in Chapter 1 with tailward flow in the distant magnetotail with a dawn-dusk electric field. On the other hand, these later findings suggest that a tailward extension of the LLBL is continuously present instead, as indicated in Fig. 6.33.

A transient process that is reminiscent of reconnection does occur, but in fact is very different from steady-state reconnection because of the nature of the electric field involving a curl (Heikkila, 1987b). Of the two connotations of the concept of reconnection, it is unlikely that the one regarding two-dimensional steady-state theories can explain three-dimensional, localized, time-dependent effects.

The magnetic field in the distant magnetotail shown by the ISEE-3 observations is northward (Heikkila, 1988; Pulkkinen et al., 1996) showing no indication of an X-line; this is shown by the bottom panel of Fig. 6.33. The LLBL should be visualized as being wrapped around the plasma sheet, mostly on closed field lines.

The main purpose of this book is to review these observations and to explain their meaning. In short, the simple Dungey model, based on steady-state reconnection, is invalidated by these observations. The criticism is meant to be direct but scholarly, not personal. Alfvén (1975) has given the best apology for this: "To criticize is often simpler than to construct, but in the present situation it is also more important as a first phase."

A parallel might perhaps be drawn from another field in geophysics, that of continental drift.

Eminent geophysicists voiced strong opposition. ... The slow acceptance of what is actually a very old idea provides a good example of the intensive scrutiny to which scientific theories are subjected, particularly in the Earth sciences where the evidence is often conflicting and where experimental demonstrations are usually not possible.

Hurley (1968)

6.12.9 What is really happening?

Poynting's theorem (Chapter 5) provides a guide. In the steady state (no time dependence) this is:

$$\iiint_{\text{vol}} \mathbf{E} \cdot \mathbf{J} d\tau = -\frac{1}{\mu_0} \oiint_{\text{surf}} \mathbf{E} \times \mathbf{B} \cdot d\mathbf{S} \qquad (6.25)$$

The left-hand side represents the dissipation rate, that is, the power due to reconnection. The dissipation with higher electron temperature (Fig. 6.6) requires a source of energy, a dynamo. That dynamo must be the term on the right. Instead of the electrons going transverse to the electric field with no energy gain, they go uphill to supply the power for the dissipation (due to the time-dependent terms). This point is where the dashed line is located in Fig. 6.7; the error is not *due to an artifact*, but it is due to having no dynamo in SMR.

Most of the theories proposed to explain the interaction between the solar wind and the geomagnetic field are stationary descriptions based on ideal MHD. In this review an alternative, non-stationary description is discussed. According to this description, most of the plasma-field irregularities, i.e., plasmoids, detected in the solar wind can penetrate inside the geomagnetic field beyond what is considered to be the mean position of the magnetopause. It is the patchy solar wind plasma impinging on the geomagnetic field which imposes rapidly changing and non-uniform boundary conditions over the whole outer magnetospheric surface. ... The emergence of the idea of "impulsive penetration" (IP) of solar wind plasma irregularities into the magnetosphere is emphasized.

Lemaire and Roth (1991, p. 59)

The boundary surface separating the magnetosheath from the magnetosphere can be either a tangential discontinuity or a rotational discontinuity, depending on the value of B_n. If $B_n \sim 0$ then we have a tangential discontinuity; there is no magnetic connection between the two. However, a finite value of B_n means rotational discontinuity; the magnetic field lines penetrate the surface of separation (called *open*). It is through these positions that the solar wind plasma particles can cross the magnetopause, as illustrated in Chapter 1 Fig. 1.18 by Lemaire and Roth (1978), and conversely magnetospheric particles can be guided along these channels and escape out of the magnetosphere.

6.13 Summary

Phan et al. (1994) have performed a revealing survey of dayside magnetopause crossings by the AMPTE/IRM satellite. They studied the variations of key plasma parameters and magnetic field in the magnetosheath region adjacent to the dayside magnetopause. A total of 38 AMPTE/IRM passes were selected as appropriate for the study. They find that the structures of the key plasma parameters and magnetic field and the dynamics of plasma flows in this region depend strongly on the magnetic shear across the magnetopause, that is, on the angle between the magnetosheath magnetic field and the geomagnetic field.

- With low shear, ending with the low value of 30 degrees, the conditions begin to change well outside the magnetopause indicating no solar wind interaction on the dayside magnetopause.
- There is a quantum jump to high shear beginning already at 60 degrees, all the way to 180 degrees, with hardly any change. It is almost as if the plasma was unaware of any obstacle at all!
- The fact that the magnetic field remains almost constant on approach to the high-shear magnetopause is an indication of a significant violation of the frozen-in magnetic field condition

somewhere at the magnetopause that presumably leads to the transfer of flux and plasma across the magnetopause.

- It is remarkable that all the high-shear magnetopauses examined in this study seem to be open, even though a signature normally associated with the openness of the magnetopause is absent. This feature can be explained by a PTE. A dynamo is involved, with $\mathbf{E}\cdot\mathbf{J}<0$. This region is a different spatial region from dissipation with $\mathbf{E}\cdot\mathbf{J}>0$.

Heikkila et al. (2006) have observed a PTE with Cluster data. CIS/CODIF data show penetration of solar wind plasma lasting for over 1 min at the time of the event. A dynamo is involved, with $\mathbf{E}\cdot\mathbf{J}<0$. This region is a different spatial region from dissipation with $\mathbf{E}\cdot\mathbf{J}>0$. This latter feature can be explained by a PTE.

- The presence of He^{++}, and the similar shape of the H^+ profile, verifies this identification of solar wind plasma. The plasma density is increased, quadrupled in the middle, for over 1 min, in a burst of plasma entering through the magnetopause.
- PEACE electron data show several aspects in agreement with a PTE.
- The pitch angle distribution for the higher-energy channels has a maximum near 90 degrees, indicating field lines are closed on either side.
- In the heart of the PTE, these fluxes are greatly reduced but still show some maximum at 90 degrees.
- Since B_n is small compared to the magnetic field on the either side (magnetosheath and LLBL) there will be some trapping along B_n in the current layer. This result may indicate open magnetic field lines, a firm requirement for the PTE process.
- At lower energies, intense fluxes are observed, with a maximum at 0 degrees at 07:30 and 180 degrees at 07:42.
- WHISPER instrument showed two bursts of plasma emissions at the beginning and end of the PTE. It is likely that these bursts coincided with the parallel component of the induction electric field as shown in Fig. 6.20. These bursts probably produced the emissions.
- EFW instrument also showed the bursty structure. The DC electric field showed that the sense reversed in E_y, and it was strong in magnitude ($\sim$5 v/km).
- Fig. 6.14 includes a possible trajectory of C3 through the PTE. GMR in three dimensions is based on the *breakdown of magnetic connection*. It opens the door so that the push of solar wind plasma is able to penetrate through the magnetopause current, and into the LLBL, onto closed magnetic field lines.

6.14 Problems

6.1 Assume that the electric field experiment on the Polar satellite can identify a spatial structure if it can take 10 samples while inside it. Assuming the fastest sample rate (and slowest orbital velocity), what is the smallest structure that the electric field experiment can identify.

6.2 A plasmoid in the solar wind travels with velocity $v_{sw} = 440$ km/s, but with an excess density dn over the average density of $n = 4.2$ particles/cm^3. If the velocity of the plasmoid is $v_e = 3$ km/s at the mean magnetopause position (where the background plasma velocity becomes zero), what is the excess density dn?

6.3 What would be the Bohm diffusion rate for a plasma of temperature $T_e = 1.0 \times 10^6$ K in a magnetic field of $B = 60$ nT?

6.4 How would you define the "key time"? Explain why it is important in superposed epoch analysis. Find a graph online of the atmospheric temperature profile for a planet. What point along the curve would make a good "key time"?

6.5 How fast would a 4 km thick electron diffusion region at the magnetopause have to be moving before MMS would not be able to capture a single 3D electron distribution function while imbedded in the region?

References

Alfvén, H., 1975. Electric current structure of the magnetosphere. In: Hultqvist, B. (Ed.), Nobel Symposium No. 30, Physics of the Hot Plasma in the Magnetosphere. Plenum Press, New York, NY, pp. 1–22.

Aubry, M.P., Russell, C.T., Kivelson, M.G., 1970. Inward motion of the magnetopause before a substorm. J. Geophys. Res. 75, 7018–7031.

Baker, D.A., Hammel, J.E., 1962. Demonstration of classical plasma behavior in a transverse magnetic field. Phys. Rev. Lett. 8 (4), 157.

Baker, D.A., Hammel, J.E., 1965. Experimental studies of the penetration of a plasma stream into a transverse magnetic field. Phys. Fluids 8 (4), 713–722.

Berkowitz, R., 2019. Satellites glimpse the microphysics of magnetic reconnection. Phys. Today 72 (2), 20. https://doi.org/10.1063/PT.3.4129.

Birn, J., Drake, J.F., Shay, M.A., Rogers, B.N., Denton, R.E., Hesse, M., et al., 2001. Geospace environmental modeling (GEM) magnetic reconnection challenge. J. Geophys. Res. 106, 3715.

Bohm, D., Burhop, E.H.S., Massey, H.S.W., 1949. The use of probes for plasma exploration in strong magnetic fields. In: Guthrie, A., Wakerling, R.K. (Eds.), The Characteristics of Electrical Discharges in Magnetic Fields. McGraw-Hill, New York, NY.

Bostick, W.H., 1957. Experimental study of plasmoids. Phys. Rev. 106, 404–412.

Bullard, E., 1975. The emergence of plate tectonics: a personal view. Annu. Rev. Earth Planet. Sci. 3, 1–130.

Burch, J.L., Moore, T.E., Torbert, R.B., Giles, B.L., 2016a. Magnetospheric multiscale overview and science objectives. Space Sci. Rev. 199 (1–4), 5–21. https://doi.org/10.1007/s11214-015-0164-9.

Burch, J.L., et al., 2016b. Electron-scale measurements of magnetic reconnection in space. Science 352, aaf2939. https://doi.org/10.1126/science.aaf2939.

Cao, D., Fu, H.S., Cao, J.B., Wang, T.Y., Graham, D.B., Chen, Z.Z., Peng, F.Z., Huang, S.Y., Khotyaintsev, Y.V., André, M., Russell, C.T., Giles, B.L., Lindqvist, P.-A., Torbert, R.B., Ergun, R.E., Le Contel, O., Burch, J.L., 2017. MMS observations of whistler waves in electron diffusion region. Geophys. Res. Lett. 44, 3954–3962. https://doi.org/10.1002/2017GL072703.

Carlson, C.W., Torbert, R.B., 1980. Solar wind ion injections in the morning auroral oval. J. Geophys. Res. 85 (A6), 2903–2908.

Chen, F.F., 1984. Introduction to plasma physics and controlled fusion. In: Plasma Physics. second ed. vol. 1. Plenum Press, New York, NY.

De Keyser, J., 2005. The Earth's magnetopause: reconstruction of motion and structure. Space Sci. Rev. 121 (1), 225–235.

Demidenko, I.I., Lomino, N.S., Padalka, V.G., Safronov, B.G., Sinel'nikov, K.D., 1966. The mechanism of capture of a moving plasma by a transverse magnetic field. J. Nucl. Eng. Part C 8, 433.

Dungey, J.W., 1961. Interplanetary magnetic field and the auroral zones. Phys. Rev. Lett. 6, 47–48.

Dunlop, M.W., Zhang, Q.-H., Bogdanova, Y.V., Lockwood, M., Pu, Z., Hasegawa, H., Wang, J., Taylor, M.G.G.T., Berchem, J., Lavraud, B., Eastwood, J., Volwerk, M., Shen, C., Shi, J.-K., Constantinescu, D., Frey, H., Fazakerley, A.N., Sibeck, D., Escoubet, P., Wild, J.A., Liu, Z.-X., 2011. Extended magnetic reconnection across the dayside magnetopause. Phys. Rev. Lett. 107, 025004.

Echim, M.M., Lemaire, J.F., 2000. Laboratory and numerical simulations of the impulsive penetration mechanism. Space Sci. Rev. 92, 565–601. https://doi.org/10.1023/A:1005264212972.

Echim, M.M., Lemaire, J.F., Roth, M., 2005. Self-consistent solution for a collisionless plasma slab in motion across a magnetic field. Phys. Plasmas 12 (7), 072904–072911.

Emslie, A.G., Miller, J.A., 2003. Particle acceleration. In: Dwivedi, B.N. (Ed.), Dynamic Sun. Cambridge University Press, Cambridge, UK, pp. 262–287.

Fälthammar, C.-G., 2004. Magnetic-field aligned electric fields in collisionless space plasmas—a brief review. Geofis. Int. 43, 225–239.

Frank, L.A., Ackerson, K.L., Paterson, W.R., Lee, J.A., English, M.R., Pickett, G.L., 1994. The comprehensive plasma instrumentation (CPI) for the GEOTAIL spacecraft. J. Geomagn. Geoelectr. 46, 23–37.

Fuselier, S.A., Vines, S.K., Burch, J.L., Petrinec, S.M., Trattner, K.J., Cassak, P.A., et al., 2017. Large-scale characteristics of reconnection diffusion regions and associated magnetopause crossings observed by MMS. J. Geophys. Res. Space Physics 122, 5466–5486. https://doi.org/10.1002/2017JA024024.

Gunell, H., Nilsson, H., Stenberg, G., Hamrin, M., Karlsson, T., Maggiolo, R., Andre', M., Lundin, R., Dandouras, I., 2012. Plasma penetration of the dayside magnetopause. Phys. Plasmas 19, 072906. https://doi.org/10.1063/1.4739446.

Gunell, H., et al., 2014. Waves in high-speed plasmoids in the magnetosheath and at the magnetopause. Ann. Geophys. 32, 991–1009. https://doi.org/10.5194/angeo-32-991-2014.

Gustafsson, G., André, M., Carozzi, T., Eriksson, A.I., Fälthammar, C.-G., Grard, R., et al., 2001. First results of electric field and density observations by Cluster/EFW based on initial months of operation. Ann. Geophys. 19, 1219–1240.

Haaland, S., Reistad, J., Tenfjord, P., Gjerloev, J., Maes, L., DeKeyser, J., Maggiolo, R., Anekallu, C., Dorville, N., 2014. Characteristics of the flank magnetopause: cluster observations. J. Geophys. Res. 119, 9019–9037. https://doi.org/10.1002/2014JA020539.

Haaland, S., Runov, A., Artemyev, A., Angelopoulos, V., 2019. Characteristics of the flank magnetopause: THEMIS observations. J. Geophys. Res. 124, 3421–3435. https://doi.org/10.1029/2019JA026459.

Haerendel, G., Paschmann, G., 1982. Interaction of the solar wind with the dayside magnetosphere (A83-48551 23-46). In: Magnetospheric Plasma Physics. Center for Academic Publications/D. Reidel, Tokyo, Japan/Dordrecht, Netherlands, pp. 49–142.

Hapgood, M.A., Bryant, D.A., 1990. Re-ordered electron data in the low-latitude boundary layer. Geophys. Res. Lett. 17 (11), 2043–2046.

Heikkila, W.J., 1982a. Impulsive plasma transport through the magnetopause. Geophys. Res. Lett. 9 (2), 159–162.

Heikkila, W.J., 1982b. Inductive electric field at the magnetopause. Geophys. Res. Lett. 9 (8), 877–880.

Heikkila, W.J., 1983. Comment on 'The causes of convection in the Earth's magnetosphere: a review of developments during the IMS' by S. W. H. Cowley. Rev. Geophys. 21 (8), 1787–1788.

Heikkila, W.J., 1987a. Neutral sheet crossings in the distant magnetotail. In: Lui, A.T. (Ed.), Magnetotail Physics. Johns Hopkins University Press, Baltimore, MD, pp. 65–71.

Heikkila, W.J., 1987b. Dialog on the phenomenological model of substorms in the magnetotail. In: Lui, A.T. (Ed.), Magnetotail Physics. Johns Hopkins University Press, Baltimore, MD, p. 416.

Heikkila, W.J., 1988. Current sheet crossings in the distant magnetotail. Geophys. Res. Lett. 15 (4), 299–302.

Heikkila, W.J., 1998. Cause and effect at the magnetopause. Space Sci. Rev. 83, 373–434.

Heikkila, W.J., Canu, P., Dandouras, I., Keith, W., Khotyaintsev, I., 2006. Plasma transfer event seen by cluster. In: Cluster and Double Star Symposium—5th Anniversary of Cluster in Space ESA SP-598.

Hudson, P.D., 1970. Discontinuities in an anisotropic plasma and their identification in the solar wind. Planet. Space Sci. 18, 161.

Hurley, P.D., 1968. The confirmation of continental drift. Sci. Am. 218, 52–64.

Johnstone, A.D., Alsop, C., Burge, S., Carter, P.J., Coates, A.J., Coker, A.J., et al., 1997. PEACE: a plasma electron and current experiment. Space Sci. Rev. 79, 351.

Kawano, H., Kokubun, S., Yamamoto, T., Tsuruda, K., Hayakawa, H., Nakamura, M., et al., 1994. Magnetopause characteristics during a four-hour interval of multiple crossings observed with GEOTAIL. Geophys. Res. Lett. 21 (25), 2895–2898. https://doi.org/10.1029/94GL02100.

Koga, J., Geary, J.L., Fujinami, T., Newberger, B.S., Tajima, T., Rostoker, N., 1989. Numerical investigation of a plasma beam entering transverse magnetic fields. J. Plasma Phys. 42 (Pt 1), 91–110.

Landau, L.D., Lifshitz, E.M., 1959. Fluid Mechanics. Translated from the Russian by J. B. Sykes and W. H. Reid Addison-Wesley: Pergamon Press, London, UK: Reading, MA.

Lemaire, J., 1977. Impulsive penetration of filamentary plasma elements into the magnetospheres of the Earth and Jupiter. Planet. Space Sci. 25, 887.

Lemaire, J., 1985. Plasmoid motion across a tangential discontinuity (with application to the magnetopause). J. Plasma Phys. 33 (3), 425.

Lemaire, J., Roth, M., 1978. Penetration of solar wind plasma elements into the magnetosphere. J. Atmos. Terr. Phys. 40 (3), 331–335.

Lemaire, J., Roth, M., 1991. Non-steady-state solar wind-magnetosphere interaction. Space Sci. Rev. 57, 59–108.

Livesey, W.A., Pritchett, P.L., 1989. Two dimensional simulations of a charge-neutral plasma beam injected into a transverse magnetic field. Phys. Fluids B 1, 914.

Lockwood, M., Hapgood, M.A., 1997. How the magnetopause transition parameter works. Geophys. Res. Lett. 24, 373.

Lundin, R., 1997. Observational and theoretical aspects of processes (other than merging and diffusion) governing plasma transport across the magnetopause. Space Sci. Rev. 80, 269–304.

Lundin, R., Dubinin, E., 1984. Solar wind energy transfer regions inside the dayside magnetopause—I. Evidence for magnetosheath plasma penetration. Planet. Space Sci. 32, 745–755.

Lundin, R., et al., 2003. Evidence for impulsive solar wind plasma penetration through the dayside magnetopause. Ann. Geophys. 21, 457–472. https://doi.org/10.5194/angeo-21-457-2003.

Lyatsky, W., Pollock, C., Goldstein, M.L., Lyatskaya, S., Avanov, L., 2016a. Penetration of magnetosheath plasma into dayside magnetosphere: 1. Density, velocity, and rotation. J. Geophys. Res. 121, 7699–7712. https://doi.org/10.1002/2015JA022119.

Lyatsky, W., Pollock, C., Goldstein, M.L., Lyatskaya, S., Avanov, L., 2016b. Penetration of magnetosheath plasma into dayside magnetosphere: 2. Magnetic field in plasma filaments. J. Geophys. Res. 121, 7713–7727. https://doi.org/10.1002/2015JA022120.

McDiarmid, I., Burrows, J., Budzinski, E., 1976. Particle properties in the day side cleft. J. Geophys. Res. 81 (1), 221–226.

Morse, P.M., Feshbach, H., 1953. Methods of Theoretical Physics. McGraw-Hill, New York, NY.

Mozer, F.S., Pritchett, P.L., 2010. Spatial, temporal, and amplitude characteristics of parallel electric fields associated with subsolar magnetic field reconnection. J. Geophys. Res. 115, A04220. https://doi.org/10.1029/2009JA014718.

Mozer, F.S., Torbert, R.B., Fahleson, U.V., Fälthammar, C.-G., Gonfalone, A., Pedersen, A., 1978. Electric field measurements in the solar wind bow shock, magnetosheath, magnetopause and magnetosphere. Space Sci. Rev. 22, 791.

Nakamura, R., Baumjohann, W., Mouikis, C., Kistler, L.M., Runov, A., Volwerk, M., et al., 2004. Spatial scale of high-speed flows in the plasma sheet observed by cluster. Geophys. Res. Lett. 31, L09804. https://doi.org/10.1029/2004GL019558.

Parker, E.N., 1996. The alternative paradigm for magnetospheric physics. J. Geophys. Res. 10 (10), 587.

Paschmann, G., Baumjohann, W., Sckopke, N., Phan, T.-D., Lühr, H., 1993. Structure of the dayside magnetopause for low magnetic shear. J. Geophys. Res. 98 (A8), 13409–13422.

Peter, W., Rostoker, N., 1982. Theory of plasma injection into a magnetic field. Phys. Fluids 25 (4), 730–735.

Phan, T.-D., Paschmann, G., 1996. The low-latitude dayside magnetopause and boundary layer for high magnetic shear: 1. Structure and motion. J. Geophys. Res. 101, 7801. https://doi.org/10.1029/95JA03752.

Phan, T.-D., Paschmann, G., Baumjohann, W., Sckopke, N., Lühr, H., 1994. The Magnetosheath region adjacent to the dayside magnetopause: AMPTE/IRM observations. J. Geophys. Res. 99 (A1), 121–141. https://doi.org/ 10.1029/93JA02444.

Phan, T.-D., Paschmann, G., Sonnerup, B.U.Ö., 1996. The low-latitude dayside magnetopause and boundary layer for high magnetic shear: 2. Occurrence of magnetic reconnection. J. Geophys. Res. 101 (A4), 7817–7828. https://doi. org/10.1029/95JA03751.

Phan, T., Kistler, L., Klecker, B., Haerendel, G., Paschmann, G., Sonnerup, B.U.O., Baumjohann, W., Bavassano-Cattaneok, M.B., Carlson, C.W., DiLellisk, A.M., Fornacon, K.-H., Frank, L.A., Fujimoto, M., Georgescu, E., Kokubun, S., Moebius, E., Mukai, T., Øieroset, M., Paterson, W.R., Reme, H., 2000. Extended magnetic reconnection at the Earth's magnetopause from detection of bi-directional jets. Nature 404, 848–850.

Phan, T.-D., Oieroset, M., Fujimoto, M., 2005. Reconnection at the dayside low-latitude magnetopause and its nonrole in low-latitude boundary layer formation during northward interplanetary magnetic field. Geophys. Res. Lett.. 32, L17101. https://doi.org/10.1029/2005GL023355.

Phan, T.-D., Eastwood, J.P., Shay, M.A., Drake, J.F., Sonnerup, B.U.O., Fujimoto, M., Cassak, P.A., Oieroset, M., Burch, J.L., Torbert, R.B., Rager, A.C., Dorelli, J.C., Gerchman, D.J., Pollock, C., Pyakurel, P.S., Haggerty, C.C., Khotyaintsev, Y., Lavraud, B., Saito, Y., Oka, M., Ergun, R.E., Rentino, A., LeContel, O., Argall, M.R., Giles, B.L., Moore, T.E., Wilder, F.D., Strangeway, R.J., Russell, C.T., Lindqvist, P.A., Magnes, W., 2018. Electron magnetic reconnection without ion coupling in Earth's turbulent magnetosheath. Nature 557 (2), 202–206.

Pickett, J.S., Menietti, J.D., Gurnett, D.A., Tsurutani, B., Kintner, P.M., Klatt, E., et al., 2003. Solitary potential structures observed in the magnetosheath by the cluster spacecraft. Nonlinear Process. Geophys. 10, 3–11.

Pulkkinen, T.I., Baker, D.N., Owen, C.J., Slavin, J.A., 1996. A model for the distant tail field: ISEE-3 revisited. J. Geomagn. Geoelectr. 48, 455.

Rème, H., Aoustin, C., Bosqued, J.M., Dandouras, I., Lavraud, B., Sauvaud, J.A., et al., 2001. First multispacecraft ion measurements in and near the Earth's magnetosphere with the identical Cluster ion spectrometry (CIS) experiment. Ann. Geophys. 19, 1303.

Russell, C.T., 1978. The ISEE 1 and 2 fluxgate magnetometers. IEEE Trans. Geosci. Electron. GE-16 (3), 239.

Russell, C.T., Elphic, R.C., 1978. Initial ISEE magnetometer results—magnetopause observations. Space Sci. Rev. 22, 681–715.

Schindler, K., Hesse, M., Birn, J., 1988. General magnetic reconnection, parallel electric fields, and helicity. J. Geophys. Res. 93, 5547–5557. https://doi.org/ 10.1029/JA093iA06p05547.

Schmidt, G., 1960. Plasma motion across magnetic fields. Phys. Fluids 3, 961.

Schmidt, G., 1979. Physics of High Temperature Plasmas, second ed. Academic Press, New York, NY.

Song, P., Sonnerup, B.U.Ö., Thomsen, M.F., (Eds.), 1995. Physics of the Magnetopause. AGU Geophysical Monograph 90. American Geophysical Union, Washington, DC.

Sonnerup, B.U.Ö., 1985. Magnetic field reconnection in cosmic plasma. In: - Kundu, M.R., Holman, G.D. (Eds.), Unstable Current Systems and Plasma Instabilities in Astrophysics. D. Reidel, Dordrecht, Netherlands.

Sonnerup, B.U.O., Paschmann, G., Phan, T.-D., 1995. Fluid aspects of reconnection at the magnetopause: in situ observations. In: Song, P., Sonnerup, B.U.O., Thomsen, M.F. (Eds.), Physics of the Magnetopause. AGU Geophysical Monograph 90. American Geophysical Union, Washington, DC, pp. 167–180.

Sonnerup, B.U.Ö., Hasegawa, H., Paschmann, G., 2004. Anatomy of a flux transfer event seen by cluster. Geophys. Res. Lett. 31, L11803. https://doi.org/10.1029/2004GL020134.

Spreiter, J.R., Alksne, A.Y., 1968. Comparison of theoretical predictions of the flow and magnetic field exterior to the magnetosphere with the observations of Pioneer 6. Planet. Space Sci. 16, 971–979.

Spreiter, J.R., Alksne, A.Y., 1969. Plasma flow around the Magnetosphere. Rev. Geophys. 7 (1,2), 11–50.

Webster, J.M., Burch, J.L., Reiff, P.H., Daou, A.G., Genestreti, K.J., Graham, D.B., et al., 2018. Magnetospheric multiscale dayside reconnection electron diffusion region events. J. Geophys. Res. Space Physics 123, 4858–4878. https://doi.org/10.1029/2018JA025245.

Woch, J., Lundin, R., 1992. Signatures of transient boundary layer processes observed with Viking. J. Geophys. Res. 97, 1431.

Zhang, X.-J., Angelopoulos, V., Artemyev, A.V., Hartinger, M.D., Bortnik, J., 2020. Modulation of Whistler waves by ultra-low-frequency perturbations: the importance of magnetopause location. J. Geophys. Res. 125, e2020JA028334. https://doi.org/10.1029/2020JA028334.

Zhukovskii, N.E., 1906. De la chute dans l'air de corps légers de forme allongée, animés d'un mouvement rotatoire. Bulletin de l'Institute Aérodynamique de Koutchino, Fascicule 1, St. Petersbourg, Russie.

7

High-altitude cusps

Chapter outline

> *The reason why new concepts in any branch of science are hard to grasp is always the same; contemporary scientists try to picture the new concept in terms of ideas which existed before.*
>
> **Dyson (1958)**

Earth's Magnetosphere. https://doi.org/10.1016/B978-0-12-818160-7.00007-7

7.1 Introduction

The Earth's magnetospheric high-altitude cusp is a region of weak magnetic fields with a funnel-shaped geometry, one in the north and one in the south, centered at about noon where the fields converge, as can be seen in Fig. 7.1. The magnetospheric cusps permit for the most direct entry of shocked solar wind plasma into the magnetosphere. This was discovered by Heikkila et al. (1970), Heikkila and Winningham (1971), and Frank (1971). These magnetic field lines at high latitudes map to the entire magnetopause surface depending on location (Chapman and Ferraro, 1931; Roederer, 1970). The cusps can respond in different ways depending on the interplanetary magnetic field (IMF) and the dipole tilt.

For decades the cusp was considered as only a sink of energy. No significant energetic particle fluxes were expected to be detected there (except passing through). Therefore it came as a big surprise when the Polar spacecraft measured cusp energetic particles (CEP) in the high-altitude cusp region (Chen et al., 1997, 1998; Sheldon et al., 1998). Subsequent satellite measurements revealed that the Earth's magnetospheric cusps are in fact broad and dynamic regions having a size of several R_E with charged particles of high energies from 20 keV up to 20 MeV (Chen and Fritz, 2002). The increased intensities of charged particles were associated with deep diamagnetic cavities.

In order to maintain a pressure balance with adjacent magnetospheric regions, the magnetic pressure in the exterior cusp must decrease to account for the increase in thermal pressure from the

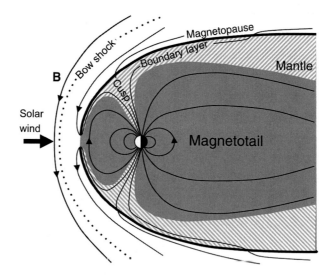

Fig. 7.1 Schematic of the magnetosphere in the noon-midnight plane showing the magnetospheric cusps in the north and south. From Mandea, M., Korte, M., Yau, A., Petrovsky E. (Eds.), 2019. Geomagnetism, Aeronomy and Space Weather: A Journey From the Earth's Core to the Sun. Cambridge University Press, Cambridge, UK.

newly injected plasma (Walsh et al., 2010). The result is a diamagnetic cavity in the exterior cusp; in some cases, this can bring the magnetic field strength close to zero nT. The whole region was not small, rather it seemed to be a broad cleft rather than a sharp cusp (Heikkila, 1972).

> *Energetic particles are a consistent and common feature of the high-altitude dayside cusp. Observing these particles in a region where they cannot be stably trapped is one the most striking findings of the Polar and Cluster satellites. The source of these cusp energetic particles (CEP) has centered on the possible role of the bow shock, leakage from the magnetosphere, and local acceleration within the cusp itself.*
>
> **Fritz (2009)**

Observationally, a CEP event is defined as follows:
- More than one order of magnitude increase in intensity of 1–10 keV ions.
- More than three sigma increase above background for >40 keV ion (dominated by proton) intensity.
- A decrease in local magnetic field.

CEP events are shedding light on the long-standing unsolved fundamental issue about the origins of the energetic particles in the magnetosphere (Fritz et al., 2000, Fritz, 2009). They may hold the key to understanding how the solar wind transfers mass, momentum, and energy into the Earth's magnetosphere. Despite a decade-long dispute about the origin of these CEP particles, how they are energized remained unknown until Chen (2008) uncovered the secret.

7.2 The magnetosheath

The magnetospheric cusps are located where magnetosheath plasma and momentum enter into the magnetosphere. A large number of low-altitude polar orbiting satellites have allowed their precise characterization in terms of plasma properties and motional behavior.

The cusp plasma is made of low-energy ions and electrons of magnetosheath origin from the shocked solar wind, and ionospheric plasma from below. It is located near magnetic noon and extends 1–2° in latitude and up to 4 h in local time. Early statistics from the DMSP satellites have shown it moves equatorward (poleward) in response to a solar wind pressure increase (decrease). For southward (northward) IMF orientation it moves equatorward (poleward) and for dawnward (duskward) IMF orientations it is displaced toward dawn (dusk) (Newell et al., 1989). Depending on these conditions, the cusp region can be found in the range 73–80° in magnetic latitude and 10:30–13:30 in magnetic local time (MLT). Surrounding the cusp, the cleft is a broader region characterized by substantially lower magnetosheath ion fluxes that probably map to the low-latitude boundary layers (LLBL) at higher altitudes.

The ions entering the cusp are subjected to a large-scale "convection" electric field. That convection field is a shortened and confusing form of the *real* electric field that includes charge separation and induction components. The plasma mantle (Rosenbauer et al., 1975) at the poleward edge of the cusp constitutes the tail of the dispersed cusp. Axford and Hines (1961) dubbed it a "zone of confusion." The early evidence for the existence of a slow-flow region of magnetosheath plasma outside of a probable magnetopause indentation was outlined by Paschmann et al. (1976) and Haerendel et al. (1978) using the HEOS-2 data. They proposed the existence of a "stagnation region," and of an adjacent "entry layer," which could permit plasma and momentum entry into the magnetosphere through eddy diffusion at their boundary. Their cusp picture was therefore close to an aerodynamic view. Although the term "stagnant" is often used in the literature since these early studies, plasma stagnation is qualitatively not obvious to occur in such a dynamic region.

Low- and mid-altitude satellites have provided major advances in the understanding of the role and characteristics of the cusp, but such spacecraft only remotely sense the solar wind interaction with the cusp, which actually occurs at higher altitudes. The Interball spacecraft (Fedorov et al., 2000; Merka et al., 1999) and the Polar spacecraft (Fuselier et al., 2000; Russell et al., 2000) observed the high-altitude regions of the dayside magnetosphere with high data sampling rates. The Interball spacecraft predominantly passed at very high latitudes and altitudes in the cusp and plasma mantle region but did not sample the most central part of the exterior cusp. On the other hand, the Polar spacecraft has an apogee of $9 R_E$ (toward the cusp) and only rarely had access to the magnetosheath. Thus it did not permit extensive study of the very distant exterior cusp and its interface with the magnetosheath. This boundary is indeed shown to be a sharp boundary, rotational in nature, and it allows plasma entry into the cusp and subsequently the magnetosphere. Currents within the magnetosheath induce the $\mathbf{J} \times \mathbf{B}$ forces to slow down and divert the flow, then speed it up again about the magnetospheric obstacle. Only a small amount of the energy that is associated with this flow enters the magnetosphere, so that we can neglect this interaction as a first approximation when considering the magnetosheath itself.

Figs. 7.2 and 7.3 are drawn for the case of a southward IMF for simplicity, recognizing its effectiveness in promoting substorms. A dawnward current near the subsolar regions as in Fig. 7.2 will produce a sunward force needed to slow down the plasma. Since the electric field is duskward (by assumption) this current acts as a generator with $\mathbf{E}\cdot\mathbf{J}<0$. Farther downstream the plasma flow speeds up; since the magnetic field will have a southward component this requires a dawn-dusk electric current. Here the plasma acts as a load on the electromagnetic system. Thus the magnetosheath behaves as a generator-motor pair. Of course, the plasma flow is not simply $\mathbf{v}^{MHD}=\mathbf{E}\times\mathbf{B}/B^2$; gradient and curvature drifts are involved, making a complicated problem.

7.2.1 The turbulent magnetosheath

Phan et al. (2018) have reported observations made by the magnetospheric multiscale (MMS) mission of an electron-scale current sheet in the turbulent magnetosheath downstream of the bow shock in which diverging bidirectional super-ion-Alfvénic electron jets, parallel electric fields, and enhanced energy conversion from magnetic fields to particles were detected. This observation is counter to Standard Magnetic Reconnection since the thin reconnecting current sheet was not embedded in a wider ion-scale current layer and no ion jets were seen. This, along with other, unidirectional events without evidence of ion reconnection

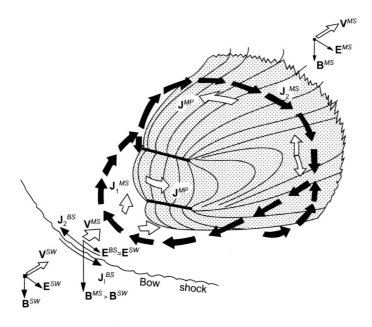

Fig. 7.2 The plasma flow in the magnetosheath is first slowed down in the subsolar regions, while farther back it is speeded up again. Thus the magnetosheath current circuit behaves as a generator-motor pair. This view holds for a southward IMF. From Heikkila, W. J., 1979. Energy budget for solar wind-magnetospheric interactions. In: Olson, W.P. (Ed.), Quantitative Modeling of Magnetospheric Processes. AGU Geophysical Monograph 21. American Geophysical Union, Washington, DC, p. 395.

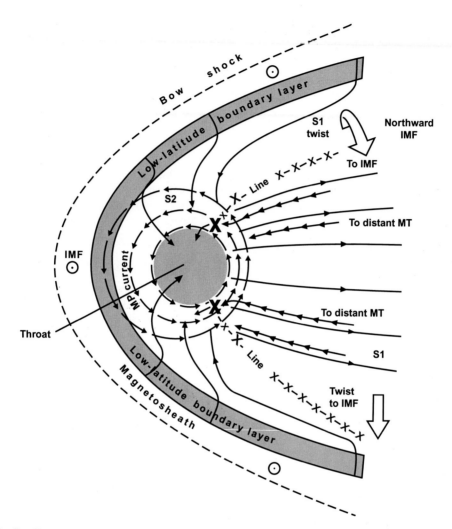

Fig. 7.3 The first line of defense is the low-latitude magnetopause with its PTEs, as discussed in Chapter 6. The next is the high-altitude cusps, the cusp throats shown here in a three-dimensional view. It was thought the solar wind plasma would meet little resistance as the plasma dips down into the lower levels. Individual particles not in the loss cone would be reflected by the mirror force; this involves considerable expense with regard to total momentum and energy. It is now known that the plasma regurgitates! From Heikkila and Brown.

reveals a process that can drive turbulent energy transfer and dissipation in electron-scale current sheets without ion coupling.

The four MMS spacecraft were in a 7–10 km tetrahedral formation at this time, which is approximately the electron skin depth scale. The high time resolution of their electron and ion spectrometers (400 and 80 times better resolved, respectively, than previous missions) allows many past ambiguities in the electron and ion behavior in turbulent current sheets to be resolved. At around 09:03:54 UT on December 9, 2016, two spacecraft (MMS3 and MMS1) detected simultaneous, oppositely directed plasma outflows (Fig. 7.4). These data are presented in a current-sheet (*LMN*) coordinate system, where the current sheet normal points along *N*, *L* is along the antiparallel magnetic field, and *M* is in the out-of-plane (X-line) direction. This ensures that the main current is in the *M* direction, while bidirectional outflows will be in the $\pm L$ directions. The duration of the current sheet crossing was 45 ms (vertical dashed lines in Fig. 7.4), which corresponds to a thickness of only 4 km, or about 4 electron skin depths. Timing analysis between the spacecraft shows a convection speed of the current sheet of 95 km/s. Both MMS3 (left) and MMS1 (right) observed fast electron flows along the *M* direction of about 900 km/s (Fig. 7.4C) that produce the main current (Fig. 7.4D and N) and the associated reversal of B_L (Fig. 7.4A and K). At the same time, the two spacecraft observed oppositely direct electron jets in the *L*, or outflow, direction with increases in speed of about +250 km/s at MMS3 (Fig. 7.4C) and —50 km/s at MMS1 (Fig. 7.4M) relative to the external electron flow in the *L* direction of about +150 km/s. These electron outflow jets had speeds of 10–18 times the ion-Alfvén speed of about 25 km/s. MMS3 was located 7.1 km in the +*L* direction from MMS1, indicating that the observations are consistent with diverging jets from an X-line located between the two spacecraft as they pass through the current sheet. There was no evidence found for ion jets at the ion-Alfvén speed (Fig. 7.4B and L). This is not surprising given that the current sheet was only 0.08 of an ion gyroradius. Importantly, though, this electron-scale current sheet was not embedded inside a much larger ion-scale current sheet, as is expected in Standard Magnetic Reconnection. This absence can be seen in Fig. 7.4A and K, which shows B_L reaching it asymptotic values immediately outside the thin current sheet. This suggests that in these turbulent magnetosheath plasmas, there is insufficient space and/or time for the ions to couple to the magnetic structures. Most of the energy is carried by the ions in the supersonic stream. To negate that would require an impossible current in response. The current that the reconnection can provide is far less, that of the turbulent plasma.

7.3   The cusp throat

At the low-latitude magnetopause the solar wind plasma encounters a barrier, with some penetration dependent upon plasma transfer events (PTEs) as discussed in Chapter 6. At higher latitudes illustrated in Fig. 7.3 the plasma can go farther in the high-altitude cusps dipping down in the cusp geometry. It was

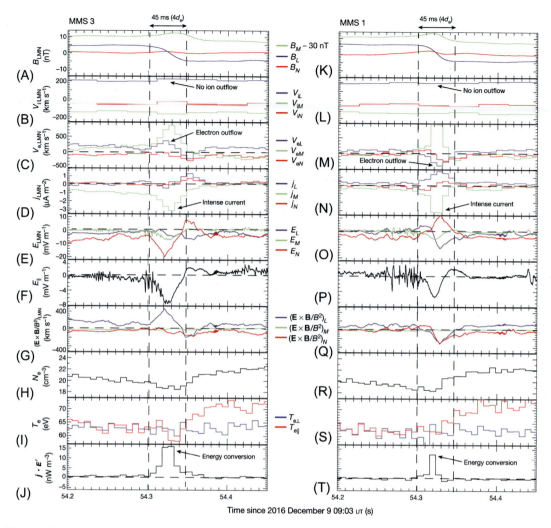

Fig. 7.4 Simultaneous MMS1 and MMS3 detections of oppositely directed superion-Alfvénic electron jets, parallel electric fields, and enhanced magnetic-to-electron energy conversion in an electron-scale current sheet. The data for both spacecraft (MMS3, (A–J); MMS1, (K–T)) are shown in a common current-sheet (*LMN*) coordinate system, determined for the MMS3 crossing of the current sheet at 09:03:54.270–09:03:54.365 UT, with $L=(-0.091, 0.87, 0.49)$ GSE, $M=(-0.25, -0.49, 0.83)$ GSE and $N=(0.96, -0.05, 0.27)$ GSE. The *vertical dashed lines* mark the left and right edges of the current sheet. (A, K) Magnetic field B_{LMN} at 8196 samples per second (from merged fluxgate and search-coil magnetometer measurements), with B_M shifted by -30 nT. (B, C, L, M) Ion and electron velocity. The 7.5 ms electron and 37.5 ms ion data products were generated by separating the individual energy sweeps that were used to form the nominal burst-mode distribution functions. These data maintain sufficient angular coverage to recover accurate plasma moments at four times the nominal temporal resolution. (D, N) Current density from plasma measurements. (E, O) Electric field in the spacecraft frame at 8196 samples per second. (F, P) Electric-field component parallel to the magnetic field. (G, Q) $(\mathbf{E} \times \mathbf{B}/B^2)$ velocity. (H) Electron density. (I) Electron temperature. (J, T) $\mathbf{j} \cdot (\mathbf{E}+\mathbf{V}_e \times \mathbf{B})=\mathbf{j} \cdot \mathbf{E}'$. The *LMN* coordinate system was determined using a hybrid minimum-variance method, which often works best in low-magnetic-shear current sheets. The current-sheet normal direction N was determined from $\mathbf{B}_1 \times \mathbf{B}_2/|\mathbf{B}_1 \times \mathbf{B}_2|$, where $\mathbf{B}_1$ and $\mathbf{B}_2$ are the fields at the two edges of the current sheet. We define $\mathbf{M}=\mathbf{L}' \times \mathbf{N}$, where $\mathbf{L}'$ is the direction of maximum variance of the magnetic field; $\mathbf{L}=\mathbf{N} \times \mathbf{M}$. MMS3 was located at $L=+7.1$ km, $M=+3.3$ km, and $N=+1.6$ km relative to MMS1. From Phan, T.-D., Eastwood, J.P., Shay, M.A., Drake, J.F., Sonnerup, B.U.O., Fujimoto, M., Cassak, P.A., Oieroset, M., Burch, J.L., Torbert, R.B., Rager, A.C., Dorelli, J.C., Gerchman, D.J., Pollock, C., Pyakurel, P.S., Haggerty, C.C., Khotyaintsev, Y., Lavraud, B., Saito, Y., Oka, M., Ergun, R.E., Rentino, A., LeContel, O., Argall, M.R., Giles, B.L., Moore, T.E., Wilder, F.D., Strangeway, R.J., Russell, C.T., Lindqvist, P.-A., Magnes, W., 2018. Electron magnetic reconnection without ion coupling in Earth's turbulent magnetosheath. Nature 557(2), 205.

expected that the plasma would meet little resistance as it goes down to the lower altitudes. Individual particles outside the loss cone would simply be reflected by the mirror force and then continue in the tailward flow.

But there is a catch to this argument: the reversal involves a reversal of its momentum. The entry is independent of the IMF, being determined by the local conditions. The cusp throats are shown in Figs. 7.3 and 7.5. Some solar wind plasma enters the converging magnetic field lines to low altitudes, partly on open field lines, partly on closed field lines. The full force of the (shocked) solar wind plasma is observed in this interaction. It is necessary to consider this reaction by Newton's third law. What happens to the energy the particles carry? With the solar wind plasma coming in nonstop, the energy and momentum they carry has to find somewhere to go!

Fig. 7.5 is a sketch for the interaction pattern of magnetosheath plasma flow with the outer cusp. The local velocity becomes a

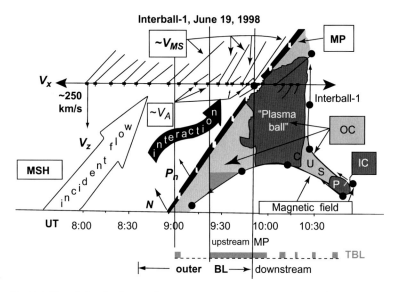

Fig. 7.5 Sketch for the interaction pattern of magnetosheath (MSH) plasma flow with outer cusp on June 19, 1998; spacecraft orbit (Interball-1 moves from left to right), with characteristic ion velocity vectors in the x, z GSE plane; N—normal to magnetopause in GSE frame $\sim$(0.7, 0.07, −0.71); P_n—projection of Poynting vector on N; V_{MS}, V_A—magnetosonic and Alfvén speeds; magnetopause is shown by the thick broken curve; OC, IC—outer and inner cusps. From Savin, S., Slassky, A., Zeleny, L., Avanov, L., Borodkova, N., Klimov, S., et al., 2005. Magnetosheath interaction with the high latitude magnetopause. In: Fritz, T.A., Fung, S.F. (Eds.), The Magnetospheric Cusps. Springer, Dordrecht, Netherlands, p. 101.

stagnant plasma ball as depicted in the throat. The cusp throat is closed by the smooth magnetopause at a larger distance. The observations of the high altitude and exterior (outer) cusp, and adjacent regions were reviewed in the book *The Magnetospheric Cusps: Structure and Dynamics* edited by Fritz and Fung (2005). Several articles presented various features of the regions sampled, boundary dynamics, and the electric current signatures observed.

> *The exterior cusp, in particular, is highly dependent on the external conditions prevailing. The magnetic field geometry is sometimes complex, but often the current layer has a well defined thickness ranging from a few hundred (for the inner cusp boundaries) to 1000 km. Motion of the inner cusp boundaries can occur at speeds up to 60 km/s, but typically 10–20 km/s. These speeds appear to represent global motion of the cusp in some cases, but also could arise from expansion or narrowing in others. The mid- to high-altitude cusp usually contains enhanced ULF wave activity, and the exterior cusp usually is associated with a substantial reduction in field magnitude.*
>
> **Dunlop et al. (2005)**

Fig. 7.6 is a sketch of the dayside boundary regions during southward and northward IMF. The solar wind is of course flowing antisunward at all times. Fig. 7.2 assumes a southward IMF, while Fig. 7.3 is drawn for a northward IMF with the twist in the distant tail for accommodation to the IMF (Laitinen et al., 2006). There are differences in the entry to the high-altitude cusp as shown in Fig. 7.6. It can be difficult to sort out the closure of the magnetic field lines in a given situation, but energetic electrons are the best solutions.

> *Energetic electrons (e.g., 50 keV) travel along field lines with a high speed of around 20 R_E s^{-1}. These swift electrons trace out field lines in the magnetosphere in a rather short time, and therefore can provide nearly instantaneous information about the changes in the field configuration in regions of geospace.*
>
> **Zong et al. (2005)**

7.3.1 Mixing of solar wind and magnetospheric plasmas

The processes that transfer solar wind mass, momentum, and energy to the magnetosphere are known to be strongly dependent on the orientation of the IMF, that is, the clock angle. The clock angle is defined as:

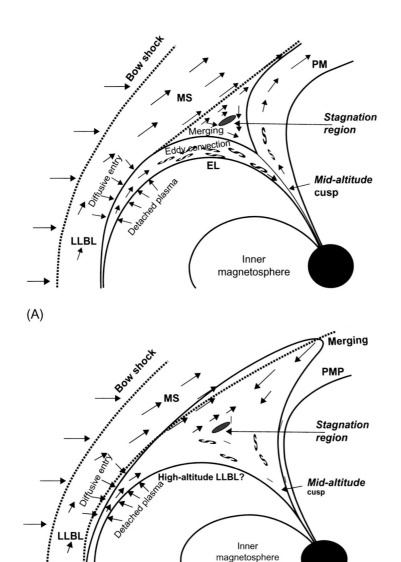

(A)

(B)

Fig. 7.6 Sketch of the dayside boundary regions related to the polar cusp field lines during (A) southward and (B) northward IMF. *MS*, magnetosheath; *PM*, plasma mantle; *LLBL*, low-latitude boundary layer; *EL*, entry layer; *S*, flow eddy. From Zong, Q.G., Fritz, T.A., Korth, A., Daly, P.W., Dunlop, M., Balogh, A., et al., 2005. Energetic electrons as a field line topology tracer in the high latitude boundary/cusp region: cluster rapid observations. In: Fritz, T.A., Fung, S.F. (Eds.), The Magnetospheric Cusps. Springer, Dordrecht, The Netherlands, p. 217.

$$\theta = \tan^{-1}\left(|B_y/B_z|\right) \ \text{ for } \ B_z > 0 \qquad (7.1)$$

$$\theta = 180° - \tan^{-1}\left(|B_y/B_z|\right) \ \text{ for } \ B_z < 0 \qquad (7.2)$$

where B_y and B_z are the components of the IMF. Values of $0° < \theta < 90°$ are indicative of positive (northward) B_z and $90° < \theta < 180°$ of negative (southward) B_z.

For the case of northward IMF, the distribution looks very different. No particular feature appears near the equatorward cusp boundary. The average tailward convection speed in most of the cusp region is lower than 100 km/s, so that the region may be considered as stagnant for the given northward IMF conditions.

Lavraud et al. (2004a,b, 2005a) have studied the plasma boundaries at the edges at mid- to high-altitudes consistent with a cusp "throat." The well-defined magnetic boundaries at high altitude are consistent with a funnel geometry. The direct control of the cusp position, and its extent, are controlled by the IMF; the north/south directions determine the effectiveness of the coupling, while the east/west components produce distortions in the dawn/dusk directions. The exact location of the cusps depends on several factors. The seasonal variation between summer and winter affects the location of the cusps in relation to the antisunward solar wind in an obvious way. The variation can be aggravated by the dipole tilt relative to the rotational axis to a maximum approaching 35°. This has a dramatic effect on the difference between the summer and winter cusps. Note that a sunward flow occurs in the plasma mantle and high-altitude cusp with northward IMF.

Fig. 7.7 shows the spatial distributions of the x component of plasma convection for southward and northward IMF, respectively. In both IMF cases the magnetosheath always shows a large tailward convection on average, which is expected. As also expected, the dayside plasma sheet, equatorward of the cusp, shows basically no flows in the x direction. Within the exterior cusp and adjacent magnetospheric regions, clear differences are observed between the southward and northward IMF cases.

7.3.2 Statistical properties of the plasma

Lavraud et al. (2005b) report on the statistical properties of the plasma flows measured by the Cluster spacecraft in the high-altitude cusp region of the northern hemisphere as a function of the IMF orientation. The technique uses a magnetic field model, taking into account the actual solar wind conditions and level of geomagnetic activity, in order to model the magnetopause and cusp displacements as a function of these conditions. The distributions of the magnetic field vector show a clear consistency with the IMF clock angle intervals chosen. They demonstrate that their technique fixes the positions of the cusp boundaries adequately.

The magnetic field vector structure clearly highlights the presence of an intermediate region between the magnetosheath and

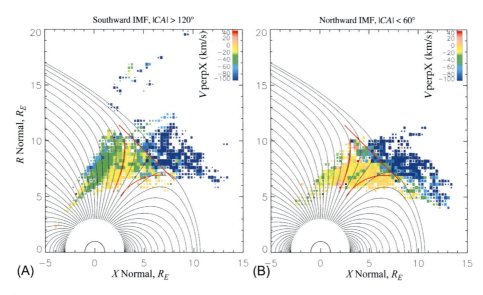

Fig. 7.7 Spatial distributions of the *x* component of the convection velocity perpendicular to the magnetic field in the high-altitude cusp and surrounding regions for (A) southward clock angle ($|CA| > 120$ degrees) and (B) northward ($|CA| < 60$ degrees) IMF. Observations have been collected and then averaged over spatial bins of 0.3 R_E. Corresponding color scale is shown in the top right-hand corner of each distribution, with *blue (red) color* indicating tailward (sunward) convection. Background magnetic field lines have been computed from the T96 model for the reference conditions. Note that a sunward flow occurs in the plasma mantle and high-altitude cusp with northward IMF. From Lavraud, B., Fedorov, A., Budnik, E., Thomsen, M.F., Grigoriev, A., Cargill, P.J., et al., 2005b. High altitude cusp flow dependence on IMF orientation: a 3 year Cluster statistical study. J. Geophys. Res. 110, 5. https://doi.org/10.1029/2004JA010804.

the magnetosphere, the exterior cusps shown by Fig. 7.7 (Lavraud et al., 2005b). This region is large and characterized by the presence of cold dense plasma near the null point of the traditional Chapman and Ferraro (1931) model and is diamagnetic in nature.

> *The density, temperature, and velocity distributions allow us to establish the presence of three distinct boundaries surrounding the exterior cusp region: with the lobes, the dayside plasma sheet, and the magnetosheath. While the two inner boundaries are well known, the average position of the external boundary with the magnetosheath is most accurately obtained through the statistical distribution of the bulk velocity. This study further demonstrates that this external boundary is characterized by a density decrease and a temperature increase, from the magnetosheath to the exterior cusp.*
>
> **Lavraud et al. (2005c)**

These studies looked at the statistical extent of the cusp region by cusp characteristics, rather than direct identification of the boundary, and found no clear magnetosheath indentation. Zhang et al. (2007) have shown some evidence of the control of cusp geometry by the IMF orientation. The pressure distributions further illustrate that the exterior cusp region is in equilibrium with its surroundings in a statistical sense. The bulk flow magnitude distributions suggest that the exterior cusp is overall stagnant under northward IMF conditions, but more convective under southward IMF conditions. The outer cusp-magnetosheath boundary appears to be a possible extension of the magnetopause boundary layer into the high-latitude and cusp region, whereas the inner boundaries define entry into the closed (equatorial) and open (polar) magnetospheric field regions.

7.4   Transfer events

Fig. 7.8 shows a type of magnetic record often seen near the magnetopause; this was first noticed by Russell and Elphic (1978) in a classic paper which called it a *flux transfer event*

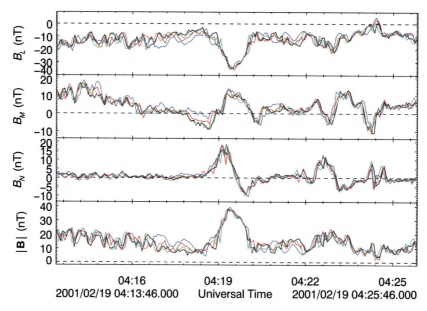

Fig. 7.8 A sample Cluster flux transfer event (FTE) observed at ∼0420 UT on February 19, 2001 at (5.8, 0.8, 10.2) R_E in GSM coordinates. Different line colors stand for different Cluster satellites. This FTE shows clear B_n bipolar signature and $|B|$ enhancement. We contend that it is also the proper signal for a PTE, and use the shorter name of transfer event (TE). From Wang, Y.L., Elphic, R.C., Lavraud, B., Taylor, M.G.G.T., Birn, J., Raeder, J., et al., 2005. Initial results of high latitude magnetopause and low latitude flank flux transfer events from 3 years of Cluster observations. J. Geophys. Res. 110, 3. https://doi.org/10.1029/2005JA011150.

(FTE). We noted in Chapter 6 that this is likely a PTE. These transfer events (TEs) are defined to be the results of temporally and spatially varying patchy impulsive processes of solar wind plasma at the magnetopause. Wang et al. (2005) have compiled a review of TEs observed by Cluster. It has been shown from observations that TEs contain a mixture of magnetospheric and magnetosheath plasmas. Thus they are important for the coupling of mass, momentum, and energy between the solar wind and the Earth's magnetosphere. This is easily understood with a PTE.

Wang et al. (2005) present initial results from a statistical study of Cluster multispacecraft transfer event observations at the high-latitude magnetopause and low-latitude flanks from February 2001 to June 2003. Most of the previous studies concentrated on low- and mid-latitude magnetopause observations and ground observations. There has been a scarcity in studies examining the statistical properties of high-latitude processes. Cluster observations provide a great opportunity to advance the understanding of PTEs in this region. This is due to the fact that Cluster has a trajectory that encounters the high-latitude magnetopause, but also because Cluster consists of four spacecraft allowing detailed study of PTE structure and motion.

Fig. 7.9, from top to bottom, shows the locations of the TEs without thresholds in the GSM x-z, y-z, and x-y planes, respectively. From left to right, Fig. 7.9 shows the TEs during all, southward, and northward IMF orientations. From the figure, we see that many of the TEs are observed at the high-latitude magnetopause near the cusps. However, there are also a considerable number of low-latitude TEs near the magnetopause flanks. Transfer events are observed at both the high-latitude magnetopause and low-latitude flanks, for both southward and northward IMF conditions.

They found that in the normalized TE MLT distribution, more TEs are observed from dawn to dusk from $\sim$9 to $\sim$17 MLT. Also, they found that TE occurrence is reduced when the dipole tilt is close to zero. Also, when dipole tilt is positive (negative), Cluster observes more TEs in the southern (northern) hemisphere. Wang et al. (2005) make use of the large Cluster TE data set to perform a detailed study of some important properties of Cluster high-latitude magnetopause and low-latitude flank TEs.

Fig. 7.10 shows the TE separation time distribution, TE $|\mathbf{B}|_{\text{Peak-Surrounding}}$ distribution, TE B_n peak-peak time distribution, and TE B_n peak-peak magnitude distribution, all with no thresholds. From the figure, we see that all these distributions show more or less smooth profiles. The mean TE separation time is 37.15 min (median: 12.12 min), the mean TE $|\mathbf{B}|_{\text{Peak-Surrounding}}$ is

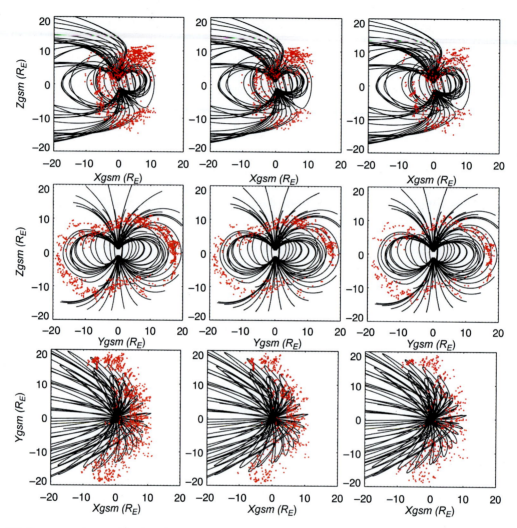

Fig. 7.9 From top to bottom: The locations of the TEs in the GSM *x-z*, *y-z*, and *x-y* planes, respectively. From left to right: TEs during all, southward, and northward IMF orientations, respectively. T96 model (Tsyganenko, 1995) magnetic field lines are shown in each panel as the background for reference. From Wang, Y.L., Elphic, R.C., Lavraud, B., Taylor, M.G.G.T., Birn, J., Raeder, J., et al., 2005. Initial results of high latitude magnetopause and low latitude flank flux transfer events from 3 years of Cluster observations. J. Geophys. Res. 110, 5. https://doi.org/10.1029/2005JA011150.

13.13 nT (median: 11.07 nT), the mean B_n peak-peak time is 25.80 s (median: 20.07 s), and the mean B_n peak-peak magnitude is 25.36 nT (median: 22.09 nT).

There has been a scarcity in studies examining the statistical properties of high-latitude TEs, possibly due to the complication of the vicinity of the cusp. Cluster observations provide a great

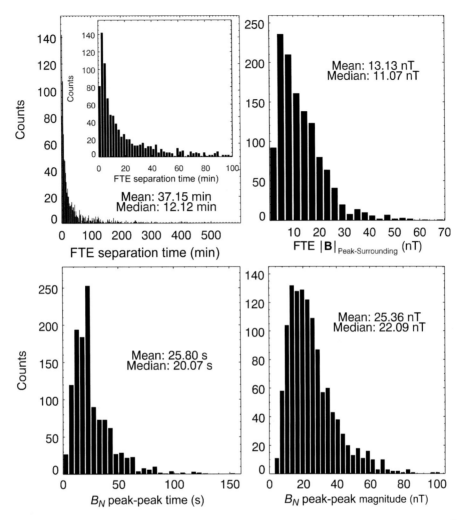

Fig. 7.10 The TE separation time distribution, FTE | **B** |$_{Peak-Surrounding}$ distribution, TE B_n peak-peak time distribution, and TE B_n peak-peak magnitude distribution, all with no thresholds. Note here that each TE separation time in the upper left panel is calculated between two contiguous TEs without requiring that they correspond to the same Cluster magnetopause crossing. From Wang, Y.L., Elphic, R.C., Lavraud, B., Taylor, M.G.G.T., Birn, J., Raeder, J., et al., 2005. Initial results of high latitude magnetopause and low latitude flank flux transfer events from 3 years of Cluster observations. J. Geophys. Res. 110, 6. https://doi.org/10.1029/2005JA011150.

opportunity to advance the understanding of TEs in this region, not only because Cluster has a trajectory that encounters the high-latitude magnetopause, but also because Cluster consists of four spacecraft allowing detailed study of TE structure and motion. From this large-scale study they reach the following conclusions:

- Cluster TEs are observed at both the high-latitude magneto-pause and low latitude flanks for both southward and northward IMF.
- There are 73% (27%) of the Cluster TEs observed outside (inside) of the magnetopause.
- Average TE separation time of 7.09 min, which is at the lower end of previous results.
- The mean B_n peak-peak magnitude of Cluster TEs is significantly larger than that from low-latitude TE studies, and clearly increases with increasing absolute magnetic latitude (MLAT).
- The B_n peak-peak magnitude dependence on Earth dipole tilt is more complex with a peak at around zero Earth dipole tilt.
- TE periodic behavior is found to be controlled by MLT.
- Further confirmation that TE statistical results do not change in a significant way by using different TE criteria.

7.5 Cusp energetic particles

It is worth repeating that energetic particles are a consistent and common feature of the high-altitude dayside cusp.

Observing these particles in a region where they cannot be stably trapped is one the most striking findings of the Polar and Cluster satellites.

Fritz (2009)

The source of these CEPs has centered on the possible role of the bow shock, leakage from the magnetosphere, and local acceleration within the cusp itself. The Polar satellite has documented that the shocked solar wind plasma enters the weak geomagnetic field of the polar region and produces cusp diamagnetic cavities (CDC) of apparent tremendous size ($\sim$6 R_E) well within the traditional magnetosphere.

7.5.1 Polar data

Cusp energetic particles were first noticed in the Polar data by Chen et al. (1997). An example from August 27, 1996 is shown in Fig. 7.11 (Chen, 2009). Polar was launched into a 1.8 by 9 R_E orbit on February 24, 1996 with an inclination of 86°. On board Polar, the Magnetospheric Ion Composition Spectrometer (MICS), a part of the Charge and Mass Magnetospheric Ion Composition Experiment (CAMMICE), is a one-dimensional time-of-flight electrostatic analyzer with post acceleration measuring ions with an energy/charge of 1 to 220 keV/e with very good angular

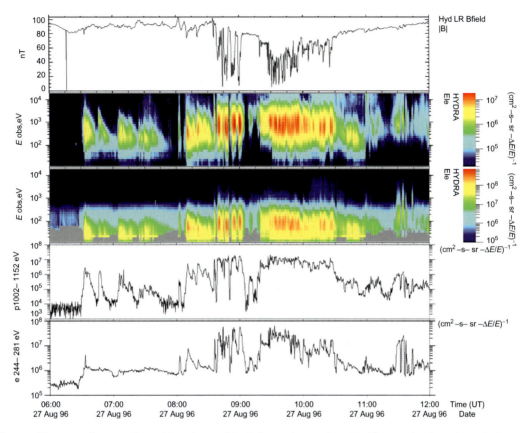

Fig. 7.11 The cusp diamagnetic cavities observed by Polar on August 27, 1996. The panels show the variation of the local magnetic field strength, the energy spectra of the ions, the energy spectra of the electrons, the 1–1.15 keV proton energy flux, and the 244–281 eV electron energy flux versus time, respectively. From Chen, J., 2009. In: Johannson, H. (Ed.), Solar Wind, Large Diamagnetic Cavities, and Energetic Particles, Handbook on Solar Wind: Effects, Dynamics and Interactions. Nova Science, Hauppauge, NY, p. 294.

resolution. The MICS, mounted perpendicular to the spin axis, is able to obtain a two-dimensional distribution at one energy per charge during each 6-s spin period. A complete energy spectrum is obtained in 32 spin periods.

Fig. 7.11 shows the time profiles of the thermalized solar wind plasma (bottom four panels) and the local geomagnetic field strength (top panel) measured by the Polar spacecraft in the high-altitude dayside cusp region on August 27, 1996. The panels from top to bottom are the local magnetic field strength, the energy spectra of the ions (mostly protons), the energy spectra of the electrons, the proton energy flux, and the electron energy flux, respectively. It shows the solar wind plasma has a peak value occurring at about 1 keV for protons and at about 100 eV for electrons. Fig. 7.11 further shows that the depressed magnetic field strength is associated with the increased intensities of the charged particles, suggesting a particle-field interaction.

At 8:00–10:36 UT on August 27, 1996, the Polar spacecraft was about 9 R_E from the Earth at ∼65° MLAT and ∼15 h MLT. Fig. 7.11 indicates Polar observed the CDCs during this period. Notice the CDCs just before 09:00 UT: both the ions and electrons show no dispersion over their whole energy bursts. This feature is common throughout all the cusp observations, arguing strongly in support of local acceleration within the cusp itself.

7.5.2 Cluster observations

This feature has also been confirmed by the Cluster mission. The Cluster mission consists of four identical satellites (C1, C2, C3, and C4). Cluster orbits take the four spacecraft in a changing formation out of the magnetosphere, on the northern leg, and into the magnetosphere, on the southern leg. During February to April the orbits are centered on a few hours of local noon and, on the northern leg, generally pass consecutively through the northern lobe and the cusp at mid- to high-altitudes. Depending upon conditions, the spacecraft often sample the outer cusp region, near the magnetopause, and the dayside and tail boundary layer regions adjacent to the central cusp (Fig. 7.12). On the southern, inbound leg the sequence is reversed. Cluster has therefore sampled the boundaries around the high-altitude cusp and nearby magnetopause under a variety of conditions. The instruments onboard provide unprecedented resolution of the plasma and field properties of the region. The spacecraft array forms a nearly regular tetrahedral configuration in the cusp and already the mission has covered this region on multiple spatial scales (100–2000 km). This multispacecraft coverage allows spatial and temporal features to be distinguished to a large degree and, in particular, enables the macroscopic properties of the boundary layers to be identified: the orientation, motion and thickness, and the associated current layers.

Fig. 7.12 shows a northern cusp crossing on March 17, 2001 that occurred during quiet external conditions, and during essentially northward IMF-B_z. The crossing occurred during the exit of the spacecraft from the magnetosphere on the outbound, northern leg of the Cluster orbits. The dots on the orbit are hours of the day, the plane of the orbit lies almost at magnetic noon.

Fig. 7.13 shows the ion spectrograms and the density and velocity moments, measured by the CIS-HIA instrument, together

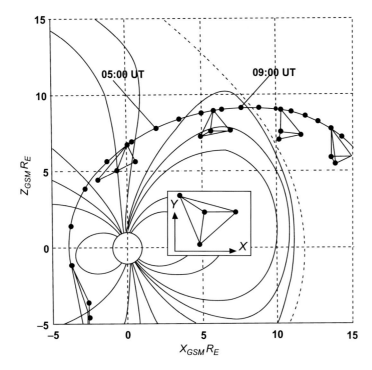

Fig. 7.12 The event of March 17, 2001 corresponds to a northern cusp crossing during the outbound leg of the orbits. The spatial configurations of the spacecraft are shown at a sequence of times around the orbit and scaled (enlarged) by a factor of 20. The inset shows the corresponding *x, y* projection. Also shown are model field lines, taken from the Tsyganenko (1989) model. From Dunlop, M.W., Lavraud, B., Cargill, P., Taylor, M.G.G.T., Balogh, A., Réme, H., et al., 2005. Cluster observations of the cusp: magnetic structure and dynamics. In: Fritz, T.A., Fung, S.F. (Eds.). The Magnetospheric Cusps. Springer, Dordrecht, Netherlands, p. 14.

with the FGM magnetic field, from spacecraft 1. Also shown is the ACE-IMF, suitably lagged for solar wind convection (Lepping et al., 1995), as taken from the observed bulk velocity (not shown). The lag times are all given in the figure captions. The plasma data mainly show an entry into and through the cusp throat and then dayside plasma sheet (closed magnetosphere) and out into the magnetosheath, confirming the interpretation from the magnetic field data. After about 05:05 UT and before 06:25 UT, the ions show a broadband signature at magnetosheath energies (top panel). We interpret this as corresponding to the passage through the cusp throat. The band fades through a boundary region, starting from about 06 UT, where the density begins to fall from $10 \, cm^{-3}$ to the low densities expected for the dayside plasma sheet $0.2-0.3 \, cm^{-3}$.

It is noteworthy that the velocities are quite low from less than 10 km/s to not more than 20 km/s. This stagnant flow is commonly referred to as a plasma ball, shown in Fig. 7.5. At the end of the plot in the magnetosheath, the velocity exceeds 100 km/s.

During this transition, there is an onset of a high-energy band of ions (between 5 and 10 keV). The spacecraft therefore appear to traverse the dayside region after about 06:25 UT, which is filled with trapped plasma sheet ions on closed field lines. After 07:00

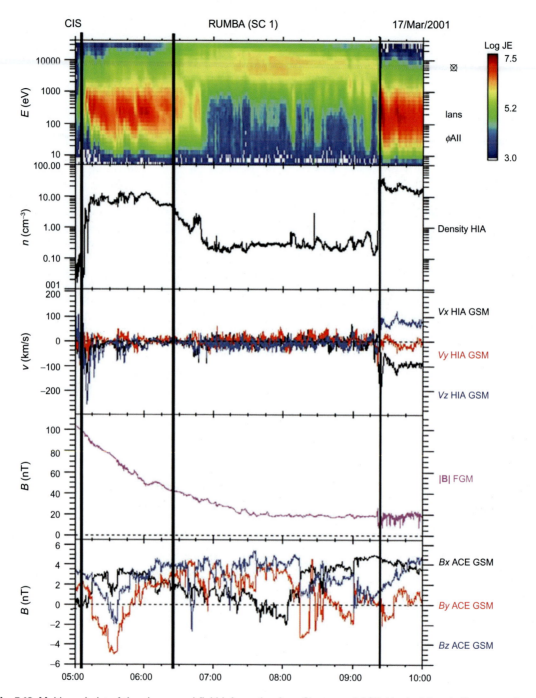

Fig. 7.13 Multipanel plot of the plasma and field information from Cluster and ACE. *Vertical lines* indicate certain features such as cusp throat, the dayside plasma sheet, and the magnetopause. From top to bottom these are: the ion energy spectrogram, density, velocity, and magnetic field from Cluster, and the lagged ACE magnetic field representing the IMF. The lag time is 82 min in this case. From Dunlop, M.W., Lavraud, B., Cargill, P., Taylor, M.G.G.T., Balogh, A., Réme, H., et al., 2005. Cluster observations of the cusp: magnetic structure and dynamics. In: Fritz, T.A., Fung, S.F. (Eds.). The Magnetospheric Cusps. Springer, Dordrecht, Netherlands, p. 16.

UT and until the magnetopause exit at 09:20 UT the ion density remains low with no significant bulk flow (during this period the GSM magnetic field turns from southward to northward). Entry into the cusp throat at 05:05 UT (5 R_E) is seen as an increase in ULF fluctuations on all spacecraft traces. The character of these fluctuations changes as the passage proceeds through the region.

Within the interval through the throat (between 05:15 and 06:15 UT) there are a number of additional transient signatures. Some of these relate to dispersive signatures in the plasma data, and some to transient, impulsive signatures, which can be correlated with plasma flows in the ionosphere. The latter are associated with brief, southward (and dawnward) turnings of the IMF during the pass. On one occasion (06:40 UT) the spacecraft appear to be taken back into the throat region from the dayside (plasma sheet). Such an occurrence would be expected for a southward (or sunward) motion of the equatorward edge of the cusp, perhaps due to erosion of the dayside MP. The event here, however, has few clear boundary crossings within the cusp structure to confirm such motions or to confirm spatial as opposed to temporal effects. Nevertheless, the spacecraft sequence through a number of features in the time series data gives timing information, which confirms the expected order in passing from one region into another.

This event therefore corresponds to a traversal across the mid- to high-altitude throat region, followed by a passage through magnetospheric field lines near the dayside boundary, before a final exit into the magnetosheath. The event is chosen because these quiet, external conditions produce a classic pass through the region, where the cusp location appears to be close to that predicted by the Tsyganenko model and the slow change in magnetic field topology can clearly be seen. The spacecraft pass out into the magnetosheath as indicated in Fig. 7.12, drawn relative to the model field lines, which apparently change orientation from the southward tailward throat alignment to the northward, dayside alignment, along the orbit. The event thus serves as a good template for the other events, some of which occurred during more dynamic conditions, often with repeated large-scale cusp motion, which adds to the observed signature.

On March 5, 2001 Cluster detected simultaneous enhancements of the energetic electron and ion fluxes (Chen, 2008). Fig. 7.14 shows the measurements of the >100 keV electrons, 30–100 keV electrons, 30–100 keV protons, and 50–150 keV helium ions (top three panels) with the local magnetic field strength (bottom panel) by the C4 satellite over 07:30–11:00 UT on March 5, 2001. It shows that the increased intensities of the charged particles were associated with the large diamagnetic cavities. Since the

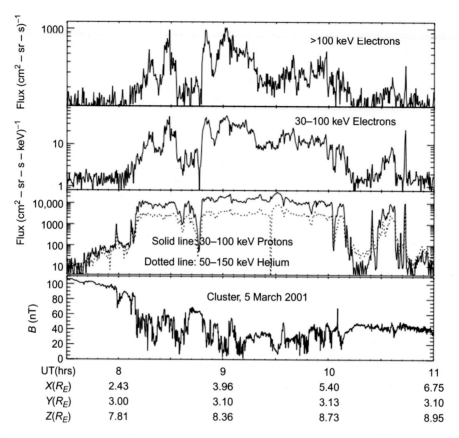

Fig. 7.14 Time variations of the cusp energetic particles and magnetic field strength. Panels from top to bottom are the time-intensity profiles of the >100 keV electrons, the 30–100 keV electrons, the 30–100 keV protons, and 50–150 keV helium ions, and the magnetic field strength measured by the Cluster satellite (C4) on March 5, 2001. The distance of Cluster from the Earth (in R_E) is shown at the bottom in the Geocentric Solar Ecliptic coordinates. From Chen, J., 2008. Evidence for particle acceleration in the magnetospheric cusp. Ann. Geophys. 26, 1994.

30–100 keV and > 100 keV relativistic electrons move at 32%–55% and > 55% of the speed of light, respectively, and since inside the geomagnetic field the electron and the proton drift in opposite directions, it is unexpected to detect them simultaneously in the CDCs for hours. Note that at 08:50–10:00 UT the 30–100 keV proton flux in the cusp was about three to four orders of magnitude higher than that before 07:40 UT and after 10:45 UT when Cluster was outside the cusp region. The top two panels show no obvious energy dispersion signatures for the relativistic electrons, suggesting a local dynamic process.

7.5.3 ISEE-1 and ISEE-2 observations

This dichotomy in the impression of the energetic particles within the cusp gained from the Polar and Cluster missions suggested that an examination of data from the older International Sun Earth Explorer (ISEE-1 and ISEE-2) satellites may indicate encounters with the cusp. This would allow the very good energy and angular resolution of the energetic particle experiment on these satellites to be used.

Two events where ISEE-1 and ISEE-2 were in the cusp have been reported by Whitaker et al. (2006, 2007), Walsh et al. (2007), and Fritz (2009). On September 29, 1978, the two ISEE satellites were outbound in near identical orbits close to local noon and each encountered a CDC. The event happened during a major magnetic storm where the upstream B_z was about -30 nT, D_{st} went to -224 nT, the hourly AE index reached 775 nT, and the dynamic pressure increased by a factor of 3. Surprisingly, ISEE-2, slightly ahead of ISEE-1, entered the CDC almost 30 min before ISEE-1 and stayed in the cavity for almost 1 h. Both satellites exited the CDC within 6 min of one another. This implies that the boundary of the CDC maintained a location between the two satellites for 27 min and must therefore have been in motion. Fig. 7.15 shows ISEE-1 data for this period. Although the time resolution of the composition data is poorer than the particle and magnetic field data, it is clear that the CDC was filled with a combination of solar (He++) and ionospheric (He+, O+) ions.

There appear to have been multiple cavities that ISEE-1 encountered with the energetic electrons defining each cavity exactly. This is true of the fluxes at all energies shown from 20 to 190 keV. The variation of the ions is similar to the electrons with the ions' intensities remaining high and extending beyond the sharp cavity boundary locations shown in the magnetic field and electron data. There is an absence of any velocity dispersion in the case of both the electrons and the ions, as seen in Fig. 7.16, indicating there was essentially no magnetic gradient or curvature drift from the source responsible for their energization.

7.6 Exterior cusp

Recognizing that the cusp also possesses three adiabatic invariants of the motion, the trapped plasma in the high-altitude cusp can be like the well-known geomagnetic trap. Low energy plasma is dominated by $\mathbf{E} \times \mathbf{B}$ drift, which because of the lack of an analogous corotation field, distorts the drift orbits and sweeps away plasma below some threshold energy, <30 keV. High-energy

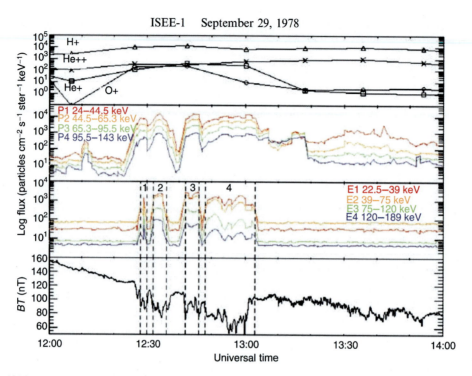

Fig. 7.15 ISEE-1 data on September 29, 1978. The top panel shows data from the Plasma Experiment (40 eV–17.36 keV), where *triangles, Xs, open squares, and diamonds* correspond to H⁺, He⁺⁺, He⁺, and O⁺ ions, respectively. The time resolution for the plasma data is 18 min. The second and third panels contain the proton and electron fluxes, respectively, integrated over all pitch angles for four energy channels measured by ISEE-1. The proton and electron flux are presented with a time resolution of 36 s. Panel four shows the total local magnetic field strength. The local magnetic field strength is displayed with 4 s resolution. Four zones of increased flux are marked with *vertical dotted lines* and labeled sequentially. From Fritz, T.A., 2009. Perspectives gained from a combination of polar, cluster and ISEE energetic particles measurements in the dayside cusp. In: Laakso, H., Taylor, M., Escoubet, C.P. (Eds.), The Cluster Active Archive—Studying the Earth's Space Plasma Environment. Springer, Dordrecht, Netherlands, p. 407.

plasma is dominated by $\nabla \mathbf{B}$ drift; the finite size of the cusp produces energy- and mass-dependent spatial limits on trapping, analogous to the inner and outer edge of the radiation belts (Fig. 7.17). The limits on the strength of the mirror force arising from topological considerations produce analogous pitch angle loss cones. The cusp loss cone loses particles to the dipolar magnetosphere, so in one sense, the cusp trap is half-embedded within the dipole trap.

All these analogies are true for a static magnetic cusp geometry, but are strongly modified by time-variable fields. The dipole trap is extremely stable, with $\nabla \mathbf{B}/\mathbf{B} < 1$ over the majority of the trap volume. The dipole trap has roughly three orders of magnitude

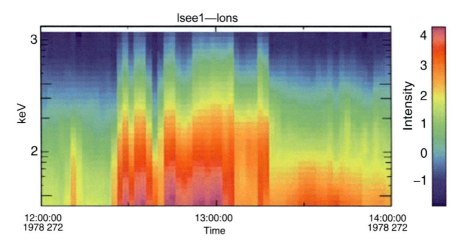

Fig. 7.16 ISEE-1 energetic ion spectrum for the event period of Fig. 7.15. Note the lack of any energy dispersion with time in the intensity of the ions during this encounter with a CDC. From Fritz, T.A., 2009. Perspectives gained from a combination of polar, cluster and ISEE energetic particles measurements in the dayside cusp. In: Laakso, H., Taylor, M., Escoubet, C.P. (Eds.), The Cluster Active Archive—Studying the Earth's Space Plasma Environment. Springer, Dordrecht, Netherlands, p. 408.

separating the gyration, bounce, and drift time scales, whereas in the cusp trap these motions may be separated by less than an order of magnitude. Induction electric fields may be quite strong.

> *Since the conservation of the adiabatic invariants depends upon the separation of timescales, the cusp trap is expected to be much more diffusive in energy and space than the dipole trap. This greater inherent diffusivity, coupled with the large perturbative power available would make the cusp an ideal location for resonance-broadened, chaotic acceleration.*
>
> **Sheldon et al. (1998)**

Diffusion in pitch angle as well as in L-shell would be required to transport these particles from the outer cusp to the radiation belts. Since the radiation belt is at a higher magnetic latitude, the minima between these trapping regions indicate that any transport between them is either taking a circuitous route or is necessarily time dependent.

The magnetospheric cusps allow for the most direct entry of shocked solar wind plasma into the magnetosphere. In order to maintain a pressure balance with adjacent magnetospheric regions, the magnetic pressure in the exterior cusp must decrease to account for the increase in thermal pressure from the newly injected plasma. The result is a diamagnetic cavity in the exterior

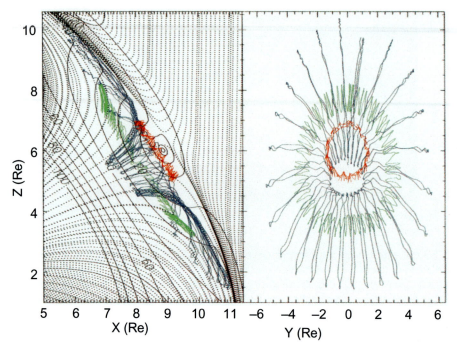

Fig. 7.17 Trajectories of trapped 1 MeV electrons in the Earth's outer cusp, projected into the GSM *x-z* and *y-z* planes. *Dashed lines* are field lines from the T96 magnetic field model (Dipole: June 21, 1996, 1300 UT; Solar Wind: +10 nT B_z, l/cm^3, and l000 km/s V_{sw}). *Black lines* are contours of |**B**| in nT. *Green, blue, and red trajectories* correspond to 1000 keV particles with various initial locations and pitch angles. From Sheldon, R.B., Spence, H.E., Sullivan, J.D., Fritz, T.A., Chen, J., 1998. The discovery of trapped energetic electrons in the outer cusp. Geophys. Res. Lett. 25(11), 1827.

cusp. In some cases this can bring the magnetic field strength close to 0 nT.

> *Through analyzing properties of energetic (E > 40 keV) electrons in the exterior cusp measured by Cluster on 27 February 2005, we have determined local energization is the primary source of this population. … Local energization is the only source that is fully consistent with the measurements.*
>
> **Walsh et al. (2010)**

Walsh et al. (2010) compared spacecraft observations by C3 (Fig. 7.18) with expected measurements for (1) a solar energetic electron source, (2) an equatorial outer ring current or radiation belt source, and (3) local energization. It is expected that all three sources have access to the cusps; however, each one would show different characteristics in the cusp and adjacent magnetospheric regions. Although other sources may contribute to the energetic

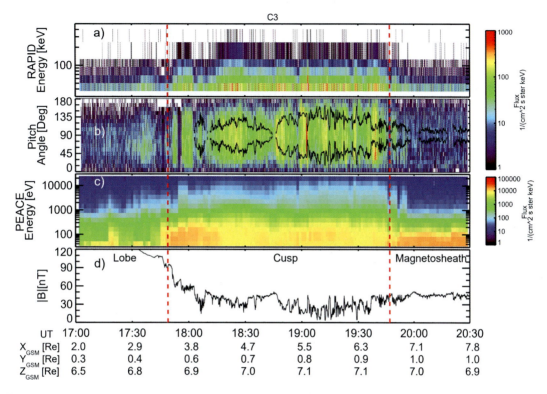

Fig. 7.18 Electron and magnetic field measurements by C3. Panel (A) is an omnidirectional flux of energetic electrons from 37 to 400 keV measured by RAPID. Panel (B) gives flux of electrons from 37 to 51 keV with pitch angle as measured with the L3DD mode on RAPID. The *black line* overplotted in panel (B) shows the cutoff pitch angle which would trap electrons within the cusp. Panel (C) is omnidirectional electron flux from 22 eV to 26.4 keV as measured by PEACE. The magnetic field strength is given in panel (D). The *vertical dashed lines* identify where the spacecraft transitions between the different magnetospheric regions. From Walsh, B.M., Fritz, T.A., Klida, M.M., Chen, J., 2010. Energetic electrons in the exterior cusp: identifying the source. Ann. Geophys. 28, 983–992, doi:10.5194/angeo-28-983-2010.

population within the cusp under certain conditions, this event serves as evidence that the cusp is capable of accelerating large amounts of particles to energies of several hundred keV.

7.6.1   Shell degeneracy

Only in the case of perfect azimuthal symmetry (as in the pure dipole) will these surfaces intersect exactly on the same line. In the general case, particles starting on the same field line at a given longitude will populate different shells, according to their initial mirror point fields or equivalently, according to their initial equatorial pitch angles. This is called *shell degeneracy*.

The quiet time Mead-Williams model has been used by Roederer (1967) to compute magnetic shells. Fig. 7.19 shows how particles, starting on a common line in the noon meridian, do indeed drift on different shells which intersect the midnight meridian along different field lines. The dots represent particles' mirror points. Curves giving the position of mirror points for constant equatorial pitch angles are traced for comparison (in a dipole field they would be constant latitude lines).

7.6.2 Diamagnetic cavities

An indicator of the dynamic processes is the magnetic fluctuations that are associated with the charged particles. As shown in Figs. 7.11, 7.15, and 7.18, strong electromagnetic fluctuations are found in the large cusp diamagnetic cavities.

7.6.3 Resonant acceleration in diamagnetic cavities

The magnetic power spectra with peaks at the ion gyrofrequencies shown in Fig. 7.20 suggest a cyclotron resonant acceleration of the field with the charged particles. On another Polar orbit, it detected MeV charged particles and large diamagnetic cavities (Chen et al., 1998). Fig. 7.21 is a hodogram of the two perpendicular components of the cusp electric field within this event period. The z-axis (out of the page) points along the local magnetic field direction and x- and y-axes complete the right-handed system. In each panel, the time period, the start point, and the end

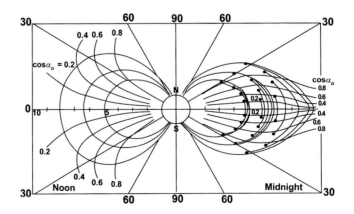

Fig. 7.19 Computed shell splitting for particles starting on common field lines in the noon meridian. *Dots* represent particles' mirror points. Curves giving the position of mirror points for constant equatorial pitch angle α_o are shown. From Roederer, J.G., 1970. Dynamics of Geomagnetically Trapped Radiation. Springer-Verlag, Berlin, Germany, p. 63.

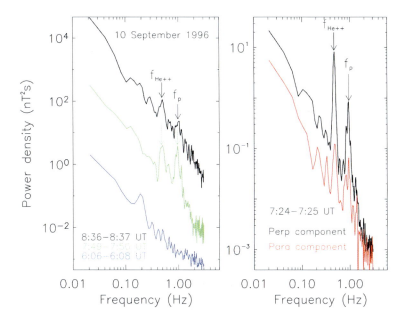

Fig. 7.20 The left panel is the total magnetic power spectra for fluctuations in the ULF range at three different 60-s periods on September 10, 1996. The right panel is the perpendicular and parallel power spectra at 07:24–07:25 UT on that day. From Chen, J., 2008. Evidence for particle acceleration in the magnetospheric cusp. Ann. Geophys. 26, 1995.

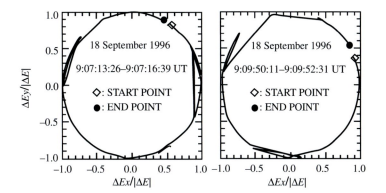

Fig. 7.21 Hodograms of the two perpendicular components of the cavity electric field at two different time periods on September 18, 1996, where $|\Delta E| = |E - <E>|$, $\Delta E_X = E_X - <E_X>$, $\Delta E_Y = E_Y - <E_Y>$ with $<>$ being the average over the time period. From Chen, J., 2008. Evidence for particle acceleration in the magnetospheric cusp. Ann. Geophys. 26, 1995.

point are shown. This hodogram shows a left-hand circular polarization of the cusp electric field in the mean magnetic field-aligned coordinates over a period near the He⁺ gyroperiod (left panel) and the 0^{+6} gyroperiod (right panel). It can energize the helium and oxygen ions, and also electrons, by a cyclotron resonant acceleration mechanism (CRAM).

Using the development of Chen (2008), the energization rate by the cyclotron resonant acceleration can be expressed analytically by measurable terms. By definition, the changing rate of the kinetic energy of a charged particle with velocity $\mathbf{v}$ in electric field $\mathbf{E}$ and magnetic field $\mathbf{B}$ is

$$dK/dt = \mathbf{F} \cdot \mathbf{v} \qquad (7.3)$$

where $\mathbf{F} = q(\mathbf{E} + \mathbf{v} \times \mathbf{B})$ is the Lorentz force with q being the particle charge. Since $(\mathbf{v} \times \mathbf{B}) \cdot \mathbf{v} = 0$ (the magnetic force is always perpendicular to the $\mathbf{v}$ and the latter term does no work on the particle), Eq. (7.3) is equivalent to

$$dK/dt = q\mathbf{E} \cdot \mathbf{v} \qquad (7.4)$$

Eq. (7.4) contains both perpendicular and parallel components. For particle cyclotron resonant acceleration, one only needs to consider the perpendicular component; that is,

$$dK_\perp/dt = qEv_\perp \qquad (7.5)$$

where E is the left-hand polarization perpendicular electric field with frequency the same as the ion gyrofrequency, and $v_\perp$ is the particle's perpendicular velocity. For nonrelativistic ions,

$$v_\perp = (2K_\perp/m)^{1/2} \qquad (7.6)$$

so that Eq. (7.5) becomes

$$dK_\perp/dt = (2K_\perp/m)^{1/2}qE \qquad (7.7)$$

or

$$dK_\perp/(K_\perp)^{1/2} = (2/m)^{1/2}qEdt \qquad (7.8)$$

where m is the ion mass. Integration of Eq. (7.8) gives the solution

$$(K_\perp)^{1/2} = (K_\perp(0))^{1/2}(2m)^{-1/2}qEt \qquad (7.9)$$

where $K_\perp(0)$ is the ion perpendicular kinetic energy at time $t = 0$. Substituting this into Eq. (7.7) yields the changing rate of the ion perpendicular kinetic energy by the ion cyclotron resonant acceleration:

$$dK_\perp/dt = (2K_\perp(0)/m)^{1/2}qE + q^2E^2t/m \qquad (7.10)$$

Over an ion gyroperiod (T_i) the energy increased due to cyclotron resonant acceleration is

$$\Delta K_\perp = (2K_\perp(0)/m)^{1/2}qET_i + q^2E^2T_i^2/(2m) \qquad (7.11)$$

The significance of Eq. (7.11) is that all terms on its right side are measurable and are independent of models and simulations. This equation indicates that an ion energy enhanced by the gyroresonant acceleration is a function of the initial perpendicular kinetic energy, charge/mass ratio, gyroperiod of the ion, and the left-hand polarization perpendicular electric field.

Assuming its initial perpendicular kinetic energy = 2 keV (the solar wind as seed population), the enhanced energy by the gyro-resonant acceleration for an alpha particle (He^{++} ion) will be 101.4 keV, 0.54 MeV, and 2.04 MeV over a 2-s gyroperiod for the perpendicular electric field of 20, 50, and 100 mV/m, respectively.

Peak power spectral densities at the ion gyrofrequencies have been measured. Over the ULF range, the power spectral densities are dominated by the perpendicular component of the local magnetic field. The cavity electric fields have an amplitude of up to 350 mV/m, and its perpendicular component varies from −300 to 350 mV/m. The measured left-hand polarization of the cusp electric field at ion gyrofrequencies demonstrates that the CRAM occurs in the diamagnetic cavities. The CRAM can energize ions from keV to MeV in seconds.

In summary, satellite observations from Polar and Cluster reveal a broad and dynamic region centered at the Earth's magnetospheric cusp. This region is filled with the energetic charged particles, large diamagnetic cavities, and strong electromagnetic fluctuations. Inside the cavities, both the energetic particle intensity and the power spectral density of the magnetic fluctuations show increases by up to four orders of magnitude in comparison to an adjacent region.

Chen (2008)

7.7 Discussion

The shocked solar wind plasma enters the throat in the geomagnetic field unimpeded (see Fig. 7.3) and produce cusp diamagnetic cavities of tremendous size ($\sim 6 \ R_E$). Inside the cavities, both the energetic particle intensity and the power spectral density of the magnetic fluctuations show increases by up to four orders of magnitude in comparison to an adjacent region. The cavity electric fields have amplitudes of up to 350 mV/m. Chen (2008) has shown that the measured left-hand polarization of the cusp electric field at ion gyrofrequencies can energize the local ions and electrons in the diamagnetic cavities. The CRAM can energize electrons and ions from keV to MeV in seconds. *This is an impressive result with a very simple calculation!*

Fig. 7.18 shows that cusp energetic particles could be trapped in this diamagnetic cavity rather than just around the high-latitude minimum, as predicted by Sheldon. The top panel is energetic electrons, the second panel is their pitch angle, and the bottom panel is the magnetic field strength. The black lines overplotted in the second panel shows which particles will be trapped within the cusp if B_{max} is the sheath field strength. The majority of the flux lies within these black lines indicating a trapped population.

7.8 Summary

The density, temperature, and velocity distributions establish the presence of three distinct boundaries surrounding the exterior cusp region: with the lobes, the dayside plasma sheet, and the magnetosheath. This external boundary is characterized by a density decrease and a temperature increase, from the magnetosheath to the exterior cusp (Lavraud et al., 2006). The antisunward convection observed in the exterior cusp suggests that this region is statistically quite convective under southward IMF. For northward IMF the region appears more stagnant, and convection in the LLBL is sunward (which will be addressed in Chapter 9).

7.8.1 Transfer events

Wang et al. (2005) presented results from a statistical study of Cluster multispacecraft transfer events. They identified these as FTEs acting under the influence of the *convection electric field*, but they are probably PTEs acting under the influence of the *real field* $\mathbf{E} = -\partial \mathbf{A}/\partial t - \nabla \phi$. The entry is independent of the IMF, being determined by the local conditions. An average transfer event separation time of 7.09 min is obtained.

The presence of large parallel (downward) flows at the equatorward edge of the cusp shows that plasma penetration occurs preferentially at the dayside low-latitude magnetopause for southward IMF conditions; in contrast, under northward IMF the results are suggestive of plasma penetration from the poleward edge of the cusp, combined with a substantial sunward convection. The transverse plasma convection in the exterior cusp appears to be controlled by the IMF B_y component as well; for dawnward (duskward) IMF orientations the convection is preferentially directed toward dusk (dawn).

7.8.2 Cusp energetic particles

The CAMMICE experiment onboard Polar has observed cusp energetic ions that were associated with a dramatic decrease and large fluctuations in the local magnetic field strength.

- The cusp helium ions had energies up to 8 MeV.
- The ions that originated from the ionosphere were also observed in the high-altitude dayside cusp.
- The magnetic moment spectra of these low charge state energetic oxygen ions showed a higher amplitude of the spectrum of ion flux in the cusp than in the equatorial ring current.
- The measured energetic ions in the cusp cannot be explained by the substorm (or the nightside plasma sheet) source.

7.8.3 New energetic particle source

We need to invoke a new energization source. Evidence presented by Chen (2008) reveals that the charged particles can be energized locally in the magnetospheric cusp. The power spectral density of the cusp magnetic fluctuations shows increases by up to four orders of magnitude in comparison to an adjacent region. Large fluctuations of the cusp electric fields have been observed with an amplitude of up to 350 mV/m. The measured left-hand polarization of the cusp electric field at ion gyrofrequencies indicates that the CRAM is working in this region. CRAM can energize ions and electrons from keV to MeV in seconds.

These ions and electrons will have easy access to the equatorial magnetosphere via their gradient and curvature drifts following a trajectory described initially by Shabansky (1970). They will form a layer of energetic particles on the magnetopause as shown by Fritz et al. (2000) and Zhou et al. (2000).

7.9 Problems

7.1. Show that you can obtain Eq. (7.10) from the previous equations in the section. Show that you can obtain Eq. (7.11) from Eq. (7.10) OR from Eq. (7.9).

7.2. The solar wind IMF has the following x-y-z components: $\mathbf{B} = (1, -3, -5\,\text{nT})$. What is the magnitude of the IMF and the clock angle? Is this IMF considered "northward" or "southward?"

7.3. Using Fig. 7.7, determine the approximate distance from the Earth to the centroid of the stagnant plasma in the cusps for both northward and southward IMF conditions.

7.4. Demonstrate that the enhanced energies by gyroresonant acceleration discussed at the end of Section 7.6.3 agree with your answers using the given values in Eq. (7.11).

7.5. Estimate the values for the ion velocity x and z components and ACE magnetic field components from the end (10:00 UT) of the time period displayed in Fig. 7.13. Calculate the magnitude and angle (relative to $-x$) of the ion flow velocity in the x-z plane, and the clock angle of the IMF at ACE.

References

Axford, W.I., Hines, C.O., 1961. A unifying theory of high-latitude geophysical phenomena and geomagnetic storms. Can. J. Phys. 39, 1433.

Chapman, S., Ferraro, V.C.A., 1931. A theory of magnetic storms. Terr. Magn. Atmos. Electr. 36, 77–97.

Chen, J., 2008. Evidence for particle acceleration in the magnetospheric cusp. Ann. Geophys. 26, 1993–1997.

Chen, J., 2009. Johannson, H. (Ed.), Solar Wind, Large Diamagnetic Cavities, and Energetic Particles, Handbook on Solar Wind: Effects, Dynamics and Interactions. Nova Science, Hauppauge, NY.

Chen, J., Fritz, T.A., 2002. Multiple spacecraft observations of energetic ions during a major geomagnetic storm. Adv. Space Res. 30 (7), 1749–1755.

Chen, J., Fritz, T.A., Sheldon, R.B., Spence, H.E., Spjeldvik, W.N., Fennell, J.F., et al., 1997. A new, temporarily confined population in the polar cap during the August 27, 1996 geomagnetic field distortion period. Geophys. Res. Lett. 24, 1447–1450.

Chen, J., Fritz, T.A., Sheldon, R.B., Spence, H.E., Spjeldvik, W.N., Fennell, J.F., et al., 1998. Cusp energetic particle events: implications for a major acceleration region of the magnetosphere. J. Geophys. Res. 103, 69–78.

Dunlop, M.W., Lavraud, B., Cargill, P., Taylor, M.G.G.T., Balogh, A., Réme, H., et al., 2005. Cluster observations of the cusp: magnetic structure and dynamics. In: Fritz, T.A., Fung, S.F. (Eds.), The Magnetospheric Cusps. Springer, Dordrecht, Netherlands, pp. 5–55.

Dyson, F., 1958. Innovation in physics. Sci. Am. 199, 76.

Fedorov, A., Dubinin, E., Song, P., Budnick, E., Larson, P., Sauvaud, J.-A., 2000. Characteristics of the exterior cusp for steady southward interplanetary magnetic field: interball observations. J. Geophys. Res. 105 (A7), 15945–15957.

Frank, L.A., 1971. Plasma in the Earth's polar magnetosphere. J. Geophys. Res. 76, 5202.

Fritz, T.A., 2009. Perspectives gained from a combination of polar, cluster and ISEE energetic particles measurements in the dayside cusp. In: Laakso, H., Taylor, M., Escoubet, C.P. (Eds.), The Cluster Active Archive—Studying the Earth's Space Plasma Environment. Springer, Dordrecht, Netherlands.

Fritz, T.A., Fung, S.F. (Eds.), 2005. The Magnetospheric Cusps. Springer, Dordrecht, Netherlands.

Fritz, T.A., Chen, J., Sheldon, R.B., 2000. The role of the cusp as a source for magnetospheric particles: a new paradigm? Adv. Space Res. 25 (7–8), 1445–1457.

Fuselier, S., Trattner, K., Petrinec, S., 2000. Cusp observations of high- and low-latitude reconnection for northward interplanetary magnetic field. J. Geophys. Res. 105 (A1), 253–266.

Haerendel, G., Paschmann, G., Sckopke, N., Rosenbauer, H., Hedgecock, P., 1978. The frontside boundary layer of the magnetosphere and the problem of reconnection. J. Geophys. Res. 83 (A7), 3195–3216.

Heikkila, W.J., 1972. The morphology of auroral particle precipitation. In: Heikkila, W.J. (Ed.), Space Research XII. Akademie-Verlag, Berlin, Germany, pp. 1343–1355.

Heikkila, W.J., Winningham, J.D., 1971. Penetration of magnetosheath plasma to low altitudes through the dayside magnetospheric cusps. J. Geophys. Res. 76, 883–891.

Heikkila, W.J., Smith, J.B., Tarstrup, J., Winningham, J.D., 1970. The soft particle spectrometer in the ISIS-I satellite. Rev. Sci. Instrum. 41, 1393–1402.

Laitinen, T.V., Janhunen, P., Pulkkinen, T.I., Palmroth, M., Koskinen, H.E.J., 2006. On the characterization of magnetic reconnection in global MHD simulations. Ann. Geophys. 24, 3059.

Lavraud, B., Fedorov, A., Budnik, E., Grigoriev, A., Cargill, P.J., Dunlop, M.W., et al., 2004a. Cluster survey of the high-altitude cusp properties: a three-year statistical study. Ann. Geophys. 22 (8), 3009–3019.

Lavraud, B., Dunlop, M.W., Phan, T.D., Rème, H., Taylor, M., 2004b. The exterior cusp and its boundary with the magnetosheath: cluster multi-event analysis. Ann. Geophys. 22 (8), 3039–3054.

Lavraud, B., Rème, H., Dunlop, M., Bosqued, J.M., Dandouras, I., Savaud, J.-A., et al., 2005a. Cluster observes the high-altitude cusp regions. In: Fritz, T.A., Fung, S.F. (Eds.), The Magnetospheric Cusps. Springer, Dordrecht, Netherlands, pp. 135–174.

Lavraud, B., Fedorov, A., Budnik, E., Thomsen, M.F., Grigoriev, A., Cargill, P.J., et al., 2005b. High altitude cusp flow dependence on IMF orientation: a 3 year Cluster statistical study. J. Geophys. Res. 110, A02209. https://doi.org/10.1029/2004JA010804.

Lavraud, B., Thomsen, M.F., Taylor, M.G.G.T., Wang, Y.L., Phan, T.D., Schwartz, S.J., et al., 2005c. Characteristics of the magnetosheath electron boundary layer under northward interplanetary magnetic field: implications for high-latitude reconnection. J. Geophys. Res. 110, A06209. https://doi.org/10.1029/2004JA010808.

Lavraud, B., Thomsen, M.F., Lefebvre, B., Budnik, E., Cargill, P.J., Fedorov, A., et al., 2006. Formation of the cusp and dayside boundary layers as a function of IMF orientation: cluster results. In: Cluster and Double Star Symposium—5th Anniversary of Cluster in Space ESA SP-598.

Lepping, R.P., Acũna, M.H., Burlaga, L.F., Farrell, W.M., Slavin, J.A., Schatten, K.H., et al., 1995. The WIND magnetic field investigation. Space Sci. Rev. 71 (1), 207–229.

Merka, J., Safrankova, J., Nemecek, Z., 1999. Interball observations of the high-altitude cusp-like plasma: a statistical study. Czechoslov. J. Phys. 48 (4a), 695.

Newell, P.T., Meng, C.I., Meng, C.-I., Sibeck, D.G., Lepping, R., 1989. Some low-altitude cusp dependencies on the interplanetary magnetic field. J. Geophys. Res. 94, 8921.

Paschmann, G., Haerendel, G., Sckopke, N., Rosenbauer, H., Hedgecock, P., 1976. Plasma and magnetic field characteristics of the distant polar cusp near local noon: the entry layer. J. Geophys. Res. 81 (16), 2883–2899.

Phan, T.-D., Eastwood, J.P., Shay, M.A., Drake, J.F., Sonnerup, B.U.O., Fujimoto, M., Cassak, P.A., Oieroset, M., Burch, J.L., Torbert, R.B., Rager, A.C., Dorelli, J.C., Gerchman, D.J., Pollock, C., Pyakurel, P.S., Haggerty, C.C., Khotyaintsev, Y., Lavraud, B., Saito, Y., Oka, M., Ergun, R.E., Rentino, A., LeContel, O., Argall, M.R., Giles, B.L., Moore, T.E., Wilder, F.D., Strangeway, R.J., Russell, C.T., Lindqvist, P.A., Magnes, W., 2018. Electron magnetic reconnection without ion coupling in Earth's turbulent magnetosheath. Nature 557 (2), 202–206.

Roederer, J.G., 1967. On the adiabatic motion of energetic particles in a model magnetosphere. J. Geophys. Res. 72 (3), 981–992.

Roederer, J.G., 1970. Dynamics of Geomagnetically Trapped Radiation. Springer-Verlag, Berlin, Germany.

Rosenbauer, H., Grünwaldt, H., Montgomery, M.D., Paschmann, G., Sckopke, N., 1975. HEOS-2 plasma observations in the distant polar magnetosphere: the plasma mantle. J. Geophys. Res. 80, 2723.

Russell, C.T., Elphic, R.C., 1978. Initial ISEE magnetometer results—magnetopause observations. Space Sci. Rev. 22, 681–715.

Russell, C.T., Le, G., Petrinec, S.M., 2000. Cusp observations of high-and low-latitude reconnection for northward IMF: an alternate view. J. Geophys. Res. 105, 5489.

Shabansky, V.P., 1970. Some processes in the magnetosphere. Space Sci. Rev. 12, 299–418.

Sheldon, R.B., Spence, H.E., Sullivan, J.D., Fritz, T.A., Chen, J., 1998. The discovery of trapped energetic electrons in the outer cusp. Geophys. Res. Lett. 25 (11), 1825–1828.

Tsyganenko, N.A., 1989. A magnetospheric magnetic field model with a warped tail current sheet. Planet. Space Sci. 37 (1), 5–20.

Tsyganenko, N.A., 1995. Modeling the Earth's magnetospheric magnetic field confined within a realistic magnetopause. J. Geophys. Res. 100 (A4), 5599–5612.

Walsh, B.M., Fritz, T.A., Lender, N.M., Chen, J., Whitaker, K.E., 2007. Energetic particles observed by ISEE-1 and ISEE-2 in a cusp diamagnetic cavity on September 29, 1978. Ann. Geophys. 25, 2633–2640 doi:0.5194/angeo-25-2633-2007.

Walsh, B.M., Fritz, T.A., Klida, M.M., Chen, J., 2010. Energetic electrons in the exterior cusp: identifying the source. Ann. Geophys. 28, 983–992. https://doi.org/10.5194/angeo-28-983-2010.

Wang, Y.L., Elphic, R.C., Lavraud, B., Taylor, M.G.G.T., Birn, J., Raeder, J., et al., 2005. Initial results of high latitude magnetopause and low latitude flank flux transfer events from 3 years of Cluster observations. J. Geophys. Res. 110, A11221. https://doi.org/10.1029/2005JA011150.

Whitaker, K.E., Chen, J., Fritz, T.A., 2006. CEP populations observed by ISEE 1. Geophys. Res. Lett. 33, L23105. https://doi.org/10.1029/2006GL027731.

Whitaker, K.E., Fritz, T.A., Chen, J., Klida, M., 2007. Energetic particle sounding of the magnetospheric cusp with ISEE-1. Ann. Geophys. 25 (5), 1175–1182.

Zhang, S.R., Holt, J.M., McCready, M., 2007. High latitude convection based on long-term incoherent scatter radar observations in North America. J. Atmos. Terr. Phys. 69 (10 – 11), 1273–1291.

Zhou, X.W., Russell, C.T., Le, G., Fuselier, S.A., Scudder, J.D., 2000. Solar wind control of the polar cusp at high altitude. J. Geophys. Res. 105 (A1), 245–251.

Zong, Q.G., Fritz, T.A., Korth, A., Daly, P.W., Dunlop, M., Balogh, A., et al., 2005. Energetic electrons as a field line topology tracer in the high latitude boundary/cusp region: cluster rapid observations. In: Fritz, T.A., Fung, S.F. (Eds.), The Magnetospheric Cusps. Springer, Dordrecht, Netherlands.

8

Inner magnetosphere

Chapter outline

Common sense has the very curious property of being more correct retrospectively than prospectively; unfortunately, they seldom are prospectively.

Ackoff (1968, p. 96)

8.1   Introduction

Although the Earth's radiation belts were among the first discoveries of orbiting satellites (Van Allen, 1959), our understanding of them has undergone a revolution since the launch, in August of 2012, of the Van Allen Probes (previously, the Radiation Belt Storm

Probes, still abbreviated as RBSP) (Reeves, 2015). Improvements in energetic particle measurements, the development of more realistic global magnetic field models, and advances in computational power have greatly enhanced our ability to study the radiation belts. That these belts are imbedded in a relatively cold and dense inner magnetospheric region named the plasmasphere has also been known since early on (Carpenter, 1963). Modern missions such as IMAGE, Cluster, and THEMIS have helped to clarify the structure and dynamics of this region as well, leading to an entirely new set of nomenclature (Fig. 8.1). An excellent review of plasmaspheric density structures and dynamics by Darrouzet et al. (2009) details many of the advancements in understanding provided by IMAGE and Cluster.

8.1.1 The radiation belts take shape

After their initial discovery, the radiation belts were the subject of much study in the following years. During these early days, much was learned about the radiation belts, including the inner

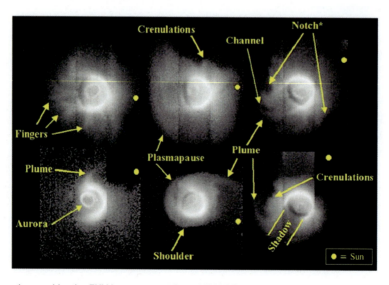

Fig. 8.1 Structures observed by the EUV instrument onboard IMAGE and new morphological nomenclature: examples of shoulders, plumes, fingers, channels, crenulations, and notches. The direction to the Sun is shown as a *yellow dot* for each image. From Darrouzet, F., Gallagher, D.L., André, N., Carpenter, D.L., Dandouras, I., Décréau, P.M.E., et al., 2009. Plasmaspheric density structures and dynamics: properties observed by the CLUSTER and IMAGE missions. Space Sci. Rev. 145, 58. https://doi.org/10.1007/s11214-008-9438-9.

and outer zone spatial structure, spectral information, pitch angle distributions, and the characteristic response of the radiation belts to geomagnetic storms. The typical storm response was found to be a rapid decrease in electron fluxes during the main phase followed by a several-day reintensification. Two primary sources for radiation belt electrons were considered: either a solar wind source or an internal acceleration process. Since little was known about the structure of the magnetosphere at that time, "internal" was naturally taken to mean internal to the radiation belts themselves (as in the early review article by Farley (1963).

8.1.2 Electron acceleration mechanisms

Eventually, two primary methods for electron acceleration were proposed. In 1965, Fälthammar and others showed that the acceleration of radiation belt electrons could occur through radial diffusion when field fluctuations occur on the time scale of electron drifts. He concluded that any radial motion of the electrons resulted in an energy change when the first and second adiabatic invariants (see Sections 1.4.5–1.4.7) were conserved. The source population in such a model is the magnetotail. Alternatively, Kennel and Petschek (1966) (extending the work of Dungey (1963)) showed that the very low frequency (VLF) whistler-mode waves that pitch angle scattered electrons into the atmosphere could be produced by plasma distributions that had temperature anisotropies unstable to the growth of whistler-mode waves. In the late 1990s Summers et al. (1998) proposed that cyclotron resonant interactions at much higher (relativistic) energies can change electron momentum. At those energies gradients in the energy dimension implied that the flow of energy is from the waves to the particles providing a mechanism for accelerating electrons in situ from the local population of lower-energy electrons.

8.1.3 New measurements needed

The two proposed acceleration methods, radial diffusion and local wave-particle coupling, can be distinguished by looking at the radial gradients of the phase space density (Fig. 8.2, after Reeves (2015)). This involves transforming the data from its natural coordinates of position, velocity, and pitch angle, into phase space density as a function of the three adiabatic invariants described in Chapter 1. At any given point, the conversion

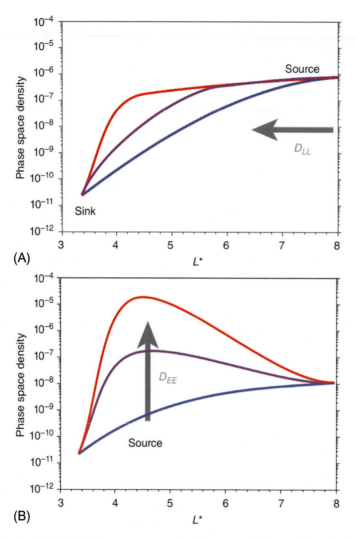

Fig. 8.2 (A) Schematic of radial diffusion from a source region (the magnetotail) to a sink region (the slot). An increase in radial diffusion will increase the transport rate and increase the phase space densities in the heart of the belts ($L^* \approx 4$–5). (B) Schematic of local acceleration. Wave-particle interactions produce energy diffusion that accelerates electrons. In this case the source of the relativistic electrons is the local lower-energy populations. From Reeves, G.D., 2015. Radiation belt electron acceleration and role of magnetotail. In: Keilling, A., Jackman, C.M., Delamere, P.A. (Eds.), Magnetotails in the Solar System. AGU Monograph 207. John Wiley & Sons, Inc., Washington, DC, pp. 351–352.

depends on a realistic global magnetic field model and computational power sufficient to solve for the unique, time-dependent solution. Finally, a sufficient quantity and quality of local plasma measurements are needed to enable the production of the radial gradients needed to distinguish between the acceleration mechanisms.

8.2   Radiation belts

8.2.1   Emerging consensus

The storm of January 1997 observed by Polar and other satellites, and the October 2013 storm observed by RBSP, have led to an emerging consensus that local acceleration can be the dominant process (Reeves, 2015).

8.2.2   January 1997 storm

The storm beginning on January 10, 1997 was observed by a number of spacecraft over a period of ten days (January 8–18). Analysis of the storm benefited from the growing network of satellites able to make energetic particle measurements in the radiation belts. Geostationary Operational Environmental Satellites (GOES) -8 and -9 observed the storm, along with Los Alamos National Laboratory (LANL) satellites 1990-095, 1991-080, and 1994-084. These five, in geosynchronous orbits, measured the electron flux at an L-shell of 6.6. Three Global Positioning System (GPS) satellites in lower orbits were able to measure electron fluxes at an L-shell of 4.6. The Polar spacecraft also made periodic observations as it passed through the radiation belts at $L=4.6$ every 18 h (Fig. 8.3).

The 2 MeV radiation belt electrons at $L=6.6$ showed a typical behavior of a gradual increase in flux over about four days. The behavior at $L=4.6$, however, was startling. The 2 MeV electrons in this inner region increased by over two orders of magnitude in less than 12 h, then remained remarkably constant. They showed none of the gradual buildup seen at the higher geosynchronous orbit. The acceleration was so rapid that the radial profiles seen by the GPS satellites as they were moving away from the equator to higher L values were actually significantly different from the fluxes measured less than an hour later during the inbound part of the orbit. It also appeared that that the gradual flux increase seen at geosynchronous orbit was a result of outward radial diffusion from the newly accelerated population at $L=4.6$ (Fig. 8.3B). In contrast to these high time resolution measurements from multiple satellites, the Polar spacecraft measurements (white diamonds in Fig. 8.3A) captured the state of the belts before and after the acceleration but could not resolve the event itself.

This event showed that the electrons appear to be accelerated in a narrow range of L-shells, well inside geosynchronous orbit in the heart of the belts. If this is the case, then the source of the MeV radiation belt electrons would not be from the magnetotail, but

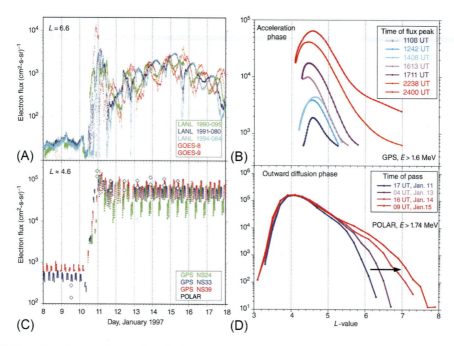

Fig. 8.3 Multisatellite electron observations during the January 1997 radiation belt event. Rapid acceleration in the heart of the (C) radiation belts ($L \approx 4.6$) and (A) geosynchronous orbit ($L \approx 6.6$) suggested local acceleration and subsequent outward radial diffusion but, without phase space density observations, were inconclusive. (B) Radial profiles from January 10 show that acceleration occurs throughout the radiation belts ($L = 4-7$) in approximately 12 h. (D) Four POLAR passes show the fluxes increasing outside $L = 5.5$ while they remain relatively stable inside $L = 5.5$. From Reeves, G.D., et al., 1998. The global response of relativistic radiation belt electrons to the January 1997 magnetic cloud. Geophys. Res. Lett. 25, 3266. https://doi.org/10.1029/98gl02509.

from the lower-energy electrons in that same inner region, somewhere between the plasmasphere and geosynchronous orbit (Reeves et al., 1998).

8.2.3 October 2013 storm

The Van Allan Probes were launched in 2012 to overcome the limitations of previous missions and clearly differentiate between the two acceleration mechanisms. The two spacecraft orbit near the equator every 9 h with an apogee of 5.8 R_E and perigee of only 1.1 R_E. The orbits are slightly different so that their relative phase slowly changes over time. When out of phase, their apogees are separated by 4.5 h, with passages through the heart of the radiation belts every 2-3 h. These orbits, paired with the comprehensive set of particle and field instruments they carry, have allowed unprecedented views of radiation belt dynamics.

Reeves et al. (2013) analyzed phase space density profiles for a storm that occurred on October 8–9, 2013, early in the Van Allen Probes mission. This storm produced a rapid flux enhancement in the outer belt and was the first test of the Van Allan Probes' ability to distinguish between the two acceleration mechanisms. The 2 MeV electron fluxes at $L=4.2$ increased by nearly three orders of magnitude in about 12 h (similar to the January 1997 event). The improved measurements allowed for conversion to phase space density for detailed analysis.

The profiles of phase space density as a function of L^* are shown in Fig. 8.4. L^* is equivalent to L in a dipolar field, but with a more realistic field model electrons with different pitch angles follow different drift paths (drift shell splitting). In the absence of acceleration or loss, L^* defines a surface of constant phase space density. The figure labels each radial pass by the time at which the satellite crossed $L^*=4.2$. From 23:17 UT on October 8

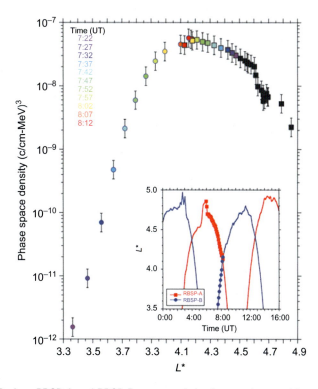

Fig. 8.4 Two Van Allen Probes, RBSP-A and RBSP-B, measured simultaneously, a positive radial gradient at $L^* < 4.2$ and a negative radial gradient at $L^* > 4.2$. Points are color coded by the time of each measurement and error bars are given. From Reeves, G.D., 2015. Radiation belt electron acceleration and role of magnetotail. In: Keilling, A., Jackman, C.M., Delamere, P.A. (Eds.), Magnetotails in the Solar System. AGU Monograph 207. John Wiley & Sons, Inc., Washington, DC, p. 356.

to 03:12 on October 9, the phase space densities increased by about two orders of magnitude at this point, and continued to increase until 13:02 UT, when the IMF turned northward. During this event, the two spacecraft were almost perfectly out of phase, providing investigators with the best possible time resolution. Also, RBSB-A was inbound from apogee and RBSB-B was outbound from perigee (red and blue lines, respectively, in the Fig. 8.4 inset). In Fig. 8.4, the measurements are shown every 5 min, color coded to show the relative positions of the two spacecraft at a given time. It can be seen that the outbound RBSB-B measured a positive gradient up to the peak at about $L^* = 4.2$, while the inbound RBSB-A was measuring a negative radial gradient at higher values. This demonstrates that the radial peak in phase space density cannot be an artifact of spatial-temporal aliasing. The two spacecraft coincidentally reached 4.2 at the same time (red points in Fig. 8.4), even though they were at quite different local times. The slight mismatch at this point was used to calculate worst case and average uncertainties as shown in the figure.

By itself, the data in Fig. 8.4 cannot completely rule out the possibility of a variable source population at high L^* (see Fig. 8.2B). However, Reeves et al. (2013) went further by plotting phase space density over time of the two Van Allen Probes along with five LANL geosynchronous satellites. The Van Allen Probes were always at similar or lower L^* values, but measured similar or higher phase space density. This rules out, at least for this particular event, a variable source at high L^* as an explanation for the peaks in phase space density, thus leaving local wave-particle acceleration as the only viable process. The specific wave-particle interaction mechanism is not yet known for certain; however, Thorne et al. (2013) have shown that chorus waves (which were present during this period) can produce the observed spectral and pitch angle distributions.

8.2.4 Implications

Even though the electrons from the magnetotail are not directly accelerated in the local wave-particle acceleration mechanism, they are still the source of the thermal and energetic electrons in the inner magnetosphere, and thus are still important to our understanding of the available seed population. The transport and subsequent trapping of electrons into the inner magnetosphere via convection and also by substorms produce distributions that are unstable to the growth of whistler-mode chorus and other waves which can cause local acceleration.

The suprathermal distributions are also responsible for Landau damping and thus the spatial extent of the chorus waves (Bortnik et al., 2007). This spatial distribution may determine which storms produce a net enhancement within the radiation belts, and which produce a net decrease (Reeves et al., 2011).

The earlier observations of the January 1997 and October 2013 storms show that the electrons accelerated in the heart of the belts later diffuse outward. Radial diffusion is enhanced during storms and always acts to redistribute electrons within the inner magnetosphere. The possibility therefore exists that electrons deep in the inner magnetosphere may actually be a source for energetic particles in the near tail.

8.3 Transient penetration

Transient penetration of plasma with magnetosheath origin is frequently observed with the hot plasma experiment onboard various spacecraft, including DE and Viking at auroral latitudes in the dayside magnetosphere. Injected magnetosheath ions exhibit a characteristic pitch angle/energy dispersion pattern, earlier reported for solar wind ions accessing the magnetosphere in the cusp regions as in Fig. 8.5. Events show temporal features which suggest a connection to transient processes at or in the vicinity of the magnetospheric boundary.

A single event study confirms previously published observations that the injected ions flow essentially tailward with a velocity comparable to the magnetosheath flow and that the energy spectra inferred for the source population resemble magnetosheath spectra. Those ion injection structures, which were resolved by the Viking mass spectrometer, consist of protons. Based on a statistical study, it is found that these events are predominantly observed around 0800 and 1600 MLT, in a region populated both by ring current/plasma sheet particles and by particles whose source is the magnetosheath plasma.

Woch and Lundin (1992)

8.3.1 Transient auroral event

On December 5, 1986, high-latitude magnetometer stations in Greenland showed a strong perturbation lasting for about 10 min beginning at 0930 UT in an otherwise quiet period with a northward IMF. These particles of magnetosheath origin (electrons and ions) can be transferred down to the ionosphere along the

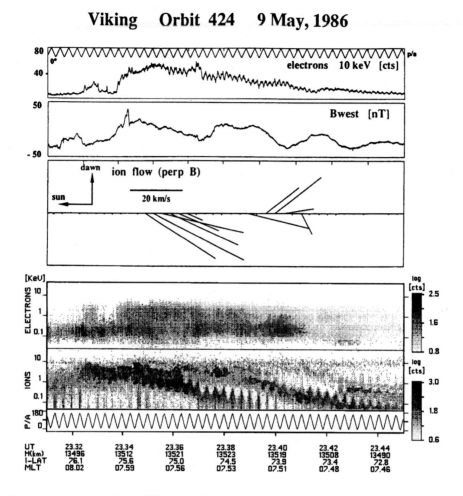

Viking Orbit 424 9 May, 1986

Fig. 8.5 Injection events observed on a Viking pass through the prenoon high-latitude region. Top to bottom: Counts of electrons at energies close to 10 keV (the variation of the pitch angle is shown at the top); the west component of the magnetic field; ion flow velocities perpendicular to the magnetic field obtained from moment calculations (direction to the Sun is toward left, dawn at the top); energy-time spectrograms for electrons and ions with energies between 40 eV and 40 keV. From Woch, J., Lundin, R., 1992. Signatures of transient boundary layer processes observed with Viking. J. Geophys. Res. 97, 1433.

magnetic field, undoubtedly affected by electric, gradient, and curvature drift terms. One can expect an ionospheric signature of a burst of particle precipitation in the cusp region which is expanding northward as well as eastward/westward (Sandholt et al., 1986; Sandholt and Farrugia, 2003).

The soft electron precipitation produces auroral emissions which were observed by all-sky photographs at the radar site.

Emission in the north began after 0430 UT; starting at 0630 UT a cleft feature was especially prominent in 630 nm emissions to the north. It is clear that the location of the cleft did not move from the far northern sky for the whole hour, observed at low elevation angles below 30 degrees. The first sign of the disturbance occurred at 0935 UT slightly to the south, followed by overhead emissions to as far north as 50 degrees elevation angle (Fig. 8.6, taken at 2-min intervals). Throughout this interval the cleft intensity peak remains at 30 degrees, with a region of decreased emission at a higher elevation angle of 40 degrees so that the cleft peak remains distinguishable at all times.

8.3.2 Plasma transfer event

The equivalent convection diagram is shown in Fig. 8.7; this diagram was drawn by rotating the magnetic perturbation vectors counterclockwise by 90 degrees and assuming a westward motion of the entire disturbance. A pair of field-aligned currents separated in the east-west sense and moving westward (tailward on the dawn side) at 4–5 km/s is consistent with the data, producing a twin vortex pattern of Hall currents. There are no discrete auroral emissions coincident with the downward Birkeland current. The first auroral forms are faintly shown to the east at this time, becoming quite intense at 0936 UT at the time of the upward current, and are overhead and to the west by 0938 UT. The photographs of Fig. 8.6 indicate a speed of travel of 3–4 km/s.

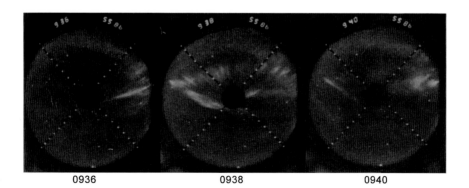

0936 0938 0940

Fig. 8.6 All-sky photographs at Sondrestrom on December 5, 1986, centered on the event. Auroras are clearly seen in northern sky at all times. No auroras were seen overhead at the time of the downward current, but a westward traveling emission is seen coincident with the upward current. This was carried by precipitating electrons in the keV range, which was recorded (not shown). From Heikkila, W.J., Jorgensen, T.S., Lanzerotti, L.J., Maclennan, C.G., 1989. A transient auroral event on the dayside. J. Geophys. Res. 94, 15299.

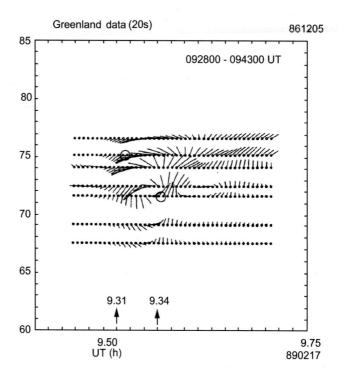

Fig. 8.7 The equivalent convection diagram drawn by rotating the magnetic perturbation vectors counterclockwise by 90 degrees. From Heikkila, W.J., Jorgensen, T.S., Lanzerotti, L.J., Maclennan, C.G., 1989. A transient auroral event on the dayside. J. Geophys. Res. 94, 15294.

The perturbation was accompanied by auroral forms overhead at Sondrestrom that also traveled westward. Meridian scanning photometer recordings at the radar site showed the cleft, located about 3–5 degrees poleward in latitude; the cleft did not move from the far northern sky for several hours, even while the disturbance was observed. The perturbation was also observed with the incoherent scatter radar at Sondrestrom; the small circles in Fig. 8.7 indicate regions sampled by the radar, these data agreeing with the twin vortex pattern.

The Viking and Polar Bear satellites passed just before the disturbance over Greenland, and DMSP encountered the disturbance near Baffin Island a few minutes later. These spacecraft observations increased confidence in the interpretation of the data. ISEE-1, ISEE-2, and IMP-8 recorded a magnetic disturbance in the solar wind, the likely cause of this event. Similar perturbations, but with reduced intensity, were also recorded on the afternoon side at Heiss Island, and at Dixon in northern Siberia. These were presumably caused by the flux tube wiping the afternoon side of the magnetopause, as shown in Fig. 8.8.

Fig. 8.8 shows magnetogram records from Dixon and Heiss on the dusk sides and Svalbard from the dawn side. The simultaneous

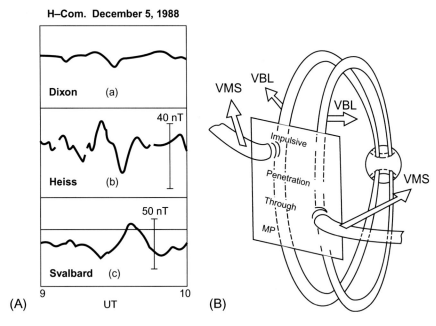

Fig. 8.8 (A) Magnetogram records from Dixon and Heiss on the dusk sides and Svalbard from the dawn side. (B) With impulsive penetration of solar wind plasma through the magnetopause onto closed field lines, two separate disturbances would be on closed field lines on the morning side as well as on the afternoon side. From Heikkila, W.J., Jorgensen, T.S., Lanzerotti, L.J., Maclennan, C.G., 1989. A transient auroral event on the dayside. J. Geophys. Res. 94, 15303.

impulsive plasma penetration on the dawn and dusk sides can only be explained as shown at the right.

8.4 Ionospheric outflow and coupling

8.4.1 Winter polar ionosphere

The winter polar ionosphere is an interesting subject for study because of the long periods without sunlight, thus providing insight into magnetospheric processes that then control it. Much of our knowledge of the shape and size of the convection electric field has come from measurements that covered only a small portion of high-latitude ionosphere at any one time. Only a few studies before the late 1970s had been able to capture the full pattern at a moment in time (STARE and SuperDARN radars were developed starting in the late 1970s).

Simultaneous ISIS 1 and 2 passes on December 14, 1971, shown in Fig. 8.9, provided an unusually comprehensive set of magnetic, ionospheric, particle, and optical measurements

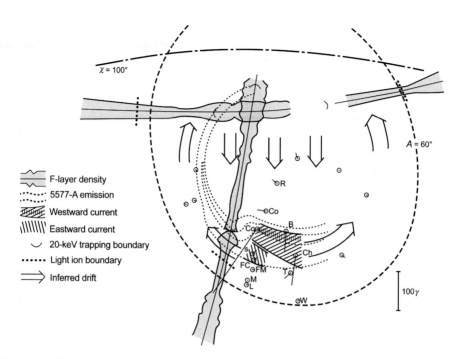

Fig. 8.9 F-layer densities, ionospheric currents, and inferred drift pattern after 0713 UT. Along the spacecraft tracks, the width of the outlined region is proportional to F-layer density. Magnetic observatories are indicated by circles, and the horizontal components of the disturbance vectors are indicated by lines extending from the circles. The diffuse 557.7 nm oval is outlined, and the approximate extent of the westward electrojet is indicated. Here, the most intense part of the westward current is shaded more heavily. Across each spacecraft track the 20-keV trapping boundary is indicated with an arc, and the light ion boundary is indicated with a *short dotted line*. The eastward current is much weaker than the westward current and may not be real. The magnetic observatories are noted in the article. From Whitteker, J.H., Shepherd, G.G., Anger, C.D., Burrows, J.R., Wallis, D.D., Klumpar, D.M., et al., 1978. The winter polar ionosphere. J. Geophys. Res. 83(A4), 1515.

(Whitteker et al., 1978). The northern magnetic polar ionosphere was in darkness to almost the maximum possible extent. During the first part of the day, magnetic activity was very low, but substorm activity began between the two sets of observations; a westward surge is seen in the optical data of the second set.

The instrumentation of the two spacecraft was described in detail by Franklin and Maclean (1969). Briefly, the instruments available for this study include topside sounders, energetic particle detectors on both spacecraft, a soft particle spectrometer, and photometers at 630.0, 557.7, and 391.4 nm but only on ISIS 2. Data were obtained also from ground-based ionosondes and magnetometers. Transverse electric fields, and the $\mathbf{E} \times \mathbf{B}$ drifts resulting from them, were not measured directly, but the nature of the drift patterns can be inferred from indirect evidence, involving the distribution of ionization and magnetic disturbances.

Before the substorm activity there was a region of low F-layer density in the central polar cap which must have been in a stagnant part of the convection pattern with no source for ionization. This region was subsequently filled in with ionization, evidently by convection that penetrated more poleward than before. During the first of the two observations, almost all high-latitude magnetograms were quiet; during the second observation the second of two closely spaced substorms was in progress.

On both occasions the two spacecraft cross the polar cap at about the same time, but on the second occasion the coincidence is particularly good, with a time separation at the crossing point of only 3 min. The ISIS 1 spacecraft was near apogee during these passes, that is, the altitude varied between 2600 and 3500 km, while ISIS 2 was in a circular orbit at 1400 km. What is plotted here is not the satellite paths themselves but these paths projected down the magnetic field to the altitude from which the optical emissions are assumed to originate (250 and 100 km).

Along the spacecraft tracks, the width of the outlined region is proportional to F-layer density. From the lower altitude of ISIS 2 (noon to midnight) the F-layer density is better defined than it is from ISIS 1 (dusk to dawn). The topside data are complete enough so that electron isodensity contours can be drawn. Magnetic observatories are indicated by circles; the horizontal components of the disturbance vectors are indicated by lines extending from the circles. The trapping boundaries derived from ISIS 2 data are marked. The 0.15- and 3.0-keV channels represent the electrons that give rise to the observed aurora, and the 0.15-keV channel indicates the fluxes of electrons that disturb the F layer and the topside ionosphere. The diffuse 557.7 nm oval is outlined, and so is the extent of the westward electrojet. Here the most intense part of the westward current is shaded more heavily.

The arcs drawn across the tracks in the neighborhood of 75 degrees invariant latitude (Λ) indicate the 20-keV electron trapping boundary. The short dotted lines across the spacecraft tracks at lower latitudes indicate points at which light ion densities (as inferred from vertical electron density distributions) increase sharply toward lower latitudes.

8.4.1.1 Soft particle data

The ISIS 2 soft particle spectrometer is shown in Fig. 8.10. This instrument measures the fluxes of electrons and positive ions over the energy range 5 eV to 15 keV. The spectrogram presented was

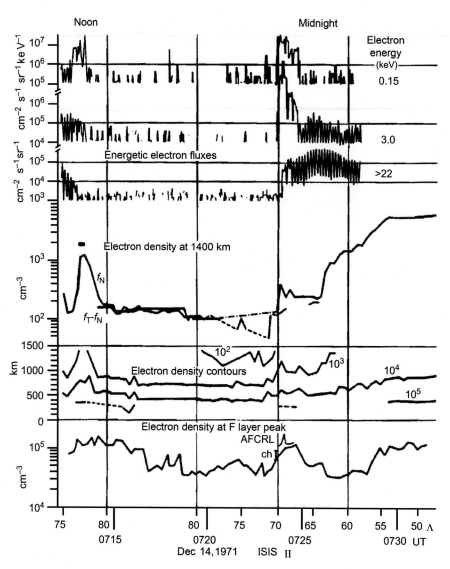

Fig. 8.10 Electron data are in energy-time spectrogram form. The ordinate is log 10 of the particle energy, and the abscissa is UT. The intensity represents counts per 11.1 ms accumulation period at a given energy. The middle and lower panels are the number and energy fluxes obtained by histogram integration of the observed spectrum from 5 eV to 15 keV. The identification of photoelectrons at the lowest energies shows a decrease near the highest invariant latitude. The designations Ch and AFCRL stand for Churchill and the AFCRL aircraft. From Whitteker, J.H., Shepherd, G.G., Anger, C.D., Burrows, J.R., Wallis, D.D., Klumpar, D.M., et al., 1978. The winter polar ionosphere. J. Geophys. Res. 83(A4), 1510.

discussed by Heikkila et al. (1970). These data are particularly valuable because they provide the information on the fluxes of low-energy electrons (~100 eV) necessary to account for both the intensity of 630.0 nm emission and the effects on the F layer and the topside ionosphere.

8.4.1.2 Optical emissions

The optical emissions are shown in Fig. 8.11 (Anger et al., 1973) for the wavelengths 630.0 and 557.7 nm. The emission rate for 391.4 nm is not shown but has an appearance very similar to that for 557.7 nm. In the quiet time data the greatest 630.0 nm intensities come from the magnetospheric cleft region on the dayside, with emissions more intense than 1 kilorayleigh (kR) originating

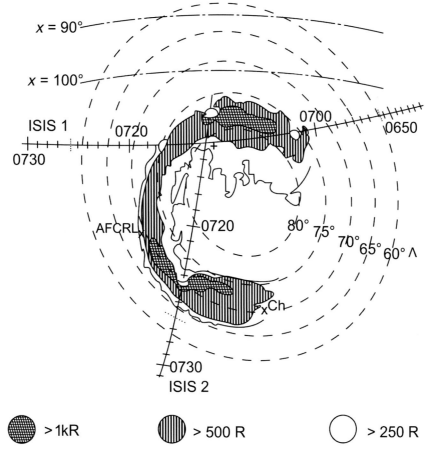

Fig. 8.11 Optical emission rates at 630.0 nm as observed by the ISIS 2 spacecraft. Data are plotted in geographic coordinates (polar projection onto a plane); the geographic North Pole is indicated with a plus sign. Only invariant latitude lines are shown, and the diagram is oriented so that noon is at the top. The solar terminator ($x=90°$) is indicated at the top of the diagram. Latitude lines, optical data, and spacecraft tracks are all projected to an altitude of 250 km. The 20-keV trapping boundary is indicated with an arc across each spacecraft track, and the light ion boundary is indicated with a *short dotted line*. From Whitteker, J.H., Shepherd, G.G., Anger, C.D., Burrows, J.R., Wallis, D.D., Klumpar, D.M., et al., 1978. The winter polar ionosphere. J. Geophys. Res. 83(A4), 1504.

from an area 4 degrees in latitude by about 4 h in magnetic local time. This bright emission is surrounded by a halo of less intense emission (>0.5 kR). The whole auroral oval is visible above the 250 R level; it is quite broad, with a jagged high-latitude edge.

In the magnetically active period, the cleft region has become narrower, about 2.5 degrees in latitude but is not significantly brighter than before. However, the rest of the oval visible on this pass has become brighter. (The morningside of the auroral oval is missing, being beyond the morning horizon, which is a line approximately parallel to the ISIS 2 path and about 20 degrees of arc from it.) The 557.7 nm data (not shown) show a smooth, almost featureless pattern of emission in the oval, with a region of minimum occurrence near the geomagnetic pole at very high latitudes in the midnight sector.

8.4.1.3 Particle-optical comparisons

There is a close correlation between the energetic particle fluxes and the optical emissions observed from ISIS 2 (Hays and Anger, 1978). On the dayside of both passes, electrons precipitated in the magnetospheric cleft region give rise to 630.0 nm emission with intensity of >1 kR. In the following pass there is a narrow region of very intense precipitation from 77 to 78 degrees, with a corresponding region of >1 kR emission.

8.4.1.4 Particle-ionosphere comparisons

The greatest effects by particles in the F layer and the topside ionosphere are caused by the intense soft electron fluxes associated with the magnetospheric cleft (Whitteker et al., 1972). The topside ionosphere expands owing to heating; the scale height increases, and there is also some F-layer ionization. The region of topside expansion coincides approximately with the region of >1 kR emission of 630.0 nm. It is remarkable how little effect the morning and afternoon auroral precipitation has on the ionospheric densities seen by ISIS 1 at its higher altitude.

8.4.1.5 Bottomside measurements

Densities at the peak of the F layer are available not only from the spacecraft, but also from a small number of bottomside sounders, one of which was carried by an Air Force Cambridge Research Laboratory (AFCRL) aircraft (Whalen and Pike, 1973). The F-layer densities are probably not unusual for quiet winter conditions. In the polar cap, in the absence of solar radiation and with only weak particle precipitation, the density of the F layer at any point must depend primarily on how long it has been since that particular element of ionosphere was last in a region where there was an ionization source and on the decay rate.

8.4.1.6 Summary of dark winter polar ionosphere

The ionosphere during the ISIS passes may be divided into four regions: the cleft, polar cap, nightside auroral region, and the plasmasphere.

- The polar cap: we contend it is mostly the LLBL.
- The cleft ionosphere: exhibits high densities at all altitudes.
- A region of stagnation: the low-density F layer was replaced by a high-density layer by the time of the second pass.
- Light ions are dominant down to much lower altitudes in the winter polar ionosphere than in the sunlight.
- Characteristic feature: almost no precipitation in the polar cap; there is almost no precipitation, no ionization, no excitation, no auroras in the polar cap during southward IMF.

8.4.2 TORDO UNO ion streak

In January 1975, two barium plasma injection experiments were carried out with rockets launched from Cape Parry, Northwest Territories, Canada. The launch was from a temporary facility located where field lines from the dayside cusp region intersect the ionosphere. One experiment, TORDO 1, took place near the beginning of a worldwide magnetic storm. It became a polar cap experiment almost immediately as convection perpendicular to **B** moved the fluorescent plasma jet away from the cusp across the polar cap in an antisunward direction. Convection across the polar cap with an average velocity of more than 1 km/s was observed for nearly 40 min (Fig. 8.12).

The three-dimensional observations of the plasma orientation and motion give an insight into convection from the cusp region across the polar cap, the orientation of the polar cap magnetic field lines out to several Earth radii, the causes of polar cap magnetic perturbations, and parallel acceleration processes. Prior to the encounter with the aurora near Greenland there is evidence of upward acceleration of the barium ions while they were in the polar cap.

A substorm began during the interval, and the barium flux tube encountered large **E** fields associated with a poleward bulge of the auroral oval near Greenland. The experiment shows that the development of the *substorm took place with constant external inputs*, in other words, it was not triggered from outside.

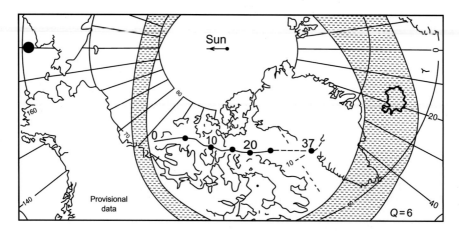

Fig. 8.12 TORDO UNO ion streak in solar magnetic coordinates. The electric field deduced represents the 100 km dawn-to-dusk electric field of the polar cap as observed for nearly 40 min, nearly 100 kV across the polar cap. Near the end of the track, large **E** fields and rapid space or time variations were observed as the barium streak encountered the evening aurora expanding poleward over Godhavn, Greenland due to an ongoing substorm. From Wescott, E., Stenbaek-Nielsen, H., Davis, T., Jeffries, R., Roach, W., 1978. The Tordo 1 polar cusp barium plasma injection experiment. J. Geophys. Res. 83(A4), 1569.

8.4.3 Theta aurora

Northward IMF produces emissions in what is called the trans-polar cap arc or theta aurora (Frank et al., 1986). Broad emission regions (containing Sun-aligned arcs) form on the dusk side (dawn side) of the auroral region in the northern hemisphere when B_y is positive (negative) and result in a "tear drop" shaped region near the pole that is void of auroral emissions when B_y is near zero. As the IMF becomes more strongly northward, the poleward edge of the auroral oval brightens.

Observations show that theta aurora can form during strictly northward IMF with its motion consistent with a change in sign of IMF B_y. Cumnock et al. (2002) consider the entire evolution of the theta aurora and the changing IMF conditions as shown in Fig. 8.13. The influence of IMF B_y is best illustrated by examples that occur during steady northward IMF as compared to times when the IMF is northward on average. For one case, DMSP F13 and F14 provide in situ measurements of precipitating particles, ionospheric plasma flows, and ion density. This unique data set enabled them to analyze in detail the evolution of a theta aurora, in one case crossing the entire polar region.

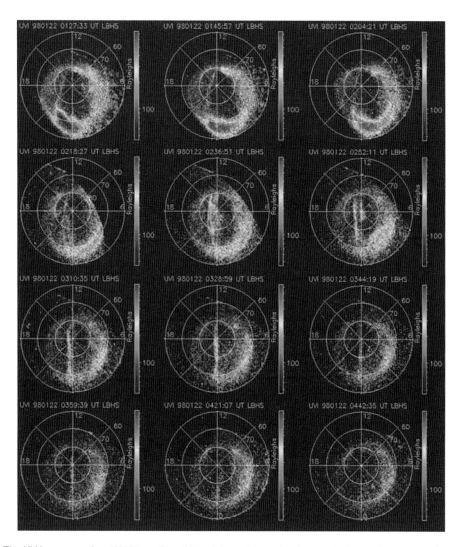

Fig. 8.13 The UV imager on day 980122 produced 36-s integration period images in the northern hemisphere using the Lyman-Birge-Hopfield short mode (LBHs) filter. Images have been selected from this day to illustrate the theta aurora development at different stages. From Cumnock, J.A., Sharber, J.R., Heelis, R.A., Blomberg, L.G., Germany, G.A., Spann, J.F., et al., 2002. Interplanetary magnetic field control of theta aurora development. J. Geophys. Res. 107(A7), 4. doi:10.1029/2001JA009126.

8.4.4   Four-cell convection pattern

The two-cell pattern of convection or equipotentials in the high-latitude regions has been clearly demonstrated in the observations under many circumstances. In addition to the two-cell convection pattern, Hardy et al. (1979) showed that when the z component of the IMF is northward, the convective pattern may have four cells, as in Fig. 8.14.

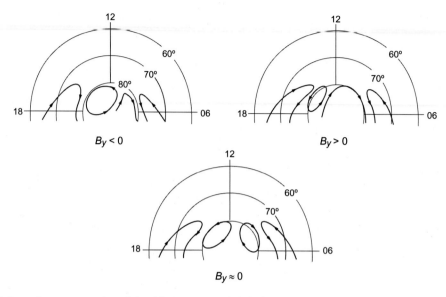

Fig. 8.14 Schematic representation of dayside patterns and their dependence on B_y during times of northward IMF. From Heelis, R.A., Reiff, P.H., Winningham, J.D., Hanson, W.B., 1986. Ionospheric convection signatures observed by DE 2 during northward interplanetary magnetic field. J. Geophys. Res. 91(A5), 5828.

Heelis et al. (1986) examined DE-2 observations of the ionospheric convection signature at high latitudes during periods of prolonged northward IMF.

> *The data from Dynamics Explorer 2 show that a four-cell convection pattern can frequently be observed in a region that is displaced to the sunward side of the dawn–dusk meridian regardless of season. In the eclipsed ionosphere, extremely structured or turbulent flow exists with no identifiable connection to a more coherent pattern that may simultaneously exist in the dayside region.*

Heelis et al. (1986)

Derivation of the electrostatic potential distribution along the satellite track, together with the convection velocity itself, allows the geometry of the convection cells, and their relative orientation, to be inferred. When examining the available data, one frequently encounters ambiguities and other difficulties in describing the most likely coherent convection pattern. The reason why it does not show in averages (e.g., Rich and Hairston, 1994) is that the averaging process removes the variations.

The two high-latitude cells show a dependence on the *y* component of the IMF that in the northern hemisphere leads to a domination of the anticlockwise circulating dusk cell when B_y is negative. When B_y is positive, clockwise circulation in the dawn cell becomes more significant. In the eclipsed ionosphere the ion drift can be extremely structured reflecting a turbulent motion rather than a coherent convection pattern.

8.5 Summary

The Van Allen Probes have given us new insight into the workings of the radiation belts. Of the two main theories proposed for electron acceleration, radial diffusion and local wave-particle interactions, the local acceleration via waves appears to be the primary mechanism in the storms studied so far. Much remains to be learned about the details of this acceleration mechanism and the waves that enable it, but as the quantity and quality of our orbital satellite network improves, more and more of these details will come to light.

The winter polar ionosphere is mostly in darkness, allowing interactions between the ionosphere and the inner magnetosphere to be studied independently of the ionizing effects of sunlight. A substorm occurring during these conditions was studied by the ISIS satellites and ground stations, providing a relatively comprehensive set of measurements. The ionosphere in this data set can be divided into four regions: the cleft, polar cap, nightside auroral region, and the plasmasphere. A region of stagnation in the low-density F layer was replaced by a high-density layer by the time of the second pass. It was also seen that light ions are dominant down to much lower altitudes in the winter polar ionosphere than in the sunlight. The polar cap in darkness is very different than when sunlit; there is almost no precipitation, no ionization, no excitation, and no auroras in the polar cap during southward IMF.

Observations of the ionospheric convection signature at high latitudes during periods of prolonged northward IMF show that a four-cell convection pattern can frequently be observed (Heelis et al., 1986). This exists in a region that is displaced to the sunward side of the dawn-dusk meridian regardless of season.

8.6 Problems

8.1. Define "L" and "L^*" and explain how they are different.

8.2. Sketch an example of the Van Allen Probes orbits in the plane perpendicular to the Earth-Sun line, including the approximate positions of the radiation belts. Why were these orbits chosen for this mission?

8.3. Explain the unit of photon flux, the Rayleigh, and calculate its value when the photon radiance is 2.7×10^{12} photons/m^2 s str.

8.4. What velocity corresponds to electrons with a kinetic energy of 2 MeV?

8.5. Fig. 8.11 shows lines of constant invariant latitude (Λ). For each value shown, calculate the corresponding L-shell and magnetic latitude at the altitude of ISIS 2 (see Section 1.12).

References

Ackoff, R.L., 1968. De Reuck, A., Goldsmith, M., Knight, J. (Eds.), Decision Making in National Science Policy. Little, Brown and Co, Boston, MA.

Anger, C.D., Fancott, T., McNally, J., Kerr, H.S., 1973. Isis-II scanning auroral photometer. Appl. Opt. 12, 1753–1766.

Bortnik, J., et al., 2007. Modeling the propagation characteristics of chorus using CRRES suprathermal electron fluxes. J. Geophys. Res. 112, A08204. https://doi.org/10.1029/2006JA012237.

Carpenter, D.L., 1963. Whistler evidence of a "knee" in the magnetospheric ionization density profile. J. Geophys. Res. 68 (6), 1675–1682.

Cumnock, J.A., Sharber, J.R., Heelis, R.A., Blomberg, L.G., Germany, G.A., Spann, J.F., et al., 2002. Interplanetary magnetic field control of theta aurora development. J. Geophys. Res. 107 (A7), 1108. https://doi.org/10.1029/2001JA009126.

Darrouzet, F., Gallagher, D.L., André, N., Carpenter, D.L., Dandouras, I., Décréau, P.M.E., DeKeyser, J., Denton, R.E., Foster, J.C., Goldstein, J., Moldwin, M.B., Reinisch, B.W., Sandel, B.R., Tu, J., 2009. Plasmaspheric density structures and dynamics: properties observed by the CLUSTER and IMAGE missions. Space Sci. Rev. 145, 55–106. https://doi.org/10.1007/s11214-008-9438-9.

Dungey, J.W., 1963. Loss of Van Allen electrons due to whistlers. Planet. Space Sci. 11, 591–595.

Farley, T.A., 1963. The growth of our knowledge of the Earth's outer radiation belt. Rev. Geophys. 1, 3–34.

Frank, L.A., Craven, J.D., Gurnett, D.A., Shawhan, S.D., Weimer, D.R., Burch, J., 1986. The theta aurora. J. Geophys. Res. 91 (A3), 3177–3224.

Franklin, C.A., Maclean, M.A., 1969. The design of swept-frequency topside sounders. Proc. IEEE 57, 897–929.

Hardy, D.A., Gussenhoven, M.S., Huber, A., 1979. The precipitation electron detectors (SSJ/3) for the block 5D/flights 2-5 DMSP satellites calibration and data presentation. Interim Report Air Force Geophysics Lab., Hanscom AFB, MA. Space Physics Division, September.

Hays, P.B., Anger, C.D., 1978. The influence of ground-scattering on satellite auroral observations. Appl. Opt. 17, 1898–1904.

Heelis, R.A., Reiff, P.H., Winningham, J.D., Hanson, W.B., 1986. Ionospheric convection signatures observed by DE 2 during northward interplanetary magnetic field. J. Geophys. Res. 91 (A5), 5817–5830.

Heikkila, W.J., Smith, J.B., Tarstrup, J., Winningham, J.D., 1970. The Soft particle spectrometer in the ISIS-I satellite. Rev. Sci. Instrum. 41, 1393–1402.

Kennel, C.F., Petschek, H.E., 1966. Limit on stably trapped particle fluxes. J. Geophys. Res. 71 (1), 1–28.

Reeves, G.D., 2015. Radiation belt electron acceleration and role of magnetotail. In: Keilling, A., Jackman, C.M., Delamere, P.A. (Eds.), Magnetotails in the Solar System. AGU Monograph 207. John Wiley & Sons, Inc., Washington, DC, pp. 345–359.

Reeves, G.D., et al., 1998. The global response of relativistic radiation belt electrons to the January 1997 magnetic cloud. Geophys. Res. Lett. 25 (17), 3265–3268. https://doi.org/10.1029/98gl02509.

Reeves, G.D., et al., 2011. On the relationship between relativistic electron flux and solar wind velocity: Paulikas and Blake revisited. J. Geophys. Res. 116. https://doi.org/10.1029/2010ja015735.

Reeves, G.D., et al., 2013. Electron acceleration in the heart of the Van Allen radiation belts. Science. https://doi.org/10.1126/science.1237743.

Rich, F.J., Hairston, M., 1994. Large-scale convection patterns observed by DMSP. J. Geophys. Res. 99 (A3), 3827–3844.

Sandholt, P.E., Farrugia, C.J., 2003. The aurora as monitor of solar wind-magnetospheric interactions. In: Newell, P.T., Onsager, T. (Eds.), Earth's Low-Latitude Boundary Layer. AGU Geophysical Monograph 133. American Geophysical Union, Washington, DC, pp. 335–349.

Sandholt, P., Deehr, C., Egeland, A., Lybekk, B., Viereck, R., Romick, G., 1986. Signatures in the dayside aurora of plasma transfer from the magnetosheath. J. Geophys. Res. 91 (A9), 10063–10079.

Summers, D., Thorne, R.M., Xiao, F., 1998. Relativistic theory of wave-particle resonant diffusion with application to electron acceleration in the magnetosphere. J. Geophys. Res. 103 (A9), 20487–20500. https://doi.org/10.1029/98JA01740.

Thorne, R.M., et al., 2013. Rapid local acceleration of relativistic radiation-belt electrons by magnetospheric chorus. Nature 504, 411–414.

Van Allen, J.A., 1959. The geomagnetically trapped corpuscular radiation. J. Geophys. Res. 64, 1683–1689.

Whalen, J., Pike, C., 1973. F-layer and 6300-A measurements in the day sector of the auroral oval. J. Geophys. Res. 78 (19), 3848–3856.

Whitteker, J.H., Brace, L.H., Burrows, J.R., Hartz, T.R., Heikkila, W.J., Sagalyn, R.C., Thomas, D.M., 1972. Isis 1 observations of the high-latitude ionosphere during a geomagnetic storm. J. Geophys. Res. 77 (31), 6121–6128.

Whitteker, J.H., Shepherd, G.G., Anger, C.D., Burrows, J.R., Wallis, D.D., Klumpar, D.M., Walker, J.K., 1978. The winter polar ionosphere. J. Geophys. Res. 83 (A4), 1503–1518.

Woch, J., Lundin, R., 1992. Signatures of transient boundary layer processes observed with Viking. J. Geophys. Res. 97 (A2), 1431–1447.

9

Low-latitude boundary layer

Chapter outline

Earth's Magnetosphere. https://doi.org/10.1016/B978-0-12-818160-7.00009-0

A new theory is seldom or never just an increment to what is already known. Its assimilation requires the reconstruction of prior theory and the re-evaluation of prior fact, an intrinsically revolutionary process that is seldom completed by a single man and never overnight.

Kuhn (1970, p. 7)

9.1 Introduction

The magnetospheric low-latitude boundary layer (LLBL) is a layer of plasma lying immediately earthward of the magnetopause (Fig. 9.1). The most important point to remember: a lot of plasma is flowing tailward for a southward interplanetary magnetic field (IMF), mostly on closed field lines. There must exist an electric field in the Earth's frame of reference given by $\mathbf{E} = -\mathbf{v} \times \mathbf{B}$ to produce a drift $\mathbf{v} = \mathbf{E} \times \mathbf{B}/B^2$ (other terms are usually smaller). To conserve momentum, the plasma will do anything, *anything at all possible*, to maintain that electric field.

Schmidt (1960) was the first one to make that point, further elaborated in his textbook in 1979. He uses an example of a gun with a magnetized plasma in which an electric field is applied by electrodes.

It is of interest to investigate what happens at the "muzzle" of this "plasma gun". Will the plasma continue to move across the magnetic field or will it stop at the end of the electrodes? A single charged particle would simply decelerate as the electric field decreases to zero, but the collective action of particles in the plasma leads to a qualitatively different behavior. The decelerating particles undergo an inertial or polarization drift resulting in the formation of charged surface layers in the electrodeless y > 0 region. The electric field set up by these layers facilitates the continued motion of the rest of the plasma across the magnetic field.

Schmidt (1979)

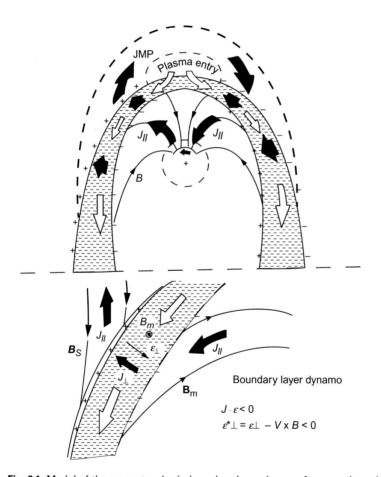

Fig. 9.1 Model of the magnetospheric boundary layer dynamo for a southward interplanetary magnetic field (IMF). *Open arrows* represent plasma flow, *solid arrows* are currents. Field-aligned Region 1 currents connect the inside of the layer to the auroral ionosphere. The lower panel shows the polarization current in the LLBL, necessary for the replenishment of charges to cause tailward flow and a braking action of the plasma. From Lundin, R., Evans, D.S., 1985. Boundary layer plasmas as a source for high-latitude, early afternoon, auroral arcs. Planet. Space Sci. 33, 1397.

This phenomenon has been observed in the laboratory. Baker and Hammel (1965) summarized a long series of experiments as follows.

These experiments have demonstrated that plasma streams from coaxial guns can enter and cross transverse fields up to 9 kG. In all cases the plasma which enters the field is observed to be only slightly diamagnetic and the motion of the plasma stream through the field is by means of an $E \times B$ drift resulting from self-polarization of the plasma stream.

Baker and Hammel (1965)

In Fig. 9.1 the plasma transfer event (PTE) supplies the plasma by a localized pressure gradient from the magnetosheath. Open arrows are plasma flows, solid arrows represent currents. Region 1 field-aligned currents connect the inside of the layer to the auroral ionosphere. The lower panel shows the polarization current in the LLBL to establish the voltage required for convection; the replenishment of charges causes a braking action of the plasma. Schmidt further stated:

It should not be thought, however, that the plasma can move indefinitely across a magnetic field. Particles drifting out into the surface layer experience a smaller electric field and hence a smaller drift than the bulk of the plasma and are consequently left behind. Since the total electric field in the plasma cannot change, the particles lost from the surface layer are continuously replaced from the plasma interior. Thus the plasma velocity remains unchanged while the mass is gradually decreasing.

Schmidt (1979)

9.1.1 Direct support for the low-latitude boundary layer

In particular, Wang et al. (2007) have investigated statistically 11 years of Geotail observations of the distributions of ion and electron density, temperature, pressure, and magnetic fields, and how these distributions respond to a change in solar wind density, speed, and magnitude (Fig. 9.2). The observed ion and electron energy spectra (0.1–100 keV for ions and 0.04–40 keV for electrons) were averaged in each 2.5×2.5 R_E region. Two-component kappa distributions (one to represent cold population and one to represent hot population) were assumed. A computed phase space density as a function of the first adiabatic invariant μ is defined as E_k/B where E_k is particle's kinetic energy. A normalized phase space density was calculated so that the

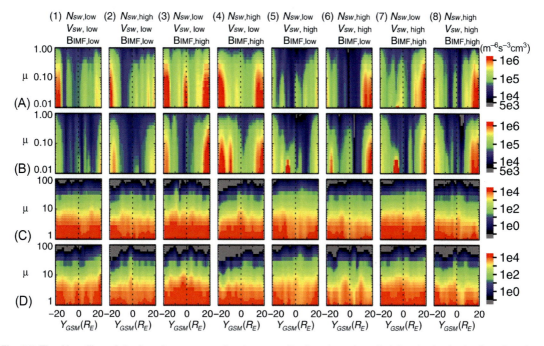

Fig. 9.2 The Y-profiles of the ion phase space density normalized to the solar wind density (*color bar* in units of m^{-6} s^{-3} cm^{-3}) for the eight interplanetary conditions. (A) For $\mu = 0.01$–1 keV/10 nT at $X = -15$ to -30 R_E, (B) for $\mu = 0.01$–1 keV/10 nT at $X = 0$ to -15 R_E, (C) for $\mu = 1$–100 keV/10 nT at $X = -15$ to -30 R_E, (D) for $\mu = 1$–100 keV/10 nT at $X = 0$ to -15 R_E. From Wang, C.-P., Lyons, L.R., Nagai, T., Weygand, J.M., McEntire, R.W., 2007. Sources, transport, and distributions of plasma sheet ions and electrons and dependences on interplanetary parameters under northward interplanetary magnetic field. J. Geophys. Res. 112, 6, A10224. https://doi.org/10.1029/2007JA012522.

changes of different interplanetary parameters can give the efficiency of the solar wind entry. The f_n versus μ for $\mu < 1$ keV and $\mu > 1$ keV across the tail in the near-Earth and midtail plasma sheet is shown for ions in Fig. 9.2 for northward IMF.

Fig. 9.2 is striking in its appearance; it divides the phase space density and particle energy into the two groups very clearly. For northward IMF we can understand how magnetosheath plasma in the LLBL (the upper panels) acts as driver causing plasma sheet plasma (the lower panels) to respond.

For both ions and electrons, the contribution of the cold population significantly enhances and becomes dominant toward the flanks. ... The temperature for the cold population near the dawn flank is similar to that near the dusk flank when V_{SW} is low. However, the dawn flank temperature increases much more with increasing V_{SW} than does the dusk flank temperature, resulting in higher temperature near the dawn than the dusk flank when V_{SW} is higher. This is seen in both ions and electrons.

Wang et al. (2007)

They found that the phase space density for <1 keV ions along the flanks is significantly larger than can be transported from the tail. This indicates that a cold particle source along the flanks could be provided by direct entry of the magnetosheath particles, as in a PTE.

In the near-Earth central plasma sheet, number density is mainly from the hot population. The temperature for the hot ion (electron) population is higher in the premidnight (postmidnight) sector. This shows that the temperature asymmetry shown in Fig. 9.2 is contributed by the high-energy particles. The distribution of the plasma sheet therefore can be understood as a mixture of a cold and a hot population. This compares the trajectories of ions of different energies coming from the same location at $X=-28.5$ R_E and $Y=0$ R_E following the drift paths $\mu B + e\Phi$ obtained from their analysis. This clearly shows how large the effect of magnetic drift can become, as the energy of particles become higher in diverting particles away from their electric drift paths. For typical thermal energy ions in the plasma sheet, their magnetic drift has become as strong as electric drift in determining their transport and energization.

This topic will now be discussed with the LLBL in mind. Fig. 9.1 shows several features of interest in solar wind interaction with the magnetosphere for a southward IMF (conditions are quite different for a northward IMF). Analyzing data from the ISEE satellites, Sckopke et al. (1981) demonstrated that the LLBL earthward of the magnetopause on the flanks is often divided into two distinct parts. The outer boundary layer (OBL) (called the "boundary layer proper" by Sckopke) is filled with dense magnetosheath-like plasma moving tailward at speeds comparable to the magnetosheath flow. The hot, tenuous plasma in the inner boundary layer (IBL) (they called it "halo") is a mixture of solar wind and magnetospheric particles; it is moving more slowly, often sunward. Fig. 9.3 illustrates their conclusion showing uniform, wavy, and patchy structures. The wavy structure is at the inner edge where the velocity reverses leading to Kelvin-Helmholtz instability.

A similar structure of the flank boundary layer was reported by Fujimoto et al. (1998a,b). They pointed out that the flow in the inner part, which they called the "mixing region," was often directed sunward. From this, they concluded that the plasma in the mixing region is most likely on closed field lines. The mixing often fills a large portion of the magnetotail.

Other early observations of an LLBL containing magnetosheath-like plasma in its outer parts and hot, tenuous plasma in its inner parts, respectively, have been presented by

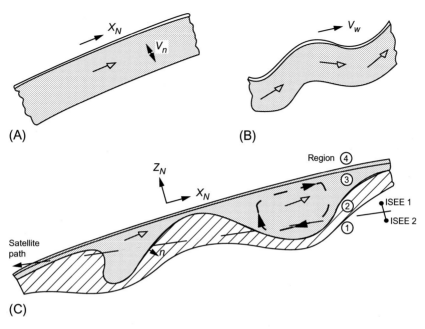

Fig. 9.3 (A) A uniform boundary oscillating at speed v_n. (B) A uniform boundary disturbed by surface waves.
(C) A boundary of nonuniform thickness attached to a smooth magnetopause. In this model, region 4 denotes the magnetosheath, region 3 is the outer boundary layer (OBL) (called the "boundary layer proper" by Sckopke), filled with dense magnetosheath-like plasma moving tailward, region 2 contains the hot, tenuous plasma of the inner boundary layer (IBL) (they called it "halo"), which is a mixture of solar wind and magnetospheric particles, and region 4 is the outer magnetosphere. The *open-headed arrows* indicate plasma flow in the spacecraft frame, while the flow vortex *(dashed line)* and satellite path are shown for a moving system in which the boundary layer structure is approximately at rest. Here z_N is the model magnetopause normal, and n is the boundary layer normal pointing into region 2. From Sckopke, N., Paschmann, G., Haerendel, G., Sonnerup, B., Bame, S., Forbes, T., et al., 1981. Structure of the low-latitude boundary layer. J. Geophys. Res. 86(A4), 2106.

Eastman et al. (1976), Eastman (1979), Williams et al. (1985), and Mitchell et al. (1987). For a review of the later boundary layer observations, see the book edited by Newell and Onsager (2003).

9.2 Comprehensive investigation of low-latitude boundary layer

A comprehensive investigation of the LLBL has been undertaken by Bauer et al. (2001) using data from AMPTE/IRM spacecraft. They studied all IRM passes through the dayside (08:00–16:00 LT) magnetopause region for which the following relevant data were available: magnetometer measurements, plasma moments at spin resolution, ion and electron distribution functions of the energy per charge, and electric wave spectra. They analyzed 40 dayside crossings that fulfilled the following criteria:

- The crossing was a complete crossing from the magnetosheath to the magnetosphere proper, or vice versa
- The duration, Δt_{BL}, of the boundary layer was at least 30 s
- At least two electron distribution functions could be measured in the boundary layer
- The time intervals in the magnetosheath before (after) the boundary layer and the time intervals in the magnetosphere after (before) the boundary layer are so long that an unambiguous identification of the magnetopause and the earthward edge of the boundary layer was possible

The second and third criteria are required in order to resolve the internal structure of the boundary layer, that is, to distinguish gradual time profiles from step-like profiles. Sharp steps in the profiles may mark topological boundaries or discontinuities associated with magnetopause processes. Gradual transitions are expected for a boundary layer formed by diffusion.

They divided the 40 crossings into four classes: for 22 crossings, 2 plateaus of the density can be distinguished. One plateau (class 1) has a density comparable to the magnetosheath density and is identified as an OBL. The other plateau has a distinctly lower density and is identified as an IBL. For 6 of the 40 crossings they observe only a high-density plateau and identify this as an OBL (class 2). For 2 of the 40 crossings they observe only a low-density plateau and identify this as an IBL (class 3). The remaining 10 crossings (class 4) do not show any plateaus or pronounced steps in the boundary layer and are classified as crossings with gradual profiles.

Table 1 in Bauer et al. (2001) gives the occurrence rates of the different particle populations in the OBL, the IBL, and the adjacent magnetosphere. Hot ring current particles are always observed in all three regions. Solar wind electrons or warm electrons are always observed in the OBL and IBL, respectively, and they are never observed in the magnetosphere proper. Solar wind ions are always observed in the OBL and never in the magnetosphere. However, for each of the 22 magnetopause crossings, there exists at least 1 distribution measured in the IBL that does show solar wind ions. This is what the value of 100% means in Table 1.

The authors always show great attention to spectral details in their work, trying to tell what happens to each constituent. Fig. 9.4 presents one-dimensional cuts through the distributions measured on September 17, 1984. Let us first have a look at the phase space density, f_e, in a cut along **B** (left diagram). Typical

Electrons ampte-IRM

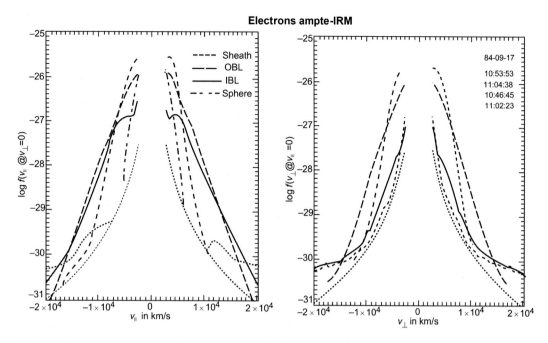

Fig. 9.4 One-dimensional cuts through the electron distributions measured on September 17, 1984. The measurements are taken in the magnetosheath at 10:46:45 *(short dashed line)*, in the OBL at 10:53:53 *(long dashed line)*, in the inner boundary layer at 11:02:23 *(solid line)*, and in the magnetosphere proper at 11:04:38 *(dash-dotted line)*. The left diagram shows the phase space density f_e (in cm^{-6} s^3) as a function of $v_\parallel$ for $v_\perp = 0$, and the right diagram shows f_e as a function of $v_\perp$ for $v_\parallel = 0$. The *dotted line* gives the detection threshold. From Bauer, T.M., Treumann, R.A., Baumjohann, W., 2001. Investigation of the outer and inner low-latitude boundary layers. Ann. Geophys. 19, 1069.

field-aligned velocities, $v_\parallel$, of the warm electrons are 6000–15,000 km/s. In this range, the value of $f_e(v_\parallel, v_\perp = 0)$ in the IBL (solid line) is comparable to that in the OBL (long dashed line). The fact that the curves of $f_e(v_\parallel, v_\perp = 0)$ in the IBL and OBL are pretty close to one another in the range of 6000–15,000 km/s suggests that the electrons in the IBL and OBL have a common source: the shocked solar wind plasma of the magnetosheath. For $v_\parallel < 6000$ km/s, the phase space density in the IBL is considerably less than that in the OBL, not very different from that in the magnetosphere proper.

9.2.1 Superposed epoch analysis

In previous studies of IRM data (Phan et al., 1994; Phan and Paschmann, 1996), the magnetopause was identified with the rotation of the magnetic field in the case of high shear crossings, and with a change in the thermal properties of the plasma in the case of low shear crossings. Since defining the magnetopause in a different manner for low and high shear might introduce some bias into the statistical analysis, Bauer et al. (2001) avoid defining the magnetopause with magnetic field data. Instead, the magnetopause is identified in both for high and low shear as a change in the distribution functions of solar wind ions and electrons. Although this change is theoretically not well understood, it provides a common criterion that can be applied to most observational data sets.

Progressing from the magnetosheath to the magnetosphere, they define the inner edge of the boundary layer as the point where solar wind electrons or warm electrons disappear. For the superposed epoch analyses, the interface between OBL and IBL is used as a key time. There are crossings in the data set during which IRM moves back and forth between the OBL and IBL. In this case, only the first (last) OBL and last (first) IBL of an inbound (outbound) crossing are used for the superposed epoch analyses.

All crossings occurred near the equatorial plane at latitudes less than 30 degrees. They distinguish between low and high magnetic shear. For low (high) shear crossings, the angle, $|\Delta\varphi_B|$ between the magnetic fields in the magnetosheath and in the magnetosphere is less (greater) than 40 degrees. Furthermore, they distinguish between "Walén events" and "non-Walén events" where they checked if the plasma moments measured across the magnetopause satisfy the tangential stress balance (Walén relation)

$$\mathbf{u}'_p = \mathbf{u}_p - \mathbf{v}_{\mathrm{HT}} = \pm\mathbf{c}_A \qquad (9.1)$$

of a rotational discontinuity. Here, $\mathbf{u}_p$ is the proton bulk velocity, $\mathbf{v}_{HT}$ is the velocity of the de Hoffmann-Teller frame, $\mathbf{u}_p{}'$ is the proton bulk velocity in the de Hoffmann-Teller frame, and $\mathbf{c}_A$ is the Alfvén velocity. The + sign (− sign) is valid when the normal component u_{pn} of the proton bulk flow has the same (opposite) direction as B_n.

It was found for 11 of the 22 selected IRM crossings that have both an OBL and IBL, a linear relation is fulfilled. For non-Walén events, the linear relation cannot be satisfied. The disagreement with the tangential stress balance in the other 11 cases might be due to reconnection being time dependent and patchy (easily understood with a PTE, see the Discussion sections in Chapters 5 and 6).

The north-south component of the proton bulk velocity in the boundary layer is, on average, directed toward high latitudes for both low and high magnetic shear in agreement with tailward flow. "Warm," counterstreaming electrons that originate primarily from the magnetosheath and have a field-aligned temperature that is higher than the electron temperature in the magnetosheath by a factor of 1–5 are a characteristic feature of the IBL. Profiles of the proton bulk velocity and the density of hot ring current electrons provide further evidence that the IBL is on closed field lines.

9.2.2 Low versus high magnetic shear

Fig. 9.5 presents a superposed epoch analysis of the 7 IRM crossings with low magnetic shear ($|\Delta\varphi_B < 40^0|$) and the 15 crossings with high magnetic shear ($|\Delta\varphi_B > 40^0|$). Two key times are used: the time when the magnetopause is crossed, and the time when the interface between the OBL and the IBL is crossed. The time of the magnetopause crossing is set to zero and the order of the time series is reversed for outbound crossings. Next, they normalize the time axis so that all OBLs have the same normalized duration. Thus normalized time $t=0$ corresponds to the magnetopause and normalized time $t=1$ corresponds to the interface between the OBL and the IBL for each event. Vertical error bars indicate the standard deviation of the average of each parameter. Finally, the traces of each parameter as a function of the normalized time t are superposed and averaged.

The panels of Fig. 9.5 include data from $t=-1$ to $t=2$. This means that all measurements are used in the magnetosheath that are obtained with less than one OBL in duration, Δt_{OBL}, sunward of the magnetopause and all measurements that are obtained with less than Δt_{OBL} earthward of the interface between the OBL and the IBL. The data are sorted into 18 bins and averaged within each bin. Parameters that can change sign are averaged linearly, and parameters that are positive by definition are averaged logarithmically.

The length of the error bars in Fig. 9.5 gives the standard deviation of the average, that is, the standard deviation, σ, of the respective parameter or its logarithm divided by the square root of the number, n, of events. The trace of φ_B in the second panel shows the rotation of the magnetic field tangential to the magnetopause. The sign of φ_B is reversed for events for which the change $\Delta\varphi_B$ from the magnetosphere to the magnetosheath is negative. As expected from the respective definition of low and high shear, φ_B changes only slightly across the low shear boundary layer, whereas it rotates, on average, by about 90 degrees across the high shear boundary layer. The third panel of Fig. 9.5 shows clearly that the interface between the OBL and the IBL is the location where most of the density decreases from the magnetosheath level of

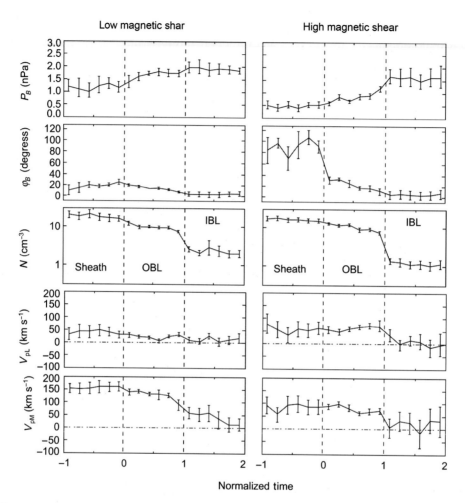

Low magnetic shar

High magnetic shear

Normalized time

Fig. 9.5 Superposed epoch analysis of the magnetic pressure, P_B, the magnetic field rotation angle, φ_B, the total plasma density, N, and the components V_{pL}, V_{pM} of the proton bulk velocity across the OBL. The left panels display data from seven low shear crossings and the right panels display data from 15 high shear crossings. From Bauer, T.M., Treumann, R.A., Baumjohann, W., 2001. Investigation of the outer and inner low-latitude boundary layers. Ann. Geophys. 19, 1070.

about $20\,\mathrm{cm}^{-3}$ to the magnetospheric level of about $2\,\mathrm{cm}^{-3}$ takes place. Phan et al. (1994) showed that for low shear, a plasma depletion layer evolves in the magnetosheath adjacent to the magnetopause. In the plasma depletion layer the magnetic field piles up and the plasma is squeezed out. Since ions with high field-aligned velocities leave the plasma depletion layer earlier than ions with low field-aligned velocities, the solar wind ions in the plasma depletion layer exhibit a strong perpendicular temperature anisotropy. For these events, the average density in front of the low shear magnetopause is not lower than in front of the high shear magnetopause. However, the first panel of Fig. 9.5 shows that the magnetic pressure, P_B, in front of the low shear magnetopause is higher by more than a factor of 2 due to its pileup in the plasma depletion layer. Moreover, the proton temperature anisotropies (not shown) are clearly different for low and high shear.

The last two panels of Fig. 9.5 display the components v_{pL} and v_{pM} of the proton bulk velocity. The plasma flow in the magnetosheath and boundary layer is directed poleward and toward the flanks. The magnitudes of v_{pL} and v_{pM} are roughly the same in the magnetosheath and in the OBL, but they drop clearly at the interface between the OBL and the IBL.

The proton and electron temperatures (not shown) exhibit a slight increase at the magnetopause and a strong increase from the OBL to the IBL. On average, the thermal energy, $KT_p(KT_e)$, of the protons (electrons) is about 500 eV (60 eV) in the OBL and about 1.5 keV (100 eV) in the IBL. The electrons show a perpendicular temperature anisotropy in the magnetosheath, while in both parts of the boundary layer field-aligned anisotropies, $T_{e\parallel}/T_{e\perp} \approx 1.2$ are observed.

9.2.3 Dawn side versus dusk side

Fig. 9.6 compares magnetopause crossings on the morning side (08:00–12:00 LT) with crossings on the evening side (12:00–16:00 LT). Since this section is focused on the IBL and on the interface between the IBL and magnetosphere proper, the parameters are now plotted as a function of a normalized time, which is defined such that $t=0$ corresponds to the interface between OBL and IBL and $t=1$ corresponds to the inner edge of the boundary layer. With this choice of the time axis, the magnetopause corresponds for each event to a different position $t<0$ and changes of the parameters occurring at the magnetopause are washed out. The averaging is performed in the same manner as for Fig. 9.5, and the data are plotted for $-1<t<2$.

The first panel of Fig. 9.6 shows again that the interface between the OBL and IBL is the location where most of the density decreases from the magnetosheath level to the magnetospheric level of about $1\,cm^{-3}$ takes place. At the inner edge of the boundary layer, N changes by a factor of 2 on the dawn side but there is no significant change on the dusk side. In the magnetosphere proper, the average of the north-south velocity component, v_{pL}, is not significantly different from zero, and it is also not significantly different from zero in the IBL on the dusk side. v_{pL} is

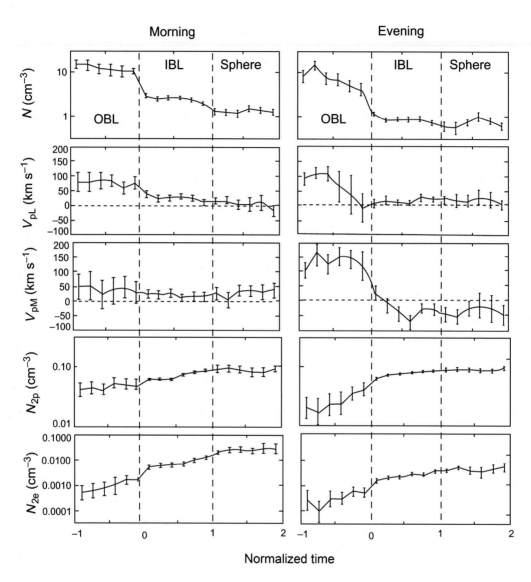

Fig. 9.6 Superposed epoch analysis of the components V_{pL}, V_{pM} of the proton bulk velocity, the total plasma density N, the partial density N_{2p}, of protons above 8 keV, and the partial density N_{2e}, of electrons above 1.8 keV across the inner boundary layer. The left panels display data from 13 crossings dawnward of local noon and the right panels display data from 9 crossings duskward of local noon. Normalized time $t=0$ corresponds to the inner edge of the boundary layer. From Bauer, T.M., Treumann, R.A., Baumjohann, W., 2001. Investigation of the outer and inner low-latitude boundary layers. Ann. Geophys. 19, 1075.

directed poleward away from the equatorial plane in the IBL on the dawn side. The dawn-dusk component, v_{pM}, exhibits a strong change at the interface between the OBL and the IBL on the dusk side.

While the flow in the OBL and magnetosheath is directed toward the dusk flank, the plasma flow in the IBL and magnetosphere is directed sunward toward the subsolar point. However, on the dawn side, the component v_{pM} of the plasma flow is directed toward the dawn flank in all four plasma regions. "We do not know the reason for this dawn-dusk asymmetry of v_{pM}" (Bauer et al., 2001).

The PTE concept provides some answers. The current carriers that are required to provide the electric field for momentum conservation, particularly the ions, may not be sufficient to carry the field-aligned current.

Another important feature is the profile of the partial density, N_{2p}, of protons above 8 keV, dominated by ring current ions, and N_{2e}, of electrons above 1.8 keV. The last two panels show the drop in N_{2p} and N_{2e} from the IBL to OBL. There is also a reduction of N_{2e}, from magnetosphere proper to the IBL. On the dawn side, N_{2e} is almost one order of magnitude higher than on the dusk side.

The average proton temperature (not shown) displayed increases at the interface between the OBL and the IBL and again at the inner edge of the boundary layer both for the dawn side and dusk-side crossings. The thermal energy of electrons, KT_e, on the dawn side increases continuously from about 60 eV sheathward of the IBL to 300 eV in the magnetosphere proper. On the dusk side, the average electron temperature in the IBL is comparable to its value on the dawn side. However, the electron temperature in the magnetosphere proper on the dusk side is considerably less than on the dawn side. The proton temperature anisotropy, $T_{p\parallel}/T_{p\perp}$, does not vary significantly in the plotted interval $-1 < t < 2$. The electron temperature is roughly isotropic in the magnetosphere, while it exhibits a strong field-aligned anisotropy, $T_{e\parallel}/T_{e\perp} \approx 1.2$ in the boundary layer.

Densities at the time resolution of one IRM spin period are available in three energy bands (not shown). The fraction of protons below 400 eV and the fraction of electrons below 60 eV, respectively, are roughly constant in the interval plotted. This implies that there are about as many ring current ions as cold ions both in the dawn-side and dusk-side magnetosphere. Note, however, that there may be a large number of cold ions that are not detected if their energy is below 20 eV.

9.2.4 Wave spectra and diffusion

To evaluate average spectra of electric and magnetic fluctuations in the vicinity of the magnetopause, the electric power spectral density is directly measured by the IRM wave experiment in 16 frequency channels, and the magnetic power spectra are obtained by Fourier analysis of the high-resolution magnetometer data.

1. **Electric fluctuations**: For each of the 16 frequency channels, the ELF/VLF spectrum analyzer provides mean values, S_E, and also peak values, $\hat{S}_E$, of the electric power spectral density with a time resolution of 1 s. In Fig. 9.7 these quantities are enhanced at the interface between the OBL and the IBL.

2. **Magnetic fluctuations**: The average spectra in the magnetosheath follow power laws, $S \propto f^\varepsilon$, with spectral slopes $1.2 < \varepsilon < 1.6$, considerably flatter than the electric spectra discussed in the previous section. Above the proton gyrofrequency, the average spectra in the magnetosheath become distinctly steeper than f^{-2}.

9.3 Studies with better resolution

Particle measurements have been limited by the spatial resolution they could achieve; this depends on the energy sweep rates of the instruments. Many detectors have sweep rates on the order of a second, which for many low altitudes may allow for only a few energy spectrums to be taken if the structure is small. AMPTE/IRM had a resolution of 4.3 s, the satellite rotation period. Later instruments, however, have much shorter sweep rates; for example, the Astrid-2 instrument has a 32-step energy sweep in 1/16th

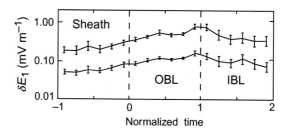

Fig. 9.7 Superposed epoch analysis of the root-mean-square amplitudes of electric fluctuations in the range of 25–800 Hz, across the OBL. From Bauer, T.M., Treumann, R.A., Baumjohann, W., 2001. Investigation of the outer and inner low-latitude boundary layers. Ann. Geophys. 19, 1075.

of a second for electrons and 1/8th of a second for ions, thus allowing resolutions down to about 400 m (Keith et al., 2001).

9.3.1 Cusp passage of DE-2

The DE-2 event on September 6, 1982 in the southern hemisphere shows a "V"-shaped structure uncovered by Keith et al. (2001, 2005). The magnetic field lines go from open/closed to closed geomagnetic through the cusp in an active geomagnetic period with $K_p = 8$. The data in the top two panels of Fig. 9.8 were taken by the Low Altitude Plasma Instrument (LAPI); the instrument consists of 15 parabolic electrostatic analyzers covering 180 degrees for ions and electrons from 5 eV to 32 keV. One 32-step spectrum is taken each second from each sensor. The data in the third panel are from two Geiger-Mueller counters which measure >35 keV electrons at 0 and 90 degrees pitch angles. The data were taken on an equatorward pass, the cusp being located at an invariant latitude of 63 degrees and a prenoon MLT of 10:30. The bottom panel of Fig. 9.8 is the AC electric wave power from the VEFI instrument from 4 to 1000 Hz.

A dispersion signature in the ion data (second panel) starts out rather constant at about 1 keV from 8:37:00 to 8:37:30 UT. The feature of interest is the last third of the dispersion pattern, at the equatorward edge from 8:37:30 to 8:37:40. This 10-s interval clearly has a "V"-shaped ion structure with higher energies on the edges and lower energies toward the center. This "V" is at unusually high energies, with peaks at around 20 keV. We need to rely on the analysis of Bauer et al. (2001) earlier to see what is going on. Comparing to Fig. 9.5, the DE-2 encounter with the magnetopause current sheet ends at 08:37:30 in Fig. 9.8. The electrons show an earlier response by about 2 s. The ions at the lowest energies show a short burst beginning at 08:37:30; this is probably explained by Sauvaud et al. (2001) in Fig. 9.9. These authors discovered that the ions are locally accelerated perpendicularly to the local magnetic field in a region adjacent to the magnetopause on its magnetospheric side. The ionospheric ions are affected by the inward push of the magnetopause responsible for the PTE.

Recurrent motions of the magnetopause are unexpectedly associated with the appearance inside closed field lines of recurrent energy structures of ionospheric ions with energies in the 5 eV to over 1000 eV range.

Sauvaud et al. (2001)

From that point, the solar wind plasma is in the OBL, on closed magnetic field lines. The heaviest ions were detected with the

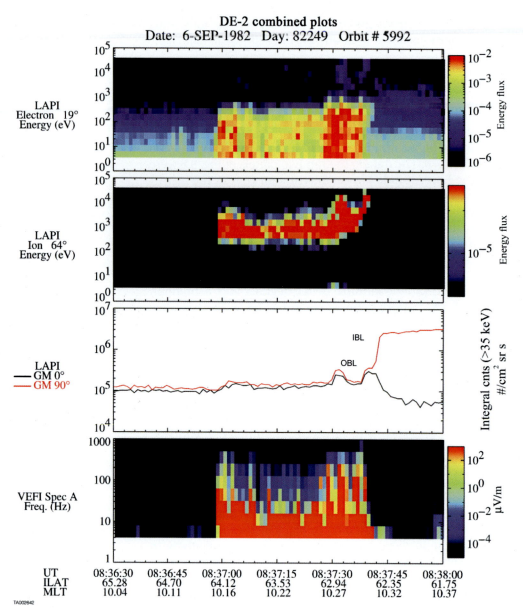

Fig. 9.8 DE-2 data from September 6, 1982 showing a cusp dispersion signature in the ions with a small "V" at the equatorward end. Top panel is electron energy flux, the second is ion energy flux. The line plots in the third panel are counts of >35 keV electrons in the parallel *(black)* and perpendicular *(red)* directions. The bottom spectrogram is electric field wave power up to 1 kHz. Note the ions below 10 eV: these mark the energization of ionospheric plasma due to motions of the magnetopause after Sauvaud et al. (2001). From Heikkila added OBL, IBL to Keith, W.R., Winningham, J.D., Norberg, O., 2001. A new, unique signature of the true cusp. Ann. Geophys. 19, 613.

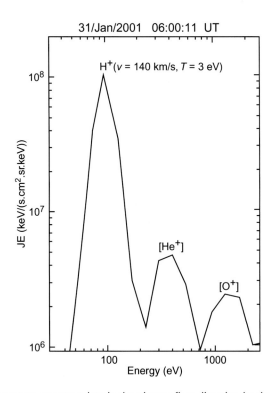

31/Jan/2001 06:00:11 UT

Fig. 9.9 Ion energy spectra taken in the plasma flow direction by the nonmass resolving CIS-2 spectrometer. Note the appearance of peaks corresponding to H+, He+, and 0+ with a velocity of $\sim$140 km/s, i.e. an electric field of $\sim$2.8 mV/m. The H+ temperature estimated from a Maxwellian fit is close to 3 eV. From Sauvaud, J.A., Lundin, R., Rème, H., McFadden, J.P., Carlson, C., Parks, G.K., et al., 2001. Intermittent thermal plasma acceleration linked to sporadic motions of the magnetopause, first cluster results. Ann. Geophys. 19(10/12), 1529.

highest energies. The ion behavior was interpreted as resulting from local electric field enhancements/decreases which adiabatically enhance/lower the bulk energy of a local dense thermal ion population.

On exit from the OBL there is some energization in the IBL. Most of the plasma in OBL goes antisunward with little energization, represented by the dip at 08:37:40.

9.3.2 Cusp passage of Astrid-2

Data from the MEDUSA and EMMA instruments aboard Astrid-2 are shown in Fig. 9.10 (Keith et al., 2005). The data were taken on January 13, 1999 during a pass over the southern auroral zone. The spacecraft was traveling equatorward at an altitude of

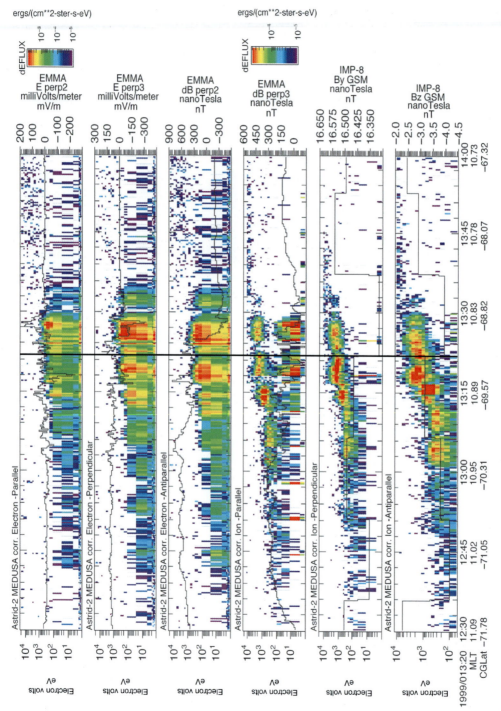

Fig. 9.10 Astrid-2 data from January 13, 1999 during an equatorward pass through the southern cusp region. The top three spectrograms are electrons in the parallel, perpendicular, and antiparallel directions, and the bottom three spectrograms are the same directions for the ions. The line plots from top to bottom are the electric field calculated for the eastward and equatorward directions, delta magnetic field data for the same two directions, and the IMF in the y and z directions as measured by IMP-8. The vertical line marks the location of the dynamo (deep minimum of precipitation). From Keith, W.R., Winningham, J.D., Goldstein, M.L., Wilbur, M., Fazakerley, A.N., Rème, H., et al., 2005. Observations of a unique cusp signature at low and mid altitudes. In: Fritz, T.A., Fung, S.F. (Eds.), The Magnetospheric Cusps. Springer, Dordrecht, The Netherlands, p. 319.

1029 km. The top three panels are electrons shown in the parallel, perpendicular, and antiparallel directions relative to the local magnetic field. The second three panels are the same directions for ion data. There are line plots in each panel, the electric field calculated for the eastward and equatorward directions (assuming E is zero along the magnetic field), and two panels below it containing delta magnetic field data for the same two directions. The bottom two panels contain the IMF in the y and z directions as measured by IMP-8.

The Miniaturized Electrostatic DUal-top-hat Spherical Analyzer (MEDUSA) was flown aboard the Swedish Astrid-2 microsatellite (Marklund et al., 2001). The instrument is composed of two spherical top-hat analyzers placed top to top with a common 360 degrees field of view. Each detector is divided into 16 azimuthal sectors of 22.5 degrees, with an elevation acceptance of about 5 degrees. The instrument has an energy per charge range of about 1 eV to 20 keV for electrons and positive ions. The details of the spectrogram have been well described by Keith et al. (2001, 2005). Here we note two additional aspects.

The first thing to note is the deep minimum in precipitation at 13:22 UT (marked with vertical line); we contend it is located at the dynamo with $\mathbf{E} \cdot \mathbf{J} < 0$, following the application of Poynting's theorem as described by Richmond (2010). The particle flux is at its minimum, but something important is happening here since the wave power is maximum. This is understandable as it is at the dynamo where the plasma gives up energy to the electromagnetic field.

A dynamo is evident as the drop in the downward flux (panels 3 and 6) in both electrons and ions at 13:22. The wave power is the highest, even higher than in the load regions on either side. The low energy upgoing ions in the load regions are well known in proton auroras. The protons alternate between the charged and uncharged states as they randomly walk through the atmosphere, some escaping. The load region 13:24–13:28 is clearly on closed field lines (the transition in magnetic field signal is sharp at 13:24). The electrons are bouncing between hemispheres in the OBL after 13:22, whereas they are absent before at 13:16–13:18. This is due to the continuity of $\mathbf{E} \times \mathbf{B}$ in Fig. 9.10, referred to earlier.

Second, there is a second maximum that follows; note that while $\mathbf{E}$ and $\mathbf{B}$ both reverse, their cross-product does not. The particles are in the LLBL, traveling antisunward. This has been observed repeatedly (e.g., Woch and Lundin, 1992).

The fact that the narrow "V" feature is consistent throughout all of the data sets from different satellites and plasma analyzer types, and the fact that the energetic particle and field instruments see a feature at the same time with a comparable scale size, points to this being physical.

*This feature has a consistent double-peaked or "V"-shaped structure
at the equatorward edge of high-latitude particle precipitation flux
… whose geometry depends greatly on the dawn/dusk component of
the IMF. Various observations are presented at low altitudes (DE-2,
Astrid-2, Munin, UARS, DMSP) and at mid altitudes (DE-1, Cluster)
that suggest a highly coherent cusp feature that is consistent with
the narrow, wedge-shaped cusp of Crooker (1988), and contains
persistent wave signatures that are compatible with previously
reported high-altitude measurements.*

Keith et al. (2005, p. 307)

9.4   Plasma transfer event

It is necessary to consider both the induction field with a finite curl and the conservative electrostatic field with a finite divergence to understand the PTE

$$\mathbf{E} = \mathbf{E}^{\mathrm{ind}} + \mathbf{E}^{\mathrm{es}} \qquad (9.2)$$

The order of these terms expresses the relative importance of the two; induction electric field can be much stronger, the organizer. However, an electrostatic field will have no effect on the electromotive force of the inductive field because its curl vanishes; any reduction in the net $E_\parallel$ must involve enhancement of the perpendicular component $E_\perp$ in any arbitrary closed contour, otherwise the curl would be affected. That brings some new aspects into the picture, some surprising.

This point is especially clear for a PTE if considered in the frame of reference of the moving or meandering magnetopause, as shown in Fig. 5.6 of Chapter 5 (at the middle of the localized perturbation). Locally, there would be no inductive electric field produced. But now, because of the localization, the magnetopause current away from the perturbation is receding (going sunward), and an inductive electric field is present there. It is in the dusk-dawn sense on the sunward side and opposite on the earthward side for a sunward recession of the magnetopause current. In the deforming part (at the edges of the localized perturbation) there again will be a normal component of the inductive electric field exactly as before. The electrostatic field will drive charges within the perturbation. We have a finite tangential component of the electric field in the wavy magnetopause frame that reverses.

In summary, we conclude that both *impulsive penetration* (Lemaire and Roth, 1991) and *PTE* (Heikkila, 1997, 1998) combine to enable solar wind plasma to cross the magnetopause current sheet into the LLBL, eventually to closed field lines. As both fields reverse on either side of the magnetopause, their product $\mathbf{E} \times \mathbf{B}$ does not reverse; the plasma can drift through the magnetopause, and onto closed field lines in the OBL. The former amounts to tapping electric energy (i.e., net transport of the plasma), while the latter taps magnetic energy (i.e., organized charge in motion) through time-dependent induction.

9.5 Identification of cusp and cleft/low-latitude boundary layer

For many years the terms "cusp" and "cleft" were used interchangeably. More recently, there has been a tendency to consider the cusp to be a subset of the cleft, with the cusp confined close to noon (Fig. 9.11). The region was called by Axford and Hines (1961) the "zone of confusion" and should be carefully defined. Chapter 7 on the high-altitude cusp shows that a lot of physics is involved in this region.

From that point the solar wind plasma is in the OBL, on closed magnetic field lines. Heikkila (1985) has offered the following definition (drawing upon the observations of G. Shepherd):

The cleft is the low altitude region around noon of about 100 eV electron precipitation associated with 630.0 nm emission, but containing also structured features of higher energy. The cusp is a

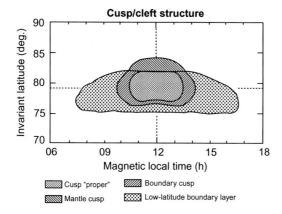

Fig. 9.11 Summary plot of the polar cusp and cleft from a statistical study using Viking particle data. Four different regions are distinguished: the cusp "proper," the "active" boundary cusp, the mantle cusp, and the dayside auroral region connected to the LLBL. From Lundin, R., 1988. On the magnetospheric boundary layer and solar wind energy transfer into the magnetosphere. Space Sci. Rev. 48, 315.

more localized region near noon within the cleft characterized by low energy precipitation only, having no discrete auroral arcs, often displaying irregular behavior, presumably associated with the magnetic cusp.

Heikkila (1985)

Newell and Meng (1988) start by proposing a conceptual definition:

The low-altitude cusp is the dayside region in which the entry of magnetosheath plasma to low altitudes is most direct. Entry into a region is considered more direct if more particles make it in (the number flux is higher) and if such particles maintain more of their original energy spectral characteristics.

Newell and Meng (1988)

The "true cusp" is characterized by high fluxes of isotropic magnetosheath electrons and ions with no signatures of particle acceleration. The "active cusp" has lower magnetosheath plasma densities and moderate acceleration of electrons and ions. The "mantle cusp" is the region with tailward flowing ions connected to the plasma mantle. The auroral region is characterized by time-dependent magnetosheath plasma injection and strong plasma acceleration (Newell and Meng, 1988).

9.6 Qualitative description of low-latitude boundary layer

There is increasing interest in the LLBL as shown by the conference held at New Orleans in 2001 (Newell and Onsager, 2003). Many observations have been reported over the years but their interpretation has been wanting, at best. It must be admitted that the theory of coupling between convective plasma motions in the magnetosphere and the footprints of those motions in the ionosphere is complicated. For tutorial purposes we start by discussing a particularly simple model following Sonnerup (1980) and Sonnerup and Siebert (2003).

9.6.1 Sonnerup's tutorial

Considering only zero IMF, Sonnerup (1980) and Sonnerup and Siebert (2003) discuss a thin layer of plasma moving tailward just inside the magnetopause. This LLBL model is shown in Fig. 9.12, a view from the Sun of the dawn-dusk meridional plane. It is in a steady state and two-dimensional. A thin layer of plasma moves

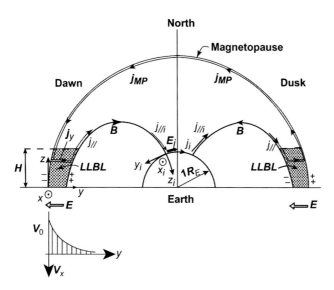

Fig. 9.12 Schematic drawing of cross section of northern hemisphere closed magnetosphere (zero IMF) at $X_{GSE}=0$. Dawn and dusk LLBLs are shown along with the flow velocity profile, field-aligned currents, and current closure over the high-latitude magnetopause. Equatorial coordinate system (x, y, z) and ionospheric system (x_i, y_i, z_i) are shown for the dawn side. From Sonnerup, B.U.Ö., Siebert, K.D., 2003. Theory of the low latitude boundary layer and its coupling to the ionosphere: a tutorial review. In: Newell, P.T., Onsager, T. (Eds.), Earth's Low-Latitude Boundary Layer. AGU Geophysical Monograph 133. American Geophysical Union, Washington, DC, p. 14.

tailward in the equatorial region, just inside the magnetopause. This tailward flow is driven by viscous forces, mass-diffusive inertia forces, and/or pressure forces. These forces are balanced by a sunward $\mathbf{J} \times \mathbf{B}$ force associated with a current $\mathbf{J}$, flowing across the layer in the dawn to dusk sense. This current is coupled to the equatorial ionosphere by field-aligned currents, an effect that has been described as "foot dragging" of the field lines.

The plasma in the equatorial dawn/dusk LLBL is moving tailward at speed $-v_x(y)$. The layer is assumed one-dimensional, in the sense that $\partial/\partial x \simeq 0$, but at the equator $\partial/\partial z \simeq 0$. The plasma moves across Earth's magnetic field, $\mathbf{B} = B\hat{\mathbf{z}}$, thereby producing a motional electric field $E_y = v_x(y)B$ pointing from dusk to dawn. An associated electric field appears across the LLBL, an electrostatic potential difference.

Because the magnetospheric plasma has high electrical conductivity along $\mathbf{B}$, the magnetic field lines are nearly equipotentials. The electric potential distribution in the LLBL is partly, if not completely, impressed at the ionospheric footprints of those field lines that thread the equatorial LLBL. The resulting horizontal ionospheric electric field, E_i, is directed poleward on the dawn side and equatorward on the dusk side. This electric field drives horizontal ionospheric Pedersen currents, but entirely on open magnetic field lines by Fig. 9.13.

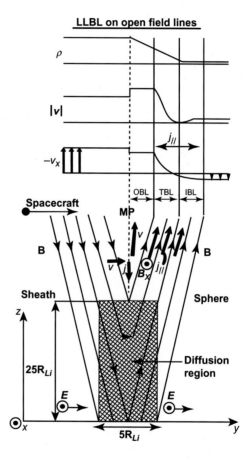

Fig. 9.13 Schematic drawing of the LLBL during quasistatic standard reconnection. The wedge-shaped outer boundary layer (OBL) is where plasma jetting and various kinetic signatures of reconnection are mainly seen. The transition boundary layer (TBL) is threaded by *open field lines* that pass through the ion diffusion region, while the inner boundary layer (IBL) is threaded by *closed field lines* that pass through the diffusion region. Signatures in plasma density ρ, plasma speed $|v|$, and tailward velocity component $-v_x$ are shown. However, this does not confront the closure of field lines in the cusp, a major finding first noted over four decades ago (see Fig. 1.28). From Sonnerup, B.U.Ö., Siebert, K.D., 2003. Theory of the low latitude boundary layer and its coupling to the ionosphere: a tutorial review. In: Newell, P.T., Onsager, T. (Eds.), Earth's Low-Latitude Boundary Layer. AGU Geophysical Monograph 133. American Geophysical Union, Washington, DC, p. 29.

Similar behavior occurs in the dusk-side LLBL, where the field-aligned current from the ionosphere feeds a duskward cross-field current. This current is deflected to flow northward along the magnetopause. It flows across the polar region and then southward along the magnetopause to feed the cross-field LLBL current on the dawn side, thus completing the current loop, as shown in Fig. 9.12.

9.6.2 Alternative description

The earlier description of the electrodynamic coupling between the LLBL and the ionosphere is often given in an alternative non-MHD way (Sonnerup and Siebert, 2003, p. 14); the electric charge separation that generates the electric fields plays a center-stage role. For example, in the dawn-side LLBL, the plasma motion leads to polarization charges, caused by minute deviations from exact charge neutrality. The motion of the plasma constitutes a dynamo, as depicted by Figs. 9.14 and 9.15 with $\mathbf{E} \cdot \mathbf{J} < 0$.

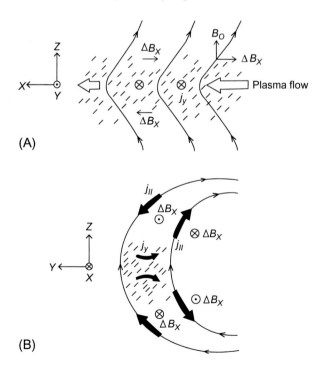

Fig. 9.14 Draping of magnetic field lines as a result of a magnetic perturbation caused by transverse and field-aligned currents in the (A) noon-midnight and (B) dawn-dusk planes. B_o marks the background magnetic field and ΔB_x the perturbation field induced by local currents (j_y and $j_{\parallel}$). From Lundin, R., 1988. On the magnetospheric boundary layer and solar wind energy transfer into the magnetosphere. Space Sci. Rev. 48, 311.

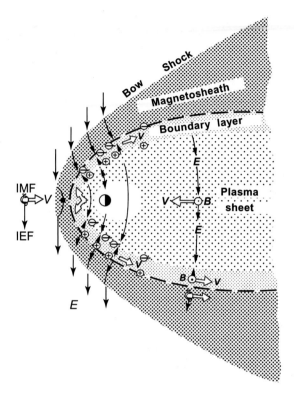

Fig. 9.15 An equatorial view of the magnetosphere with a southward IMF. The dawn-side magnetopause must carry a negative charge as a terminus of the interplanetary electric field (IEF), by Gauss's law; the charge is positive at dusk for the continuation. The signs of charge are the opposite at the inside edges of the LLBL giving a dusk-dawn sense for a tailward flow $\mathbf{V} \approx \mathbf{E} \times \mathbf{B}$ with the northward sense of closed geomagnetic field lines. The dipole charged layer has little effect on the IEF, allowing its effective transfer, as shown. From Heikkila, W.J., 1978. Electric field topology near the dayside magnetopause. J. Geophys. Res. 83 (A3), 1076.

The potential at high altitudes will necessitate a field-aligned drop that has been observed using particle distribution functions, for example, by Reiff et al. (1985) with the DE satellites. The charge density is negative at the dawn magnetopause and positive at dusk, as it must be with southward IMF. The opposite happens at the inner edges of the LLBL. The upward current on the inner edge of the dusk LLBL causes the evening auroras (Lyons and Williams, 1984). The downward current on the morning side is not conducive to electron auroras. The charges are continually drained along field lines into the conducting ionosphere, where a steady state is maintained by a current that flows over the polar cap. This pattern of field-aligned currents amounts to an electric dipolar layer.

9.7 Topology of the magnetosphere

Fig. 9.16 from Heikkila (1984) is a projection of field lines in the dawn-dusk meridian plane for a strictly southward IMF. In that special case we will have an X-line completely around the entire magnetosphere. The field lines in one branch will go out from the X-line into the magnetosheath labeled $S1$, and into the ionosphere in the other branch $S2$. All open field lines must be within the narrow strip in the lower part of Fig. 9.16. Fig. 9.17 shows a projection of field lines

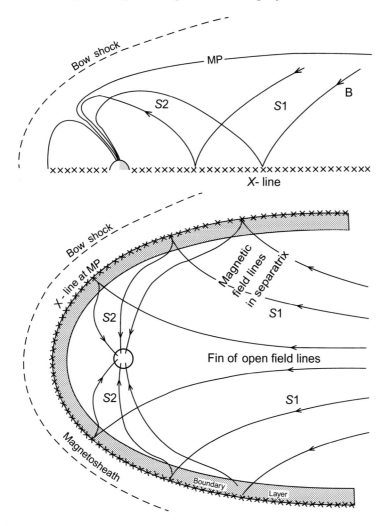

Fig. 9.16 A projection of field lines in the dawn-dusk meridian plane for a strictly southward IMF. An X-line is completely around the entire magnetosphere. Magnetic field lines form two sheets of separatrix $S1$ and $S2$. All open field lines must be within the narrow strip in the lower part. From Heikkila, W.J., 1984. Magnetospheric topology of fields and currents. In: Potemra, T.A. (Ed.), Magnetospheric Currents. AGU Geophysical Monograph 28. American Geophysical Union, Washington, DC, p. 219.

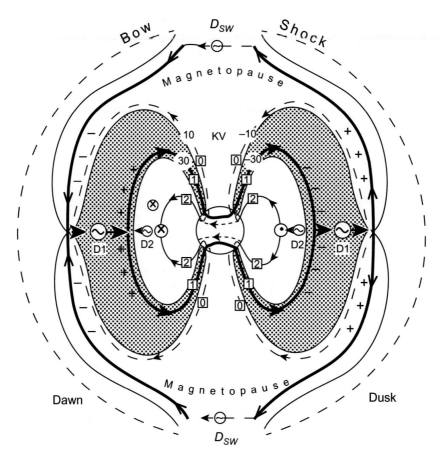

Fig. 9.17 A projection of field lines in the dawn-dusk meridian plane for a strictly southward IMF. The field lines in one branch will go out from the X-line into the magnetosheath labeled S1 in Fig. 9.16, and into the ionosphere in the other branch S2. From Heikkila, W.J., 1984. Magnetospheric topology of fields and currents. In: Potemra, T.A. (Ed.), Magnetospheric Currents. AGU Geophysical Monograph 28. American Geophysical Union, Washington, DC, p. 219.

in the dawn–dusk meridian plane, again for a strictly southward IMF. The field lines in one branch will go out from the X-line into the magnetosheath labeled S1 in Fig. 9.16, and into the ionosphere in the other branch S2. Evening auroras occur on the inside edge of the dusk boundary layer in the upward current region with the precipitating electrons as the current carriers. Of course, the results for a general orientation of the IMF will differ.

Cowley (1982) presented a highly regarded paper on the physical processes which give rise to convection in the Earth's magnetosphere.

Most of the discussion has centered on two basic pictures, one in which closed magnetospheric flux tubes are transported from dayside to nightside in a boundary layer around the flanks of the magnetosphere by a "viscouslike" process occurring at the magnetopause and the other in which open flux tubes are transported over the poles of the Earth after reconnection has taken place with the interplanetary magnetic field. These processes may coexist on a continuous basis, and the question then arises as to their relative contributions to usual total cross-magnetospheric voltages of ~40–100 kV.

Cowley (1982)

9.8 ISEE observations

The geotail mission of ISEE-3 presented a good opportunity for a critical test of the two main processes that have been advanced for solar wind-magnetosphere interaction, standard magnetic reconnection, and viscous interaction.

9.8.1 Tailward-moving vortex pattern

The model of the near-Earth magnetosphere sketched in Fig. 9.18 (Hones et al., 1981) has vortices, centered near the sides of the tail, moving tailward. These tailward-moving vortex patterns could be caused by the Kelvin-Helmholtz instability. The placement of the vortex axes at the inner edge of the boundary layer is inspired by several observations, in the morning sector, of flow rotation reversals from clockwise to counterclockwise coincident with the appearance of weak boundary layer plasma, followed, after a few minutes, to a change back to the original conditions.

9.8.2 Average conditions in the distant tail

Average and substorm conditions in the lobe and plasma sheet regions of the Earth's magnetotail were first studied by Zwickl et al. (1984) and Slavin et al. (1985) using ISEE-3 magnetopause and plasma analyzer measurements (Fig. 9.19). On the basis of 756 magnetopause crossings, a low-latitude magnetotail diameter of $60 \pm 5\ R_E$ at $|X| = 130$–$225\ R_E$ was determined. The plasma sheet magnetic field intensity and electron temperature decrease with increasing downstream distance, while flow speed, density, and Alfvénic Mach number all increase.

Fig. 9.19 shows both earthward and tailward flows in the near tail, changing to be only tailward at the greatest distance. Strong

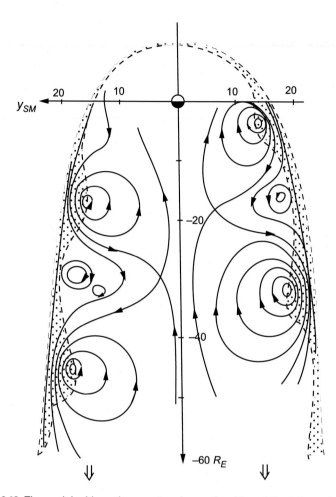

Fig. 9.18 The model with vortices centered near the sides of the tail, moving tailward. The placement of the vortex axes at the inner edge of the boundary layer, and specifically at crests in waves on the plasma sheet-boundary layer interface, is inspired by observations with the appearance of weak boundary layer plasma. From Hones, E.W., Jr., Birn, J., Bame, S., Asbridge, J., Paschmann, G., Sckopke, N., et al., 1981. Further determination of the characteristics of magnetospheric plasma vortices with Isee 1 and 2. J. Geophys. Res. 86(A2), 818. https://doi.org/10.1029/JA086iA02p00814.

density and weak velocity and temperature gradients are observed as ISEE-3 moves from the center of the lobes out toward the magnetopause (Zwickl et al., 1984). In particular, factors of 3–6 increases in plasma density were observed as the spacecraft moves from the center of the tail at $|Y'| < 10$ R_E (Y' refers to the aberrated GSM system) toward the dawn and dusk portions of lobes at $|Y'| > 20$ R_E.

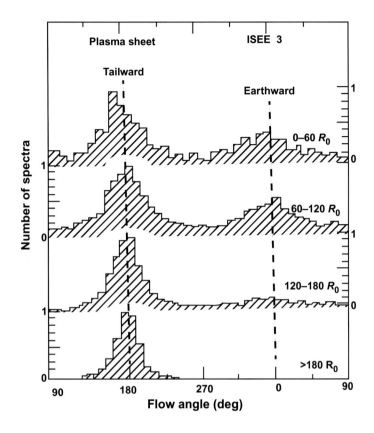

Plasma sheet flow angle

	Radial range, R_E			
	0–60	**60–120**	**120–180**	**>180**
Number spectra	680	4158	2013	4442
Velocity cutoff, km/s	180	150	120	90
Tailward flow,* %	72	70	88	97
Earthward flow,* %	28	30	12	3
Statistical error, %	±4	±1.5	±2	±1.5

*Based on number of spectra within ±36° of peak, minus background.

Fig. 9.19 Histograms of the electron velocity flow angle for plasma sheet plasma. A bimodal distribution is evident out to 120 R_E, beyond which the earthward flowing peak decreases rapidly to near zero. From Zwickl, R.D., Baker, D.N., Bame, S.J., Feldman, W.C., Gosling, J.T., Hones, E.W., Jr, et al., 1984. Evolution of the earth's distant magnetotail: ISEE 3 electron plasma results. J. Geophys. Res. 89(A12), 11011.

The distribution of temperatures is seen to move to lower values at larger distances with the plasma sheet temperatures decreasing more rapidly as shown in Fig. 9.20. Tailward flow speed and electron temperature exhibit maxima in the central portion of the plasma sheet where $B_x < 0$. At $|Y'| > 15\ R_E$, where $B_z > 0$, the flow direction remains tailward, albeit at a reduced speed. Earthward of $|X| = 100\ R_E$, the average B is northward and the flow is on average sub-Alfvénic. Between $|X| = 100$ and $180\ R_E$ the flow becomes predominantly tailward and super-Alfvénic $M_A \sim 1$–2, across the entire width of the tail. Along the flanks, closed field lines are apparently still being swept tailward at $|X| = 200\ R_E$. This result is important as it shows that $\mathbf{E} \cdot \mathbf{J} < 0$; it is a dynamo. The plasma loses energy as it travels further downstream.

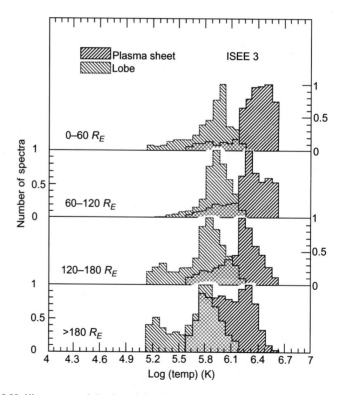

Fig. 9.20 Histograms of the log of the electron temperature for the plasma sheet and lobe distributions. The distribution of temperatures is seen to move to lower values at larger distances with the plasma sheet temperatures decreasing more rapidly. From Zwickl, R.D., Baker, D.N., Bame, S.J., Feldman, W.C., Gosling, J.T., Hones, E.W., Jr, et al., 1984. Evolution of the earth's distant magnetotail: ISEE 3 electron plasma results. J. Geophys. Res. 89(A12), 11009.

9.8.3 Analysis of B_z data from ISEE-3

The most obvious test for the Dungey (1961) model is to search for an average southward component of the magnetic field in the distant plasma sheet, indicating that the steady-state reconnection line, or X-line, is earthward of the spacecraft location. Indeed, such a result would be, to many persons, overwhelming proof that the Dungey model is appropriate. On the other hand, failure to find $B_z < 0$ should mean that the X-line (if any) would be more distant; however, such a failure has not been taken too seriously.

ISEE-3 provided an excellent opportunity during 1983 for a comprehensive evaluation of conditions in the distant magnetotail. Heikkila (1988) chose a 16-day period near the first deep tail apogee of ISEE-3 at $X = -220 R_E$ for Fig. 9.21. A total of 166 points yielded B_z (average) $= 0.27 \pm 0.05$ $(x^2 = 3)$ with the probability of 99%. Selection criteria were used to weed out points in the magnetosheath, and magnetopause crossings, however, the conclusion is not sensitive to the exact numbers used. This result suggests that the field lines are closed out to $X = -220$ at all levels of the AL geomagnetic index, in the absence of plasmoids. In the regions with northward B_z, the electric field must be directed from dusk to dawn for a tailward velocity, reverse of its sense within the plasma sheet.

9.8.4 Analysis for B_z for the whole year 1983

The average B_z at the tail current sheet is an interesting question that bears significance also to the global topology of the field. Pulkkinen et al. (1996) analyzed this important topic; the top panel of Fig. 9.22 shows the measured B_z values within the plasma sheet and the averages (20 R_E bins in the X direction, averaged over all Y). The averages decrease as a function of increasing radial distance and reach a local minimum close to the flow reversal region at $\sim$100 R_E. Tailward of that, B_z is small but slightly positive, reaching values very close to zero tailward of $\sim$200 R_E. Note that the variability is quite large, and that negative B_z values are frequently observed at all radial distances. Slavin et al. (1987) found that the probability for negative B_z increases substantially with increasing magnetic activity: in a subset of observations covering only the magnetically active periods, the B_z averages were negative tailward of $\sim$100 R_E. The middle panel concentrates on B_z values exactly at the current sheet crossings ($|B_x| < 0.5$ nT). The averaged results are virtually the same for both data sets, suggesting that the current sheet structure is relatively homogeneous. Because B_z is generally small in this region, there could also be

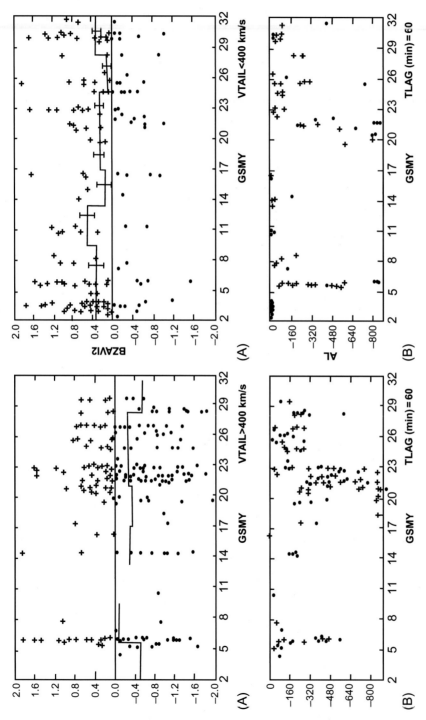

Fig. 9.21 Left: (A) B_z for $V > 400$ km/s, and AL for each point, plotted versus GMSY; the middle of the tail is about 15–20 R_E due to aberration. (B) The high velocity points are well correlated with magnetic activity indicated by AL. Right: (A) Similar data for $V < 400$ km/s. (B) These low-velocity points are present at all levels of geomagnetic activity, with averages positive everywhere. From Heikkila, W.J., 1988. Current sheet crossings in the distant magnetotail. Geophys. Res. Lett. 15(4), 301.

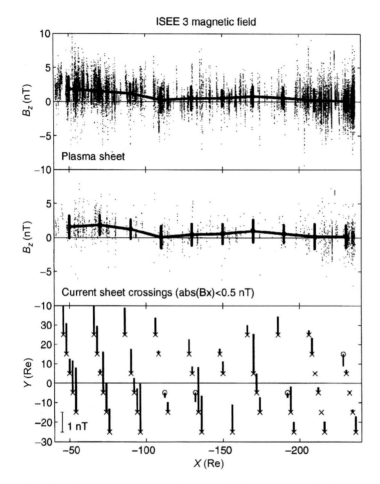

Fig. 9.22 (Top) Plasma sheet B_z as a function of X. The averages (20-R_E bins) and variances are shown with the heavy lines. (Middle) Plasma sheet B_z at current sheet crossings ($B_x < 0.5$ R_E). The averages (20-R_E bins) and variances are shown with the heavy lines. (Bottom) A quasi-three-dimensional view of the B_z at the plasma sheet. The *vertical bars* indicate the average B_z, the scale is given in the lower left corner. From Pulkkinen, T.I., Baker, D.N., Owen, C.J., Slavin, J.A., 1996. A model for the distant tail field: ISEE-3 revisited. J. Geomagn. Geoelectr. 48, 461.

coupling between B_z and other field components due to tail flapping and twisting. Whereas in individual cases this has to be analyzed in detail, we have here assumed that any such effects average out in the relatively large data set.

The bottom panel shows the B_z distribution in the XY plane in a quasi-three-dimensional representation. The data were divided into 20 R_E by 5 R_E bins. The thick vertical bars indicate the averaged B_z values within each bin (the scale is given in the lower left corner). B_z is slightly larger in the flanks than near the tail center; also the only four bins where the average was negative (marked with circles instead of crosses) were all near the central part of the tail (see also Slavin et al., 1985).

> *This model does not include a distant tail neutral line, as the averaging procedure resulted in weakly positive* B_z *throughout the region where measurements were made. The variability in the data is sufficient (the mean is much smaller than the variance) to suspect that at least part of the time the average* B_z *could have been negative in the distant part of the tail (especially near the tail center and during magnetically active periods). This is a topic that needs to be further investigated.*
>
> **Pulkkinen et al. (1996)**

Splitting the results on the basis of tailward velocity (noted in Fig. 9.21) might help to resolve the matter.

9.9   Massive flow in the boundary layer

The flow in the boundary layer is massive enough that it must keep on going because of momentum considerations; most of this flow never comes back (again, we must note that the flow is super-Alfvénic, beyond the control of the magnetic field). The boundary layer flow is not cyclic, as has been proposed (Cowley, 1982), except for a fraction that is left behind to feed the plasma sheet.

When Axford and Hines (1961) advanced their theory, the boundary layer had not been discovered; Fig. 9.23 adds this feature to their model. The boundary layer acts as a dynamo, with the plasma losing energy, as discovered by Zwickl et al. (1984) (see Fig. 9.20). The plasma supplies energy to the electromagnetic field (Eastman et al., 1976). The massive flow is able to enforce earthward plasma circulation within the plasma sheet, energize particles, generate field-aligned currents, and produce auroral phenomena. It appears that it is this massive boundary layer flow that maintains the magnetotail (see Section 9.15).

Fujimoto et al. (1998a) have used Geotail observations of the LLBL to study the near-Earth tail flanks. They find that cold-dense stagnant ions, which are likely to be of magnetosheath origin, are detected in this region of the magnetosphere. Charge neutrality is maintained by accompanying dense thermal ($<300\,\mathrm{eV}$) electrons presumably also from the magnetosheath. Compared to the magnetosheath component, however, the electrons are anisotropically heated to have enhanced bidirectional flux along the field

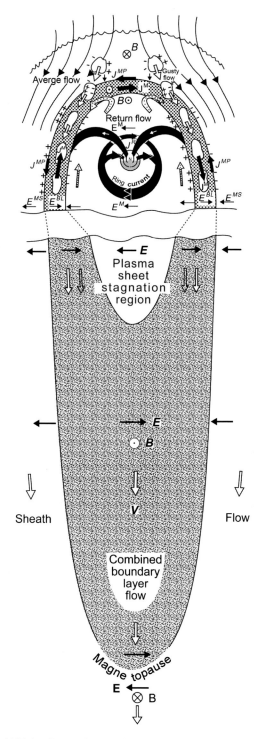

Fig. 9.23 The LLBL has been added to the model of Axford and Hines. The plasma sheet is a small cavity of energized low density plasma surrounded by higher density solar wind plasma. This plasma forms a dynamo with $\mathbf{E} \cdot \mathbf{J} < 0$, dictated by the lack of precipitation over the polar caps. From Heikkila, W.J., 1984. Magnetospheric topology of fields and currents. In: Potemra, T.A. (Ed.), Magnetospheric Currents. AGU Geophysical Monograph 28. American Geophysical Union, Washington, DC, p. 216.

lines. The enhanced bidirectional flux is well balanced, suggesting a closed topology of the field lines.

In addition to these common characteristics, a dawn-dusk asymmetry is observed in data for several keV ions, which is attributed to the dawn-to-dusk cross-tail magnetic drift of the plasma sheet ions. We also show a case that strongly suggests that this entry of cold-dense plasma from the magnetosheath via near-Earth tail flanks can be significant at times. In this case, the cold-dense plasma is continuously detected as the spacecraft moves inward from the magnetospheric boundary to deep inside the magnetotail. By referring to the solar wind data showing little dynamic pressure variation during the interval,

We interpret the long duration of the cold-dense status as indicative of a large spatial extent of the region: The cold-dense plasma is not spatially restricted to a thin layer attached to the magnetopause (LLBL) but constitutes an entity occupying a substantial part of the magnetotail, which we term as the cold-dense plasma sheet. The continuity of the cold-dense plasma all the way from the boundary region supports the idea that the magnetosheath plasma is directly supplied into the cold-dense plasma sheet through the flank.

Fujimoto et al. (1998a)

9.10 Observational summary of the low-latitude boundary layer

A PTE permits solar wind plasma to cross the (moving) magnetopause, as described in Chapter 6, to create the LLBL. We need 3D to describe the transfer, and an explicit dependence on time, to investigate both the induction and polarization electric fields.

- **Injection from rocket flights**: The earliest reports of impulsive magnetosheath plasma injection come from sounding rocket observations. Rockets launched from Greenland in 1974–5 discovered time-of-flight dispersion in sheath plasma, as discussed by Carlson and Torbert (1980). Clemmons et al. (1995) presented the results of later rocket flights.
- The most compelling work on impulsive sheath injections comes from midaltitude observations. Woch and Lundin (1992) used Viking data to study time-dependent injections of sheath plasma into the LLBL (see Fig. 8.4). These results are consistent with, but more definitive than, the rocket experiments.
- Stenuit et al. (2001) studied 131 instances using the Interball-Auroral satellite. They present a $1/v$ spectrogram of an

injection, with two bounce echoes (the multiple bounces indicate closed field lines). The temporal narrowness of the injections is clear from the $1/v$ time dependence of arrival.

- **Low-altitude observations**: Newell et al. (1989) and Newell and Meng (1991), using low-altitude data from the DMSP satellites, presented instances of sheath plasma extending largely unmodified beyond the convection reversal boundary.

- Heikkila et al. (1989), using multiple data sources including all-sky cameras, magnetometer stations, a global UV (still) image from Polar Bear, and others, investigated a single large-scale dayside auroral transient. The transient included a very large-scale convection vortex which formed (in conjunction with the auroral transient) around 0730 MLT.

- Moen et al. (1994) examined a series of multiple discrete arcs using meridian scanning photometers and radar (Eiscat) observations. The multiple transient arcs were thin (0.2 degrees MLAT), and occurred predominately equatorward of the convection reversal boundary. They suggested that magnetosheath plasma is injected into the magnetosphere through the LLBL.

- Fujimoto et al. (1998a) have presented reports in several papers on cold-dense stagnant ions, likely to be of the magnetosheath origin, detected in this region of the magnetosphere. Compared to the magnetosheath component, however, the electrons are anisotropically heated to have enhanced bidirectional flux along the field lines.

- Masson and Nykyri (2018) used Cluster and THEMIS data to find that the Kelvin-Helmholtz instability can lead to the development of rolled-up vortices along the magnetopause, allowing plasma transport and also significant ion heating due to enhanced ion-scale wave activity. This result persisted under all IMF conditions, although it was predominantly seen when the field was northward. They identified three main physical mechanisms to explain the cross-scale energy transport within a vortex.

In summary, the large number of observations spanning four decades indicates the introduction of sheath plasma through the flank boundary layers deep into the magnetosphere. These injections can extend beyond the convection reversal boundary onto sunward convecting field lines. The presence of hot magnetospheric electrons, spectrally complete ions, and other data establishes that these injections reach closed field lines. Despite the significance of these observations for understanding solar wind-magnetosphere coupling, they have received comparatively little attention. The earliest reports come from rocket flights around 0900 MLT. Further reports of injection events appear in recent extensive observations from Interball, Geotail, Cluster, THEMIS, MMS, and other spacecraft.

9.11 Study with southward interplanetary magnetic field

Øieroset et al. (1997) considered a typical dusk-dawn satellite orbit through the cusp and LLBL (see Fig. 9.24) showing a characteristic sequence of poleward moving auroral forms. The F11 satellite of the Defense Meteorological Satellite Program (DMSP) followed an essentially east-to-west, circular (apogee/perigee = 848 ± 22 km) geomagnetic path, crossing the meridian scanning photometer located at Svalbard at 0735:45 UT. The spacecraft reached a maximum latitude of about 75.7° MLAT (Fig. 9.24). A polar dial at the right side of the plots shows the spacecraft orbit track in MLT and MLAT.

Each DMSP spacecraft carries a set of three space environment sensing instruments: Special Sensor for Ionospheric Electrodynamics and Scintillation (SSIES), precipitating energetic particle spectrometer (SSJ/4), and Special Sensor Magnetometer (SSM). Various regions of interest can be seen in the electron and ion spectra derived from the particle detectors aboard the satellite in Fig. 9.24. At the two extremities of the plot there are increased fluxes of keV electrons, mapping from the plasma sheet (we use the energy flux scale rather than the particle flux used by Øieroset et al. (1997)). The top panel of the plots shows the integral precipitating energy flux for both ions and electrons in ergs/cm^2 s sr.

The second and third panels show the color spectrogram of the precipitating electron and ion energy flux, respectively (note that the scale of ion and electron spectrograms is reversed). The fourth panel shows the ion drift in the plane perpendicular to the spacecraft velocity vector, plotted in the Earth's corotation frame. The bottom panel shows the H$^+$, O$^+$, and total ion concentration.

9.11.1 Open geomagnetic field lines

There is little doubt that the magnetosphere is partly open (Heikkila and Winningham, 1971; Frank, 1971). The region of openness is marked by two dashed lines in Fig. 9.24, defined by the SSIES sensor in the next subsection.

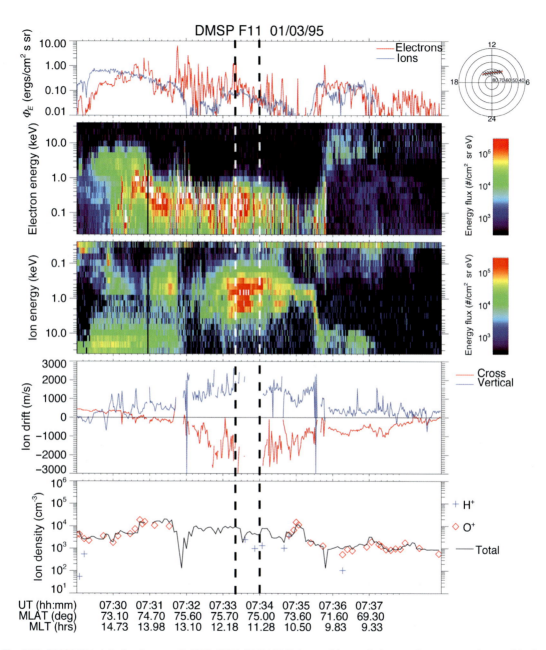

Fig. 9.24 DMSP F11 data for January 3, 1995, 0725–0755 UT. Enhanced ion and electron fluxes were observed in the 0731:00–0732:00 and the 0732:40 0734:45 UT interval in the LLBL. The ionospheric convection was strongly antisunward (∼1–2 km/s) during the latter interval. From Heikkila and Anderson, after Øieroset, M., Sandholt, P., Lühr, H., Denig, W., Moretto, T., 1997. Auroral and geomagnetic events at cusp/mantle latitudes in the prenoon sector during positive IMF by conditions: signatures of pulsed magnetopause reconnection. J. Geophys. Res. 102(A4), 7199.

9.11.2 Ionospheric convection

The cross-track and vertical ion drifts are shown in Fig. 9.25. There is a gap in the data nearly a minute near 0734 UT which represents magnetosheath plasma; plasma densities are too low here which is thought to be on open field lines. Nevertheless, the horizontal drift measurements show clear indication of antisunward flow of several hundred m/s after 0732 UT, reaching up 2 km/s. The LLBL is a reminder of the context.

9.11.3 Poleward moving auroral forms

Ground magnetic signatures associated with a sequence of poleward moving auroral forms are shown in Fig. 9.26 (Øieroset et al., 1997). We have already met this in Fig. 8.7, where it was closely related to PTEs. Previous studies have indicated that precipitation in the prenoon sector is favored for positive IMF B_y conditions.

The occurrence, structure, and motion pattern of these events are found to be strongly regulated by the IMF B_z and B_y components. The present cases are observed in the prenoon sector during positive IMF B_y and negative B_z conditions. ... The current system consists of a central filament or sheet-like field-aligned current (FAC) ... with oppositely directed return currents on the poleward and equatorward sides. The ground magnetic deflections can be explained by ionospheric Hall currents generated by this FAC system and the associated pattern of ionospheric electric fields. The propagation speed of the events (auroral and magnetic signatures) are found to be ... 1–2 km/s.

Øieroset et al. (1997)

- Sequences of poleward moving auroral forms have been interpreted as evidence for particle precipitation (Sandholt et al., 1986; Sandholt and Farrugia, 2003).
- The event recurrence time in these sequences is variable, but the auroral observations indicate that there is a subclass of major poleward moving auroral forms recurring at 5–8 min during IMF $B_z < 0$.
- A qualitatively similar current system with opposite current polarity has recently been proposed for the corresponding auroral and magnetic signatures observed during negative IMF B_y conditions.
- The horizontal drift measurements show clear indication of antisunward flow of several hundred m/s on the afternoon side

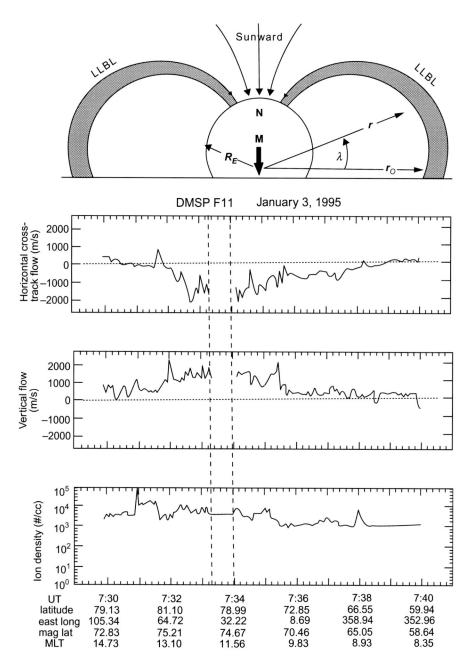

Fig. 9.25 The cross-track and vertical ion drifts are shown, and the relatively constant ion density. Note that the plasma flows tailward and raises in altitude in agreement with the observations of Fig. 9.24. From Heikkila and Hariston.

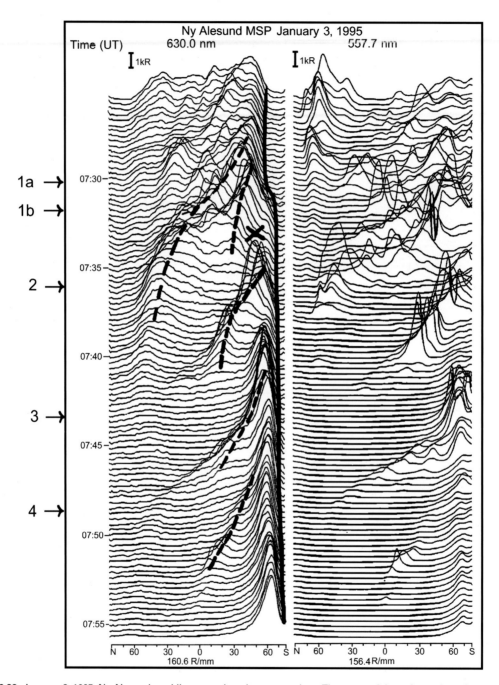

Fig. 9.26 January 3, 1995, Ny Alesund meridian scanning photometer data. The onset of the poleward moving events is marked by *arrows*. The equatorial boundary of the aurora and the optical events are given by the lines, and the intersection point with the DMSP satellite at 0734:45 UT is marked by the cross. From Øieroset, M., Sandholt, P., Lühr, H., Denig, W., Moretto, T., 1997. Auroral and geomagnetic events at cusp/mantle latitudes in the prenoon sector during positive IMF By conditions: signatures of pulsed magnetopause reconnection. J. Geophys. Res. 102(A4), 7195.

of the track (0918 UT), with sunward flow near magnetic noon (immediately prior to 0919 UT) and a substantial antisunward flow of some 2 km/s on the morning side (marginally later than 0921 UT) that diminished in magnitude with decreasing MLT, consistent with convection flow pattern illustrated in Fig. 9.25.

9.12   Polar cap during northward interplanetary magnetic field

Ionospheric convection depends strongly on the direction and strength of the IMF. For periods with southward IMF B_z, the convection configuration at high latitudes consists of a well-organized two-cell pattern with antisunward flow over the polar cap and sunward flow in the lower latitudes. The neutral winds tend to mimic the ion convection pattern because of the significance of the ion drag force on neutral momentum. When B_z is northward, the convection patterns are more complex due to the twist of the entire magnetotail as shown by Figs. 4.4 and 9.27. A feature of sunward ion flow in the central polar cap (Burke et al., 2007) is a reversal from the configuration for

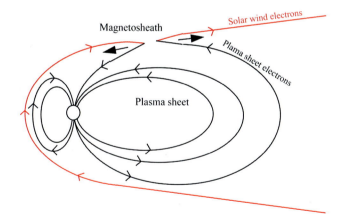

Fig. 9.27 A sketch illustrating the creation of the open plasma sheet by tailward-of-cusp reconnection between the draped northward-directed IMF and closed plasma sheet field lines. Reconnection is not shown in the two hemispheres, but the figure as drawn has opposing directions in the south which should indicate that there is a normal component. The requirement for PTE is a finite B_n which is implied by the reversal in **B** in the lower part of the plot. From Øieroset, M., Phan, T.D., Fairfield, D.H., Raeder, J., Gosling, J.T., Drake, J.F., et al., 2008. The existence and properties of the distant magnetotail during 32 hours of strongly northward interplanetary magnetic field. J. Geophys. Res. 113, 6, A04206. https://doi.org/10.1029/2007JA012679.

southward IMF. If the duration of the northward B_z is comparable to the time constant (few hours) for ion neutral momentum exchange, sunward neutral winds can also be established and maintained by ion drag forcing.

Øieroset et al. (2008) reported observations made by the Wind spacecraft in the distant magnetotail on October 22–24, 2003 when the IMF was strongly northward for more than 32 h.

> *A well-defined magnetotail was observed down to at least*
> $X_{GSM} = -125 \, R_E$ *even after 32 h of strongly northward IMF*
> *conditions. However, the observed tail properties were strikingly*
> *different from the typical magnetotail.*
>
> **Øieroset et al. (2008)**

Fig. 9.28 shows Wind observations from 02:55 to 04:05 UT on October 23, 2003. In addition to the cold, dense, and fast-flowing magnetosheath-like plasma flowing *tailward*, two other distinct plasma regions were observed. Both regions were more tenuous and much hotter than the magnetosheath. However, the two regions had distinctly different plasma properties.

The first region had a density of $0.3 \, cm^{-3}$, an ion temperature of 0.7–1.5 keV, a tailward flow speed of 100–200 km/s, low $|B_x|$ (<6.5 nT), and nearly isotropic electrons. The ion temperature of this plasma sheet during northward IMF is cooler (0.7–1.5 keV vs 1–4 keV) while the density remains similar ($\sim 0.3 \, cm^{-3}$). The lack of solar wind strahl electrons in this region (e.g., at 03:45–03:51 UT on October 23, 2003) provides additional evidence that this region is the plasma sheet on closed field lines.

While the first region (the closed plasma sheet under northward IMF) is similar to the closed plasma sheet under southward or B_y-dominated IMF, the second region (located at larger $|B_x|$) is quite different from the mantle. The electron distributions in this region consist of two populations: counterstreaming field-aligned electrons, and electrons near 90 degrees pitch angle. The counterstreaming electrons at solar wind strahl energies in this region are similar to those observed in the southward/B_y-dominated IMF mantle and indicate that this is a "mantle-like" region, which the authors said is on open field lines. However, the additional population at 90 degrees pitch angle was not seen on mantle field lines. The energy of this population is between 400 eV and 2 keV, resulting in a total electron temperature of ~ 500 eV, a factor of five higher than the conventional mantle.

The authors do not use the electric field in their deliberations. For now, let us note that the plasma sheet is a dynamo since $\mathbf{E} \cdot \mathbf{J} < 0$, a dynamo. We saw in Fig. 9.10 that a dynamo is an active

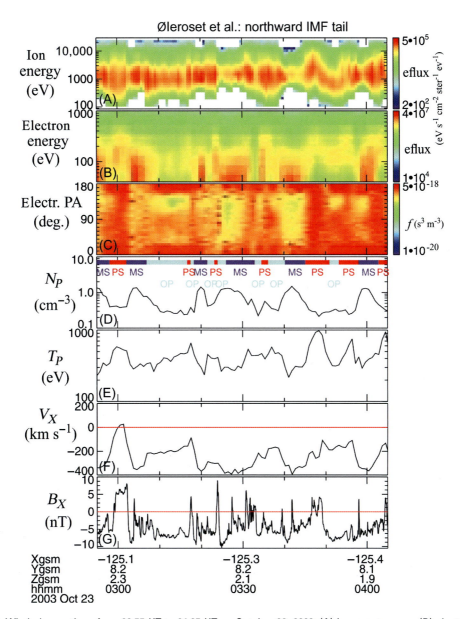

Fig. 9.28 Wind observations from 02:55 UT to 04:05 UT on October 22, 2003. (A) Ion spectrogram, (B) electron spectrogram, (C) 430 eV electron pitch angle, (D) plasma density, (E) ion temperature, (F) velocity along X_{GSM}, (G) magnetic field along X_{GSM}. The closed plasma sheet (PS), the open plasma sheet (OP), and the magnetosheath (MS) intervals are marked by *color bars* in panel (D) (according to the authors). From Øieroset, M., Phan, T.D., Fairfield, D.H., Raeder, J., Gosling, J.T., Drake, J.F., et al., 2008. The existence and properties of the distant magnetotail during 32 hours of strongly northward interplanetary magnetic field. J. Geophys. Res. 113, 5.

region with electric and magnetic activity reaching a maximum (the solid line). We shall discuss this further in the discussion section.

9.12.1 November 2004 storm

The November 2004 storm event was a superstorm with three periods of strongly southward IMF B_z that caused substantial geomagnetic disturbances. The electric potential pattern displayed the two-cell configuration with a positive potential cell on the dawn side and a negative potential cell on the dusk side.

The most interesting period is the strongly northward B_z that followed for 4 h (21 UT on November 9, to 01 UT on November 10) when B_z was larger than 20 nT for most of the time. The solar wind speed was close to 800 km/s and the dynamic pressure varied between 5 and 35 nPa. The daily solar radiation index was high; on November 9, 2004 it was 124×10^{-22} W/m^2 Hz. The electric potential pattern reversed in both hemispheres, with a negative potential cell on the dawn side and a positive potential cell on the dusk side. Reversed convection in both hemispheres for more than 1 h has rarely been observed.

Using the AMIE (Assimilative Mapping of Ionospheric Electrodynamics) outputs to drive the TIEGCM (National Center for Atmospheric Research Thermosphere Ionosphere Electrodynamics General Circulation Model), the impact of the reversed convection on the thermosphere was examined. The influence on the neutral dynamics through changing the ion drag force was found to be substantial. In order to eliminate the impact of solar EUV-induced pressure gradients on the neutral wind and to emphasize the wind variations caused by geomagnetic activities, a background condition was subtracted from the simulation results. The interval of 2000–2400 UT on November 8, 2004 was chosen to represent the background when the solar wind and IMF were relatively steady.

Fig. 9.29A shows the difference $\mathbf{E} \times \mathbf{B}$ drift and the difference neutral wind with respect to the background condition in the northern hemisphere at 2000 UT on November 9, when B_z was strongly southward ($B_z \sim -20$ nT). Fig. 9.29B shows those at 2230 UT on November 9, when B_z became strongly northward ($B_z \sim 30$ nT). The difference $\mathbf{E} \times \mathbf{B}$ drift in the polar cap region changed from the antisunward-duskward to sunward-duskward direction with reduced magnitude. The coupling between the neutral wind and $\mathbf{E} \times \mathbf{B}$ drift is rather evident.

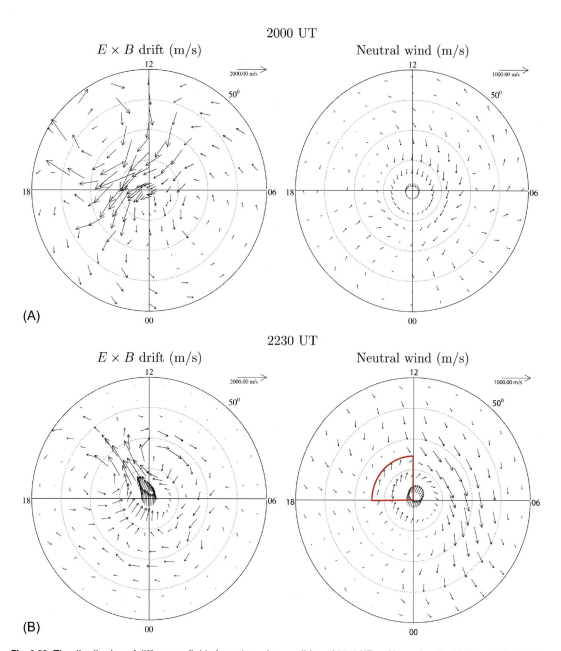

Fig. 9.29 The distribution of difference fields from the quiet condition of 2000 UT on November 8, at 400 km altitude to the fields on November 9 at (A) 2000 UT with southward IMF and (B) 2330 UT with northward IMF. The outside ring is 50 degrees and geographic coordinates are used. Note that **E** × **B** drifts are strikingly different. This feature has rarely been observed on both poles, apparently because of the long time constant involved. From Deng, Y., Lu, G., Kwak, Y.-S., Sutton, E., Forbes, J., Solomon, S., 2009. Reversed ionospheric convections during the November 2004 storm: impact on the upper atmosphere. J. Geophys. Res. 114, 5, A07313. https://doi.org/10.1029/2008JA013793.

Indeed, the AMIE outputs show strong reversed convection cells in both hemispheres for a long period (>1h), which have rarely been observed. The impact on the thermospheric neutral wind has been investigated using ... the National Center for Atmospheric Research Thermosphere Ionosphere Electrodynamics General Circulation Model (TIEGCM). After the ionospheric convection reversed, the neutral wind distribution at 400km altitude changed correspondingly, and the difference wind patterns reversed in the polar cap region. ... The neutral wind response time (e-folding time) clearly has an altitudinal dependence varying from 45min at 400km altitude to almost 1.5h at 200km.

Deng et al. (2009)

9.12.1.1 Surface singularities

Surface singularities of the charge density are of particular interest in both electrostatics and magnetostatics. Let us consider a double layer charge arrangement with a dipole moment per unit area designated by τ. The potential arising from such a distribution, in the case when r is uniform and directed along the normal to the surface outward from the observation point, reduces to (Panofsky and Phillips, 1962, p. 20):

$$\phi = \frac{-|\tau|}{4\pi\varepsilon_o}\int\frac{\mathbf{r}\cdot\mathbf{dS}}{r^3} = \frac{-|\tau|}{4\pi\varepsilon_o}\Omega \tag{9.3}$$

Here Ω is the solid angle subtended by the dipole sheet at the point of observation. The solid angle subtended by a nonclosed surface jumps discontinuously by 4π as the point of observation crosses the surface. It will have a continuous derivative at the dipole sheet.

A simple surface charge layer will not result in a discontinuity in potential, but will produce a discontinuity in the normal derivative of the potential (Fig. 9.30). Since surface charge layers and dipole layers enable us to introduce arbitrary discontinuities in the potential and its derivatives at a particular surface, we can make the potential vanish outside a given volume by surrounding the volume with a suitably chosen charge layer and dipole layer.

9.12.2 Magnetosheath flow is in control

An equatorial view of the magnetosphere with a southward IMF is shown on the left in Fig. 9.31, similar to Fig. 9.15. The dawn-side magnetopause must carry a negative charge on the surface of the magnetopause, and a positive charge at dusk; this meets the requirements for interaction with the IMF.

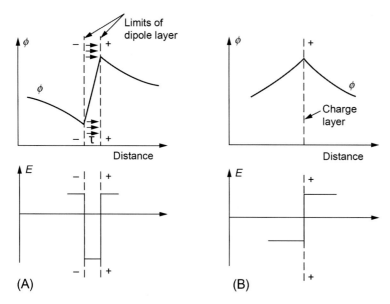

Fig. 9.30 A charged layer produces an electric field which is repulsive. (A) The left plate, being negatively charged, produces a field which points toward it, while the right plate produces a field which points away. The two fields cancel in regions on both sides; they conspire in the layer (note that the *arrows* shown are the dipole moment per area τ). If an electric field exists on one side, it is transferred to the other as shown. (B) A single charge layer would reverse the direction of the electric field. From Heikkila, after Panofsky, W.K.H., Phillips, M., 1962. Classical Electricity and Magnetism, second ed. Addison-Wesley, Reading, MA, p. 21.

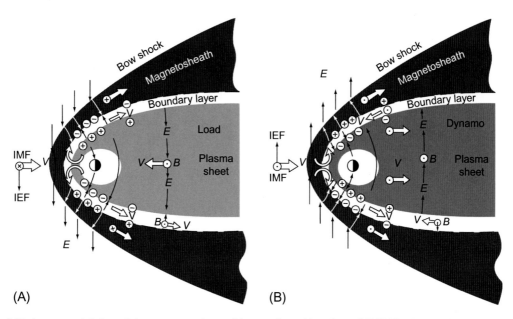

Fig. 9.31 An equatorial view of the magnetosphere with a southward (northward) IMF. The dawn-side magnetopause must carry a negative (positive) charge as a terminus of the IEF, by Gauss's law; the charge is positive (negative) at dusk. The signs of charge are the opposite at the inside edges of the LLBL giving a tailward (sunward) flow sense. Opposite charges for the plasma sheet make the difference between load and dynamo by the sign of **E · J**. The twist for northward sense is omitted. From Heikkila and Brown.

A strictly northward IMF as shown in Fig. 9.31B brings in some exciting physics. First, we again note that magnetosheath flow is in control because of the overriding principle of momentum conservation. The IEF must be in a dusk-to-dawn orientation, just the opposite for a southward IMF. That demands positive charge on the morning-side magnetopause by Gauss's law and negative on the dusk side. The electric field is oriented in the *opposite sense* to the southward IMF case, in all regions.

Now let us look at what happens in the boundary layer; there is a lot of plasma involved. It keeps its tailward momentum by a polarization current, with $\mathbf{E} \cdot \mathbf{J} < 0$ acting as a dynamo. Using the superposition principle and Gauss's theorem, the signs of charge are the opposite at the inside edges of the LLBL giving a tailward flow of the LLBL plasma. The charged dipolar layer has little effect on the IEF, allowing its effective transfer, as shown in Fig. 9.30. The geomagnetic field is northward at the equator, therefore the *plasma flow in the LLBL is sunward*, rather than tailward.

When mapped to the ionosphere, the sunward flow appears as a band through the poles, aligned in the noon–midnight sense. In fact, there are *two bands*, one from dawn and one from dusk, separated by *open field lines*.

One matter of great importance to the plasma sheet is the sign of $\mathbf{E} \cdot \mathbf{J}$. A positive sign means dissipation, common with southward IMF in the night-side plasma sheet. A negative sign indicates cooling of the plasma. With the signs of $\mathbf{E}$ being changed between by the reversal of the IMF, the flow in the plasma sheet is a generator with a northward IMF. The cold and dense plasma sheet (Fujimoto et al., 1998a) is thereby explained. The plasma has to counter the resistive force of the closed geomagnetic field (for the plasma to get past Earth we have to add the corotation electric field as well).

In summary, the AMIE outputs show strong reversed convection for a long period ($>1\,\mathrm{h}$) in both hemispheres. The sunward flow out of the dawn and dusk boundary layers will meet near noon. The only way out of a collision is a sharp earthward turn, as shown in Fig. 9.31. In the continued tailward flow $\mathbf{E} \cdot \mathbf{J} < 0$ (it is still a polarization electric field).

9.13 Penetration of interplanetary electric field into magnetosphere

The IMF always exists, so that an IEF also exists due to the solar wind flow. There are many radar observatories at high latitudes that can determine the electric field. SuperDARN results are shown later, in Chapters 10 and 11. For a southward IMF, the IEF is duskward, in the positive y direction. The opposite is true for a northward IMF; the IEF is negative, toward dawn. How is this change communicated to the magnetosphere?

Fig. 9.31B shows the situation in the equatorial plane for a strictly northward IMF. The electric field in the magnetosheath is rotated in step with diverted flow around the magnetospheric obstacle. In the LLBL the electric field has the opposite sense because the z-component of the magnetic field is positive (being on closed geomagnetic field lines), in agreement with Fig. 9.30. Gauss's law and Coulomb's law are not two independent physical laws, but the same law expressed in different ways. Symmetry is the key to the application of Gauss's law. Three kinds of symmetry are sufficient for analysis; for planar symmetry we use a Gaussian "pillbox" which straddles the surface as illustrated in Fig. 9.30. For a Gaussian charged layer, the electric field points away on both sides with a positive charge that is constant:

> *It seems surprising, at first, that the field of an infinite plane is independent of how far away you are. What about the $1/r^2$ in Coulomb's law? Well, the point is that as you move farther away from the plane, more and more charge comes into your "field of view," and this compensates for the diminishing influence of any particular piece.*

> **Griffiths (1981)**

The electric field in the magnetosphere and ionosphere can be determined in several ways. One is by radar, tracking the motion of plasma structures; Fig. 9.32 shows results of this type, with Jicamarca incoherent scatter radar in Peru (Huang et al., 2005).

These are measurements on the dayside near the equator. The IMF on 1848 UT November 9, 2004 became southward between 1848 and 2048 UT, turned northward at 2048 UT, and remained strongly northward until 0130 UT on November 10. The ionospheric electric field E_y measured by the Jicamarca radar on the dayside is presented in the bottom panel; the ionospheric electric field data are averaged over the altitude range of 248–368 km. When the IMF turned southward between 1848 and 2048 UT, the ionospheric electric field became strongly eastward (dawn to dusk). The shape of the enhanced eastward ionospheric electric field is very similar to the positive enhancement of the IEF E_y, indicating that the electric field penetration occurred during the entire interval of southward IMF. When the IMF turned northward between 2048 and 0130 UT, the ionospheric electric field became westward (dusk to dawn). In particular, the magnitude of the

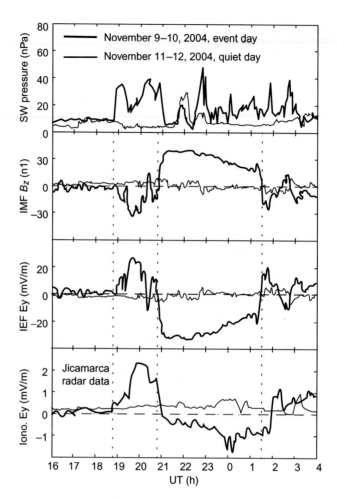

Fig. 9.32 Solar wind pressure, IMF B_z, and IEF E_y measured by the Wind satellite, and ionospheric electric field E_y measured by the Jicamarca radar on the dayside on November 9–10 and 11–12, 2004 (plotted as a quiet-time reference). The *vertical dotted lines* indicate the interval of southward IMF and eastward ionospheric electric field between 1848 and 2048 UT and of northward IMF and enhanced westward ionospheric electric field between 2048 and 0130 on November 9–10, 2004. (Local noon at Jicamarca is 1800 UT.) From Huang, C.-S., Foster, J.C., Kelley, M.1C., 2005. Long-duration penetration of the interplanetary electric field to the low-latitude ionosphere during the main phase of magnetic storms. J. Geophys. Res. 110, 4, A11309, https://doi.org/10.1029/2005JA011202.

westward electric field increased gradually without decay between 2048 and 0130 UT. After the IMF turned southward again at 0130 UT, the westward ionospheric electric field remained for about an additional 20 min. The temporal difference between the IMF southward turning and the end of the westward ionospheric electric field may be caused by an inaccurate time shift.

Fig. 9.32 represents a dilemma for the space physics community: how can the response be so rapid? It was thought that this electric field would be shielded from the low-latitude ionosphere by the action of the region-2 *field-aligned currents*. However, it takes *charge separation* to cancel an electric field or a changing current of right polarity. The efficiency of the shielding process by the region-2 field-aligned currents appears to be very limited, if it works at all.

9.14 A study with northward interplanetary magnetic field

Key observations were presented by Pryse et al. (2000) of the polar ionosphere under steady, northward IMF during winter conditions. The state of the magnetosphere requires the reality of a northward IMF as shown by Fig. 9.33; here the X-lines are over the poles, and the electrical currents are reversed from Fig. 9.17. The measurements were made by six complementary experimental techniques, including radio tomography, all-sky and meridian-scanning photometer optical imaging, incoherent and coherent scatter radars, and satellite particle detection. Each of the instruments individually provides valuable information on certain aspects of the ionosphere, but the paper demonstrates that taken together the different experiments complement each other to give a consistent and comprehensive picture of the dayside polar ionosphere.

9.14.1 Defense Meteorological Satellite Program particle and ionospheric data

The DMSP F13 satellite followed an essentially east-to-west geomagnetic path, crossing the noon meridian near 0920 UT and reaching a maximum latitude of about 78.0 degrees MLAT (Fig. 9.34). Various regions of interest can be seen in the electron and ion spectra derived from the particle detectors aboard the satellite. At the two extremities of the plot, dusk and dawn, there are increased fluxes of keV electrons, mapping from the plasma sheet. Between 0917 UT and approximately 0922 UT the satellite saw an increased flux of soft electron and ion precipitation in magnetosheath-like plasma. On either side within this central region, the ion spectrum shows the energy of the maximum flux increasing with latitude, indicative of reverse ion energy dispersion. The latitude of the equatorward edge of the dispersion region is in accord with the magnetic latitude of the density

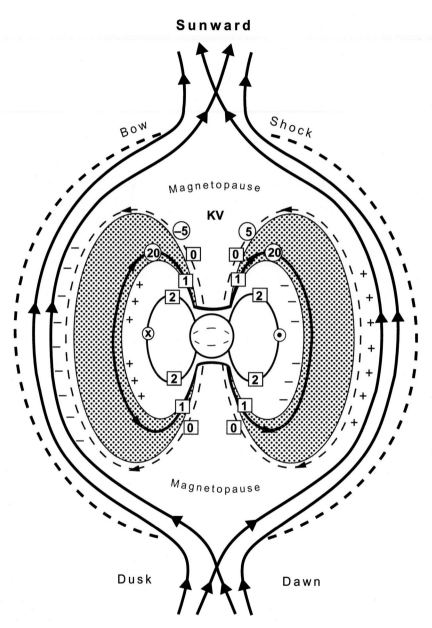

Fig. 9.33 Illustration of the separatrices projected to the dawn-dusk meridian for northward IMF, for a tailward sense revised from Fig. 9.17. The *arrows* indicate the sense of the field-aligned currents. Note that flow in the boundary layer is sunward, appropriate for theta aurora in the dusk sector. From Heikkila and Brown.

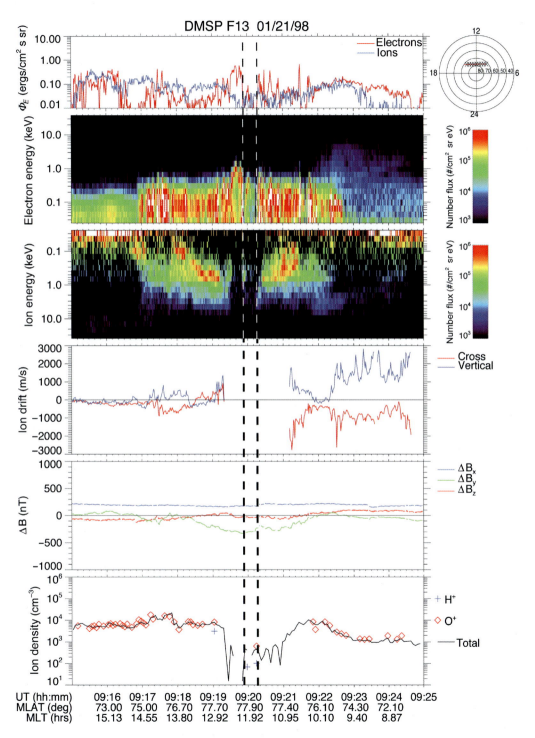

Fig. 9.34 The electron spectrogram shows theta-aurora just before 09:19:53 UT; the electrons reach 1 keV energies, then abruptly reach very low values until 09:20:17 UT. We interpret this short interval of 24 s (180 km) as open field lines. From Heikkila and Anderson.

gradient in the tomographic image and the soft, red-line precipitation indicated by the meridian scanning photometer.

The cross-track and vertical ion drifts are shown in the fourth panel of Fig. 9.34. Unfortunately, there is a gap in the data between about 0919 and 0921 UT, possibly due to low-plasma densities. Nevertheless the horizontal drift measurements show clear indication of antisunward flow of several hundred meters per second on the afternoon side of the track (0918 UT), with sunward flow near magnetic noon (immediately prior to 0919 UT), and a substantial antisunward flow of some 2 km/s on the morning side (marginally later than 0921 UT) that diminished in magnitude with decreasing MLT.

Inspect carefully the downward electron flux in Fig. 9.34: there is a short drop at 0919:53 and again at 0920:17 UT. With only this as a guide we must imagine what the complete three-dimensional picture would be. We assume that somewhere solar wind plasma would be on open field lines, butting up to closed geomagnetic field lines in this region of space (look at the plot in Fig. 9.35 between the dashed lines). These downgoing electrons or ions might be diverted at the magnetopause (after Willis, 1975), causing the current carriers in the ionosphere. Therefore we postulate that this region, marked by dashed vertical lines in Figs. 9.34 and 9.35, is on open field lines, lasting for 24 s or 180 km.

There is a very slight current in this region as verified by ΔB_y in Fig. 9.34. Some extra current, as shown by the top panel of Fig. 9.35, might be the deflected magnetopause current.

Fig. 9.36 is a series of plots of the ion density that gives an indication that this is not ionospheric plasma. There is a very weak Solar Wind plasma presence on "open" field lines. The weakness could be related to an upward ion flow (which is not measured; it is measured at 0919 and after 0921 UT). An upward flow was measured in the article by Whitteker et al. (1978) and Wescott et al. (1978).

9.14.2 Aurora observations

DMSP F13 recorded a moderate electron flux up to 1 keV between 09:19 and 09:19:55 UT at 13 MLAT slightly on the dusk side. Observations of plasma flows during the time interval of interest were made by the HF Collaborative UK SuperDARN radar (CUTLASS) with transmitters in Finland and Iceland (Fig. 9.37). The component of the plasma drift along the beam is relatively strong and away from the radar on the western side of the field of view. It is toward the radar in the central beams, and away

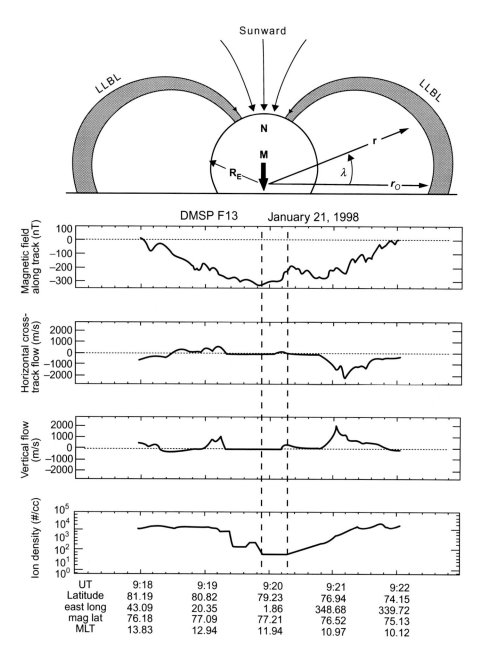

Fig. 9.35 Plots of the three components of the magnetic field and the plasma density. The *vertical dashed lines* indicate where the region of open field lines exists, according to Fig. 9.34. From Heikkila and Hairston.

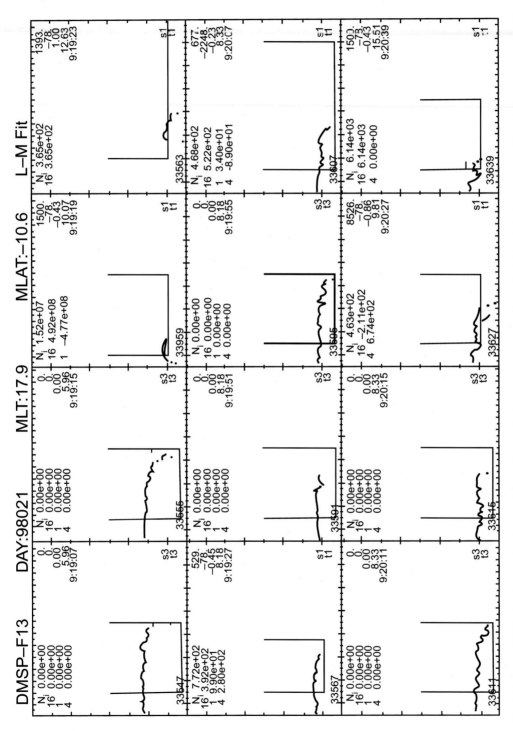

Fig. 9.36 Around 0920 the plasma density was low, and most likely at a temperature of 0.1 eV. Solar wind plasma is considerably hotter, near 100 eV. This figure shows the absence of the retarding potential effectiveness at modulating the stream. From Heikkila and Coley.

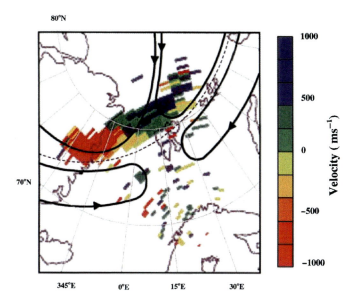

Fig. 9.37 Line-of-sight velocities measured by the Finland radar of the CUTLASS Superdarn facility at 0852 UT on a geographic grid. The positive values represent flows toward the radar and negative values flows away. Superimposed are flow stream lines *(solid curves)* and the adiaroic boundary *(dashed curve)*. From Pryse, S.E., Smith, A.M. ,Kersley, L., Walker, I.K., Mitchell, C.N., Moen, J., Smith, R.W., 2000. Multi-instrument probing of the polar ionosphere under steady northward IMF. Ann. Geophys. 18(1), 92.

on the eastern side at the equatorward edge of signal return, confirming the existence of the theta aurora.

9.14.3   Ionospheric tomography

The tomography chain of receivers operated by the University of Wales, Aberystwyth, comprises four stations in northern Scandinavia from Ny Alesund (78.9°N, 12.0°E) to Tromsø (69.8°N, 19.0°E). These monitor the phase coherent signals from the polar orbiting satellites in the Navy Ionospheric Monitoring System (NIMS) and enable the measurement of total electron content along a large number of intersecting satellite-to-receiver raypaths. Inversion of the data yields the distribution of electron density over a meridional section of the polar ionosphere. Fig. 9.38 shows the plasma structure determined from a north-to-south satellite pass on the morning of January 21, 1998, with the satellite crossing latitude 75.0°N at 0845 UT. The trajectory was essentially aligned along the noon magnetic meridian.

The striking feature of the image is the sharp gradient near 79.5°N, with densities decreasing abruptly to the north. The

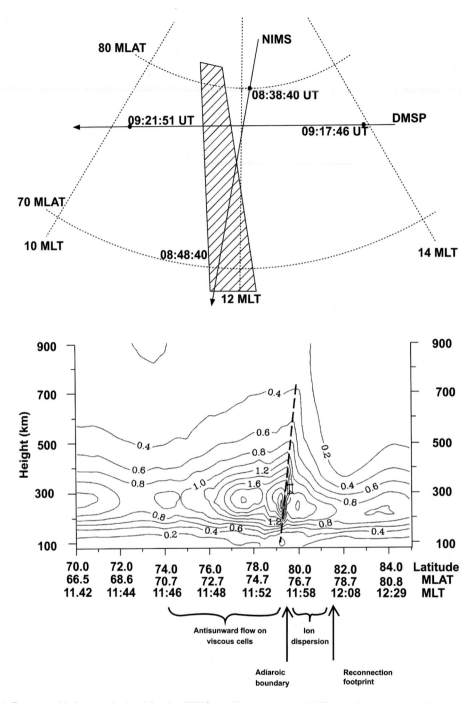

Fig. 9.38 Tomographic image obtained for the NIMS satellite pass at 0845 UT showing contours of electron density. (Top) The trajectory in terms of the magnetic coordinates of the 250 km intersection of the satellite-to-receiver ray paths was aligned along the noon magnetic meridian. (Bottom) The striking feature of the image is the sharp gradient near 79.5°N, with densities decreasing abruptly to the north. The transition is borne out, not only near the peak of the F-region, but also at altitudes extending into the topside ionosphere as indicated by the height change in the contours between 79.8°N and 79.9°N. From Pryse, S.E., Smith, A.M., Kersley, L., Walker, I.K., Mitchell, C.N., Moen, J., Smith, R.W., 2000. Multi-instrument probing of the polar ionosphere under steady northward IMF. Ann. Geophys. 18(1), 95.

transition is borne out not only near the peak of the F-region, but also at altitudes extending into the topside ionosphere (alignment of the feature with the geomagnetic field is not obvious at first sight). The dotted line drawn along the maximum gradient indicates the position of the field line that intersects an altitude of 250 km at 76.2 degrees MLAT. The gradient was interpreted by the authors as being at the *adiaroic boundary* separating LLBL plasma on closed flux tubes from ionization circulating on open magnetic flux in the polar lobe cells. This concept is irrelevant to the idea of a PTE, as in Chapters 5 and 6.

A change in the flow properties occurs as shown in Fig. 9.29 and explained by Fig. 9.31 near the subsolar region. A bright spot at 15 MLT was common in the 70s; curl **E** in the vorticity in the flow involves field-aligned acceleration leading to auroras. With northward IMF to the right there is a similar curl on the afternoon side of correct polarity; the NIMS data was lucky to find that situation. The all-sky camera did observe a spotty record shown by Pryse et al. (2000, p. 93).

9.14.4 Synopsis of daytime auroras

According to the classification of dayside auroral forms as shown in Fig. 9.39 (Sandholt et al., 1986), for large clock angles the cusp auroral form is typically located at ~72–74 degrees MLAT, and is characterized by a sharp equatorward boundary subject to intensifications (Sandholt and Farrugia, 2003). This aurora, called

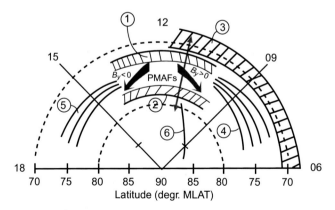

Fig. 9.39 Schematic overview of dayside auroral forms for different regimes of IMF orientation. From Sandholt, P.E., Farrugia, C.J., 2003. The aurora as monitor of solar wind-magnetospheric interactions. In: Newell, P.T., Onsager, T. (Eds.), Earth's Low-Latitude Boundary Layer. AGU Geophysical Monograph 133. American Geophysical Union, Washington, DC, p. 337.

type 1, is accompanied by poleward (antisunward) convection. Forms usually propagate poleward from the equatorward boundary intensifications, and fade at 3–5 degrees higher latitudes, in the regime of mantle precipitation. The aurora is accompanied by equatorward convection and is called type 2 (Murphree et al., 1993).

An additional dayside emission, strongly dominated by the green line at 557.7 nm, is often observed equatorward of the cusp, particularly in the prenoon sector. This so-called type 3 emission is caused by precipitating particles of plasma sheet energies. During southward IMF orientation the type 1 and 3 auroral emissions are typically separated by a 50–100 km wide latitudinal gap in the emission profile, which corresponds to a precipitation regime called void (Newell and Meng, 1994). The auroral/precipitation gap is associated with the "true cusp" (Keith et al., 2001, 2005).

Poleward moving auroral forms (see Figs. 8.5 and 9.26) must be associated with the injection events illustrated by Fig. 8.4. They tend to expand northwest into the prenoon sector if IMF $B_y > 0$ (east), and northwest if IMF $B_y < 0$. This motion asymmetry corresponds to a similar pattern observed in cusp region field-aligned currents and mantle precipitation.

During southward turnings of the IMF, type 1 forms have been observed to expand in the east-west direction, reaching beyond the 0900 and 1500 MLT meridians (Sandholt and Farrugia, 2003). During strongly northward IMF orientation, polar arcs, called type 6 in the schematic, are observed to emanate from the poleward boundary of the type 2 cusp aurora. During east-west oriented IMF conditions both types 1 and 2 may coexist as two latitudinally separated forms in a bifurcated or double cusp. When this happens, types 1 and 2 are often subject to coordinated brightenings.

Types 4 and 5 are multiple arcs in the pre- and postnoon sectors, which correspond to current filaments in the large-scale Region 1 field-aligned current system and particle precipitation classified as boundary plasma sheet. The type 4 arcs in the prenoon sector are observed to be strongly intensified associated with substorm activity expanding into this sector from the nightside. They are also activated in relation to the phenomenon of traveling convection vortices (Fig. 8.6) (Heikkila et al., 1989). At midmorning local times cusp-type auroral forms, associated with IMF B_y-related field-aligned currents, are observed to be located poleward of the Rl system/type 4 aurora.

9.15 Discussion

It has often been assumed that the polar cap must be on open field lines in the far magnetotail, connected to the swept-back IMF with the plasma mantle on open field lines. First, there are hardly any auroras above the auroral zone; the implication is that the solar wind particles must simply follow the field lines in the tailward motion. Second, a weak flux of soft electrons called polar rain was thought to be a simple mapping of the solar wind plasma. A third fact is that energetic particles from the Sun have almost instant access to the polar cap. Consequently, the model shown in Fig. 9.23 in the distant magnetotail (with the distant LLBL being on closed field lines) seems, at first glance, not a plausible solution.

We suggest that it is the proper and perfect model to explain a variety of observations. However, there are a number of problems to be dealt with; putting this thought into a positive sense, knowledge of the importance of the LLBL should shed new possibilities for realistic solutions.

9.15.1 Polar rain

Polar rain was discovered by Heikkila (1972) and Winningham and Heikkila (1974); it is apparent in the electron spectra to the extreme right of Fig. 9.42A and B. According to the usual interpretation based on the Dungey model, polar rain streams on open field lines from the solar wind (Fairfield and Scudder, 1985; Sotirelis et al., 1997). However, with no reconnection line in the near tail by Fig. 9.22, the model of Fig. 9.23 with distant magnetotail on closed field lines is appropriate. Now the reversal of the electric field (not the magnetic field) at about 100 R_E is the one that matters.

Polar rain has strong hemispherical asymmetry (Fig. 9.40), with the northern (southern) hemisphere favored for an away (toward) IMF sector structure (Yeager and Frank, 1976; Hardy et al., 1979; Gussenhoven and Madden, 1990). It has a dawn-dusk gradient controlled by IMF B_y (Meng and Kroehl, 1977). Fairfield and Scudder (1985) proposed that the suprathermal solar wind heat flux, or strahl, is the source of polar rain (see also Wing et al., 2005). For these reasons, details of particle fluxes have long been regarded as an important diagnostic of open-closed field lines: those with polar rain are *surely* (!) open.

Not so true! Winningham and Heikkila (1974) commented on polar rain being stronger during magnetic storms than during quiet conditions. This has been confirmed by Sotirelis et al.

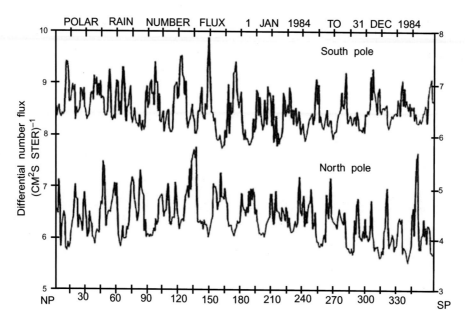

Fig. 9.40 Polar rain number flux calculated according to the polar rain index in the (top) southern hemisphere and (bottom) northern hemisphere and plotted against day number for 1984. The North Pole (South Pole) flux scale is on the left (right). From Gussenhoven, M.S., Madden, D., 1990. Monitoring the polar rain over a solar cycle: a polar rain index. J. Geophys. Res. 95(A7), 10406.

(1997); they have found that the relationship between the presence of bright polar rain in the polar cap and the IMF conditions preceding its observation "occurs almost exclusively during times of rapid dayside merging."

> *The occurrence of bright (>0.0016 ergs cm^{-2} s^{-1} sr^{-1}) polar rain is found to be determined by interplanetary magnetic field (IMF) conditions. An automated search of 11 years of DMSP particle precipitation data (1984–94) was used to identify polar rain. Comparison was made with 30-min segments of appropriately lagged 15-s IMP-8 magnetic field data. Bright polar rain away from the dayside merging line occurred almost exclusively under conditions favorable for rapid merging ($B_Z < 0$ or $|B_Y| > 2.5|B_Z|$).*
>
> **Sotirelis et al. (1997)**

This finding that polar rain *occurred almost exclusively under conditions favorable for substorms* has many consequences for the magnetospheric configuration. Polar rain is not passed on by *passive means*; it is *produced by substorm process*. Chapter 3 shows that low-energy electrons circle around the ions as Debye shielding; some higher-energy ions can escape earthwards along the field lines to form the polar rain. Newell et al. (2009) have stated:

> *The auroral oval itself lies largely on closed field lines, with the significant exception of the cusp. The region inside (poleward of) the auroral oval, which is usually called the "polar cap", is therefore a region of open field lines, connecting out to the solar wind, generally far downstream of the Earth.*
>
> **Newell et al. (2009)**

While a good source of polar rain data, their interpretation is at variance with the model of Fig. 9.23. McDiarmid et al. (1976) were the first to establish that the cusp is partly (mainly) on closed field lines. Sotirelis et al. (1997) found that polar rain occurred almost exclusively under conditions favorable for substorms; this has many consequences for the magnetospheric configuration. Newell et al., (2009) report observations of very intense polar rain with keV electrons, citing the DMSP orbit of 84/149. The 100 eV is apparently the true polar rain, while the keV electrons are produced by the substorm process. Sun-aligned arcs are connected with the low-latitude boundary as shown by Figs. 9.34 and 9.35.

9.15.2 Relativistic particle access

Gussenhoven and Mullen (1989) report observations of relativistic electrons (Fig. 9.41) made across the central polar cap using detectors onboard DMSP satellites on February 7, 1986 under different IMF conditions. While a solar proton event was still in progress, polar cap arc activity (IMF B_z large and positive) occurred prior to a prolonged period of oval arc activity (IMF B_z large and negative).

> *Throughout these periods the following are identified: the equatorward edge of the relativistic electron precipitation; the outer zone electron poleward and equatorward boundaries; the equatorward auroral electron and ion boundaries; and the auroral electron transition boundary. The history of the boundaries is presented as the transition from intense polar cap activity to intense auroral oval activity takes place. The data clearly show that the polar cap region of relativistic electron precipitation does not change significantly as the sign of B changes.*
>
> **Gussenhoven and Mullen (1989)**

Plots like those in Figs. 9.24 and 9.34 are a key ingredient for particle access showing a narrow region of open field lines. As shown by Roederer (1967), the cusp is connected to the entire magnetopause surface. For example, these energetic electrons

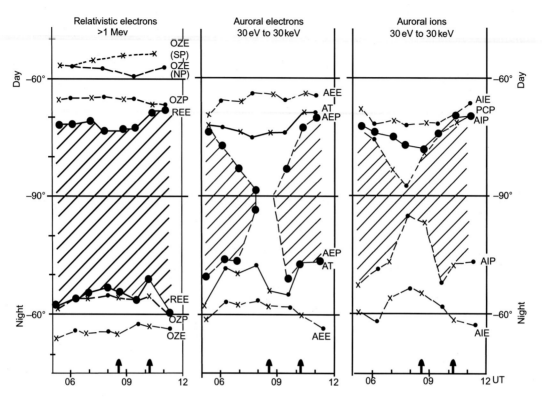

Fig. 9.41 Boundaries of precipitating electrons and ions over the southern polar cap for various energy ranges as a function of time on February 7, 1986. Cross-hatching shows the polar cap size as defined by relativistic solar electrons (*left*), polar rain (*middle*), and the absence of keV ions (*right*). OZP and OZE are outer zone poleward and equatorward boundaries, respectively; REE is the equatorward boundary of the relativistic electrons. AEP and AEE are the auroral electron poleward and equatorward boundaries, respectively. AT is the transition boundary between the hot and cool auroral electron populations. AIP and AIE are the auroral ion poleward and equatorward boundaries of the 4.5–9.6 keV population, respectively. PCP is the position of the peak in the ion polar cusp or cleft population. From Gussenhoven, M., Mullen, E., 1989. Simultaneous relativistic electron and auroral particle access to the polar caps during interplanetary magnetic field Bz northward: a scenario for an open field line source of auroral particles. J. Geophys. Res. 94(A12), 17125.

and ions will penetrate to very low altitudes in the narrow open field region beginning at 0919:53 UT in Figs. 9.34, cross into the LLBL along with solar wind plasma, aided by the reversal of the electric field in the PTE.

9.15.3   Velocity-dispersed ion structures

Beams of ions and electrons have been observed in the plasma sheet boundary layer (PSBL) for some time now (see Fig. 9.42 from Sotirelis et al. (1999), and references therein). At the outer edge of the PSBL, counterstreaming electrons are observed with the more

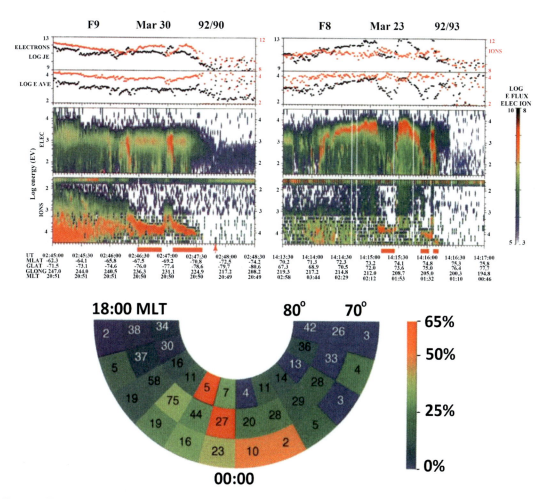

Fig. 9.42 Two velocity-dispersed ion structure (VDIS) events at top, with polar rain transitions to the right with lower energy. The top two panels of each plot show log energy flux (eV cm^{-2} sr^{-1} s^{-1}) and log average energy (eV), respectively. The bottom two panels are energy-time spectrograms of the differential energy flux in cm^{-2} sr^{-1} s^{-1}. Bottom: The percent of open-closed boundary crossings for which VDIS are observed as a function location. The total number of boundary crossings in each 5 degrees by 1 h bin is indicated by the number within. From Sotirelis, T., Newell, P.T., Meng, C.I., Hairston, M., 1999. Low altitude signatures of magnetotail reconnection. J. Geophys. Res. 104(A8), 17316.

energetic electrons headed tailward. They are dispersed in energy with depth into the PSBL, the more energetic electrons being outermost. Farther in, counterstreaming ions are observed, again dispersed in energy with depth into the PSBL and with the more energetic ions headed tailward. These observations are interpreted as the result of hot plasma streaming from a tailward reconnection site toward Earth (where it mirrors), as field lines convect from the lobe toward the plasma sheet. The observed structures are formed by the velocity filter effect: faster particles make it to the spacecraft before the field line has moved very deeply into the PSBL. Tailward traveling particles, having mirrored at Earth, have gone farther in the same time than those that are still headed earthward; hence the tailward component of a counterstreaming beam is more energetic.

These effects can also be seen where the PSBL impinges on the ionosphere. The velocity filter effect results in a sharp low-energy cutoff in the ion spectra, since only ions with sufficient velocity will have been able to travel the intervening distance in the time since the field line was reconnected and populated. If conditions remain steady for at least several minutes, then convection can effect a smooth dispersion of the cutoff energy with latitude. The events are called velocity-dispersed ion structures (VDIS) type 2 (type 2 to distinguish from a velocity dispersion of upwelling ionospheric plasma seen near the equatorward edge of the oval). These type-2 VDIS events have been the subject of several previous studies.

These observations have been interpreted as the result of hot plasma originating at a stationary X-line 50 to 100 R_E down the tail. However, as Pulkkinen et al. (1996) have observed, there is no steady-state X-line at these distances (see Fig. 9.22).

Sotirelis et al. (1999) have done a detailed study using data from DMSP spacecraft. Roughly one-third of type-2 VDIS were accompanied by a sharp transition in the polar rain near the open-closed boundary that aids in their analysis. From 886 night-side open-closed boundary crossings by DMSP spacecraft, 148 type-2 VDIS were identified.

> They were found most frequently within 2–3 hours of midnight and for 40% of the open-closed boundary crossings between 2200 and 0100 magnetic local time. Minimum variance fits to the cutoff energies and polar rain transitions are performed on 49 of these events. For four of them, the information from the minimum variance fit and observed convection velocities are used to infer distances to the reconnection site that varied from 30 to 80 R_E. In three of these four cases a sharp transition in the convection velocity is observed, coincident with the arrival of ions from the reconnection site.
>
> **Sotirelis et al. (1999)**

We focus on the two events depicted in Fig. 9.42. Their study emphasized the low-energy cutoffs in the ion energy spectra, as this is the principal consequence of the velocity filter effect. In Fig. 9.23 the plasma convection reverses near 100 R_E from earthward in the plasma sheet to tailward in the LLBL; VDIS most likely is associated with this reversal. Inspection of Fig. 9.42 shows that it is transient, perhaps related to ULF activity. It seems that the phenomenon may be related to generation of energetic particles in the PSBL.

9.15.4 ULF wave activity

Spacecraft in the cusp at a variety of altitudes have consistently found the cusp to be filled with intense, often irregular power in the upper ULF frequency range. On most days, no broadband ULF power was observed on the ground above the noise level near noon when only soft cusp precipitation or poleward moving auroral forms occurred overhead (Engebretson et al., 2006). However, several bursts of band-limited Pc 1–2 waves were observed in association with regions of intense soft precipitation that peaked near the poleward edge of the cusp. Their properties are consistent with origin in the plasma mantle (Fig. 9.43). Engebretson et al. (2006) have reviewed ground-based studies of high-latitude ULF waves on three regions: (1) near the low-altitude cusp, (2) in the polar cap, and (3) at the nightside aurora/cap boundary.

The two principal findings of this study are (1) the absence of a ground signature in the Pc 1–2 frequency range beneath the cusp, either narrowband or broadband, and (2) additional evidence in ground records for, and details of, narrowband Pc 1–2 waves that originate at the poleward edge of the cusp, in the plasma mantle.

Engebretson et al. (2009)

Geomagnetic field lines from all parts of the magnetospheric boundary reach Earth near the low-altitude northern and southern cusp regions. As a result, observations in the ULF frequency range make it possible to characterize the turbulent energy transfer from the solar wind into both dayside regions and the magnetotail, which is significant for the dynamics of the magnetosphere.

Several new types and features of ULF pulsations specific to the polar cap and dayside boundary regions have been identified using data from magnetometer arrays in Antarctica. The lowest frequency class, polar geomagnetic variations in the period range 4–20 min, is designated cap-associated $Pi_{cap}3$ pulsations. Statistical 1D and 2D spatial patterns of spectral power and coherence show the occurrence of intense variations, but with low spatial coherence, near the cusp projection and in the night-side auroral

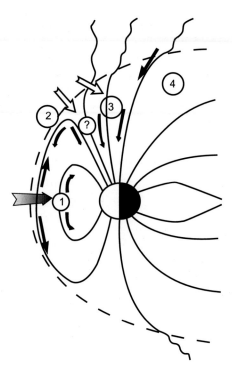

Fig. 9.43 Possible channels of Pc 3–4 wave penetration to the ground: (1) dayside equatorial magnetosphere, (2) near-cusp regions, (3) mantle/lobe, and (4) reconnected field lines. The *wide arrows* denote compressional-type disturbances, and the *thin arrows* denote Alfvén waves. The *question mark* refers to the transmission via the cusp proper. From Engebretson, M.J., Posch, J.L., Pilipenko, V.A., Chugunova, O.M., 2006. ULF waves at very high latitudes. In: Takahashi, K., Chi, P.J., Denton, R.E., Lysak, R.L. (Eds.), Magnetospheric ULF Waves: Synthesis and New Directions. AGU Geophysical Monograph 169. American Geophysical Union, Washington, DC, p. 146.

region. At the same time, low-amplitude $Pi_{cap}3$ pulsations are very coherent throughout the polar cap and are decoupled from auroral and cusp ULF activity. The primary sources of these long-period polar cap pulsations are probably related to tail lobe oscillations and turbulence.

> *Observational evidence has emerged (such as the occurrence of monochromatic Pc 3 waves deep in the polar cap, Pi1B pulsations associated with tail plasma flows, and ULF waves in the magnetospheres of giant planets) which indicates that some basic assumptions of this theory may be insufficient in regions with open or very extended field lines.*

Pilipenko et al. (2008)

The most intense F-region irregularities in the high-latitude ionosphere appear to be produced by convective plasma processes (Tsunoda, 1988). Such irregularities are produced by convectively mixing plasma across a plasma density gradient leading to the development of an irregularity spectrum that extends in scale from about 10 km down to the ion gyroradius. The mean plasma density gradient that must be present to allow irregularity production by this interchange process appears to be associated with larger-scale (>10 km) plasma structure produced by other means.

The final irregularity spectrum appears to be produced by (1) global convective processes acting on solar-produced plasma at the largest scales (>50 km), (2) particle precipitation at scales greater than 10 km, (3) perhaps some form of wave activity around 10 km, and (4) the $\mathbf{E} \times \mathbf{B}$ instability at the smaller scales (<10 km).

Tsunoda (1988)

9.16 Summary

The LLBL, the layer of plasma lying immediately earthward of the magnetopause, is mostly on closed field lines.

9.16.1 Magnetosheath flow is in control

A lot of plasma is flowing tailward in the magnetosheath; an electric field in the Earth's frame of reference given by $\mathbf{E} = -\mathbf{v} \times \mathbf{B}$ will produce an electric drift to force the plasma convection $\mathbf{v} = \mathbf{E} \times \mathbf{B}/B^2$. To conserve momentum, the plasma must maintain this electric field, especially at the magnetopause, for both southward and northward IMF. This result has emphasized by Lavraud et al. (2006) in Fig. 9.44.

- An important result of a superposed epoch analysis is that the dayside LLBL has a dense outer part that can be distinguished from its dilute inner part. Whereas the plasma in the outer boundary layer (OBL) is dominated by solar wind particles, the partial densities of solar wind and magnetospheric particles in the inner boundary layer (IBL) are lower, comparable to plasma sheet.

- Schmidt's (1960) mechanism works: Particles drifting out into the surface layer from the OBL experience smaller electric field, hence a smaller drift than the bulk of the plasma; they are consequently left behind in the IBL as well in the plasma sheet. The particles lost from the surface layer are

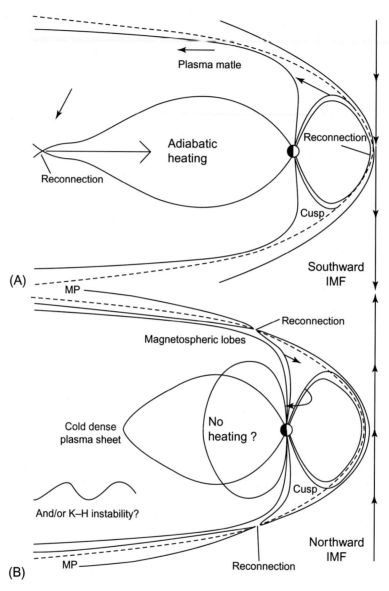

Fig. 9.44 Schematic of the solar wind circulation and magnetic topology of the magnetosphere for (A) southward and (B) northward IMF directions. The differences can be understood by Fig. 9.31. With southward IMF the negative charge on the dawn magnetopause will drive antisunward flow in the LLBL because both fields reverse. In contrast, with northward IMF the plasma flow in LLBL is sunward; the only possibility of the plasma sheet flow is tailward away from the Earth. The signs of charge are the opposite at the inside edges of the LLBL giving a tailward (sunward) flow sense. Opposite charges for the plasma sheet make the difference between load and dynamo by the sign of $\mathbf{E} \cdot \mathbf{J}$. From Lavraud, B., Thomsen, M. F., Lefebvre, B., Budnik, E., Cargill, P. J., Fedorov, A., et al., 2006. Formation of the cusp and dayside boundary layers as a function of imf orientation: cluster results. In: Cluster and Double Star Symposium—5th Anniversary of Cluster in Space, ESA SP-598, p. 6.

continuously replaced from the plasma interior, that is, from the OBL. Thus the plasma velocity remains unchanged while the mass is gradually decreasing.

- Polar cap during southward IMF: An antisunward LLBL flow component at highest latitudes produced an electric field associated with a sunward flow component at lower latitudes, both on the dawn and dusk sides.
- TORDO UNO rocket ion streak represents the 100 km dawn-to-dusk electric field of the polar cap as observed for nearly 40 min. It shows a constant electric field irrespective of the substorm development.
- Polar cap during northward IMF: with no X-line in the subsolar region, there is the possibility of B_N at neutral points poleward of the cusp, where the local rotation of the field across the magnetopause (the magnetic shear) is large. "Thus, our observations suggest that there is some secondary entry process in addition to these mechanisms … which becomes efficient on the flank sides under northward IMF" (Hasegawa et al., 2004). This process is a PTE.
- For northward IMF the flow distributions are suggestive of plasma penetration taking place at the poleward edge of the cusp, combined with a sunward convection there. "These results indicate that the exterior cusp flow structure is compatible with the preferential presence of reconnection at low-latitudes for southward IMF and at high-latitudes in the lobes for northward IMF." (Lavraud et al., 2005).
- Penetration electric fields have represented a dilemma for the space physics community. The answer is simple: this happens because the LLBL is an electric dipolar layer. This works for both southward IMF as well as northward IMF.
- The polar cap, which had long been assumed to be a relatively quiescent and uninteresting region, is now known to have its own distinct wave phenomena, which are not yet well understood. "The physics of nightside bursty/wave phenomena turns out to be much richer than traditional notions of Pi2 pulsations." (Engebretson et al., 2006).
- With a northward magnetic field, the *plasma flow in the LLBL is sunward*, rather than tailward. When mapped to the ionosphere, the sunward flow appears as a band through the poles, aligned in the sunward sense, and theta aurora.

9.16.2 Low-latitude boundary layer

The LLBL can conform to all of these observations:
- Supplies plasma to the plasma sheet

- Supplies the IEF to the plasma sheet
- LLBL is a dynamo, necessary to force plasma convection on closed magnetic field lines
- Produces no aurora over the polar cap, except theta aurora with northward IMF
- Dusk-side evening aurora arcs during southward IMF
- Polar rain with super-Alfvénic flow

9.17 Problems

9.1. Cowley (1982) said that cross-magnetospheric voltages are typically 40–100 kV. Given a typical magnetospheric width, what would be the range of electric fields in mV/m?

9.2. Consider the simple shock of Fig. 1.9. If the flow velocity $\mathbf{u}_u$ is 350 km/s and the angle of the upstream magnetic field to the normal line is 32 degrees, what Alfvén velocity would satisfy the Walén relation?

9.3. A spacecraft is traveling in a geocentric circular orbit with an altitude of 500 km. It has a plasma instrument that completes 128 energy sweeps per second. What is the maximum spatial resolution that can be expected from this spacecraft?

9.4. Using typical values for the density and velocity of the solar wind, calculate the momentum per cubic kilometer. What about the momentum per cubic R_E?

9.5. In the ideal case of an infinitely thin dipole charge layer, what would be the magnitude of the discontinuity of the potential function?

References

Axford, W.I., Hines, C.O., 1961. A unifying theory of high-latitude geophysical phenomena and geomagnetic storms. Can. J. Phys. 39, 1433.

Baker, D.A., Hammel, J.E., 1965. Experimental studies of the penetration of a plasma stream into a transverse magnetic field. Phys. Fluids 8 (4), 713–722.

Bauer, T.M., Treumann, R.A., Baumjohann, W., 2001. Investigation of the outer and inner low-latitude boundary layers. Ann. Geophys. 19, 1065–1088.

Burke, W.J., Gentile, L.C., Huang, C.Y., 2007. Penetration electric fields driving main-phase Dst. J. Geophys. Res. 112, A07208. https://doi.org/10.1029/2006 JA012137.

Carlson, C.W., Torbert, R.B., 1980. Solar wind ion injections in the morning auroral oval. J. Geophys. Res. 85 (A6), 2903–2908.

Clemmons, J.H., Carlson, C.W., Boehm, M.H., 1995. Impulsive ion injections in the morning auroral region. J. Geophys. Res. 100, 12133–12149.

Cowley, S.W.H., 1982. The causes of convection in the Earth's magnetosphere – a review of developments during the IMS. Rev. Geophys. Space Phys. 20, 531–565.

Crooker, N.U., 1988. Mapping the merging potential from the magnetopause to the ionosphere through the dayside cusp. J. Geophys. Res. 93 (A7), 7338–7344.

Deng, Y., Lu, G., Kwak, Y.-S., Sutton, E., Forbes, J., Solomon, S., 2009. Reversed ionospheric convections during the November 2004 storm: impact on the upper atmosphere. J. Geophys. Res. 114, A07313, https://doi.org/10.1029/2008 JA013793.

Dungey, J.W., 1961. Interplanetary magnetic field and the auroral zones. Phys. Rev. Lett. 6, 47–48.

Eastman, T.E., 1979. The Plasma Boundary Layer and Magnetopause Layer of the Earth's Magnetosphere (Ph.D. Thesis). Los Alamos Scientific Laboratory, Los Alamos, NM.

Eastman, T.E., Hones Jr., E.W., Bame, S.J., Asbridge, J.R., 1976. The magnetospheric boundary layer: site of plasma momentum, and energy transfer from the magnetosheath into the magnetosphere. Geophys. Res. Lett. 3, 685.

Engebretson, M.J., Posch, J.L., Pilipenko, V.A., Chugunova, O.M., 2006. ULF waves at very high latitudes. In: Takahashi, K., Chi, P.J., Denton, R.E., Lysak, R.L. (Eds.), Magnetospheric ULF Waves: Synthesis and New Directions. AGU Geophysical Monograph 169. American Geophysical Union, Washington, DC, pp. 137–156.

Engebretson, M.J., Moen, J., Posch, J.L., Lu, F., Lessard, M.R., Kim, H., et al., 2009. Searching for ULF signatures of the cusp: observations from search coil magnetometers and auroral imagers in Svalbard. J. Geophys. Res. 114(A6), A06217. https://doi.org/10.1029/2009JA014278.

Fairfield, D.H., Scudder, J.D., 1985. Polar rain: solar coronal electrons in the Earth's magnetosphere. J. Geophys. Res. 90, 4055.

Frank, L.A., 1971. Plasma in the Earth's polar magnetosphere. J. Geophys. Res. 76, 5202.

Fujimoto, M., Terasawa, T., Mukai, T., Saito, Y., Yamamoto, T., Kokubun, S., 1998a. Plasma entry from the flanks of the near-Earth magnetotail: geotail observations. J. Geophys. Res. 103, 4391.

Fujimoto, M., Mukai, T., Kawano, H., Nakamura, M., Nishida, A., Saito, Y., et al., 1998b. Structure of the low-latitude boundary layer: a case study with Geotail data. J. Geophys. Res. 103 (A2), 2297–2308.

Griffiths, D., 1981. Introduction to Electrodynamics, first ed. Prentice Hall, Englewood Cliffs, NJ.

Gussenhoven, M.S., Madden, D., 1990. Monitoring the polar rain over a solar cycle: a polar rain index. J. Geophys. Res. 95 (A7), 10399–10416.

Gussenhoven, M., Mullen, E., 1989. Simultaneous relativistic electron and auroral particle access to the polar caps during interplanetary magnetic field Bz northward: a scenario for an open field line source of auroral particles. J. Geophys. Res. 94 (A12), 17121–17132.

Hardy, D.A., Gussenhoven, M.S., Huber, A., 1979. The precipitation electron detectors (SSJ/3) for the block 5D/flights 2-5 DMSP satellites calibration and data presentation. Interim Report Air Force Geophysics Lab., Hanscom AFB, MA. Space Physics Division, September.

Hasegawa, H., Fujimoto, M., Saito, Y., Mukai, T., 2004. Dense and stagnant ions in the low-latitude boundary region under northward interplanetary magnetic field. Geophys. Res. Lett. 31, L06802. https://doi.org/10.1029/2003GL019120.

Heikkila, W.J., 1972. The morphology of auroral particle precipitation. In: Heikkila, W.J. (Ed.), Space Research XII. Akademie-Verlag, Berlin, Germany, pp. 1343–1355.

Heikkila, W.J., 1984. Magnetospheric topology of fields and currents. In: Potemra, T.A. (Ed.), Magnetospheric Currents. AGU Geophysical Monograph 28. American Geophysical Union, Washington, DC, pp. 208–222.

Heikkila, W.J., 1985. Definition of the cusp. In: Holtet, J.A., Egeland, A. (Eds.), Polar Cusp, NATO ASI Series. In: vol. 145. D. Reidel, Dordrecht, Netherlands.

Heikkila, W.J., 1988. Current sheet crossings in the distant magnetotail. Geophys. Res. Lett. 15 (4), 299–302.

Heikkila, W.J., 1997. Interpretation of recent AMPTE data at the magnetopause. J. Geophys. Res. 102 (A5), 2115.

Heikkila, W.J., 1998. Cause and effect at the magnetopause. Space Sci. Rev. 83, 373–434.

Heikkila, W.J., Winningham, J.D., 1971. Penetration of magnetosheath plasma to low altitudes through the dayside magnetospheric cusps. J. Geophys. Res. 76, 883–891.

Heikkila, W.J., Jorgensen, T.S., Lanzerotti, L.J., Maclennan, C.G., 1989. A transient auroral event on the dayside. J. Geophys. Res. 94, 15291–15305.

Hones Jr., E.W., Birn, J., Bame, S., Asbridge, J., Paschmann, G., Sckopke, N., et al., 1981. Further determination of the characteristics of magnetospheric plasma vortices with Isee 1 and 2. J. Geophys. Res. 86 (A2), 814–820. https://doi.org/10.1029/JA086iA02p00814.

Huang, C.-S., Foster, J.C., Kelley, M.C., 2005. Long-duration penetration of the interplanetary electric field to the low-latitude ionosphere during the main phase of magnetic storms. J. Geophys. Res. 110, A11309. https://doi.org/10.1029/2005JA011202.

Keith, W.R., Winningham, J.D., Norberg, O., 2001. A new, unique signature of the true cusp. Ann. Geophys. 19, 611–619.

Keith, W.R., Winningham, J.D., Goldstein, M.L., Wilbur, M., Fazakerley, A.N., Rème, H., et al., 2005. Observations of a unique cusp signature at low and mid altitudes. In: Fritz, T.A., Fung, S.F. (Eds.), The Magnetospheric Cusps. Springer, Dordrecht, Netherlands, pp. 305–337.

Kuhn, T.S., 1970. The Structure of Scientific Reductions. The University of Chicago Press, Chicago, IL.

Lavraud, B., Fedorov, A., Budnik, E., Thomsen, M.F., Grigoriev, A., Cargill, P.J., et al., 2005. High altitude cusp flow dependence on IMF orientation: a 3 year Cluster statistical study. J. Geophys. Res. 110, A02209. https://doi.org/10.1029/2004JA010804.

Lavraud, B., Thomsen, M.F., Lefebvre, B., Budnik, E., Cargill, P.J., Fedorov, A., et al., 2006. Formation of the cusp and dayside boundary layers as a function of imf orientation: Cluster results. In: Cluster and Double Star Symposium—5th Anniversary of Cluster in Space, ESA SP-598.

Lemaire, J., Roth, M., 1991. Non-steady-state solar wind-magnetosphere interaction. Space Sci. Rev. 57, 59–108.

Lyons, L.R., Williams, D.J., 1984. Quantitative Aspects of Magnetospheric Physics. Springer, New York, NY.

Marklund, G., Ivchenko, N., et al., 2001. Temporal evolution of the electric field accelerating electrons away from the auroral ionosphere. Nature 414, 724–727.

Masson, A., Nykyri, K., 2018. Kelvin–Helmholtz instability: lessons learned and ways forward. Space Sci. Rev. 214(71). https://doi.org/10.1007/s11214-018-0505-6.

McDiarmid, I., Burrows, J., Budzinski, E., 1976. Particle properties in the day side cleft. J. Geophys. Res. 81 (1), 221–226.

Meng, C.I., Kroehl, H., 1977. Intense uniform precipitation of low-energy electrons over the polar cap. J. Geophys. Res. 82 (16), 2305–2313.

Mitchell, D.G., Kutchko, F., Williams, D.J., Eastman, T.E., Frank, L.A., Russell, C.T., 1987. An extended study of the low-latitude boundary layer on the dawn and dusk flanks of the magnetosphere. J. Geophys. Res. 92, 7394. https://doi.org/10.1029/JA092iA07p07394.

Moen, J., Sandholt, P., Lockwood, M., Egeland, A., Fukui, K., 1994. Multiple, discrete arcs on sunward convecting field lines in the 14-15 MLT region. J. Geophys. Res. 99 (A4), 6113–6123.

Murphree, J.S., Elphinstone, R.D., Henderson, M.G., Cogger, L.L., Hearn, D.J., 1993. Interpretation of optical substorm onset observations. J. Atmos. Terr. Phys. 55, 1159–1170.

Newell, P.T., Meng, C.-I., 1988. The Cusp and the cleft/boundary layer: low-altitude identification and statistical local time variation. J. Geophys. Res. 93 (A12), 14549–14556. https://doi.org/10.1029/JA093iA12p14549.

Newell, P.T., Meng, C.-I., 1991. Ion acceleration at the equatorward edge of the cusp: low-altitude observations of patchy merging. Geophys. Res. Lett. 18, 1829–1832. https://doi.org/10.1029/91GL02088.

Newell, P.T., Meng, C.-I., 1994. Ionospheric projections of magnetospheric regions under low and high solar wind pressure conditions. J. Geophys. Res. 99 (A1), 273–286. https://doi.org/10.1029/93JA02273.

Newell, P.T., Onsager, T. (Eds.), 2003. Earth's Low-Latitude Boundary Layer. AGU Geophysical Monograph 133. American Geophysical Union, Washington, DC.

Newell, P.T., Meng, C.-I., Sibeck, D.G., Lepping, R., 1989. Some low-altitude cusp dependencies on the interplanetary magnetic field. J. Geophys. Res. 94, 8921.

Newell, P.T., Liou, K., Wilson, G.R., 2009. Polar cap particle precipitation and aurora: review and commentary. J. Atmos. Sol. Terr. Phys. 71 (2), 199–215.

Øieroset, M., Sandholt, P., Lühr, H., Denig, W., Moretto, T., 1997. Auroral and geomagnetic events at cusp/mantle latitudes in the prenoon sector during positive IMF By conditions: signatures of pulsed magnetopause reconnection. J. Geophys. Res. 102 (A4), 7191–7205.

Øieroset, M., Phan, T.D., Fairfield, D.H., Raeder, J., Gosling, J.T., Drake, J.F., et al., 2008. The existence and properties of the distant magnetotail during 32 hours of strongly northward interplanetary magnetic field. J. Geophys. Res. 113, A04206. https://doi.org/10.1029/2007JA012679.

Panofsky, W.K.H., Phillips, M., 1962. Classical Electricity and Magnetism, second ed. Addison-Wesley, Reading, MA.

Phan, T.-D., Paschmann, G., 1996. The low-latitude dayside magnetopause and boundary layer for high magnetic shear: 1. Structure and motion. J. Geophys. Res. 101, 7801. https://doi.org/10.1029/95JA03752.

Phan, T.-D., Paschmann, G., Baumjohann, W., Sckopke, N., Lühr, H., 1994. The Magnetosheath region adjacent to the dayside magnetopause: AMPTE/IRM observations. J. Geophys. Res. 99 (A1), 121–141. https://doi.org/10.1029/93JA02444.

Pilipenko, V., Fedorov, E., Heilig, B., Engebretson, M.J., 2008. Structure of ULF Pc3 waves at low altitudes. J. Geophys. Res. 113, A11208. https://doi.org/10.1029/2008JA013243.

Pryse, S.E., Smith, A.M., Kersley, L., Walker, I.K., Mitchell, C.N., Moen, J., Smith, R.W., 2000. Multi-instrument probing of the polar ionosphere under steady northward IMF. Ann. Geophys. 18 (1), 90–98.

Pulkkinen, T.I., Baker, D.N., Owen, C.J., Slavin, J.A., 1996. A model for the distant tail field: ISEE-3 revisited. J. Geomagn. Geoelectr. 48, 455.

Reiff, P.H., Spiro, R.W., Wolf, R.A., Kamide, Y., King, J.H., 1985. Comparison of polar cap potential drops estimated from solar wind and ground magnetometer data: CDAW 6. J. Geophys. Res. 90 (A2), 1318–1324.

Richmond, A.D., 2010. On the ionospheric application of Poynting's theorem. J. Geophys. Res. 115, A10311. https://doi.org/10.1029/2010JA015768.

Roederer, J.G., 1967. On the adiabatic motion of energetic particles in a model magnetosphere. J. Geophys. Res. 72 (3), 981–992.

Sandholt, P.E., Farrugia, C.J., 2003. The aurora as monitor of solar wind-magnetospheric interactions. In: Newell, P.T., Onsager, T. (Eds.), Earth's Low-Latitude Boundary Layer. AGU Geophysical Monograph 133. American Geophysical Union, Washington, DC, pp. 335–349.

Sandholt, P., Deehr, C., Egeland, A., Lybekk, B., Viereck, R., Romick, G., 1986. Signatures in the dayside aurora of plasma transfer from the magnetosheath. J. Geophys. Res. 91 (A9), 10063–10079.

Sauvaud, J.A., Lundin, R., Rème, H., McFadden, J.P., Carlson, C., Parks, G.K., et al., 2001. Intermittent thermal plasma acceleration linked to sporadic motions of the magnetopause, first cluster results. Ann. Geophys. 19 (10/12), 1523–1532.

Schmidt, G., 1960. Plasma motion across magnetic fields. Phys. Fluids 3, 961.

Schmidt, G., 1979. Physics of High Temperature Plasmas, second ed. Academic Press, New York, NY.

Sckopke, N., Paschmann, G., Haerendel, G., Sonnerup, B., Bame, S., Forbes, T., et al., 1981. Structure of the low-latitude boundary layer. J. Geophys. Res. 86 (A4), 2099–2110.

Slavin, J.A., Smith, E.J., Sibeck, D.G., Baker, D.N., Zwickl, R.D., Akasofu, S.-I., 1985. An ISEE 3 study of average and substorm conditions in the distant magnetotail. J. Geophys. Res. 90 (10), 10875–10895.

Slavin, J.A., Daly, P.W., Smith, E.J., Sanderson, T.R., Wenzel, K.-P., Lepping, R.P., et al., 1987. Magnetic configuration of the distant plasma sheet-ISEE 3 observations. In: Lui (Ed.), Magnetotail Physics. Johns Hopkins University Press, Baltimore, MD.

Sonnerup, B.U.Ö., 1980. Theory of the low-latitude boundary layer. J. Geophys. Res. 85, 2017. https://doi.org/10.1029/JA085iA05p02017.

Sonnerup, B.U.O., Siebert, K.D., 2003. Theory of the low latitude boundary layer and its coupling to the ionosphere: a tutorial review. In: Newell, P.T., Onsager, T. (Eds.), Earth's Low-Latitude Boundary Layer. AGU Geophysical Monograph 133. American Geophysical Union, Washington, DC, pp. 13–32.

Sotirelis, T., Newell, P.T., Meng, C.I., 1997. Polar rain as a diagnostic of recent rapid dayside merging. J. Geophys. Res. 102 (A4), 7151–7157.

Sotirelis, T., Newell, P.T., Meng, C.I., Hairston, M., 1999. Low altitude signatures of magnetotail reconnection. J. Geophys. Res 104 (A8), 17311–17321.

Stenuit, H., Sauvaud, J.-A., Delcourt, D.C., Mukai, T., Kokubun, S., Fujimoto, M., et al., 2001. A study of ion injections at the dawn and dusk polar edges of the auroral oval. J. Geophys. Res. 106, 29619–29631. https://doi.org/10.1029/2001JA900060.

Tsunoda, R.T., 1988. High-latitude F region irregularities: a review and synthesis. Rev. Geophys. 26 (4), 719–760. https://doi.org/10.1029/RG026i004p00719.

Wang, C.-P., Lyons, L.R., Nagai, T., Weygand, J.M., McEntire, R.W., 2007. Sources, transport, and distributions of plasma sheet ions and electrons and dependences on interplanetary parameters under northward interplanetary magnetic field. J. Geophys. Res. 112, A10224. https://doi.org/10.1029/2007JA012522.

Wescott, E., Stenbaek-Nielsen, H., Davis, T., Jeffries, R., Roach, W., 1978. The Tordo 1 polar cusp barium plasma injection experiment. J. Geophys. Res. 83 (A4), 1565–1575.

Whitteker, J.H., Shepherd, G.G., Anger, C.D., Burrows, J.R., Wallis, D.D., Klumpar, D.M., Walker, J.K., 1978. The winter polar ionosphere. J. Geophys. Res. 83 (A4), 1503–1518. https://doi.org/10.1029/JA083iA04p01503.

Williams, D., Mitchell, D., Eastman, T., Frank, L., 1985. Energetic particle observations in the low-latitude boundary layer. J. Geophys. Res. 90 (A6), 5097–5116.

Willis, D.M., 1975. The microstructure of the magnetopause. Geophys. J. R. Astron. Soc. 41 (3), 355–389.

Wing, S., Newell, P.T., Meng, C.-I., 2005. Cusp modeling and observations at low altitude. Surv. Geophys. 26, 341. https://doi.org/10.1007/s10712-005-1886-0.

Winningham, J.D., Heikkila, W.J., 1974. Polar cap auroral electron fluxes observed with ISIS 1. J. Geophys. Res. 79, 949. https://doi.org/10.1029/JA079i007p00949.

Woch, J., Lundin, R., 1992. Signatures of transient boundary layer processes observed with Viking. J. Geophys. Res. 97, 1431.

Yeager, D., Frank, L., 1976. Low-energy electron intensities at large distances over the Earth's polar cap. J. Geophys. Res. 81 (22), 3966–3976.

Zwickl, R.D., Baker, D.N., Bame, S.J., Feldman, W.C., Gosling, J.T., Hones Jr., E.W., et al., 1984. Evolution of the earth's distant magnetotail: ISEE 3 electron plasma results. J. Geophys. Res. 89 (A12), 11007–11012.

10

Driving the plasma sheet

Chapter outline

Earth's Magnetosphere. https://doi.org/10.1016/B978-0-12-818160-7.00010-7

In many cases the parameters that are used in the models [even the models] tend to conceal or distort some of the key physical processes that occur, a high price to pay for simplified analytical [or computer] solutions.

Richmond (1985)

10.1   Introduction

The transport of the plasma sheet plasma affects the spatial structure of the whole magnetotail. This structure, its variation in location and time, ultimately depends on changes in the solar wind. Field-aligned currents, the auroral currents, couple the magnetosphere with the ionosphere, which in turn affects plasma transport there. In this chapter we discuss the static (or nearly static) case, reserving the dynamics for Chapter 11. Happily, we have made some progress in the static case; the thermodynamics, although long neglected, are beginning to be understood. We are able to predict the structure, the transport, and the energization of solar wind plasma under different interplanetary magnetic field (IMF) conditions. Some problems do remain, both qualitative and quantitative.

The magnetosphere is constantly being stirred by momentum and energy transfer from the solar wind through the low-latitude boundary layer (LLBL). The primary process behind this inflow has some immediate effect on the aurora via initiation of flow and pressure buildup. The generation of two-cell (or sometimes four-cell) convection patterns is the result. The essential element is the formation of elongated narrow auroral arcs during the growth phase of substorms. However, we do have our work cut out for us as shown by the following quotation:

Practically all auroral theories treat only part of the total problem. For instance, most often the currents are considered as given and the attention is focused on the particle energization and the fields and waves which are instrumental in achieving this. There are models which address solely the parallel energy and mass exchange, others focus on the transverse electric fields and heating by shear flows, although parallel and transverse fields are intimately related. The microphysics of the auroral acceleration regions is often treated without reference to the source of energy. In fact, the latter one is one of the most neglected elements in the context of auroral arc formation. Even less work has been done on the impact of arc formation on the source plasma. In short, it is fair to say that we are still far from a self-consistent theory, even on the macroscopic scale.

Paschmann et al. (2003, p. 360)

The aurora is due to energetic particle impact on the upper atmosphere, but the reasons for the precipitation can differ. All auroras are not of the same kind. One may distinguish major classes: quiescent and dynamic convection-related aurora, growth-phase aurora, diffuse aurora, and very active forms of several kinds such as the westward traveling surge, omega bands, and pulsating aurora. None of these are well understood. We approach this question on first principles in order to try to engage the correct physics.

10.2   Transfer of plasma and electric field

The major entry sites for solar wind plasma into the plasma sheet are located somewhere in the distant tail and/or the flank magnetopause (Axford and Hines, 1961; Cole, 1961; Dungey, 1961). We can do better now since we know by Chapter 9 that the LLBL is intimately involved.

First, the electric field is transferred from the magnetosheath into the plasma sheet by the LLBL because of its electric dipole nature; this can be seen from the figure at the bottom part of Fig. 10.1, in the equatorial view. Second, the LLBL also supplies the plasma through the polarization drift required to maintain the electric field for tailward flow.

The LLBL, which extends from the magnetopause to closed field lines, separates the magnetosheath from direct contact with the plasma sheet. The result of Pulkkinen et al. (1996) on the variation of B_z out to 220 R_E is shown in Fig. 9.22 of Chapter 9. We reiterate this result to stress one outstanding aspect, that the average B_z is positive everywhere, that is, northward. This indicates that the field lines in the plasma sheet are closed out to 220 R_E. With the observed tailward flow, the electric field is in the dusk-dawn direction (Slavin et al., 1985). Since the cross-tail current is still from dawn to dusk, $\mathbf{E} \cdot \mathbf{J} < 0$. The vast region is a dynamo; this is logical as the plasma must do work against the resistance due to the geomagnetic field.

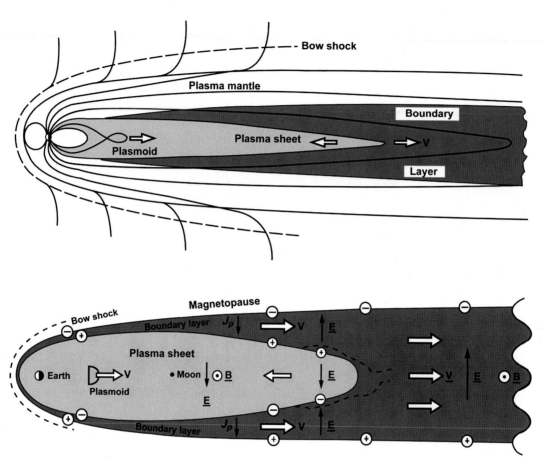

Fig. 10.1 The northward component of the magnetic field in the distant magnetotail shown by the ISEE-3 observations (in the absence of plasmoids) implies that the electric field is from dusk to dawn, reverse of its sense in the plasma sheet. These two cuts of the entire magnetosphere show that the plasma sheet is a small cavity of low-density plasma within a sea of boundary-layer and mantle plasma; the LLBL should be visualized as being wrapped around the plasma sheet, at least partly on closed field lines. From Heikkila, W.J., 1990. Magnetic reconnection, merging, and viscous interaction in the magnetosphere. Space Sci. Rev. 53, 4.

10.3 Plasma sheet from low-altitude observations

A method of inferring central plasma sheet temperature (T), density (n), and pressure (p) from ionospheric observations was developed by Wing and Newell (1998, 2002). Ionospheric polar orbiting satellites move rapidly, with a typical orbital period near 100 min, allowing them to image the plasma sheet twice per orbit, over two dozen times per day. Plasma in the plasma sheet has been observed to be nearly isotropic, and as a result T, n, and p should be conserved along the field line. For their study, they used data obtained from instruments onboard DMSP satellites F8, F9, F10, and F11 for the entire year 1992. There were at least three DMSP satellites in operation simultaneously; and for 1 month (March) all four were in operation.

Fig. 10.2 shows the two-dimensional equatorial profiles of plasma sheet ion (A and B) density and (C and D) temperature, separated by the sign of the B_Z component of the IMF. The plasma sheet ion density is higher during periods of northward IMF than southward IMF. During periods of northward IMF, density peaks appear along the plasma sheet dawn and dusk flanks. However, during periods of southward IMF, in the dawn flank the density peak is smaller and narrower (in the y direction). During periods of northward IMF, the ion temperature has minima along the dawn and dusk flanks. However, during periods of southward IMF, the temperature minima are smaller and narrower.

Using this method and DMSP observations for 1992, we present 2D profiles of the equatorial plasma sheet ion density and temperature for southward and northward IMF for the isotropic precipitation of ring current protons … for $K_p < \sim 3$. *During periods of northward IMF, cold dense ions can be found plentifully along the plasma sheet flanks. However, during periods of southward IMF, the presence of these cold dense ions is noticeably diminished, especially at the dusk flank where the density peak is less discernable. These cold dense ions have been previously interpreted in terms of magnetosheath ion entry into the plasma sheet. Our result suggests that any mechanism proposed to transport magnetosheath ions from dusk LLBL to the plasma sheet along the dusk flank during periods of northward IMF would have to be able to do so efficiently over a large spatial scale.*

Wing and Newell (2002)

Ion spectra occurring in conjunction with electron acceleration events were specifically excluded, making the results valid for quiescent conditions. Because of the variability of magnetotail stretching, the mapping to the plasma sheet was done using a modified Tsyganenko (1989) magnetic field model (T89) adjusted to agree with the actual magnetotail stretch at observation time. Because of the abundance of the ionospheric observations, inferred 2D spatial profiles of plasma sheet ion temperature, density, and pressure could be constructed at unprecedented fine spatial resolution. Instead of computing moments (which is widely used in general) each ion spectrum was fitted to distribution functions (one-component Maxwellian, two-

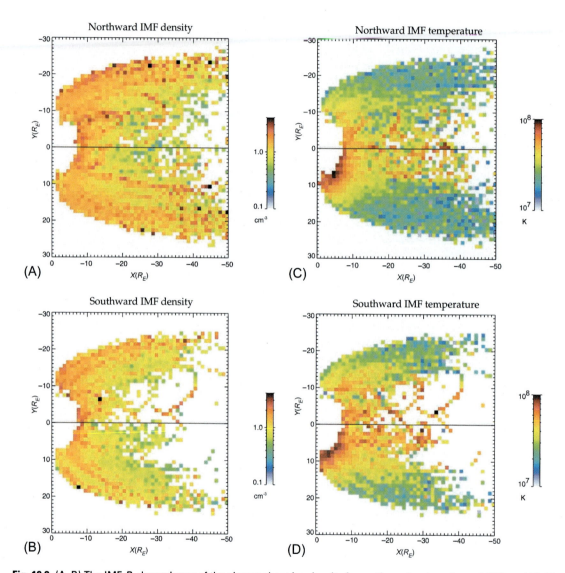

Fig. 10.2 (A, B) The IMF B_z dependence of the plasma sheet ion density for northward and southward IMF; and (C, D) their temperature profiles. From Wing, S., Newell, P.T., 2002. 2D plasma sheet ion density and temperature profiles for northward and southward IMF. Geophys. Res. Lett. 29(9), 3. https://doi.org/10.1029/2001GL013950.

component Maxwellian, and κ distribution). This takes into account ions that are outside the detectors' energy range.

In summary, for the first time, the IMF B_Z dependence of the plasma sheet ion density n and temperature T profiles was exhibited in fine spatial detail in 2D. These differ for the northward and southward IMF cases as shown in Fig. 10.2, but the most remarkable contrast occurs in the density profiles along the plasma sheet flanks. During periods of northward IMF, ions are colder and denser along both flanks of the plasma sheet, that is, adjacent to the magnetopause. However, during periods of southward IMF, this density peak is smaller along the dawn flank, and it is barely discernible along the dusk flank. Likewise, the temperature minima are smaller and narrower at the flanks, but the dawn-dusk asymmetry appears not to be as strong as in the density profile. The presence of the cold dense ions along the plasma sheet flanks has been interpreted as strong indication of magnetosheath ion entry through the magnetopause (e.g., Fujimoto et al., 1998). The plasma transfer event (PTE) process discussed in Chapters 6, 7, and 9 meets the requirements as described in Section 10.8.

10.4 Plasma sheet observations

Equatorial distributions of the plasma sheet ions have been studied by several groups including Hori et al. (2000), Wang et al. (2006, 2007), Kaufmann et al. (2005), and Kaufmann and Paterson (2006) using data from GEOTAIL, with similar results.

Fig. 10.3 shows the number density, temperature, and plasma pressure for the four IMF B_z conditions from Wang et al. (2006). They have investigated statistically the equatorial distributions of ions and magnetic fields from GEOTAIL when the IMF has been continuously northward or southward for shorter or longer than 1 h.

The perpendicular flow shows that ions divert around the Earth mainly through the dusk side in the inner plasma sheet because of westward diamagnetic drift. The magnetic fields indicate that field lines are more stretched during southward IMF. Ions' electric and magnetic drift paths evaluated from the observations show that for thermal energy ions, magnetic drift is as important as electric drift. Comparison of the distributions of the observed phase space density with the evaluated drift paths at different energies indicates that the electric and magnetic drift transport is responsible for the observed dawn-dusk asymmetries in the plasma sheet structure.

Wang et al. (2006)

The overall density can be seen to be higher (lower) during northward (southward) IMF and to increase (decrease) as the northward (southward) IMF proceeds. The IMF B_z is opposite to that of density. Across the tail at fixed X, relatively higher density and lower temperature near the flanks than near midnight are seen regardless of the IMF conditions, and the dawn flank appears

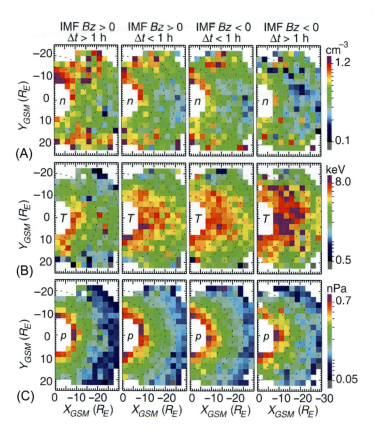

Fig. 10.3 Equatorial distributions of: (A) number density, (B) temperature, and (C) plasma pressure during the four IMF B_z conditions. From Wang, C.-P., Lyons, L.R., Weygand, J.M., Nagai, T., McEntire, R.W., 2006. Equatorial distributions of the plasma sheet ions, their electric and magnetic drifts, and magnetic fields under different interplanetary magnetic field Bz conditions. J. Geophys. Res. 111, 3. https://doi.org/10.1029/2005JA011545.

denser than the dusk flank. In general the earlier distributions and their differences between northward and southward IMF are consistent with the results of equatorial mapping of DMSP observations in the previous section.

The density and temperature distributions indicate that the pressure in the premidnight sector is mainly contributed by hotter and tenuous plasma while that in the postmidnight sector are mainly by colder but denser plasma. The pressure has relatively weaker dawn-dusk asymmetry than do the density and temperature and depends more strongly on r, with higher pressure seen at smaller r. The pressure in the region $|Y| < 2.5\ R_E$ and between $r = 8$ and $12\ R_E$ is seen to be distinctly higher than at other local times when southward IMF lasts longer than 1 h and is likely due to the contribution from energization associated with the substorm dipolarizations.

Fig. 10.4A shows the perpendicular flow velocities for the same four IMF B_z conditions. Note that only $|v_\perp| \le 200\ \text{km/s}$ has been included in this study. For northward IMF longer than 1 h, the

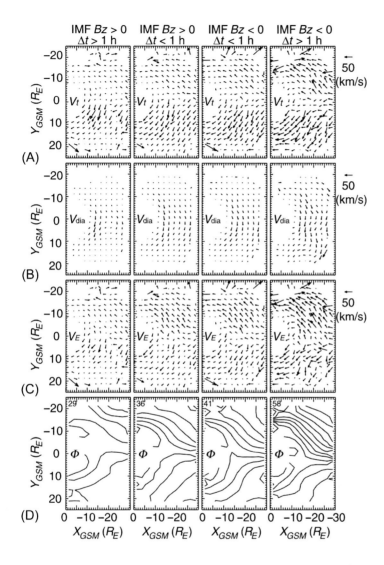

Fig. 10.4 Equatorial distributions of (A) perpendicular flow, (B) diamagnetic drift, (C) electric drift, and (D) electric potential contours (5 kV intervals). The *arrow* at the top right corner represents a unit vector of 50 km/s. The values of the cross-tail potentials are indicated at the top left corner of each electric potential plot. From Wang, C.-P., Lyons, L.R., Weygand, J.M., Nagai, T., McEntire, R.W., 2006. Equatorial distributions of the plasma sheet ions, their electric and magnetic drifts, and magnetic fields under different interplanetary magnetic field Bz conditions. J. Geophys. Res. 111, 7. https://doi.org/10.1029/2005JA011545.

overall flow speed is less than 50 km/s. Most of the flow in the $Y < 0$ region is mainly directed earthward and dawnward, while almost all the flow in the $Y > 0$ region is earthward and duskward. The flow is more earthward at larger $|Y|$. However, in the region $|Y| < \sim 10$ R_E there is a dawn-dusk asymmetry with larger flow speed at $Y > 0$ and smaller and less dawnward flow at $Y < 0$. Asymmetry becomes more significant with decreasing radial distance. The flow pattern indicates that the plasma from the tail near the midnight meridian diverts around the Earth to the dayside mainly through the dusk side.

As the IMF condition changes from being shorter northward to shorter southward and then to longer southward, it can be seen that the overall flow pattern described earlier does not change significantly. However, the flow speed everywhere increases gradually, with the overall speed for southward IMF $\Delta t > 1\,\mathrm{h}$ being about three times the speed for northward IMF $\Delta t > 1\,\mathrm{h}$. These are consistent with the flow changes with increasing AE observed by ISEE.

In the slow-flow approximation, which is valid within the plasma sheet except during dynamic periods such as the expansion phase of substorms, plasma drift results from electric drift $\mathbf{v}_E = (\mathbf{E} \times \mathbf{B})/B^2$ and diamagnetic drift $\mathbf{v}_{\mathrm{dia}}$, where $\mathbf{E}$ is electric field and $\mathbf{B}$ is magnetic field. $\mathbf{v}_{\mathrm{dia}} = (\mathbf{B} \times \nabla p)/(neB^2)$ for protons under the condition of isotropic pressure. The middle plots of Fig. 10.4 show $\mathbf{v}_{\mathrm{dia}}$ in the x-y directions at the center of the current sheet ($\mathbf{B} = B_z\hat{z}$) using the n and p distributions of Fig. 10.3 assuming all ions are protons. $\mathbf{v}_{\mathrm{dia}}$ is mainly directed westward because ∇p is mainly in the radial direction. The flow pattern for $\mathbf{v}_{\mathrm{dia}}$ is similar for different IMF B_z conditions but its overall speed during southward IMF is stronger than that during northward IMF. It is evident that the dawn-dusk asymmetry seen at $|Y| < 10 \ R_E$ in the total drift is mainly due to diamagnetic drift.

The electric drift shows that flow in the $Y < 0$ region is mainly directed earthward and dawnward, while the flow in the $Y > 0$ region is earthward and duskward. The electric drift pattern appears more dawn-dusk symmetric than does the total drift. The electric drift speed becomes larger, indicating stronger convection, as the IMF conditions change from longer to shorter northward, to shorter southward, then to longer southward. The potential difference between $Y = \pm 19 \ R_E$ at $X = -26 \ R_E$ is shown in the top left corner of each plot. The potential difference across the tail is stronger during southward than during northward IMF, as expected from the dependence of the cross-polar cap potential drop on the direction of the IMF B_z, a clear indication of the control of magnetospheric convection by the IMF B_z.

For the high-energy particles ($\mu = 52$ and $190\,\mathrm{keV}\,(10\,\mathrm{nT})^{-1}$), the drift is dominated by the westward magnetic drift. This makes the postmidnight sector accessible mostly by particles from the dawn flank and the premidnight sector accessible mostly by particles from the tail. The paths are confirmed by the general agreement with constant phase space density contours. That the number of the high-energy particles from the dawn flank is much smaller than that from the tail results in the dawn-dusk flux asymmetry for the high-energy ions.

The earlier phase density distributions as a function of ion energy invariant are in qualitative agreement with the results of Garner et al. (2003) that is based on earlier statistical studies of plasma sheet parameters. Fig. 10.5 compares the trajectories of ions of different energies coming from the same location at $X=-28.5\ R_E$ and $Y=0\ R_E$ following the drift paths $\mu B+e\Phi$, which clearly shows how large the effect of magnetic drift can become, as a particles' energy gets higher, in diverting particles away from their electric drift paths. It is important to point out that for typical thermal energy ions ($\mu=6.5\ \text{keV}\ (10\ \text{nT})^{-1}$) in the plasma sheet, their magnetic drift has become as strong as electric drift in determining their transport and energization.

In summary, the magnetic fields indicate that the field lines are more stretched during southward IMF than northward IMF, and the field lines become further (less) stretched as southward (northward) IMF proceeds. The perpendicular flow at $|Y|<\sim 10\ R_E$ is seen to be stronger in the premidnight sector, and this dawn-dusk asymmetry becomes stronger with decreasing radial distance, indicating the ions from the tail divert around the Earth mainly through the dusk side. The asymmetry results from westward diamagnetic drift.

10.4.1 Direct support for the low-latitude boundary layer

Fig. 10.3 is striking in its appearance; it divides the phase space density and particle energy into the two groups very clearly. For northward IMF we can understand how magnetosheath plasma in the LLBL acts as the driver causing plasma sheet plasma to respond.

The transport of low-energy ions is dominated by electric drift, while magnetic drift dominates the transport of high-energy ions. The phase space density for <0.5 keV ions along the flanks is significantly larger than what can be transported from the tail during northward IMF, indicating contribution from a cold particle source along the flanks at $X>-30\ R_E$ that could be provided by direct entry of the magnetosheath particles into the plasma sheet through the flanks.

10.4.2 The Rice Convection Model

Since the time scales of magnetic field changes in the magnetosphere are much longer than the typical periods of particle's gyro and bounce motion, under many circumstances it is

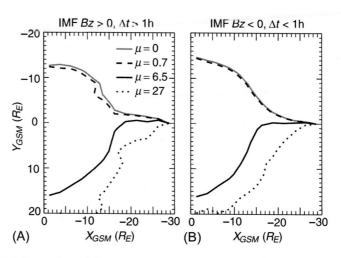

Fig. 10.5 Comparison of the trajectories of ions of different energies (μ value in keV $(10\,nT)^{-1}$) coming from the same location at $X = -28.5\ R_E$ and $Y = 0\ R_E$ following the drift paths ($\mu B + e\Phi$) during (A) northward IMF longer than 1 h and (B) southward IMF shorter than 1 h. From Wang, C.-P., Lyons, L.R., Weygand, J.M., Nagai, T., McEntire, R.W., 2006. Equatorial distributions of the plasma sheet ions, their electric and magnetic drifts, and magnetic fields under different interplanetary magnetic field Bz conditions. J. Geophys. Res. 111, 9. https://doi.org/10.1029/2005JA011545.

appropriate to base the transport equations on some form of the adiabatic guiding center drift theory.

Different assumptions are used in the procedure to simplify and complete the chain of equations. However, it should be pointed out that there are really only two different formalisms that combine plasma transport (based on adiabatic drift theory) and the electrodynamic coupling to the ionosphere, that lead to a set of closed equations and yield realistic solutions. A special form of the kinetic approach to the systematic calculation of magnetospheric convection is explained in Heinemann and Wolf (2001).

This formalism underlies the well-known Rice Convection Model (RCM) numerical code developed for decades at Rice University. This approach combines an equation of current conservation and includes field-aligned currents resulting from pressure gradients of hot magnetospheric plasma with transport equations for the distribution function assuming isotropic pitch angle distribution of particles. Due to energy dependence of drift velocities, a suitable discretization of the distribution function is used, resulting in a large number of transport equations.

The two-fluid description of Peymirat and Fontaine (1994), referred to here as "FCM1," was obtained from nearly the same physical assumptions and spatial region as the RCM. While the electrodynamics equation for the electric field is the same one, the equations replacing the drift RCM equations are just four partial differential equations for mass density and energy for each of the two species. Instead of the actual drift velocity, the formalism requires use of the diamagnetic velocity (Peymirat and Fontaine, 1994). Heinemann (1999) pointed out that an additional term accounting for the heat flux should be added, which is perpendicular to both magnetic field and the temperature gradient, since the heat flux is comparable to the retained terms in the energy equation, with the addition of the heat flux term. Under the assumption of the Maxwellian distribution function, the model of Peymirat and Fontaine (1994) is the same as that of Heinemann (1999), and is referred to here as the "FCM2" or "fluid" model. Fig. 10.6 illustrates the various assumptions.

The relationship between the two formalisms was analyzed by Heinemann and Wolf (2001). Fig. 10.6 shows that if the RCM model includes an additional assumption that the distribution function locally comes to thermal equilibrium faster than the characteristic drift times, the two sets of equations become equivalent. Without that formal assumption, a one-dimensional solution indicated dramatically different evolution of a density and pressure perturbation.

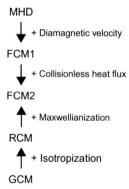

Fig. 10.6 Relationships of the different theories. FCM1 is related to MHD by the addition of the diamagnetic velocity. FCM2 is related to FCM1 by the addition of Chapman-Enskog collisionless heat flux. RCM is related to the General Circulation Model (GCM) by the additional assumption that the particle distribution is isotropized during convection. FCM2 is related to RCM by the additional assumption that particles are Maxwellianized during convection. From Heinemann, M., Wolf, R.A., 2001. Relationships of models of the inner magnetosphere to the Rice Convection Model. J. Geophys. Res. 106, 15,548.

10.4.3 Plasma irreversibly heated

Eight years of GEOTAIL particle and magnetic field measurements were separated by Kaufmann et al. (2005) into 12 data sets on the basis of the ion flow speed. The same measurements were separated into 12 other data sets using a magnetic flux transport sorting parameter. Magnetic field lines in the three-dimensional models created using these two sorting methods were dipolar when the flow or transport was fast and stretched into a tail-like configuration when the flow or transport was slow. The magnitude of B_x measured in the outer central plasma sheet decreased weakly, and B_z at the neutral sheet increased strongly as the magnetic flux transport rate increased. These observations showed that fast flow flux tubes were typically located near but earthward of the primary region in which localized Region 1 sense currents were diverted to the ionosphere. The plasma density was low and the temperature was high when the flow was fast. The particle pressure depended only weakly on flow speed. The average entropy was higher at $z = 0$ during fast flow events than it was anywhere in the region that could be studied when flows were slow or moderate.

The average entropy also decreased as $|z|$ increased, as shown in Fig. 10.7. These observations suggest that the plasma was irreversibly heated by the process that produced the fast flows.

Ions and electrons were found, on average, to be remarkably isotropic at the neutral sheet. Scattering through 90° each minute during slow and moderate flow conditions and as rapidly as every 10s during the fastest flows was needed to maintain this average degree of isotropy.

Kaufmann et al. (2005)

Scattering need not be so rapid that it gives instantaneous results; when it occurs it is likely to be at very low altitudes, in the ionosphere, once per bounce. The temperature anisotropy increased away from the neutral sheet, reaching 1.1–1.3 at some point along most field lines. This variation along field lines was attributed primarily to a parallel electric field needed to maintain charge neutrality. The average ion to electron temperature ratio was as low as 5 and as high as 10 in certain regions. These observations showed that electrons and ions were heated or cooled at different rates depending on their locations and bulk flow speeds.

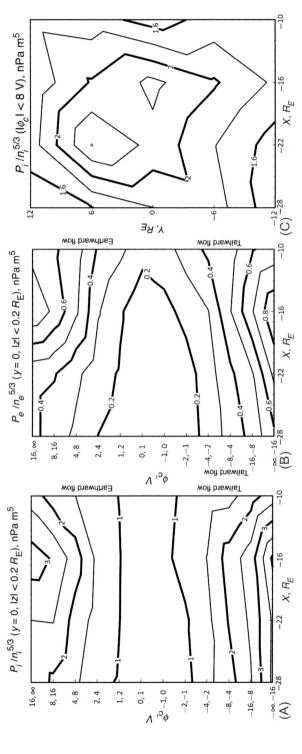

Fig. 10.7 Entropy parameter for (A) ions, (B) electrons, and (C) maximum of the entropy parameter for ions in all boxes. From Kaufmann, R.L., Paterson, W.R., Frank, L.A. (2005). Relationships between the ion flow speed, magnetic flux transport rate, and other plasma sheet parameters. J. Geophys. Res. 110, 8. https://doi.org/10.1029/2005JA011068.

10.4.4 Other fluid models

In 2010 Liu published a model that is based on the RCM-like adiabatic drift theory but is a single-fluid model (with ions carrying all of the particle pressure and electrons assumed to be cold) with a heat-flux term. With a clever choice of a Cartesian coordinate system, and by employing the concept of time-reversal mapping, it presents two partial differential equations describing the evolution of density and energy that supposedly includes the RCM drift physics formalism in an average way. Thus the model would appear to have the best of the RCM and fluid descriptions published before.

One reason for analysis is that for practical (computational) purposes, a one- or two-fluid model is simpler and faster to evolve numerically, although the full details of the distribution function must be sacrificed. A kinetic model is constructed from the guiding center drift theory and would appear to be closer to the real world, but solving equations numerically is more difficult and computationally slower.

A less obvious, but no less important reason, is related to a recent line of work building "coupled" numerical models of the magnetosphere where some form of the inner magnetospheric representation is installed, in a modular way, into a global MHD code. In this approach, the drift physics of the inner magnetospheric "hot" particle population is tracked in a kinetic model, and the results are used in a global MHD code by computing the appropriate plasma moments. The MHD code provides a time-dependent magnetic field model for the inner magnetospheric model and continuously updated boundary conditions for the electric potential and plasma. The inner magnetospheric model computes the distribution functions of particles in the inner magnetosphere and passes the calculated inner magnetospheric pressures to the MHD model.

10.5 Particle dynamics

The motion of charged particles in a given electromagnetic field is well described in many textbooks. Plasmas consist of very large numbers of particles; it would be very complicated indeed to follow the path of each of them. Actually, we are not even interested in the fate of an individual particle, but we prefer to look into the properties of the plasma as a macroscopic entity.

Fluid plasma physics (such as MHD) has had impressive results. The plasma moments such as density, temperature, pressure, and perpendicular flow are easily computed, as well as flux versus particle energy and the magnetic field. We infer the plasma sheet electric field and electric potential from the observed plasma and magnetic field data. The electromagnetic field can no longer be considered as "given" but has to be found simultaneously with the equation of motion of particles, in a self-consistent way.

A suitable way to treat the plasma as a collection of many charged particles moving in the self-consistent electromagnetic field is presented by the Boltzmann equation. Let us represent each particle by a point in the six-dimensional configuration space-velocity space-hyperspace, with the "coordinates" (x, y, z, v_x, v_y, v_z) or, briefly, $(\mathbf{r}, \mathbf{v})$. If a great number of particles are present, it suffices to know the density of points in this $(\mathbf{r}, \mathbf{v})$ space, $f(\mathbf{r}, \mathbf{v}, t)$ (note the addition of time). Consequently,

$$\frac{\partial f}{\partial t} + \nabla_{\mathbf{r}, \mathbf{v}} \cdot [f(\dot{\mathbf{r}}, \dot{\mathbf{v}})] = 0 \tag{10.1}$$

One arrives at the collisionless Boltzmann equation:

$$\frac{\partial f}{\partial t} + \frac{\partial f}{\partial x_i} \dot{x}_i + \frac{\partial f}{\partial v_i} \dot{v}_i = 0 \tag{10.2}$$

or

$$\frac{\partial f}{\partial t} + \frac{\partial f}{\partial x_i} v_i + \frac{F_i}{m} \frac{\partial f}{\partial v_i} = 0 \tag{10.3}$$

The divergence of the six-dimensional velocity vector vanishes. Since in hydrodynamics fluids with this property are described as incompressible, we conclude that whatever the behavior of the particles in real space, their image points in the $(\mathbf{r}, \mathbf{v})$ hyperspace behave as particles of an incompressible fluid. One can ascribe the following physical meaning to this equation: following a point on its motion in $(\mathbf{r}, \mathbf{v})$ space, one finds that the point density around it remains unchanged. This is again characteristic of an incompressible fluid. (Note, however, that the density need not be uniform.)

> In deriving the Boltzmann equation we have used the equation of motion... but no other physical principle. Vice versa, the equation of motion can be derived from the Boltzmann equation. Hence the two principles are equivalent; neither one contains more information than the other. The latter is, however, better suited to handle collective phenomena, especially since simpler equations can be easily deduced from it by "destroying unnecessary information."
>
> **Schmidt (1979, p. 57)**

There has been debate about the use of the **B**, **V** versus the **E**, **J** approach (Parker, 1996; Heikkila, 1997). We note that only the electric field can change the energy of plasma particles so "the particle approach is then necessary for at least some aspects of the problem" (Hines, 1963). By "destroying unnecessary information" fluid theory is forced to use some other means of particle energization, for example, define an equation of state. By definition, that is localized. That is a big problem, especially with a tensor conductivity that is necessary in space plasmas.

A collisionless plasma consists of charged particles moving freely through space, influenced only by electromagnetic fields if they happen to be present. In general, particles close in space with completely different velocity vectors going off in different directions to remote parts of the plasma. The motion of the local center of mass is a fiction that does not correspond to any real mass motion.

> *It is rather fortunate, therefore, to have found that these fictitious quantities obey some quasi-hydrodynamic equations: an equation of continuity and an equation of motion. This is about as far as the analogy can be stretched. Unfortunately these equations do not suffice to provide solutions for the great number of unknown quantities.*
>
> **Schmidt (1979, p. 72)**

10.5.1 Conservation of entropy

When a system evolves in an adiabatic manner without dissipation the entropy S is conserved. On the other hand, dissipative processes increase the entropy, relaxing the system toward a more stable state. Macroscopic fluid approaches, such as ideal MHD, usually assume an adiabatic pressure law, one that is reversible and that conserves entropy.

In the ideal MHD approach the frozen-in condition implies the conservation of the global entropy per unit flux of convecting plasma. Observations and simulations in space plasmas indicate that entropy is not conserved during steady flows in the plasma sheet, and during transitions such as substorms. Nonconservation of entropy in space plasmas can result from magnetic reconfiguration as well as nonadiabatic processes such as turbulent transport. Examination of entropy properties may provide insight into processes leading to plasma entry across the magnetopause and processes leading to the tail pressure balance crisis in the plasma sheet. It may provide constraints on which processes are most important in the dynamical transition related to substorms.

MHD models have invariably assumed that S is constant in space and time. Goertz and Baumjohann (1991) have plotted S against various quantities such as density, temperature, pressure, plasma beta, magnetic pressure, Alfvén speed, and distance from the neutral sheet. They have not found any systematic dependence on position, magnetic pressure, plasma beta, Alfvén speed, or distance from the neutral sheet. They did, however, find a remarkable correlation between S and T, which is shown in Fig. 10.8. The best fit is provided by $S \sim P/N^{5/3}$ with a correlation coefficient $r > 0.8$. The statistical results shown can be summarized as follows:

- As plasma is transported from the plasma sheet boundary layer into the central plasma sheet, it is compressed (N increases) and heated. The pressure and density are correlated. This transport is thus nearly adiabatic.
- In the central plasma sheet, the temperature and density are anticorrelated for quiet times (AE $<$ 100 nT) but weakly correlated during active times (AE $>$ 300 nT) (not shown here).

Ideal MHD predicts only that the specific entropy should be constant along the $\mathbf{E} \times \mathbf{B}$ drift path of a plasma element. Unfortunately, the equations do not suffice to provide solutions for the great number of unknown quantities; see later regarding *these fictitious quantities*.

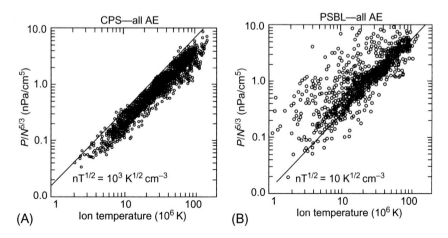

Fig. 10.8 Correlation between specific entropy S and temperature T in the (A) central plasma sheet (CPS) and the (B) plasma sheet boundary layer (PSBL). Only every 60th sample point is plotted. From Goertz, C., Baumjohann, W., 1991. On the thermodynamics of the plasma sheet. J. Geophys. Res. 96(A12), 20,995.

10.5.2 Effect of collisions

Thus far, we have not discussed the effect of collisions on particle motions. Boltzmann's equation without collisions deals with reversible physics. When we add collisions we enter the world of irreversible phenomenon. We know how to handle that with fluid physics: simply add them, as with billiard balls. With the Boltzmann equation it is not so easy:

$$\frac{\partial f}{\partial t} + \frac{\partial f}{\partial x_i}v_i + \frac{F_i}{m}\frac{\partial f}{\partial v_i} = \left(\frac{\partial f}{\partial t}\right)_{\text{coll}} \tag{10.4}$$

The right-hand side is merely an indication that collisions are important; the description of the mechanics remains to be addressed. Fig. 10.9 indicates one result. On the left we see the behavior without collisions; the volume might change but the physics is reversible, and entropy is constant. On the right, the colliding particle has lost its memory: the entropy can change. Even with a collisionless plasma the collisions can take place in the ionosphere.

10.5.3 Raising and lowering mirror points

A key concept of plasma theory is that the first adiabatic invariant, the magnetic moment, is conserved;

$$\mu_m = \frac{1}{2}mv^2/B \tag{10.5}$$

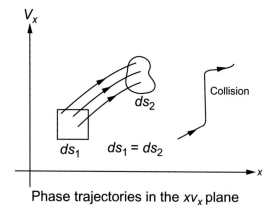

Phase trajectories in the xv_x plane

Fig. 10.9 Phase trajectories in the xv_x plane. On the left, the volume changes but the physics is reversible and entropy is constant. On the right, the colliding particle has lost its memory and the entropy can change. From Schmidt, G., 1979. Physics of High Temperature Plasmas, second ed. Academic Press, New York, NY, p. 57.

This "principle" is true for particle orbit theory using the guiding center as well as for fluid theory such as ideal MHD. Given this principle we look at the mirror points of particles, related to the longitudinal invariant

$$\oint [B_{MP} - B]^{1/2} = \text{invariant} \tag{10.6}$$

At the mirror point of a particle the pitch angle is defined to be 90 degrees; all the energy is in the transverse component. The action of the magnetic field on the particle is governed by force

$$\mathbf{F} = \mu \nabla B \tag{10.7}$$

The motion of the particle turns around at the mirror point leading to bouncing motion between hemispheres. If the particle gains or loses energy along the way, and μ_m is constant, then

$$\mu_m = \frac{1/2\left(mv_1^2\right)}{B_{MP1}} = \frac{1/2\left(mv_2^2\right)}{B_{MP2}} \tag{10.8}$$

The mirror point will come down for an increase in energy; conversely, the mirror point will be raised when the particle loses energy. To put this in another way, precipitation will occur with an electrical load with $\mathbf{E} \cdot \mathbf{J} > 0$, not with a dynamo given by $\mathbf{E} \cdot \mathbf{J} < 0$. The sign of $\mathbf{E} \cdot \mathbf{J}$ is critical.

10.5.4 Curvature drift

The good progress with fluid analysis does sound quite promising, but one thing seems to have been forgotten: only the electric field can energize charged particles. As Colin Hines said a long time ago,

> ...in general there seems to be little advantage in the further pursuit of the hydromagnetic-thermodynamic approach when a precise description is required. The particle approach is then necessary for at least some aspects of the problem, and it might as well be adopted for all.
>
> **Hines (1963)**

Let us start at the beginning without going directly into fluid plasma physics. In a magnetized plasma the particles gain or lose energy ONLY by gradient and curvature drift in the presence of an electric field $\mathbf{E}$. The steady-state electric drift $\mathbf{E} \times \mathbf{B}$ is transverse to $\mathbf{E}$, so it does not energize the plasma particles (see Hines, 1963, p. 239). Gradient drift in the far tail is not effective because the gradients are not large in the tail-like structure. Aurora particles that penetrate to the ionosphere have extremely small pitch angles in the plasma sheet, so curvature drift must be the way to energize them.

With a thin current sheet (TCS), both electrons and ions are energized dependent on the electric field at their location; their mass comes in only through the first invariant μ. Both electrons and ions should react in the same manner within the plasma sheet depending only on the magnetic moment; the energy they gain or lose by curvature drift is proportional to their magnetic moment. Since the ions have more energy in the first place, they acquire more energy each time they cross the neutral sheet where $\mathbf{E} \cdot \mathbf{J}$ is positive.

Let us take a closer look at curvature drift. Consider a coordinate system that follows the particle motion along the geomagnetic field lines.

$$-m\frac{d\mathbf{v}_D}{dt} = -\frac{mv_\parallel^2}{R_B^2}\mathbf{R}_B \tag{10.9}$$

The curvature of the magnetic field line can be defined as

$$\frac{\mathbf{R}_B}{R_B^2} = \left(\frac{\mathbf{B}}{B} \cdot \nabla\right)\frac{\mathbf{B}}{B} = (\mathbf{b} \cdot \nabla)\mathbf{b} \tag{10.10}$$

where $\mathbf{b} = \mathbf{B}/B$ is the unit vector of the magnetic field. Using some vector identities this expression can be rewritten as

$$\left(\frac{\mathbf{B}}{B} \cdot \nabla\right)\frac{\mathbf{B}}{B} = -\frac{\mathbf{B}}{B} \times \left(\nabla \times \frac{\mathbf{B}}{B}\right) \tag{10.11}$$

$$= -\frac{\mathbf{B}}{B} \times \left[\frac{1}{B}\nabla \times \mathbf{B} - \mathbf{B} \times \nabla\left(\frac{1}{B}\right)\right] \tag{10.12}$$

$$= -\frac{1}{B^2}\mathbf{B} \times (\nabla \times \mathbf{B}) + \frac{1}{B}(\nabla B)_\perp \tag{10.13}$$

The centrifugal drift can now be written as

$$\mathbf{v}_C = -\frac{\mathbf{B}}{qB^2} \times \left(-m\frac{d\mathbf{v}_D}{dt}\right) \tag{10.14}$$

$$= \frac{\mathbf{B}}{qB^2} \times mv_\parallel^2 \frac{\mathbf{R}_B}{R_B^2} \tag{10.15}$$

$$= -\frac{\mathbf{B}}{qB^2} \times mv_\parallel^2\left(-\frac{1}{B}\nabla B_\perp + \frac{\mathbf{B}}{B^2} \times (\nabla \times \mathbf{B})\right) \tag{10.16}$$

(Fälthammar, 1973, p. 132). Notice that the presence of a current $\nabla \times \mathbf{B}$ has a distinct contribution. When the current can be neglected

$$\mathbf{v} = \mathbf{v}_E + \mathbf{v}_G + \mathbf{v}_C = -\frac{\mathbf{B}}{qB^2} \times \left[q\mathbf{E} - \mu \left(1 + \frac{2v_\parallel^2}{v_\perp^2} \right) \nabla B \right] \quad (10.17)$$

(or by Chen, 1984, p. 30).

$$\mathbf{v} = \mathbf{v}_G + \mathbf{v}_C = \frac{m}{q} \frac{\mathbf{R}_B \times \mathbf{B}}{R_B^2 B^2} \left(\frac{1}{2} v_\perp^2 + v_\parallel^2 \right) \quad (10.18)$$

10.5.5 Energy of auroral particles

Fig. 10.10 illustrates the relationships between the tangential magnetic unit vector, the curvature vector, and the vector normal to the field line. Inspection of this figure suggests that a particle spiraling about a field line will interact with the distinct curvature similar to Fig. 1.6 in Chapter 1. The centrifugal force has the effect

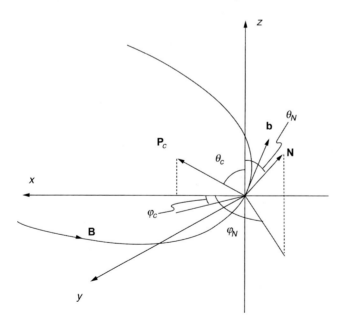

Fig. 10.10 Illustration of the relationship between the unit tangential vector **b**, curvature vector $\mathbf{P}_c$, and the normal **N** of the osculating plane of one magnetic field line in GSM coordinates. The polar angles θ_C and θ_N are from the z-axis, and the azimuthal angles φ_C and φ_N are from the x-axis. From Shen, C., Li, X., Dunlop, M., Liu, Z.X., Balogh, A., Baker, D.N., et al., 2003. Analyses on the geometrical structure of magnetic field in the current sheet based on cluster measurements. J. Geophys. Res. 108(A5), 3.

of taking it past the field line, overshooting slightly to the right, and contributing to a tailward velocity. The mathematics earlier show that it gains energy by this deflection, and if its first adiabatic invariant is maintained its mirror point will either be raised or lowered in the process.

However, the electric field has a component along the magnetic field just above the auroral arc (Evans, 1975; Reiff et al., 1985). This might be in addition to energization by curvature drift in the plasma sheet.

The final energy of the precipitated particle would be

$$W^F = gW^I + q\Delta\phi \qquad (10.19)$$

where the first term represents the gain in energy produced by plasma sheet processes (W^I represents the initial particle energy and g is a factor to be defined by the plasma sheet processes, mainly curvature drift), while the second term shows the gain in energy by a parallel electric field above the auroral form.

Pellinen and Heikkila (1984, p. 29)

Thus curvature drift yields field-aligned particles, both electrons and ions, with only a cross-tail transverse electric field. In addition, there could be acceleration due to charge separation along the magnetic field.

10.5.6 Thin current sheets

The geometrical configuration of the geomagnetic field lines plays a crucial role in magnetospheric dynamics, especially in the current sheet of the magnetotail. TCS formation is a slow process with gradual thinning during the growth phase, typically over 30–60 min. The particle velocity of curvature drift, as well as the resultant current, is reciprocal to the curvature radius of the geomagnetic field lines. They play a major role in controlling the magnetospheric dynamics.

... thin current sheets are a major part of substorm growth phases as well as of sawtooth events and steady convection intervals. Observations, empirical models, and MHD simulations suggest that thin current sheets have thickness of the order of ion gyroradius, cross-tail width about 15–25 R_E and tail dimension of about 20 R_E. The current sheet inner edge is typically at or slightly tailward of geostationary orbit; during storms it can extend around the Earth in the duskward direction. As global simulations suggest that the magnetotail flow is diverted around the thin current sheet to the

flanks in the inner part of the tail, the temporal scale associated
with the current sheet intensification and thinning may play a role
in determining the type of activity developing in the magnetotail.
Pulkkinen et al. (2006)

The geometrical configuration of geomagnetic field lines also controls the nonadiabatic and chaotic motions of particles having gyroradius comparable to or larger than the characteristic scale of the magnetic field. In a TCS, these kinds of particles may perform diverse motions, for example, the serpentine motion (Speiser, 1965). Furthermore, the spatial configuration of geomagnetic field lines can also significantly affect the properties of waves and instabilities in plasmas (Kivelson, 1995a,b; Engebretson et al., 2006).

It is important to know the actual geometrical configuration of the geomagnetic field lines in the tail current sheet and also its dynamical evolution. Shen et al. (2003) have reviewed the thickness and other features can be determined based on single- or two-satellite measurements. The half thickness of the current sheet near the end of the growth phase is about 0.1 R_E. Zhou et al. (1997) have investigated in detail the average structure of the current sheet and revealed that the current sheet could be twisted significantly due to the existence of IMF B_y.

The Cluster mission has made it possible to directly deduce the three-dimensional structure of the magnetic field in the magnetosphere from the four-point simultaneous magnetic measurements. Some methods have been developed to determine the spatial configuration and velocity of a discontinuity plane, wave vector, the gradient of magnetic field, and current density (e.g., see Dunlop et al., 2005). However, presently there is still no special stress on drawing the geometrical structure of the geomagnetic field lines in the magnetosphere based on the Cluster four-point magnetic observational data.

10.5.7 The curvature vector

In the investigation of Shen et al. (2003) the authors develop an approach to calculate the curvature vector and normal of the osculating plane of magnetic field lines based on Cluster magnetic measurements.

The curvature parameter κ, which is the squared ratio between the minimum curvature $R_{C\,\min}$ of the magnetic field and the maximum Larmor radius $\rho_{\max}$

$$\kappa = (R_{C\,\min}/\rho_{\max})^{1/2} \qquad (10.20)$$

has been put forward to determine the motion types of charged particles in the current (Büchner and Zelenyi, 1987, 1989). For $\kappa \gg 1$ the particles perform adiabatic motions with their magnetic moment conserved. However, when $\kappa \lesssim 1$ the motion of the particles will become nonadiabatic and some chaotic processes may appear. Furthermore, the spatial configuration of geomagnetic field lines can also significantly affect the properties of waves and instabilities in plasmas. It is believed that a TCS with $\kappa \lesssim 1$ favors the occurrence of the tearing mode, current instability, and even current disruption and particle energization. The geometrical structure of the magnetic field is a critical characteristic in the magnetospheric dynamics. The Cluster crossing event during September 17, 2001 is shown in Fig. 10.11 in GSM coordinates. There is a thin layer with low magnetic field strength, and the curvature radius is very small from 09:00 to 09:35 UT.

The results are (1) Inside of the tail neutral sheet (NS), the curvature of magnetic field lines points toward Earth ... and the characteristic half width (or the minimum curvature radius) of the NS is generally less than 2 R_E, for many cases less than 1600 km. (2) Outside of the NS ... the curvature radius is about 5 $R_E \sim$ 10 R_E. (3) Thin NS ... occurs more frequently near to midnight. (4) ... NS is thin during the growth and expansion phases, grows thick during the recovery phase.

Shen et al. (2003)

10.6 Auroral current circuit

Many auroral models assume a generator for the field-aligned current systems that must exist somewhere in the plasma sheet. Such a generator taps the kinetic energy of the magnetospheric plasma and converts it to the electromagnetic energy of the auroral current system. This electromagnetic energy is transmitted as a Poynting flux to the acceleration region and ionosphere. There it is either dissipated or returned to the outer magnetosphere as a reflected Alfvén wave.

One speaks of current or voltage generators. These two extremes may be visualized by an analogy in which the generator is a battery with an internal resistance. If the internal resistance is small compared to the load on the circuit, the load will receive a fixed voltage. In contrast, if the load is small compared to the internal resistance, a fixed current, given by the emf of the battery divided by the internal resistance, will be delivered to the load. An important distinction between current and voltage generators is

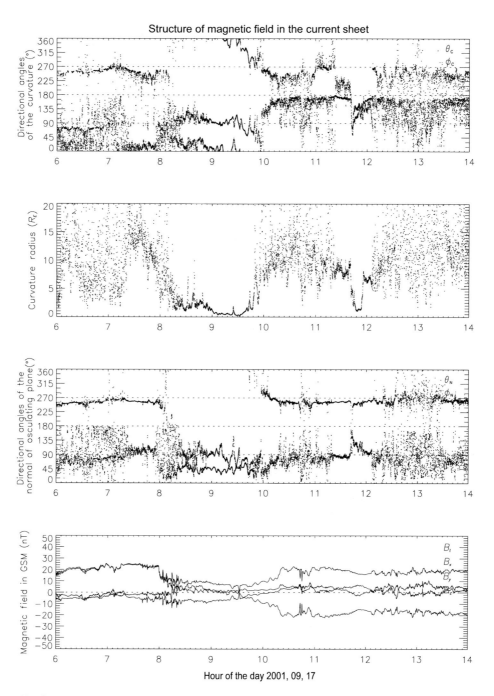

Fig. 10.11 The Cluster crossing event during September 17, 2001. The first panel shows the direction of the curvature of the geomagnetic field lines; the second panel shows the curvature radius; the third panel shows the direction of the normal of the osculating plane of the geomagnetic field lines; the last panel shows the three components and strength of the magnetic field. The GSM coordinates are used. From Shen, C., Li, X., Dunlop, M., Liu, Z.X., Balogh, A., Baker, D.N., et al., 2003. Analyses on the geometrical structure of magnetic field in the current sheet based on cluster measurements. J. Geophys. Res. 108(A5), 4.

that the ionospheric dissipation will be proportional to the Pedersen conductivity in the voltage generator case, while it is inversely proportional to the conductivity if the current is held fixed.

10.6.1 The primary circuit

The bright night-time aurora that is visible to the unaided eye is caused by electrons precipitated toward Earth. An upward-pointing electric field was the cause proposed by Evans (1974) and found by Reiff et al. (1988). The return current is probably due to an electric field accelerating electrons away from the auroral ionosphere as shown by Fig. 10.12 (Marklund et al., 2001), but there are some difficulties due to the huge amount of charge required with an intense arc.

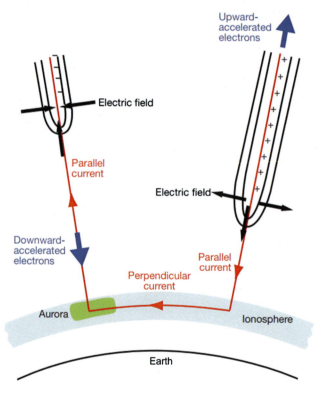

Fig. 10.12 Acceleration structures *(black)* within the auroral current circuit *(red)*. This diagram shows a negatively charged potential structure (indicated by equipotential contours) representative of the aurora (left) and a positively charged potential structure representative of the auroral return current (right). From Marklund, G., Ivchenko, N., et al., 2001. Temporal evolution of the electric field accelerating electrons away from the auroral ionosphere. Nature 414, 724.

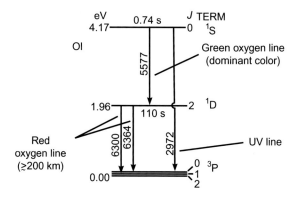

Fig. 10.13 Energy level of the oxygen atom. The terms as well as the radiative half-lives of the ^{1}S and ^{1}D levels are indicated, the half-life and the wavelengths of photons emitted in transitions between energy levels are shown. From Jones, A.V., 1974. Aurora. D. Reidel, Norwell, MA.

The brightest visible feature of the aurora, the "green line" at 557.7 nm, is due to the forbidden transition of an electron from the ^{1}S excited state to the ^{1}D state of atomic oxygen, as illustrated in Fig. 10.13. At high altitudes the collision rate is low enough to permit an emission. Another commonly observed line, particularly in the polar cusp and polar cap, is the "red line" at 630 nm as the ^{1}D state relaxes to the ground state; it is also forbidden as the half-life is almost 2 min.

It was long thought that the two branches of the equipotential contours close at typical altitudes of 5000–8000 km in the upward-current region above the aurora, and at 1500–3000 km in the auroral return-current region, respectively. Above that, they extend to very high altitudes along the geomagnetic field, forming the characteristic U-shape. The field-aligned currents are carried by the downward and upward accelerated electrons, respectively. Together with the ionospheric closure, current, and the magnetospheric generator, these form the complete auroral current circuit. The figure shows a north-south section through the structures that usually extend several hundreds of kilometers in the east-west direction.

10.6.2 Current thinning event

While there is nothing wrong in the above auroral circuit, it is not complete. What is missing is the connection to the dynamo. Everyone should know that the real electric field has induction and electrostatic components:

$$\mathbf{E} = -\partial \mathbf{A}/\partial t - \nabla \phi \qquad (10.21)$$

Since the growth-phase arcs are slowly changing toward the explosion point it is essential that they have a time dependence, making use of induction. Nearly all papers in space plasma use only one electric field, as in Fig. 10.12. Fig. 10.4 demonstrates that "the total drift is mainly due to diamagnetic drift," that is, dependent on the induction.

The plasma response to an induction electric field $\mathbf{E}^{\text{ind}} = -\partial \mathbf{A}/\partial t$ (the *cause* that can be robust) leads to the creation of an electrostatic field by charge separation $\mathbf{E}^{es} = -\nabla \phi$ (the *effect*). Of course, that response is affected by the (tensor) conductivities of the medium, as in a PTE at the magnetopause.

It is instructive to express the induction electric field in integral form:

$$\text{emf} = \varepsilon = \oint_{\text{circuit}} \mathbf{E} \cdot d\mathbf{l} = -d\Phi^{\text{mag}}/dt \qquad (10.22)$$

Φ^{mag} is the magnetic flux through the circuit used for the integration. It is only by this emf that we can tap stored magnetic energy. Comparison of Fig. 10.14 with a PTE shows that this emf results from the line integral along the perturbation current. The "convection" field has no emf, being in 2D.

The source of the aurora is in the plasma sheet. Let us consider a very simple meander in the equatorial plane, that of an

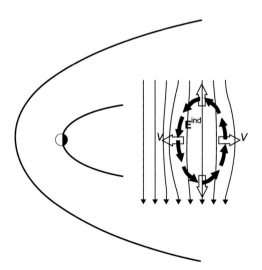

Fig. 10.14 The model of perturbation current responsible for a TCS to develop. The tailward advance of the current implies a dusk-dawn inductive electric field with tailward convection, a dynamo. On the earthward side the field is reversed for an electrical load; precipitation causes the growth-phase aurora. This is similar to a PTE at the magnetopause. From Heikkila and Brown.

elongated current in the y direction bounded at each end (dawn and dusk) with currents in the x direction (see Fig. 10.14). It is a perturbation current to the cross-tail current. A close look at Helmholtz's theorem shows that the perturbation current is closed with div $\delta \mathbf{J} = 0$ (see Chapter 3).

On the tailward side the sign is negative, $\mathbf{E} \cdot \mathbf{J} < 0$, to mark the advancing current; the result is a dynamo, and the sky is black. On the earthward side there is a decrease in current, and an opposing induction (in the direction of $-\delta \mathbf{J}$); this leads to an electrical load. The particles would precipitate as we saw in an earlier subsection, producing an auroral arc. The inductive electric field will drive the charges, possibly enhanced by the plasma. This is the same as with a PTE at the magnetopause.

To make all of this happen there must be general magnetic reconnection (GMR) both on the dawn and dusk sides of the perturbation. It is here that we have a parallel component of the inductive electric field, reversing on the two sides.

Let us now look closely at the very interesting auroral data in Fig. 10.15. This is a rare example of a satellite crossing above multiple parallel arcs and simultaneous optical coverage of these arcs from an aircraft. The arcs can be identified by peaks in energy spectra, fluxes of the downgoing electron component. The top panel shows multiple arcs seen in the all-sky image, taken by a low-light level TV camera from an aircraft; the orbit of the FAST spacecraft is mapped to 110 km altitude. The second panel shows the electron energy spectrum measured by FAST, and the third is the precipitated energy flux on a linear scale. When ion beams were present, the flux was mapped assuming a parallel potential between FAST and 110 km equal to the mean energy of the ion beam (Stenbaek-Nielsen et al., 1998).

It is noteworthy that FAST showed that plasma was present even in the gaps corresponding to the dark sky in the all-sky photograph. Only the spectrum was different, less energetic suggesting a dynamo with the plasma loosing energy. Each arc can potentially "break up" leading to a substorm, discussed in Chapter 11.

Finally, Fig. 10.16 shows the most probable location of the downtail mapping. The dark curve shows the normalized result for all 2588 mappings, and the light curve shows the normalized result for space-storm time mappings only. The most probable location of the mapped auroral brightening tends to be slightly closer to the Earth during storms.

Wanliss and Rostoker (2006) consider the list of substorm onsets from the IMAGE satellite and use Tsyganenko models (T96, T01) to map these ionospheric locations into the magnetotail. They investigate, in a statistical fashion, the source region of

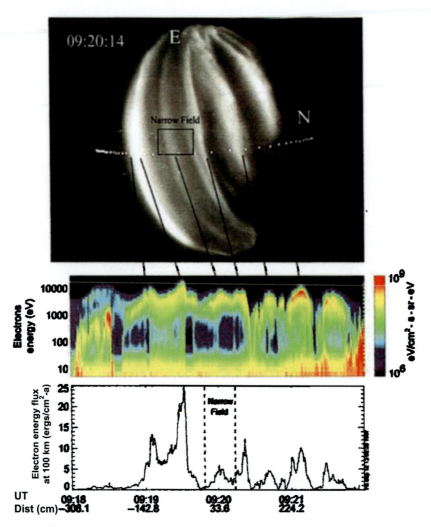

Fig. 10.15 The top panel shows multiple arcs seen in the all-sky image, taken by a low-light level TV-camera from an aircraft, and the FAST orbit mapped to 110 km altitude. The second panel shows the electron energy spectrum measured by FAST. The peaks in the bottom panel would correspond to auroral emissions. From Stenbaek-Nielsen, H., Hallinan, T.J., Osborne, T.D.L., Kimball, J., Chaston, C., McFadden, J., et al., 1998. Aircraft observations conjugate to FAST: auroral arc thicknesses. Geophys. Res. Lett. 25(12), 2074. https://doi.org/10.1029/98GL01058.

the auroral arc that brightens at the onset of expansive phase. This arc is usually identified as the ionospheric signature of the expansive phase onset that occurs in the magnetotail, which is a possible signature of the beginnings of a poleward border intensification. The auroral brightening associated with this higher latitude current system occurs near the poleward edge of the oval.

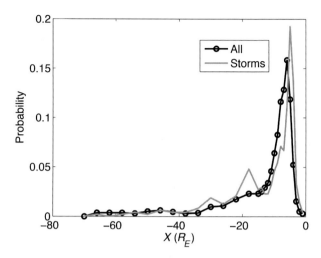

Fig. 10.16 Most probable mapped downtail location of the auroral brightenings. The *dark curve* shows the normalized result for all 2588 mappings, and the *light curve* shows the normalized result for space-storm time mappings only. From Wanliss, J.A., Rostoker, G., 2006. IMAGE analysis and modelling of substorm onsets. In: Syrjäsuo, M., Donovan, E. (Eds.), Proceedings of the Eighth International Conference on Substorms, University of Calgary, Alberta, CA, p. 333.

There are bright spots out to 70 R_E. This appears indicative of a poleward border intensification rather than an expansion phase onset.

> *Steady equatorward motion of the auroral oval during growth phase is associated with slow stretching of the inner magnetotail field. We assume that during the growth phase stretching of the tail and plasma sheet thinning take place without a major reconfiguration of magnetic field lines. This is not an unreasonable assumption, and several studies have shown how this is consistent with observations. … We exploited this loophole to map onset locations for several hundred substorms.*
>
> **Wanliss and Rostoker (2006)**

These results can be interpreted in two ways. First, the main onsets are initiated in the near-Earth magnetotail, typically at and within geostationary orbit. Second, Canadian Auroral Network for the OPEN Program Unified Study (CANOPUS) data demonstrates that the IMAGE onset list contains auroral brightenings that are not classical substorm onsets, but are actually poleward border intensifications. These occur farther out to 70 R_E. It is likely that the maxima and minima in Fig. 10.16 reach out past the moon's orbit.

10.6.3 Quasisteady state of growth

An early example of growth-phase aurora demonstrates important results on the development process, as shown in Fig. 10.17. On November 11, 1976 it began with a magnetically quiet

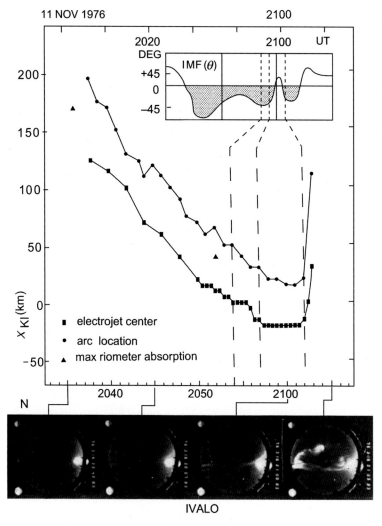

Fig. 10.17 The southward speed of both the arc and the westward electrojet was 170 km/s until they stopped at 2058 UT. Before the onset of the expansion phase the maximum riometer absorption was equatorward of the arc. From Pellinen, R.J., Baumjohann, W., Heikkila, W.J., Sergeev, V.A., Yahnin, A.G., Marklund, G., et al., 1982. Event study on pre-substorm phases and their relation to the energy coupling between the solar wind and magnetosphere. Planet. Space Sci. 30, 379.

period where the IMF was directed northward (Pellinen et al., 1982). A sudden southward turning of the IMF immediately led to a worldwide intensification of convection which was observed to start almost simultaneously at stations within the auroral zone and polar cap. The two-dimensional equivalent current system over the northern hemisphere had a typical two-cell convection pattern with a maximum disturbance (that is, change in the horizontal component of the geomagnetic field) ΔH of -300 nT observed on the morningside in the westward electrojet region. This enhancement of activity ended after 35 min in a localized substorm onset at 21:02 UT in the midnight sector over Scandinavia.

The recordings made in this area indicate large fluctuations of various ionospheric parameters starting several minutes before the substorm onset. Three subsequent stages can be resolved:

- High-energy particle precipitation recorded by balloon X-ray detectors.
- Maximum ionospheric electrojet current density increases, while the electrojet half-width shrinks.
- The auroral brightness increases, and the total electrojet current and its half-width show a growing trend prior to the final breakup.

All these developments move equatorward with the growth-phase aurora as shown in Fig. 10.17, taking tens of minutes. Several things happen as the auroral arc moves. During quiet times the equivalent current system, the solar daily variation, is nearly symmetric about the Earth-Sun line, with two cells centered near the dawn and dusk terminators. The current flows sunward across the polar cap as a broad sheet. In the auroral oval it is concentrated by the high conductivity in the auroral oval into the eastward electrojet (dusk) and westward electrojet (dawn).

Already during the IGY it was found that the two-cell pattern is also observed in the early stages of an auroral substorm. The instantaneous pattern was named DP-2 (disturbance polar of the second type) as distinct from the pattern from the single cell called DP-1. The axis of symmetry was found to differ from the solar daily pattern, tilted along a line from late morning to late evening rather than along the Earth-Sun line (Fig. 10.18, also reproduced as Fig. 13.17 in McPherron, 1995).

The main division is between upward, downward, and time-varying field-aligned current regions, as shown in Fig. 10.17. The current producing the ground disturbance is a Hall current, flowing at right angles to the ionospheric electric field, maximizing in the E-region. The existence of the DP-2 current system implies that ionospheric plasma is moving in a circulation system.

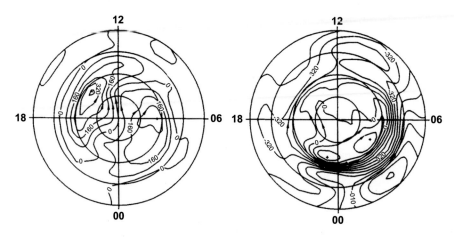

Fig. 10.18 Results from a determination of the patterns of ionospheric currents during magnetic disturbance. Closed contours show the flow lines for an equivalent ionospheric current that produces the observed ground magnetic perturbations. Left: The two-cell DP2 current system present in the substorm growth phase. Right: The single-cell DP-1 current system that dominates during the substorm expansion phase. From Clauer, C.R., Kamide, Y., 1985. DP 1 and DP 2 current systems for the March 22, 1979 substorms. J. Geophys. Res. 90(A2), 1350.

The direction of this flow is approximately from noon to midnight across the poles, and then back to the dayside through the auroral oval, both dawn and dusk.

10.6.4 Aurora study with EISCAT

It has been found that growth-phase arcs travel slowly, advancing equatorward. The study by Aikio et al. (2002) took advantage of this feature to obtain the values of all the essential electrodynamic parameters (electric fields, conductances, and horizontal and field-aligned currents) according to distance from the arc. The duration of the EISCAT measurement period varied between 16 and 36 min. During the optical observations the mean velocity of the arc in one event was 114 m/s with a standard deviation of 18 m/s. EISCAT radar was making a magnetic field-aligned measurement at Tromsø. The electric field vector was calculated from a measurement of the plasma drift velocity in an F-region volume, common for three radars located at Tromsø, Kiruna, and Sodankyla. The CP-1 measurement mode utilized for events allows the calculation of true electron densities in the E-region above 90 km. The electron densities together with the neutral atmospheres were used to calculate conductivity profiles between altitudes of 70 and 200 km. The height-integrated Pedersen and Hall

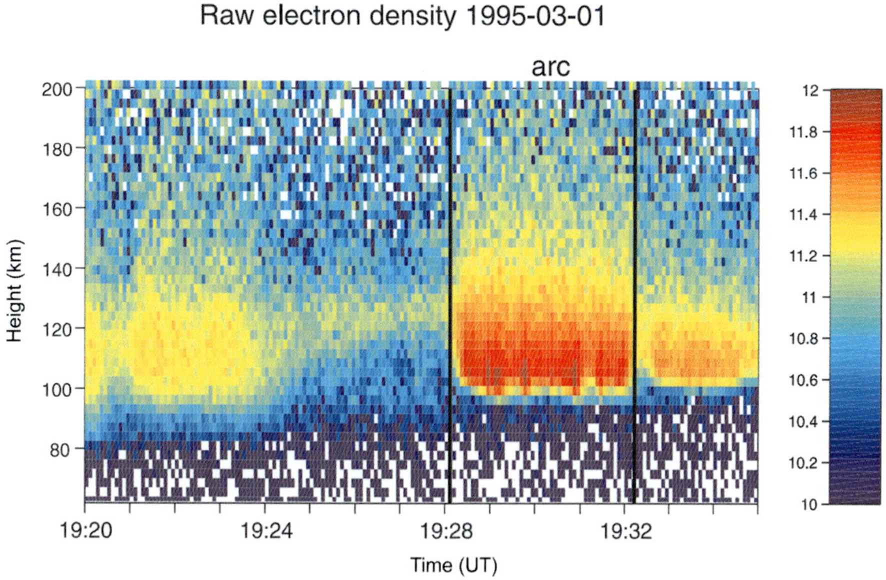

Fig. 10.19 Auroral arc by the KIL all-sky camera between 1928 and 1932 UT. The velocity is about 100 m/s, and stays about 4 min within the beam with remarkably constant characteristics. Note the ionospheric density cavity in the region of the downward current. From Aikio, A.T., Lakkala, T., Kozlovsky, A., Williams, P.J.S., 2002. Electric fields and currents of stable drifting auroral arcs in the evening sector. J. Geophys. Res., 107(A12), 4. https://doi.org/10.1029/2001JA009172.

conductivities were calculated from the profiles, and using the measured electric fields, the height-integrated Pedersen and Hall currents were derived.

An auroral arc was observed by the KIL all-sky camera (Fig. 10.19) between 1920 and 1935 UT on March 1, 1995; it was followed by other arcs, all drifting equatorward. The arc enters into the EISCAT beam at 1928 UT when the arc velocity is about 100 m/s with remarkably constant characteristics. The width of the arc that follows from these values is 24 km. High time resolution raw-electron densities indicate that the arc is very homogenous, a unique observation. To the south an auroral cavity is clearly apparent where the return downward current should be located. The article continues with a careful study on conductances and currents.

The current systems of stable arcs residing in the northward convection electric field region show a consistent pattern: currents flow downward on the equatorward side of the arcs, then poleward, and upward from the arcs. … Most of the arcs are associated with an enhanced northward-directed electric field region on the equatorward side of the arc, colocated with downward field-aligned currents (FACs) and suppressed E- *and* F-*region electron densities. The width of the region of the enhanced electric field is one to four times the width of the arc. In some cases, the electron density reduction is so pronounced that the region can be described as an auroral ionospheric density cavity.*

Aikio et al. (2002)

10.6.5 Second low-altitude generator

A phenomenological model of a quiescent auroral arc is proposed in which the magnetic field lines are not (nearly) equipotentials within the arc. The auroral plasma is negatively charged as a direct result of stopping the primary electrons (see Fig. 10.20). This net charge is that of the keV primary electrons responsible for the arc; the primary keV electrons can force their way into the negative charge region, being delivered by the upward Birkeland current. This is possible because of the high collision frequency of the secondary electrons in the region up to 200 km. This charge creates a downward electric field in the upper *E*-region and *F*-region of the ionosphere.

There is a region above the arc where the potential is a maximum; secondary electrons are trapped between two electric mirrors. Near the upper mirror point these secondaries continue to ionize and excite the atmospheric constituents to produce a high-altitude extension of the aurora. Being secondaries with low energies they excite the 630 nm emission of oxygen to cause the well-known type-A red aurora (see Figs. 10.20 and 10.21).

The ions created by fresh ionization above the arc will fall under this downward electric field. Some thermal electrons will contribute to the equatorward Pedersen current; after that they will be accelerated upward as part of the downward return current. With a large negative potential, some electrons (those at the high-energy end of the distribution function) may become energized to form runaways. The balance of the return current are the ions executing curvature drift in the thin plasma sheet.

10.7 Key results from SuperDARN, CANOPUS

A superposed epoch study of SuperDARN convection observations during substorms has been published by Bristow and Jensen (2007). Intervals included in this study were selected based on observations from the SuperDARN network, along with observations from the Alaska Magnetometer Chain, the Poker Flat all-sky

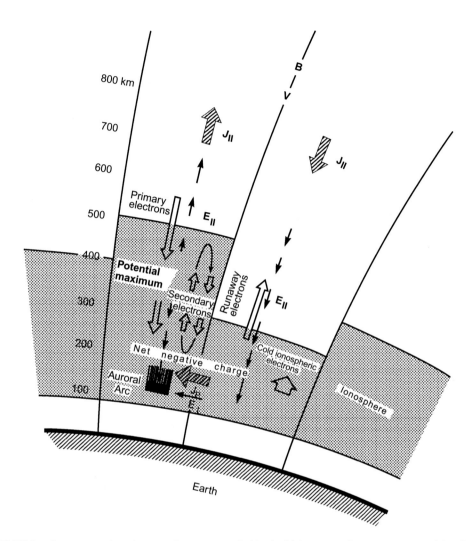

Fig. 10.20 With an intense arc there is so much current carried by the high-energy electrons to cause violet glow at low attitudes due to nitrogen. The low energy secondaries are repelled upward causing the red type-A aurora due to oxygen. They come under the influence of the upward electric field at high altitudes and are reflected. From Heikkila and Ricamore.

imager (ASTV), the Poker Flat meridian scanning photometer (MSP), the CARISMA magnetometer network, and the NORSTAR array of meridian scanning photometers (called CANOPUS in Fig. 10.22). Initial selection of intervals was based on optical or magnetometer observations providing an estimate of substorm onset time. The SuperDARN observations were examined to determine if there were sufficient data throughout the interval to provide

Fig. 10.21 A beautiful red type-A aurora, presumably at the onset of a westward traveling surge. The low energy secondaries are repelled upward causing the red type. From Finnish Meteorological Institute.

vector velocity estimates in the premidnight region starting about 1.5 to 2 h prior to onset.

10.7.1   Observations

A single event has been examined to illustrate how the data were treated, as shown in Fig. 10.23 (Bristow and Jensen, 2007). An isolated substorm on October 14, 2001 occurred with onset at about 0622 UT. The Spectroscopic Imager on the IMAGE satellite indicated that the onset occurred over central Canada. The onset was observed by the CARISMA magnetometer network and the NORSTAR meridian scanning photometers. By using the magnetometer on the ACE spacecraft it is safe to assume that the southward IMF reached the Earth well prior to the expansion onset, and the IMF remained southward throughout the interval of the substorm.

There is some time and space dispersion across the magnetometer network, with the earliest indication of the onset

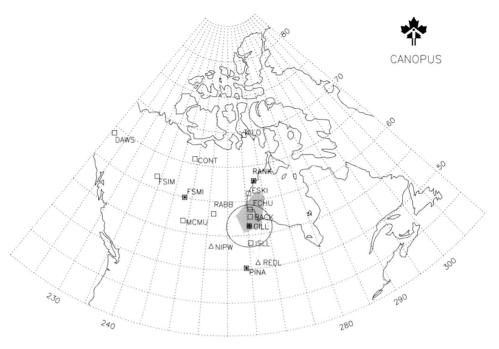

Fig. 10.22 The CANOPUS (now NORSTAR) array of ground stations. The data used in these studies were obtained at stations along the Churchill-Gillam meridian. The *circle* in the center of the map shows the field of view of the Gillam all-sky camera. The area covered by SuperDARN at Saskatoon is shown by the hatched area. The magnetic latitude at Gillam is 67.1°. From Lessard, M.R., Lotko, W., LaBelle, J., Peria, W., Carlson, C.W., Creutzberg, F., et al., 2007. Ground and satellite observations of the evolution of growth phase auroral arcs. J. Geophys. Res. 112, 3. https://doi.org/10.1029/2006JA011794.

occurring in the Gilliam observations, indicating that the onset latitude was close to Gilliam. Fig. 10.23 shows a clear and abrupt onset occurring near the center of the scan at about 0622 UT.

Fig. 10.24 shows two frames of the convection patterns chosen 10 min prior to onset (0612 UT) and at the time of the onset (0622 UT). Prior to onset, the characteristic pattern of sheared flow in the premidnight region is clear, with the convection reversal extending across the midnight meridian. The expansion onset pattern shows the velocity shear to have relaxed, and flows at midnight to be meridional out of the polar cap.

Two lines are shown in the figures, one at 2230 MLT, and another at 1030 MLT. Plots showing the velocity vectors as a function of latitude along a meridian from each interval included in the study were formed near these local times and were examined as time series (Fig. 10.25). Figures of this form are referred to as latitude-time-vector (LTV) plots.

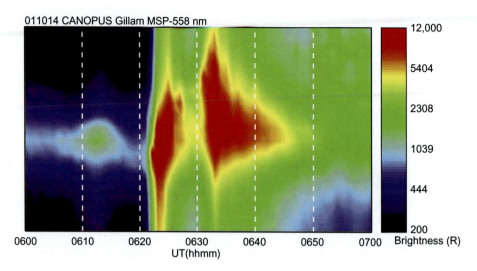

Fig. 10.23 Observations from the Gillam meridian scanning photometer for the hour of 0600 to 0700 UT on October 14, 2001. The meridian scanning photometer data format shows observations from the north of the station at the top of the figure, and observations from the south at the bottom of the figure. The horizontal center of the figure gives the observations from the zenith. From Bristow, W.A., Jensen, P., 2007. A superposed epoch study of SuperDARN convection observations during substorms. J. Geophys. Res. 112, 5. https://doi.org/10.1029/2006JA012049.

Fig. 10.25 presents the LTV plot for the superposed epoch time 2200 MLT meridian. The vertical axis of the plot is labeled "Substorm Latitude" and the horizontal axis is labeled "Substorm Time." The reference latitude was selected to be the latitude of the convection reversal at the onset time. The figure shows that prior to onset, the flows in the region equatorward of the reversal boundary were predominantly westward and were relatively uniform.

The superposition of all 10 intervals used in the study looks like a smoothed version of one of the contributing intervals such as the one shown in Fig. 10.25. Prior to expansion onset, flow vectors equatorward of the reversal boundary show a region of relatively high-velocity uniform-westward flow covering about 5° to 6° of latitude. Below that flow region, lower velocity flow was observed over an additional 4° to 5° of latitude. Poleward of the reversal, the flow vectors were comparable in magnitude to those in the equatorward region and were directed equatorward and eastward.

At onset, flows in the equatorward region decreased in magnitude and rotated from westward to equatorward. The change was abrupt and clearly tied to the expansion onset time, as is illustrated in Fig. 10.26, which was formed by averaging the peak velocities and latitude of peak from the 10 intervals. The solid lines

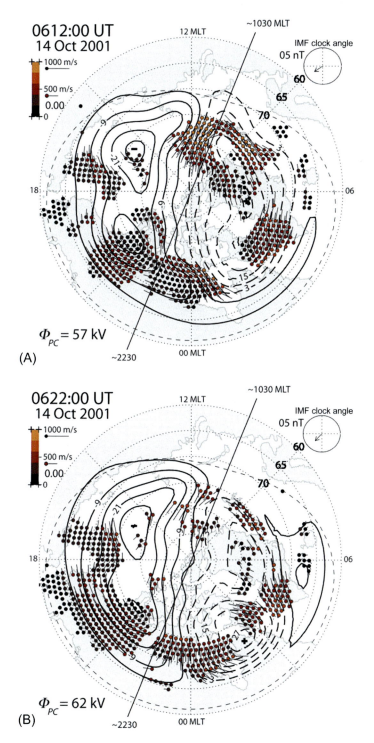

Fig. 10.24 Two frames of convection estimated from the SuperDARN observations from October 14, 2001 at (A) 0612 UT, and (B) 0622 UT time of onset. From Bristow, W.A., Jensen, P., 2007. A superposed epoch study of SuperDARN convection observations during substorms. J. Geophys. Res. 112, 6. https://doi.org/10.1029/2006JA012049.

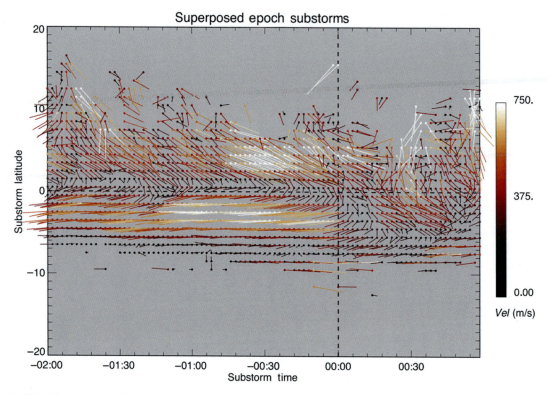

Fig. 10.25 Superposed LTV plots with substorm onset of 22:00 MLT. Flow is highly sheared to the east before onset. From Bristow, W.A., Jensen, P., 2007. A superposed epoch study of SuperDARN convection observations during substorms. J. Geophys. Res. 112, 9. https://doi.org/10.1029/2006JA012049.

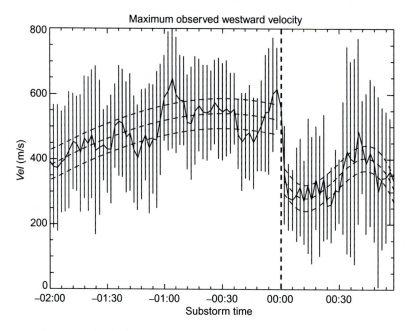

Fig. 10.26 Average peak westward velocity, formed by averaging the peak velocities from each of the 10 intervals. These velocities were observed equatorward of the convection reversal. From Bristow, W.A., Jensen, P., 2007. A superposed epoch study of SuperDARN convection observations during substorms. J. Geophys. Res. 112, 10. https://doi.org/10.1029/2006JA012049.

in the figure show the observations. On the velocity plot, the dashed lines show a third-order polynomial fit to the time series, plus and minus one standard deviation. The vertical lines at each observation point represent the standard deviation of the individual data points. The figure shows elevated velocities in the period before the onset with a significant decrease at the onset time. They first became elevated about 1 h before onset, when there was an increase to about 600 m/s, followed by a period of about 30 min with velocities of about 550 m/s. There was a decrease in the observed flow velocities to about 450 m/s about 20 min prior to onset, followed by an increase from 450 m/s beginning about 10 min prior to onset and peaking at over 600 m/s at about 2 min prior to onset. At onset, the velocity abruptly decreased to about 270 m/s.

Velocities over 600 m/s are observed throughout the interval at local times before 2000 according to convection maps. However, there is a significant area between 2000 MLT and magnetic midnight over which the velocities appear depressed after onset. The latitude of the peak westward velocity remained relatively constant at about 3° equatorward of the convection reversal. At the time of the abrupt increase of velocity prior to onset, the latitude moved to about 2° from the reversal and then at onset moved to about 4° from the reversal. This motion of the peak after onset is illustrated in the LTV plot (Fig. 10.25) as the region equatorward of the reversal showing decreased flow velocity, which is directed primarily equatorward.

10.7.2 Conclusions about current buildup

Bristow and Jensen (2007) examined average flow patterns for 10 substorm intervals in an attempt to identify common features. The features that could be identified were as follows:

- An increase of the zonal convection speed in the premidnight region equatorward of the convection reversal boundary associated with a southward turning of the IMF becoming effective at the dayside magnetosphere.
- A further enhancement of the zonal convection speed about 8 min prior to expansion onset.
- An abrupt and significant decrease of convection speed at the onset time accompanied by a rotation of the near-midnight flow from zonal to meridional out of the polar cap.

[Previous] case studies of flow patterns during substorms have revealed characteristic patterns in the period leading to expansion onset. Velocities in the premidnight region have been observed to

become strongly sheared, and the premidnight convection reversal has been observed to extend across the midnight meridian. At the onset of the expansion phase, the shear decreases and flow at midnight becomes meridional, out of the polar cap. The study presented here superposes flow patterns from 10 substorm intervals to determine the average flow patterns. The study confirms the findings of the case studies and extends them to additional local time sectors. In addition, changes to the flow pattern after expansion onset are examined.

Bristow and Jensen (2007)

Examination of dayside flows did not show any significant signature prior to substorm onset, from which it could be concluded that the substorms were not externally triggered. These observations agree with MHD simulations and empirical models of thin current sheets. They confirm that the magnetotail flow is diverted around to the flanks in the inner part of the tail, and are consistent with thin current sheets being a major part of substorm growth phase.

10.8 Large-scale flow dynamics

The convection picture is continuously interrupted by large-scale reconfiguration processes associated with changes in energy input. A substorm starts when this energy cycle is enhanced (Akasofu, 1964). This has three phases: the growth phase, the expansion phase, and the recovery phase. A fourth phase, the "trigger phase," was proposed by Pellinen et al. (1982) and Pellinen and Heikkila (1984), and was described as the slow, linear growth-phase processes that are interrupted just before the breakup at the beginning of the expansion phase.

During the growth phase, equatorward-drifting discrete auroral arcs are often observed in the evening sector as in Figs. 10.17 and 10.19. In the magnetotail, the growth phase is seen as an increase in the lobe magnetic field consistent with the polar cap area increase. In the inner tail near the geostationary orbit, the magnetic field normal to the current sheet (B_z) decreases.

It is well known that these effects are caused by the formation of an elongated thin current sheet that extends from near-geostationary orbit out to 30 or 40 R_E (Fairfield and Ness, 1970). This TCS typically has a thickness of only about 1000 km (ion Larmor radius scale), and it is embedded within the much thicker plasma sheet. This leads to a trigger phase ending with the expansion phase.

10.8.1 Auroral fading

The equatorward motion of both proton and electron aurora during the growth phase is well documented (see Liu et al., 2007 and references therein). This motion reflects the motion of the plasma in the nightside magnetosphere and the storage of energy that is subsequently released during the expansion phase. Auroral intensity also undergoes changes during the growth phase. One such effect is the fading of electron auroras and cosmic radio absorption immediately prior to substorm onset, which has been a topic of some uncertainty and dispute. Pellinen and Heikkila (1978) carried out an early study of electron auroral fading and concluded that the phenomenon was associated mostly with the onset region and was fairly common. Kauristie et al. (1997) studied carefully selected fading events from the Magnetometers-Ionospheric Radars All-Sky Cameras Large Experiment (MIRACLE) observations and found that they occurred predominantly in disturbed periods of the magnetosphere and on average lasted 2 min.

An exceptionally clear fading event occurred on October 5, 1986 (Fig. 10.27) with data taken by several different instruments, including EISCAT radar at Tromsø and all-sky cameras (Kauristie et al., 1995). The power profile along the geomagnetic field line recorded by EISCAT, and three ASC frames at Kilpisjärvi (KIL) during this event, are shown. Both EISCAT and ASC recordings show the fading simultaneously; the onset took place very close to the radar beam. The breakup precipitation after the fading increased electron density at exceptionally low altitudes, indicating electron energies by the low altitude of the ionization (85 km) exceeding 30 keV.

10.8.2 Proton aurora

The TCS is carried by particles in the plasma sheet undergoing gradient and curvature drift. Since the ions (mostly protons) have a higher mass and energy they will dominate in carrying the current; therefore we must look at them closely. Samson et al. (1992) analyzed 40 intervals with substorm intensifications seen in the CANOPUS data, ranging from about 2100 to 0200 local magnetic time. The data show a narrow (approximately 1–2° latitudinal width) band of H_P emissions, typically near 68–69° geomagnetic latitude at 0300 UT. It begins moving equatorward, reaching 65° to 66° just before a substorm intensification. The initial brightening of the electron arc associated with

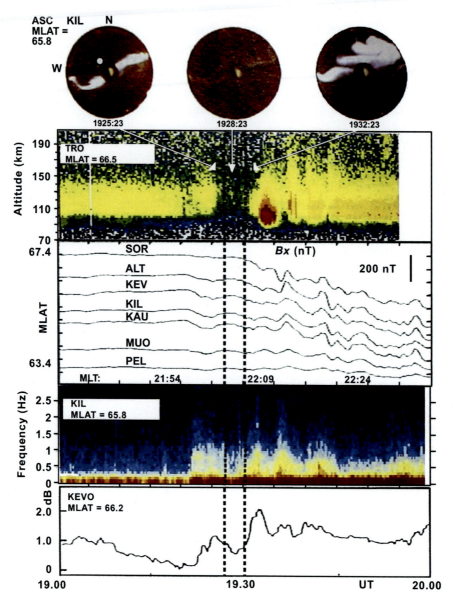

Fig. 10.27 *From top to bottom*: ASC pictures from KIL, the EISCAT power profile (*red, yellow*, and *green* correspond to electron densities of 10^{12}, 2.5×10^{11}, and 6.3×10^{10} m^{-3}, respectively), the magnetic north-south components from the EISCAT Magnetometer Cross, dynamic spectrum of magnetic pulsations at KIL (H-component, *yellow* and *red* correspond to values of 10^{-3} and 10^{-1} nT^2/Hz, respectively), and the riometer recordings at KEV (the magnetic coordinates are by Baker and Wing, 1989). From Kauristie, K., Pulkkinen, T.I., Pellinen, R.J., Janhunen, P., Huuskonen, A., Viljanen, A., et al., 1995. Analysis of the substorm trigger phase using multiple ground-based instrumentation. Geophys. Res. Lett. 22(15), 2066. https://doi.org/10.1029/95GL01800.

the substorm intensification started at 66°, near or just poleward of the maximum proton emissions.

> *The precipitating protons are from a population that is energized via earthward convection from the magnetotail into the dipolar region of the magnetosphere and may play an important role in the formation of the electron arcs leading to substorm intensifications on dipolelike field lines.*
>
> **Samson et al. (1992)**

Mende et al. (2003) surveyed 91 substorms with the IMAGE satellite far ultraviolet imagers and concluded that fading was not statistically significant in either the proton or electron aurora. However, the observation sensitive to the proton aurora showed a significant dropout about 20 min prior to onset. Using Polar Ultraviolet Imager (UVI) data, Newell et al. (2001) found that a 10% decrease in the power of electron auroral emission was common just before onsets.

During the growth phase, the proton aurora can be expected from pitch angle scattering due to nonadiabatic effects in the vicinity of a TCS, and its variation a potential diagnostic of this important configuration usually presaging the substorm onset. Thus an investigation of proton auroral brightness can shed light on proton dynamics in the magnetosphere. A partial demagnetization of protons in the current sheet is often deemed important.

Liu et al. (2007) carried out a superposed epoch analysis of the proton auroral brightness for the 50 events as a function of time relative to the proton auroral onset (determined manually in each case by the start of obvious brightening of the proton aurora). Onset so defined can be robustly determined to within 2 min. Fig. 10.28 shows the results of the analysis. The two panels consist of two-dimensional histograms (gray level with darker values indicating higher relative occurrence) and latitude-averaged values (red diamonds) of the two quantities. In the top panel are the results for the peak proton auroral intensity as a function of time. The peak intensity is the maximum value of the raw scan, and no fitting was involved. The bottom panel shows the superposed epoch analysis of the equatorward boundary of the proton auroral band, determined according to the "optical b2i" algorithm of Donovan et al. (2003). From this analysis, we see that the optical b2i moves to a lower latitude systematically during the growth phase and then retreats to a higher latitude rapidly at onset.

Liu et al. (2007) proposed that the surprising result of Fig. 10.28 is a distinct feature of magnetic field stretching (i.e., a decreasing equatorial B_z) in the central plasma sheet prior to onset. They

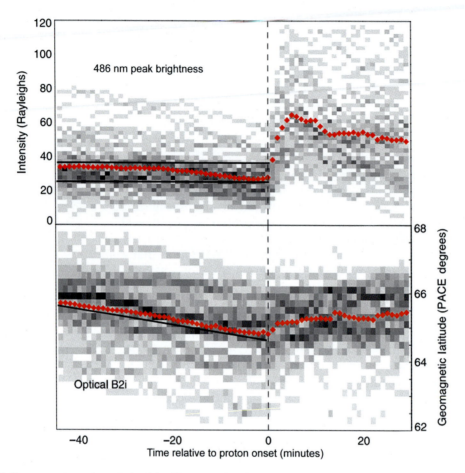

Fig. 10.28 Superposed epoch analysis of the 50 events, showing two-dimensional histograms (*grayscale*, with darker indicating higher frequency of occurrence) and average intensity and optical b2i *(red symbols)*. Time is relative to the proton auroral onset (see text). Geomagnetic latitude is assuming the coordinate system of Baker and Wing (1989) and an emission height of 110 km. The *solid lines* are visual aids, highlighting the difference between the brightness and b2i variations. From Liu, W.W., Donovan, E.F., Liang, J., Voronkov, I., Spanswick, E., Jayachandran, P.T., et al., 2007. On the equatorward motion and fading of proton aurora during substorm growth phase. J. Geophys. Res. 112, 4. https://doi.org/10.1029/2007JA012495.

used 50 high-quality events from the CANOPUS Gillam meridian scanning photometer over 10 years (1989–98).

We show that proton aurora in the substorm growth phase exhibits a systematic tendency to fade. The average pattern of fading consists of a period of relatively stable proton aurora brightness and then a period of 15–20 min before onset during which the brightness decreases by an average of 15%. We interpret the observed proton aurora brightness variation in terms of the magnetic field stretching in the near-Earth magnetosphere; in particular, the fading is interpreted as a result of the central plasma sheet magnetic field lines having stretched to such a degree that the loss cone closing effect dominates precipitation due to nonadiabatic proton motions.

Liu et al. (2007)

10.8.3 Growth-phase auroras

A combination of ASTV and meridian scanning photometers were used by Deehr and Lummerzheim (2001) to identify the optical signature of the growth phase, onset, expansion, and recovery of 33 auroral substorms in the Alaskan sector.

The example of the meridian scanning photometer and ASTV data shown in Figs. 10.29–10.31 was chosen because the substorm onset took place in the field of view of the instruments, and it contains all of the elements of a typical, isolated substorm. This onset occurred near the local zenith at 1145 UT (0030 MLT) on February

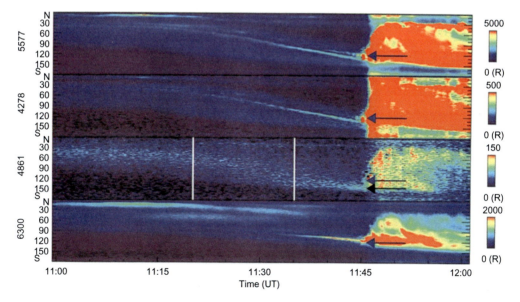

Fig. 10.29 Meridian scanning photometer plots of auroral emissions as a function of elevation angle from north to south through zenith and UT taken through four wavelength filters: 5577 Å (O I), 4278 Å (N_2^+), 4861 Å (H_β), and 6300 Å (O I). These plots show the growth phase and onset for a substorm on February 1, 1990, at Poker Flat Research Range. The area between the two vertical lines on the *H* emission panel is where the *H* emission shows increased equatorward motion. The *black arrow* indicates the location of the peak *H* emission at onset (140 degrees elevation from north), and the *blue arrows* indicate the location of the electron-induced arc at onset (120 degrees elevation from north). From Deehr, C., Lummerzheim, D., 2001. Ground-based optical observations of hydrogen emission in the auroral substorm. J. Geophys. Res. 106(A1), 35. https://doi.org/10.1029/2000JA002010.

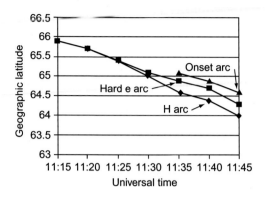

Fig. 10.30 Geographic latitude of the intensity maxima for three auroral arcs from the Poker Flat Research Range meridian scanning photometer record for February 1, 1990. The 557.7-nm [OI] emission was used for the electron-induced arcs, and the 486.1-nm H_β emission was used for the H arc. The location of the arcs was obtained graphically, with a curved Earth, and no allowance was made for atmospheric scattering or extinction. The assumed altitude in all three cases was 110 km. Note that if a higher altitude were used for the less energetic electron onset arc, the separation of the arcs would be even more apparent. From Deehr, C., Lummerzheim, D., 2001. Ground-based optical observations of hydrogen emission in the auroral substorm. J. Geophys. Res. 106(A1), 38. https://doi.org/10.1029/2000JA002010.

1, 1990, after a period of little activity indicated by the poleward position of the precipitation zone which moved equatorward starting at 1115 UT. There are three identifiable precipitation regions at different latitudes visible in Fig. 10.29:
- The first region consists of poleward arcs and bands due to electrons with a peak energy of 0.3 keV (first, second, and fourth panels before 1145).
- The second region consists of an arc excited by 10 keV electron precipitation (first panel 1135–1145).
- The third region consists of the "hydrogen arc," a diffuse arc, extending over 100-km of latitude on the equatorward side (third panel before 1145).

These three regions are summarized in Fig. 10.30. Also apparent here is the departure of the H emission from the electron arcs.

The meridian-scanning photometer data for the H_β emission during the growth phase is shown in the top panel of Fig. 10.31. The bottom panel is a plot of the hydrogen emission integrated over the meridian between 0 and 180 degrees and 20 to 150 degrees elevation angle measured from south. It is apparent that the intensity of hydrogen emission over the meridian is nearly constant, except for the period between 1115 and 1130 UT, when it

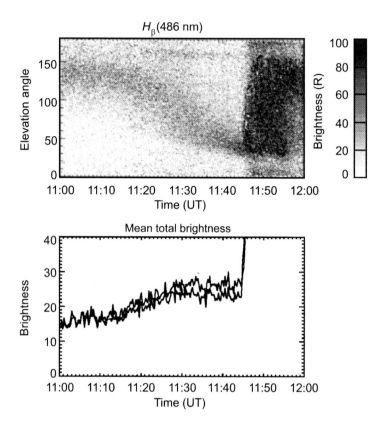

Fig. 10.31 *Top:* The intensity of 486.1-nm H_β emission as a function of time and elevation angle from Poker Flat Research Range between 1100 and 1200 UT on February 1, 1990. *Bottom:* The total brightness of the H_β emission as a function of time, integrated from horizon to horizon (upper trace) and between 20 and 150 degrees elevation angle from south. From Deehr, C., Lummerzheim, D., 2001. Ground-based optical observations of hydrogen emission in the auroral substorm. J. Geophys. Res. 106(A1), 40. https://doi.org/10.1029/2000JA002010.

increases monotonically to nearly twice its original intensity. This is also the period of the sudden equatorward movement of the hydrogen arc. Thus the change in H intensity occurs during the equatorward motion that appears faster than that of the electron induced arc.

The discrete auroral arc that brightens at auroral substorm onset was found to be poleward of the diffuse aurora that contains the H *emission by a distance of between 10 and 300 km. … The poleward crossover of the peak* H *emission occurs shortly after substorm onset.*

Deehr and Lummerzheim (2001)

The peak H emission crosses back to the equatorward position during substorm recovery between 2100 and 0100 MLT. After 0100 MLT the peak H emission remains poleward of the electron-trapping boundary for the rest of the night.

10.8.4 Average trend at onset

From a survey of the cross-tail current disruption events in the near-Earth plasma sheet collected from the Time History of Events and Macroscale Interactions during Substorms (THEMIS) mission, Liang et al. (2009) identified a highly repeatable class of event occurring at the current sheet boundary in the few minutes before the local current disruption onset. Fig. 10.32 shows the average trend at the onset of the expansion phase.

> *Salient features of this class of event include (1) a precipitous drop of the ion temperature, (2) concurrent growth of a neutral sheet-pointing electric field, and (3) ULF wave activations at Pi1/Pi2 bands. We interpret the ion temperature drop as a manifestation of the extreme thinning of the local current sheet prior to its disruption. This thinning process is inferred as nonadiabatic by nature.*
>
> **Liang et al. (2009)**

When the current sheet thickness is down to ion kinetic scales, the ions are demagnetized, and a quasielectrostatic neutral sheet-pointing electric field emerges owing to the charge separation. The ULF fluctuations of electric/magnetic field have a two-band structure. The lower-frequency band with a period of 50 to 80 s is interpreted by Liang et al. (2009) as an Alfvénic mode coupled from other preonset wave modes excited at the equatorial plasma sheet such as the ballooning. The higher-frequency wave at 10- to 20-s periods is attributed to some instability mode, directly leading to the disruption of the thin current sheet. They suggest that an extremely thinned non-Harris TCS and the emergence of quasielectrostatic field constitute the conducive conditions for a local current disruption to occur.

10.9 Discussion

In the high-latitude ionosphere during the last few minutes before the substorm breakup, the growth-phase processes sometimes show puzzling behavior. During a "trigger phase," the named coined by Pellinen and Heikkila (1978, 1984), the slow linear growth-phase processes are interrupted just before the breakup. Fading, such as in Fig. 10.27, may be observed both in

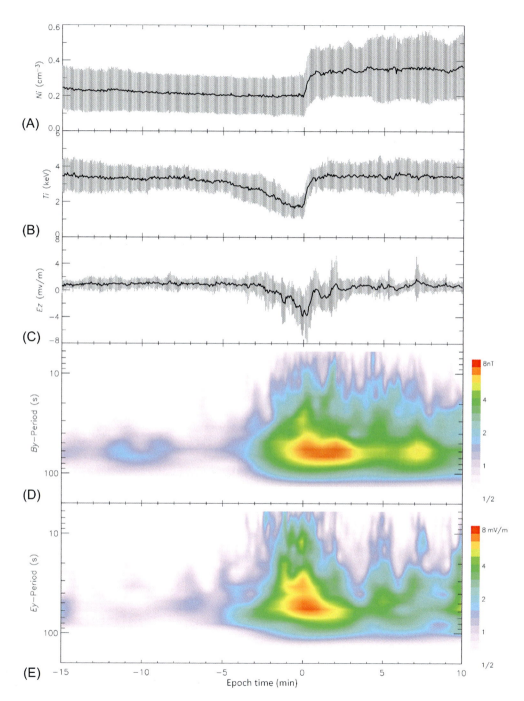

Fig. 10.32 (A) Average trend of the variation of the ion density, (B) ion temperature, and (C) proxied E_z field from −15 to +10 min around the current disruption onset for included events. *Vertical gray lines* indicate the standard deviation. (D and E) Averaged wavelet scalogram of B_y and E_{DSL_y} *Color bar* indicates the scalogram amplitude that is related to the actual wave amplitude (see article). From Liang, J., Liu, W.W., Donovan, E.F., 2009. Ion temperature drop and quasi-electrostatic electric field at the current sheet boundary minutes prior to the local current disruption. J. Geophys. Res. 114, 16. https://doi.org/10. 1029/2009JA014357.

soft and hard precipitation, in magnetic pulsation, and westward electrojet activity.

10.9.1 Current thinning event

The observational evidence shows that a neutral sheet-pointing electrostatic field frequently arises in the late growth-phase current sheet in the magnetotail. This electric field is associated with the thinning of the current sheet to the ion scale at which the electron and ion current sheets begin to separate (see Fig. 10.33 and the cited article). The attendant effect of a decreasing ion temperature, also interpreted in terms of a TCS, suggests that a cold plasma population is involved. Liu et al. (2007) review existing theories of electrostatic fields and show that they cannot explain the observations for various reasons. A particular problem is the overshielding of the electrostatic field by the cold population embedding the current sheet.

They interpret the observed proton aurora brightness variation in terms of the magnetic field stretching in the near-Earth

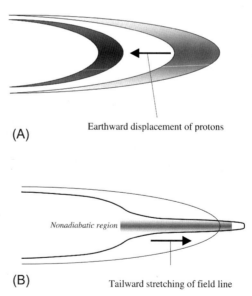

(A)

Earthward displacement of protons

Nonadiabatic region

(B)

Tailward stretching of field line

Fig. 10.33 Two scenarios of equatorward motion of proton aurora. (A) The displacement scenario involves an earthward motion of protons contained by a flux tube. (B) The field line stretching scenario involves the stretching of field lines and corresponding decrease in the equatorial magnetic field. From Liu, W.W., Donovan, E.F., Liang, J., Voronkov, I., Spanswick, E., Jayachandran, P.T., et al., 2007. On the equatorward motion and fading of proton aurora during substorm growth phase. J. Geophys. Res. 112, 5. https://doi.org/10.1029/2007JA012495.

magnetosphere; in particular, the fading is interpreted as a result of the central plasma sheet magnetic field lines having stretched to such a degree that the loss cone closing effect dominates precipitation due to nonadiabatic motions.

> *What has not been realized in previous theoretical treatments of the underlying hot-cold plasma interaction problem is that the cold plasma is a separate entity and entails a different treatment from that of the hot plasma. ... The cold plasma is dominated by the MHD dynamics while the hot TCS plasma is described in the classic Harris sheet formulism. The two components are coupled through a small correction to the cold plasma's MHD response, related to the polarization drift associated with the primary electric field due to the hot TCS plasma.*
>
> **Liu et al. (2010)**

Such an electromagnetic field is included in Fig. 10.14; to the left, since current is increasing, the field is from dusk to dawn by Faraday's law. On the right it is decreasing, and the directions are reversed. This is exactly like in a plasma transfer event (PTE) at the magnetopause (Fig. 6.21) so we call it Current Thinning Event (CTE).

- We emphasize that only the electric **E** can energize charged particles. No matter what expedients are used in fluid theory to get around this fact of life. "The particle approach is then necessary for at least some aspects of the problem" (Hines, 1963).
- Curvature drift in the plasma sheet is the only term that provides energy to any particle that can reach into the ionosphere.
- Newton's third law requires that the outgoing plasmoid experiences a reaction that is inward. This principle has not been used before. Fig. 11.2 use this idea to stress that the reaction to a plasmoid leaving (class II) is accomplished by a second entity, (class I), and that the two are equal, opposite, and collinear.
- The result is that ions (protons) will be precipitated on the west side, and electrons to the east at the poleward edge.
- The TCS means that the ions cannot remain stably trapped by the magnetic field with continued energization. This was pointed out by Heikkila et al. (2001); this adds a new electric field, and it is explosive.

The net result is that the LLBL is formed by a PTE onto closed field lines acting as a dynamo. Since the plasma is losing energy, Axford and Hines (1961) were correct with the inclusion of the LLBL as depicted by Fig. 10.34.

It is impressive that the 2D MHD code is able to produce almost identical results from the PIC code "despite the fact that

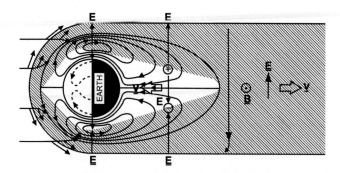

Fig. 10.34 The Axford and Hines (1961) model of the magnetotail, but with the addition of the LLBL. Since $\mathbf{E} \cdot \mathbf{J} < 0$, the plasma loses energy and momentum. This shows that it behaves as a viscous medium. From Heikkila, W.J., 1987. Neutral sheet crossings in the distant magnetotail. In: Lui, A.T. (Ed.), Magnetotail Physics. Johns Hopkins University Press, Baltimore, MD, p. 70.

kinetic approaches include anisotropy, a different dissipation mechanism, and different waves not included in MHD" (Birn and Schindler, 2002). One reason is that MHD does include conservation of mass, momentum, energy, and entropy. However, MHD leaves open a number of questions, for example, what is the exact dissipation mechanism? We should recall that by "destroying unnecessary information" fluid theory is forced to use some other means of particle energization (e.g., define an equation of state, *one that is localized*). A *precise description* is offered by the particle approach, needed to find correct answers to these complicated questions *if we can*.

> *Equilibrium configurations that satisfy the constraints cease to exist when the boundary deformation exceeds the critical value. This provides a strong argument for the onset of instability or the loss of equilibrium, regardless of the dissipation mechanism.*
>
> **Birn and Schindler (2002)**

10.10 Summary

The major entry site for solar wind plasma into the plasma sheet is the LLBL. Because of its electric dipole nature, the electric field is transferred from the magnetosheath into the plasma sheet by the LLBL. The LLBL also supplies the plasma through the polarization drift required to maintain the electric field for tailward flow.

- **Electric and magnetic drift**: With the magnetic field **B** and electric potential **E** it is possible to evaluate the magnetic drift $\mu \mathbf{B} \times \nabla B / eB^2$ and electric drift $\mathbf{B} \times \nabla \Phi / B^2$ paths of protons of different first adiabatic invariants μ assuming all ions are protons and have 90 degrees pitch angle. The drift paths are equivalent to the contours of constant $\mu B + e\Phi$ Wang et al. (2006).
- **Plasma irreversibly heated**: Ions and electrons were found, on average, to be remarkably isotropic at the neutral sheet. The average entropy was higher at $z = 0$ during fast flow events than it was anywhere in the region that could be studied when flows were slow or moderate. The plasma was irreversibly heated by the process that produced the fast flows. Scattering through 90 degrees each minute during slow and moderate flow conditions and as rapidly as every 10 s during the fastest flows was needed to maintain this average degree of isotropy. Scattering need not be so rapid that it gives instantaneous results; it is likely to be at very low altitudes, in the ionosphere (Kaufmann et al., 2005).
- **Fluids and plasmas**: A collisionless plasma consisting of charged particles moving freely through space, with completely different velocity vectors, go off in different directions. The motion of the local center of mass is a fiction that does not correspond to any real mass motion. These equations do not suffice to provide solutions for the great number of unknown quantities (Schmidt, 1979, p. 57). The effect of the magnetic field might to a certain extent replace the role of collisions perpendicular to the magnetic field lines. To force a closer analogy to fluid dynamics we have to make assumptions concerning the longitudinal motion.
- **Raising and lowering mirror points**: A key concept of plasma theory is that the first adiabatic invariant, the magnetic moment, is conserved. If the particle gains or loses energy then

$$\mu_m = \frac{1/2\left(mv_1^2\right)}{B_{MP1}} = \frac{1/2\left(mv_2^2\right)}{B_{MP2}} \qquad (10.23)$$

- The mirror point will come down for an increase in energy; conversely, the mirror point will be raised when the particle loses energy. Precipitation will occur with an electrical load with $\mathbf{E} \cdot \mathbf{J} > 0$, not with a dynamo given by $\mathbf{E} \cdot \mathbf{J} < 0$ (Pellinen and Heikkila, 1984).
- **Current thinning event (CTE)**: Thin current sheet formation is a slow process with gradual thinning during the growth phase,

typically over 30–60 min. CTE and PTE have a lot in common. Observations, empirical models, and MHD simulations suggest that TCSs have a thickness of the order of the ion gyroradius, a cross-tail width of about 15–25 R_E, and a tail dimension of about 20 R_E. Simulations suggest that the magnetotail flow is diverted around the TCS to the flanks in the inner part of the tail (Pulkkinen et al., 2006).

- **SuperDARN and CANOPUS/NORSTAR array**: Numerous superposed epoch studies of SuperDARN convection observations during substorms have been published. Case studies of flow patterns during substorms have revealed characteristic patterns in the period leading to expansion onset. Velocities in the premidnight region have been observed to become strongly sheared, and the premidnight convection reversal has been observed to extend across the midnight meridian. At the onset of the expansion phase, the shear decreases and flow at midnight becomes meridional, out of the polar cap (Bristow and Jensen, 2007).

- **Aurora study with EISCAT**: The current systems of stable arcs residing in the northward convection electric field region show a consistent pattern: currents flow downward on the equatorward side of the arcs, then poleward, and upward from the arcs. Most of the arcs are associated with an enhanced northward-directed electric field region on the equatorward side of the arc, colocated with downward field-aligned currents and suppressed E- and F-region electron densities. In some cases, the electron density reduction is so pronounced that the region is described as an auroral ionospheric density cavity (Aikio et al., 2002).

- Cosmic radio noise absorption (CNA) is generally attributed to the enhancement of D-region ionization produced by particle precipitation important because of high-collision frequency. The maximum absorption was produced at an altitude of about 80 km.

- **The auroral current circuit**: One speaks of current or voltage generators. If the internal resistance is small compared to the load on the circuit, the load will receive a fixed voltage. In contrast, if the load is small compared to the internal resistance, a fixed current, given by the emf of the battery divided by the internal resistance, will be delivered to the load. An important distinction between current and voltage generators is that the ionospheric dissipation will be proportional to the Pedersen conductivity in the voltage generator case, while it is inversely proportional to the conductivity if the current is held fixed.

- **The primary circuit**: The bright nighttime auroras that are visible to the unaided eye are caused by electrons accelerated toward Earth by an upward-pointing electric field. The return current is due to an electric field accelerating electrons away from the auroral ionosphere. The two branches of the equipotential contours close at typical altitudes of 5000–8000 km in the upward-current region above the aurora, and at 1500–3000 km in the auroral return-current region, respectively (Marklund et al., 2001). The field-aligned currents are carried by the downward and upward accelerated electrons, respectively. The source of energy is hardly ever discussed.

- **Second low-altitude generator**: There can be a region above the arc where the potential is a maximum; secondary electrons are trapped between two electric mirrors. Near the upper mirror point these secondaries continue to ionize and excite the atmospheric constituents to produce a high-altitude extension of the aurora. Being secondaries with low energies they excite the 630 nm emission of oxygen to cause the well-known type-A red aurora.

- **Ground and FAST observations**: Ground and FAST satellite observations of the evolution of growth-phase auroral arcs show evidence that the growth-phase arc is essentially stationary except perhaps for a slow drift. That suggests an electrostatic field, but that is a major mistake; there is slow change in the magnetic profile. The intensity of auroral emissions increases gradually.

- A rare example of a satellite crossing above multiple parallel arcs and simultaneous optical coverage of these arcs from an aircraft. The arcs can be identified by peaks in energy spectra and fluxes of the downgoing electron component. It is noteworthy that FAST showed that plasma was present even in the gaps corresponding to the dark sky in the all-sky photograph. Only the spectrum was different, less energetic, suggesting a dynamo with the plasma losing energy (Stenbaek-Nielsen et al., 1998)

- **Auroral fading**: An exceptionally clear fading event occurred on October 5, 1986 with data taken by several different instruments, including EISCAT radar at Tromsø and all-sky cameras. Both EISCAT and ASC recordings show the fading simultaneously. Fading is not associated with every breakup (Kauristie et al., 1997). The observational evidence presented earlier indicates that proton aurora fading is both real and statistically significant. The average brightness decrease of proton aurora is ~15% during the growth phase. The proton aurora exhibits a consistent tendency to fade during the last 20 min or so before optical onsets (Liu et al., 2007).

- **Current disruption**: Salient features of this class of event include (1) a precipitous drop of the ion temperature, (2) concurrent growth of a neutral sheet-pointing electric field, and (3) ULF wave activations at Pi1/Pi2 bands. The ion temperature drop is interpreted as a manifestation of the extreme thinning of the local current sheet prior to its disruption. This thinning process is nonadiabatic by nature (Liang et al., 2009).

10.11 Problems

10.1. Show explicitly the identities and assumptions used in the progression from Eqs. (10.11) to (10.13).

10.2. At a particular point in space, the electric and magnetic fields are perpendicular to each other, and the electric drift is 43 km/s. If the total drift at this point is zero, and ∇p is in the same direction as E, what is the pressure gradient? Assume typical values of number density and magnetic field strength of $n = 0.5\,\mathrm{cm}^{-3}$ and $B = 5\,\mathrm{nT}$.

10.3. What is the maximum perpendicular velocity a proton in the neutral sheet can have before its motion starts to become nonadiabatic? Assume the geomagnetic field has a magnitude of 6 nT and a minimum curvature of $0.2\,R_E$.

10.4. Write Boltzmann's Eq. (10.3) in vector notation. Is this equation (in either form) valid for relativistic velocities? Explain why or why not.

10.5. An auroral arc is observed to be drifting equatorward with a velocity of $80 \pm 5\,\mathrm{m/s}$, for a duration of $350 \pm 10\,\mathrm{s}$. What is the magnitude and uncertainty of the width of the arc?

References

Aikio, A.T., Lakkala, T., Kozlovsky, A., Williams, P.J.S., 2002. Electric fields and currents of stable drifting auroral arcs in the evening sector. J. Geophys. Res. 107 (A12), 1424. https://doi.org/10.1029/2001JA009172.

Akasofu, S.-I., 1964. The development of the auroral substorm. Planet. Space Sci. 12 (4), 273–282. https://doi.org/10.1016/0032-0633(64)90151-5.

Axford, W.I., Hines, C.O., 1961. A unifying theory of high-latitude geophysical phenomena and geomagnetic storms. Can. J. Phys. 39, 1433.

Baker, K.B., Wing, S., 1989. A new magnetic coordinate system for conjugate studies at high latitudes. J. Geophys. Res. 94 (A7), 9139–9143.

Birn, J., Schindler, K., 2002. Thin current sheets in the magnetotail and the loss of equilibrium. J. Geophys. Res. 107 (A7), 1117. https://doi.org/10.1029/2001J A000291.

Bristow, W.A., Jensen, P., 2007. A superposed epoch study of SuperDARN convection observations during substorms. J. Geophys. Res. 112, A06232. https://doi.org/10.1029/2006JA012049.

Büchner, J., Zelenyi, L., 1987. Chaotization of the electron motion as the cause of an internal magnetotail instability and substorm onset. J. Geophys. Res. 92 (A12), 13,456–13,466.

Büchner, J., Zelenyi, L.M., 1989. Regular and chaotic charged particle motion in magnetotaillike field reversals, 1. Basic theory of trapped motion. J. Geophys. Res. 94 (A9), 11,821–11,842. https://doi.org/10.1029/JA094iA09p11821.

Chen, F.F., 1984. Introduction to plasma physics and controlled fusion. In: Plasma Physics. second ed. vol. 1. Plenum Press, New York, NY.

Cole, K.D., 1961. On solar wind generation of polar geomagnetic disturbances. Geophys. J. R. Astron. Soc. 6, 103.

Deehr, C., Lummerzheim, D., 2001. Ground-based optical observations of hydrogen emission in the auroral substorm. J. Geophys. Res. 106 (A1), 33–44. https://doi.org/10.1029/2000JA002010.

Donovan, E.F., Jackel, B.J., Voronkov, I., Sotirelis, T., Creutzberg, F., Nicholson, N.A., 2003. Ground-based optical determination of the b2i boundary: a basis for an optical MT-index. J. Geophys. Res. 108 (A3), 1115. https://doi.org/10.1029/2001JA009198.

Dungey, J.W., 1961. Interplanetary magnetic field and the auroral zones. Phys. Rev. Lett. 6, 47–48.

Dunlop, M.W., Lavraud, B., Cargill, P., Taylor, M.G.G.T., Balogh, A., Réme, H., et al., 2005. Cluster observations of the cusp: magnetic structure and dynamics. In: Fritz, T.A., Fung, S.F. (Eds.), The Magnetospheric Cusps. Springer, Dordrecht, Netherlands, pp. 5–55.

Engebretson, M.J., Posch, J.L, Pilipenko, V.A, Chugunova, O.M., 2006. ULF waves at very high latitudes. In: Takahashi, K., Chi, P.J., Denton, R.E., Lysak, R.L. (Eds.), Magnetospheric ULF Waves: Synthesis and New Directions. AGU Geophysical Monograph 169. American Geophysical Union, Washington, DC, pp. 137–156.

Evans, D.S., 1974. Precipitating electron fluxes formed by a magnetic field aligned potential difference. J. Geophys. Res. 79, 2853. https://doi.org/10.1029/JA079i019p02853.

Evans, D.S., 1975. Evidence for the low altitude acceleration of auroral particles. In: Physics of the Hot Plasma in the Magnetosphere. Proceedings of the Thirtieth Nobel Symposium, Kiruna, Sweden, April 2–4. Plenum Press, New York, NY, pp. 319–340.

Fairfield, D., Ness, N., 1970. Configuration of the geomagnetic tail during substorms. J. Geophys. Res. 75 (34), 7032–7047.

Fälthammar, C.-G., 1973. Motions of charged particles in the magnetosphere. In: Egeland, H., Omholt, (Eds.), Cosmical Geophysics. Universitetsforlaget, Oslo, Norway.

Fujimoto, M., Terasawa, T., Mukai, T., Saito, Y., Yamamoto, T., Kokubun, S., 1998. Plasma entry from the flanks of the near-Earth magnetotail: geotail observations. J. Geophys. Res. 103, 4391.

Garner, T., Wolf, R., Spiro, R.W., Thomsen, M.F., Korth, H., 2003. Pressure balance inconsistency exhibited in a statistical model of magnetospheric plasma. J. Geophys. Res. 108 (A8), 1331.

Goertz, C., Baumjohann, W., 1991. On the thermodynamics of the plasma sheet. J. Geophys. Res. 96 (A12), 20,991–20,998.

Heikkila, W.J, 1997. Comment on "The alternative paradigm for magnetospheric physics" by E.N. Parker. J. Geophys. Res 102 (A5), 9651–9656.

Heikkila, W.J., Chen, T., Liu, Z.X., Pu, Z.Y., Pellinen, R.J., Pulkkinen, T.I., 2001. Near Earth Current Meander (NECM) model of substorms. Space Sci. Rev. 95, 399–414.

Heinemann, M., 1999. Role of collisionless heat flux in magnetospheric convection. J. Geophys. Res. 104 (A12), 28,397–28,410. https://doi.org/10.1029/1999JA 900401.

Heinemann, M., Wolf, R.A., 2001. Relationships of models of the inner magnetosphere to the Rice Convection Model. J. Geophys. Res. 106 (A8), 15,545–15,554.

Hines, C.O., 1963. The energization of plasma in the magnetosphere: hydromagnetic and particle-drift approaches. Planet. Space Sci. 10, 239.

Hori, T., Maezawa, K., Saito, Y., Mukai, T., 2000. Average profile of ion flow and convection electric field in the near-Earth plasma sheet. Geophys. Res. Lett. 27 (11), 1623–1626.

Kaufmann, R.L., Paterson, W.R., 2006. Magnetic flux and particle transport in the plasma sheet. J. Geophys. Res. 111, A10214. https://doi.org/10.1029/2006JA 011734.

Kaufmann, R.L., Paterson, W.R., Frank, L.A., 2005. Relationships between the ion flow speed, magnetic flux transport rate, and other plasma sheet parameters. J. Geophys. Res. 110, A09216. https://doi.org/10.1029/2005JA011068.

Kauristie, K., Pulkkinen, T.I., Pellinen, R.J., Janhunen, P., Huuskonen, A., Viljanen, A., et al., 1995. Analysis of the substorm trigger phase using multiple ground-based instrumentation. Geophys. Res. Lett. 22 (15), 2065–2068. https://doi.org/10.1029/95GL01800.

Kauristie, K., Pulkkinen, T.I., Huuskonen, A., Pellinen, R.J., Opgenoorth, H.J., Baker, D.N., et al., 1997. Auroral precipitation fading before and at substorm onset: ionospheric and geostationary signatures. Ann. Geophys. 15 (8), 967–983.

Kivelson, M.G., 1995a. Physics of space plasmas. In: Kivelson, M.G., Russell, C.T. (Eds.), Introduction to Space Physics. Cambridge University Press, New York, NY, pp. 27–55.

Kivelson, M.G., 1995b. Pulsations and magnetohydrodynamic waves. In: Kivelson, M.G., Russell, C.T. (Eds.), Introduction to Space Physics. Cambridge University Press, New York, NY, pp. 330–353.

Liang, J., Liu, W.W., Donovan, E.F., 2009. Ion temperature drop and quasi-electrostatic electric field at the current sheet boundary minutes prior to the local current disruption. J. Geophys. Res. 114, A10215. https://doi.org/10.1029/2009JA014357.

Liu, W.W., Donovan, E.F., Liang, J., Voronkov, I., Spanswick, E., Jayachandran, P.T., et al., 2007. On the equatorward motion and fading of proton aurora during substorm growth phase. J. Geophys. Res. 112, A10217. https://doi.org/10.1029/2007JA012495.

Liu, W.W., Liang, J., Donovan, E.F., 2010. Electrostatic field and ion temperature drop in thin current sheets: a theory. J. Geophys. Res. 115, A03211. https://doi.org/10.1029/2009JA014359.

Marklund, G., Ivchenko, N., et al., 2001. Temporal evolution of the electric field accelerating electrons away from the auroral ionosphere. Nature 414, 724–727.

McPherron, R.L., 1995. Magnetospheric dynamics. In: Kivelson, M.G., Russell, C.T. (Eds.), Introduction to Space Physics. Cambridge University Press, New York, NY, p. 400.

Mende, S.B., Carlson, C.W., Frey, H.U., Peticolas, L.M., Østgaard, N., 2003. FAST and IMAGE-FUV observations of a substorm onset. J. Geophys. Res. 108, 1344.

Newell, P.T., Liou, K., Sotirelis, T., Meng, C.-I., 2001. Polar ultraviolet imager observations of global auroral power as a function of polar cap size and magnetotail stretching. J. Geophys. Res. 106 (A4), 5895–5905.

Parker, E.N., 1996. The alternative paradigm for magnetospheric physics. J. Geophys. Res. 10, 10,587.

Paschmann, G., Haaland, S., Treumann, R., 2003. Auroral Plasma Physics. Kluwer, Norwell, MA (reprinted from Space Sci. Rev. 103, 1–485).

Pellinen, R.J., Heikkila, W.J., 1978. Energization of charged particles to high energies by an induced substorm electric field within the magnetotail. J. Geophys. Res. 83, 1544.

Pellinen, R.J., Heikkila, W.J., 1984. Inductive electric fields in the magnetotail and their relation to auroral and substorm phenomena. Space Sci. Rev. 37, 1–61. https://doi.org/10.1007/BF00213957.

Pellinen, R.J., Baumjohann, W., Heikkila, W.J., Sergeev, V.A., Yahnin, A.G., Marklund, G., et al., 1982. Event study on pre-substorm phases and their relation to the energy coupling between the solar wind and magnetosphere. Planet. Space Sci. 30, 371.

Peymirat, C., Fontaine, D., 1994. Relationships between field-aligned currents and convection observed by EISCAT and implications concerning the mechanism that produces region-2 currents: statistical study. Ann. Geophys. 12 (4), 304–315.

Pulkkinen, T.I., Baker, D.N., Owen, C.J., Slavin, J.A., 1996. A model for the distant tail field: ISEE-3 revisited. J. Geomagn. Geoelectr. 48, 455.

Pulkkinen, T.I., Palmroth, M., Tanskanen, E.I., Janhunen, P., Koskinen, H.E.J., Laitinen, T.V., 2006. New interpretation of magnetospheric energy circulation. Geophys. Res. Lett. 33, L07101. https://doi.org/10.1029/2005GL025457.

Reiff, P.H., Spiro, R.W., Wolf, R.A., Kamide, Y., King, J.H., 1985. Comparison of polar cap potential drops estimated from solar wind and ground magnetometer data: CDAW 6. J. Geophys. Res. 90 (A2), 1318–1324.

Reiff, P.H., Collin, H.L., Craven, J.D., Burch, J.L., Winningham, J.D., Shelley, E.G., 1988. Determination of auroral electrostatic potentials using high- and low-altitude particle distributions. J. Geophys. Res. 93, 7441.

Richmond, A.D., 1985. Atmospheric physics: atmospheric electrodynamics. Science 228 (4699), 572–573.

Samson, J.C., Lyons, L.R., Newell, P.T., Creutzberg, F., Xu, B., 1992. Proton aurora and substorm intensifications. Geophys. Res. Lett. 19 (21), 2167–2170. https://doi.org/10.1029/92GL02184.

Schmidt, G., 1979. Physics of High Temperature Plasmas, second ed. Academic Press, New York, NY.

Shen, C., Li, X., Dunlop, M., Liu, Z.X., Balogh, A., Baker, D.N., et al., 2003. Analyses on the geometrical structure of magnetic field in the current sheet based on cluster measurements. J. Geophys. Res. 108 (A5), 1168.

Slavin, J.A., Smith, E.J., Sibeck, D.G., Baker, D.N., Zwickl, R.D., Akasofu, S.-I., 1985. An ISEE 3 study of average and substorm conditions in the distant magnetotail. J. Geophys. Res. 90 (10), 10875–10895.

Speiser, T., 1965. Particle trajectories in model sheets, 1, analytical solutions. J. Geophys. Res. 70, 4219.

Stenbaek-Nielsen, H., Hallinan, T.J., Osborne, T.D.L., Kimball, J., Chaston, C., McFadden, J., et al., 1998. Aircraft observations conjugate to FAST: auroral arc thicknesses. Geophys. Res. Lett. 25 (12), 2073–2076. https://doi.org/10.1029/98GL01058.

Tsyganenko, N.A., 1989. A magnetospheric magnetic field model with a warped tail current sheet. Planet. Space Sci. 37, 5–20.

Wang, C.-P., Lyons, L.R., Weygand, J.M., Nagai, T., McEntire, R.W., 2006. Equatorial distributions of the plasma sheet ions, their electric and magnetic drifts, and magnetic fields under different interplanetary magnetic field Bz conditions. J. Geophys. Res. 111, A04215. https://doi.org/10.1029/2005JA011545.

Wang, C.-P., Lyons, L.R., Nagai, T., Weygand, J.M., McEntire, R.W., 2007. Sources, transport, and distributions of plasma sheet ions and electrons and dependences on interplanetary parameters under northward interplanetary magnetic field. J. Geophys. Res. 112, A10224. https://doi.org/10.1029/2007JA 012522.

Wanliss, J.A., Rostoker, G., 2006. IMAGE analysis and modelling of substorm onsets. In: Syrjäsuo, M., Donovan, E. (Eds.), Proceedings of the Eighth International Conference on Substorms. University of Calgary, Alberta, CA, pp. 331–335.

Wing, S., Newell, P.T., 1998. Central plasma sheet ion properties as inferred from ionospheric observations. J. Geophys. Res. 103 (A4), 6785–6800.

Wing, S., Newell, P.T., 2002. 2D plasma sheet ion density and temperature profiles for northward and southward IMF. Geophys. Res. Lett. 29(9). https://doi.org/10.1029/2001GL013950.

Zhou, X.-.Y., Russell, C.T., Mitchell, D.G., 1997. Three spacecraft observations of the geomagnetic tail during moderately disturbed conditions: global perspective. J. Geophys. Res. 102 (A7), 14,425–14,438.

11

Magnetospheric substorms

Chapter outline

Earth's Magnetosphere. https://doi.org/10.1016/B978-0-12-818160-7.00011-9

The trouble with ideologies that preach inevitable progressor one of the troubles, for there are many—is that they encourage linear thinking and discourage the factoring in of surprises, discontinuities, and disasters.

Harries (2005)

11.1 Introduction

Discrete auroras are usually made up of long east-west bands of ionospheric light emissions. During the International Geophysical Year of 1957, arrays of all-sky cameras were placed around the auroral oval allowing a phenomenological model to describe the exceptional development that follows, the *elementary storm* by Birkeland (1908) or *auroral substorm* by Akasofu (1964).

It was Akasofu's study of substorm aurora during the International Geophysical Year that revealed systematic auroral activity over the entire polar region, called the auroral substorm, as shown in Fig. 11.1. The first indication is a sudden brightening of the auroral curtain in the midnight or late evening sector.

The first disturbance is a sudden brightening of a portion of the most equatorward arc somewhere in the premidnight sector (panel B). This event is called the onset of the auroral substorm. The brightening expands rapidly westward and poleward (panel C). Within a short time a bright bulge of auroral disturbance forms in a broad region spanning the midnight sector close to where the aurora originally brightened. Within the bulge the aurora is very dynamic. Arcs appear and disappear; patches form and pulsate. Most arcs develop drapery-like folds that rapidly move along the arc. Lower borders of the arcs may become intensely colored. The

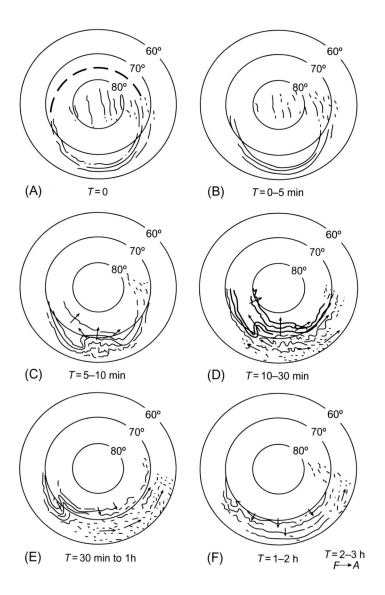

Fig. 11.1 Schematic representation of six stages in the development of an auroral substorm as determined from all-sky camera data during the IGY 1957. (A) Quiet state with multiple arcs drifting equatorward. (B) Sudden brightening at onset. (C) Rapid expansion westward and poleward. (D) Westward traveling surge and omega bands. (E) End of expansion phase, beginning of recovery. (F) Recovery phase. From Akasofu, S.-I., 1964. The development of the auroral substorm. Planet. Space Sci. 12(4), 273. https://doi.org/10.1016/0032-0633(64)90151-5.

interval of time during which the disturbed region is growing is called the expansion phase of the substorm. Eventually the auroral bulge develops a sharp kink at its westward edge, where it joins with the bright arc extending farther westward. This kink often appears to move westward, becoming more pronounced with time; hence it is called the westward travelling surge (panels C and D). At the eastern edge of the bulge, torchlike auroral forms appear, extending poleward from the diffuse aurora and drifting eastward (panel D).

These forms are called omega bands, from the shape of the dark regions defining their poleward borders. At the equatorward edge of this eastern region, dim pulsating patches of aurora appear, drifting eastward (panel E). After about 30–50 min, the auroral activity ceases to expand poleward, and the expansive phase of the substorm has ended (panel E). With the end of the expansion phase, auroral activity begins to dim at lower latitudes in the oval, and quiet arcs reappear. To the west, the westward-travelling surge degenerates, and a westward-drifting loop replaces it (panel E). In the morning sector, a pulsating aurora proceeds for some time. This phase of a substorm lasts for about 90 minutes called the recovery phase.

McPherron (1995, p. 421)

McPherron (1970) added the *growth phase*, the subject of Chapter 10; there must be a time-dependent evolution toward the explosion that follows. Pellinen et al. (1982) then added a *trigger phase*; this phase is required to make explicit the actual condition that caused the eruption.

We must ask, just what is happening at the onset of the expansion phase? A model or theory can be built by assembling different substorm mechanisms into one substorm process. This is based on the assumption that they are all being produced by the yet-to-be-determined "true" magnetospheric substorm onset. There should be differences at different places because disturbances propagate at different speeds.

Figs. 11.1 and 11.2 are used repeatedly throughout this chapter. Typical auroral forms in each phase, starting with long east-west arcs, westward traveling surge with north-south oriented streamers, and ending with bursty bulk flows and omega bands in the poleward edge of the diffuse morning sector auroral oval during the later phase of the substorm recovery. Particles become energized up to the MeV range in the plasma sheet. As observed long ago by IMP-8, in certain regions electrons and ions exhibit counterstreaming along the field lines (Kirsch et al., 1977). Taylor et al. (2006) reported Cluster data with similar results. Ions show inverse time dispersion with low-energy particles arriving before the high (Sarafopoulos and Sarris, 1988). Blake et al. (2005) showed that energization occurred over wide regions in the magnetotail. This suggests that a strong field-aligned electric field was suddenly turned on at some location. Electrons and ions reach 1 MeV energies in less than 1 min!

Only observations made local to the onset can be meaningful because of travel time limits. Auroral breakups identified with global auroral images from Polar imagers to calibrate other

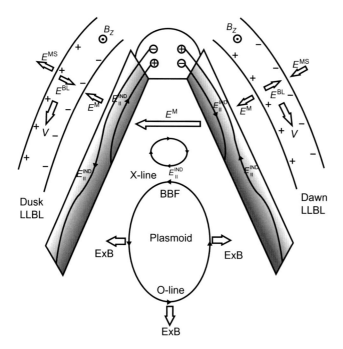

Fig. 11.2 The grand scheme of a magnetospheric substorm. Driven by the dynamo in the low-latitude boundary layer (LLBL), the plasma sheet is energized, at least for a southward interplanetary magnetic field (IMF). A local instability produces an induction electric field. That takes control, eventually to become a plasmoid traveling tailward. There is a reaction as given by Newton's third law, an earthward response compressing the inner plasma sheet. The electric field is rather simple, as shown by the cyclic pattern over the polar caps. From Heikkila and Keith.

substorm onset signatures have been carried out in the last few years (Meng and Liou, 2004). Fig. 11.3 presents a summary of these results. It is clear that all common substorm onset identifiers typically lag behind the auroral breakup, identified as the first indication of an increase in the intensity of an existing arc, except perhaps auroral kilometric radiation (AKR). The average time difference between AKR onset and auroral breakup is within ∼1 min. The small time difference indicates that when the satellite is suitably located, AKR can be a good substorm onset indicator.

11.2 Statistical description of the substorm

The substorm process has been enthusiastically investigated for at least six decades. Akasofu (1964) analyzed a large number of photographs from an array of all-sky cameras during substorms to obtain a detailed description from a global perspective of the development and decay of a typical auroral substorm. His analysis was based on the premise that substorms have common features on a global scale. The substorm concept has been improved by auroral imagers on spacecraft and additional ground-based

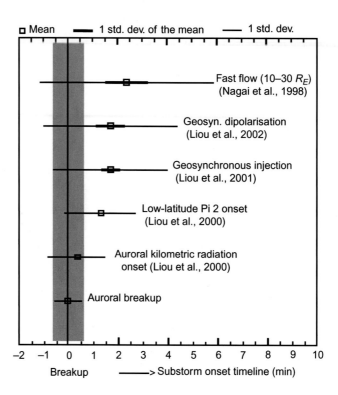

Fig. 11.3 Comparisons of auroral breakups, determined with Polar UVI images, with a number of substorm onset signatures. *Square boxes* are the averaged proxy onset times relative to auroral breakups, *thin error bars* are one standard deviation, and *thick error bars* are the standard error of the mean for the sample distribution. From Liou and Heikkila, after Meng, C.-I., Liou, K., 2004. Substorm timings and timescales: a new aspect. Space Sci. Rev. 113, 53.

observations by, for example, McPherron (1970), Kisabeth and Rostoker (1974), Galperin and Feldstein (1991), and others. Akasofu's general description still serves as a framework to which particle precipitation and electrodynamic parameters can be related.

Newell et al. (2001) performed a superposed epoch analysis of substorm auroral power; Mende et al. (2003) performed a comprehensive statistical study of the relative intensities and boundaries of proton and electron auroras during substorms using data from the IMAGE satellite.

In fact, to date there is still no quantitative description of the classical global Akasofu-type substorm beyond the initial diagrams. This is rather remarkable considering the massive amount of high quality global auroral images obtained by space borne imagers. In questioning what a substorm looks like, the consensus has been to adopt the Akasofu diagrams. What is missing is a superimposed epoch analysis of the auroral substorm that can quantitatively describe the growth, brightening, and decay of the auroral substorm.

Gjerloev et al. (2008)

The research by Gjerloev et al. (2007, 2008) has focused on particular features, such as the onset locations, to provide valuable information. Several figures from their papers on the auroral substorm in the far ultraviolet are included in this chapter.

11.2.1 Data

The primary data for their study come from the "Earth camera" images acquired by the Visible Imaging System (VIS) on the Polar satellite (Frank et al., 1995). This camera provides global auroral images in the far ultraviolet (FUV), in contrast to the more limited views but higher spatial resolution from its two visible imaging cameras. The imager obtains an average intensity over the linear size of a pixel, typically ~70 km near apogee, but with higher sensitivity than the all-sky camera pictures. The FUV image repetition rate was 1–5 min, depending upon whether the Earth camera was sharing the VIS telemetry allocation with the visible imagers. To ensure an onset timing with a precision of about 1 min, the Earth camera data were often supplemented with images from the visible imaging cameras and image data from the Ultraviolet Imager (UVI) on Polar (Torr et al., 1995). The combination of these data sets also enabled the elimination of pseudo-onsets and ensured that the substorm developed continuously out of the identified onset.

Using global auroral images during 116 substorms, they obtained quantitative measures of the key features. The selection of an event to be included in this statistical study was based both on its optical characteristic and the AL index pattern around the time of the event:

- Isolated single event, optically and magnetically
- Localized onset
- Bulge-type auroral event
- Expansion and single recovery phase
- At least the entire bulge region in darkness
- Magnetic storms excluded (Dst index > -30 nT)

The requirement of a localized onset in the night-time hours is the key requirement for the study of bulge-type auroras. Darkness requirement allows the high- and low-latitude boundaries of the aurora to be seen with minimal contamination, but biases the selection to the months around the northern winter solstice. Their selection of isolated single events and removal of events during complex activity periods provides simplicity.

The expansion period was identified solely from images and varied primarily from 10 to 40 min, with an average of 30.9 min. To avoid mixing expansion data with recovery data, they normalized the time of each substorm to one unit from onset to maximum expansion. The average onset location was 22.6 magnetic local time (MLT) and 66.8 degrees invariant latitude (ILat), in good agreement with previous analyses.

11.2.2 Three-step normalization technique

The substorm aurora is highly variable in time and space. This variability raises the fundamental question: "How can the major characteristics of individual events be maintained in a statistical average model?" They used time-honored methods and achieved outstanding results.

The events selected were fairly isolated and had to expand from a localized onset. This compilation required a three-step normalization technique, one temporal based on the expansion time, and two spatial (MLT and ILat).

11.2.2.1 Normalize substorm time

The temporal evolution of auroral substorms varies widely from some tens of minutes to several hours. Since we wish to describe the characteristics of the optical substorm as a function of time, it is necessary to normalize the time of each event. They defined the substorm time to begin at the onset of brightening at $T=0$, extending to $T=1$ at the maximum of the substorm, that is, the end of the expansion phase. Thus their selection provides a more integrated optical characteristic to the global aurora in the selection of the maximum. Fig. 11.4 contains the superposition of the AL indices as a function of normalized substorm time. The dark line is the average.

The normalized time steps used for each event during the expansion phase are $T=0.0$, $T=0.5$, $T=1.0$ (see red lines in Fig. 11.4). Prior to the event (growth phase) they use the number of minutes prior to onset: $T=-20$, $T=-15$, $T=-10$, and $T=-5$-min. During the recovery phase the normalized time scale at times $T=1.5, 2.0, 2.5$, and 3.0 are chosen to represent the sketches C–F in Fig. 11.1 (Akasofu, 1964). The average auroral boundaries for onset ($T=0$), the end of the expansion phase ($T=1$) and recovery ($T=2$) are shown in Fig. 11.5.

11.2.2.2 Normalize the magnetic local time position

The second step in the normalization scheme eliminates the varying local time position of the auroral bulge. The MLT position of the head of the surge to the west and the east end of the bulge (from the Gjerloev et al. (2007) database) is used. The red lines in Fig. 11.6A and B indicate the MLT position of these boundaries.

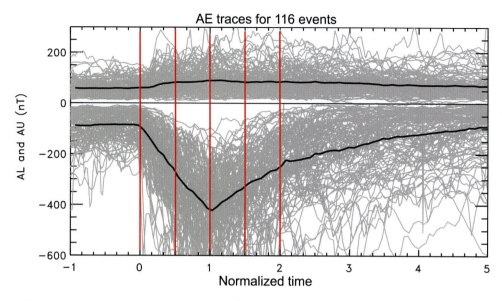

Fig. 11.4 The superposition of the AL indices from all substorms as a function of the normalized substorm time (defined in the text). From Gjerloev, J.W., Hoffman, R.A., Sigwarth, J.B., Frank, L.A., Baker, J.B.H., 2008. Typical auroral substorm: a bifurcated oval. J. Geophys. Res. 113, 4, A03211. https://doi.org/10.1029/2007JA012431.

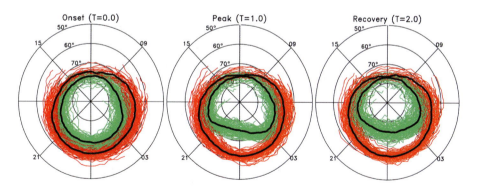

Fig. 11.5 Automated determination of auroral oval boundaries for 116 events. *Green* and *red* show the high-latitude and low-latitude boundaries, respectively, while the *black lines* are the average locations. The three normalized time steps *T* = 0.0, *T* = 1.0, and *T* = 2.0 are shown. From Gjerloev, J.W., Hoffman, R.A., Sigwarth, J.B., Frank, L.A., Baker, J.B.H., 2008. Typical auroral substorm: a bifurcated oval. J. Geophys. Res. 113, 5, A03211. https://doi.org/10.1029/2007JA012431.

Fig. 11.7A contains the distribution of onset locations in MLT/ILat in polar coordinates, while Fig. 11.7B and C shows the distribution as histograms in MLT and ILat, respectively. The average location of the onset is 22.6 MLT and 66.8 degrees ILat, with medians of 22.7 MLT and 67.0 degrees ILat. Gaussian fits seem to represent the distributions fairly well with maxima at 22.4 MLT

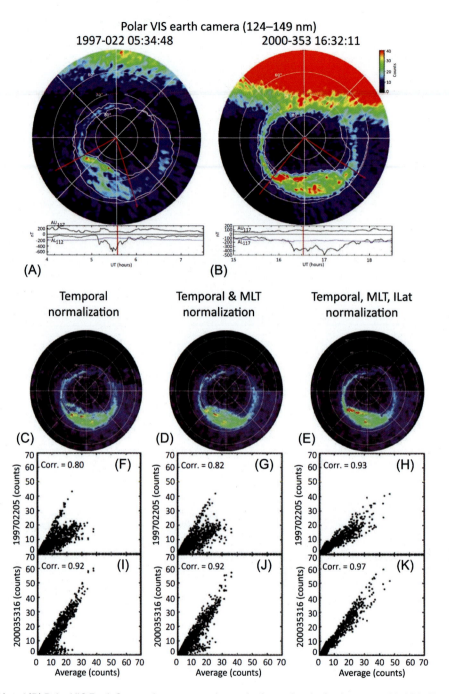

Fig. 11.6 (A) and (B) Polar VIS Earth Camera images near the peak of two classical substorms with AE indices. The *red lines* on the images indicate MLT boundaries of the poleward expansion while the *vertical lines* on the AE plots indicate times of images. Also shown are (C)–(E) average patterns of the two images using three different levels of normalization and (F)–(K) the results from a correlation analysis of images and the average patterns. From Gjerloev, J.W., Hoffman, R.A., Sigwarth, J.B., Frank, L.A., Baker, J.B.H., 2008. Typical auroral substorm: a bifurcated oval. J. Geophys. Res. 113, 4, A03211. https://doi. org/10.1029/2007JA012431.

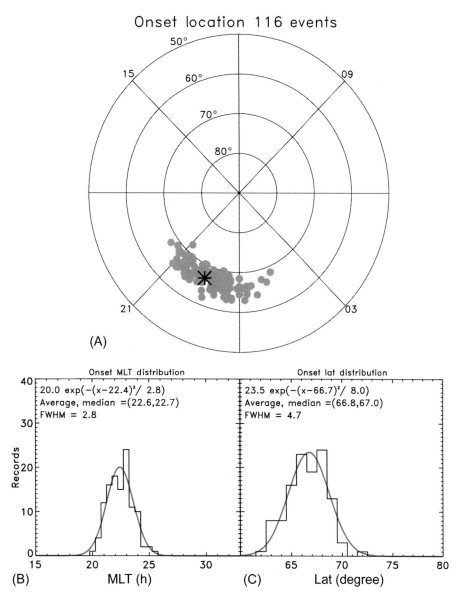

Fig. 11.7 (A) The distribution of onset locations in MLT/ILat with the average marked as an *asterisk*. (B) and (C) The distribution as histograms in MLT and ILat, respectively, of the onset locations. Gaussian curves are fitted to the data, with the equation, average, median, and full-width at half-maximum included. From Gjerloev, J.W., Hoffman, R.A., Sigwarth, J.B., Frank, L.A., 2007. Statistical description of the bulge-type auroral substorm in the far ultraviolet. J. Geophys. Res. 112, 6, A07213. https://doi.org/10.1029/2006JA012189.

and ILat of 66.7 degrees. The Gaussian curves yield full widths at half-maximum (FWHM) of 2.8 MLT, almost a quarter of the night side auroral oval, and 4.8degrees ILat, with only one event above 70degrees ILat. The widths at half-maximum argue that a simple MLT/ILat superposition of substorm intensity distributions would certainly smear or even obliterate key features of the optical substorm. These average patterns were then fit to Gaussian distributions in latitude for each of 24 MLTs.

11.2.2.3 Normalize the latitudinal extent

The third and final step eliminates the varying latitudinal location and north-south width of the disturbance. It requires a nontrivial determination of the low- and high-latitude auroral boundaries, using a software package developed for the purpose.

11.2.2.4 Effectiveness of normalization

The normalization techniques were applied to determine the average UV emission patterns as a function of substorm phase using all 116 events. An example using two sample substorms is shown in Fig. 11.6C–K. The position of each pixel in an image is normalized to a pixel in the average substorm and then all the intensities in the normalized pixel are averaged. The localized average onset is centered at 22.4 MLT and 66.7degrees ILat. The characteristic poleward expansion of the oval is accompanied by eastward and westward expansions. During expansion the poleward boundary of the bulge is nearly a straight line. The maximum extent of poleward expansion remains at approximately the MLT of the onset. The entire pattern is asymmetric around midnight. The most intense emissions are found in the western and poleward part of the bulge. Thus the key features of the individual events are maintained in the normalized substorm.

11.2.3 A clear example of expansion

An example of an optical substorm using the VIS Earth Camera taken on January 3, 2000 is shown in Fig. 11.8. It is an exceptionally clear event selected to emphasize the key positions P1 to P6. The original image on the left is mapped to an MLT ILat coordinate system on the right. By a visual inspection of the images, they have identified the position of:

- P1: westward traveling surge (MLT and ILat)
- P2: east end of the bulge (MLT and ILat)
- P3: west end of the oval aurora (MLT)
- P4: east end of the oval aurora (MLT)

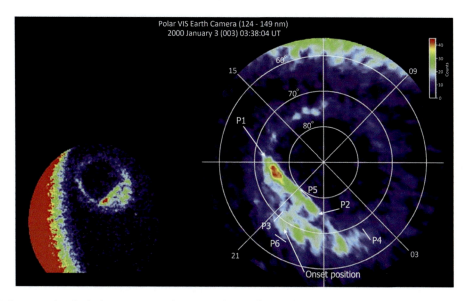

Fig. 11.8 An example of a bulge-type auroral event at the maximum of the expansion phase of a substorm in FUV wavelengths. The original image on the left, and a mapping to an MLT, ILat coordinate system on the right. From Gjerloev, J.W., Hoffman, R.A., Sigwarth, J.B., Frank, L.A., 2007. Statistical description of the bulge-type auroral substorm in the far ultraviolet. J. Geophys. Res. 112, 3, A07213. https://doi.org/10.1029/2006JA012189.

- P5: poleward boundary of the bulge at the MLT of onset (ILat)
- P6: equatorward boundary of auroral oval at MLT of onset (ILat)

11.2.4   Six key positions

The six key positions were determined for the average sub-storm. In Fig. 11.9 (left), the MLTs of the west and east ends of the bulge (P1 and P2) and the west and east ends of the oval aurora (P3 and P4) from the onset location are plotted as a function of the normalized substorm time. This plot clearly shows that the majority of expansion occurs during the first half of the expansion period (to $T=0.5$), both to the west and east. The bulge expansion is remarkably symmetric around the onset MLT, though slightly more to the east. The bulge location remains quite stable in MLT after $T=1.0$. Thus there is essentially no movement, on average, of the bulge as a whole in MLT as the substorm expands and decays, so the center of the bulge remains close to the onset MLT.

The total length of all auroral brightenings, from the west end of the bulge to the east end of the oval aurora (P1 to P4), increases to 7.1 h of MLT at $T=1.5$. Fig. 11.9 (right) shows the average

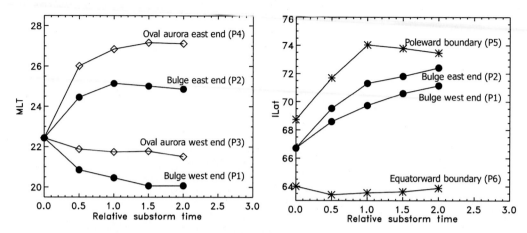

Fig. 11.9 (Left) Average MLT for the ends of the bulge and oval aurora. (Right) ILat of the ends of the bulge aurora and boundaries of the oval aurora at the MLT of the onset, as a function of substorm time. From Gjerloev, J.W., Hoffman, R.A., Sigwarth, J.B., Frank, L.A., 2007. Statistical description of the bulge-type auroral substorm in the far ultraviolet. J. Geophys. Res. 112, 11, A07213. https://doi.org/10.1029/2006JA012189.

latitudinal positions of the west and east ends of the bulge (P1 and P2) as well as the poleward boundary of the bulge (P5) and the equatorward boundary of the oval aurora (P6) at the onset MLT. For the west and east ends of the bulge, the onset point (ILat) was used for $T=0$ since the bulge expands out of the onset, not the poleward boundary of the oval at this time. The equatorward boundary is remarkably stable moving about 0.5 degrees in the early expansion phase and then slowly recovering back toward the growth-phase position. The poleward boundary moves rapidly poleward in the expansion phase after which a gradual equatorward recovery is seen.

11.2.5 Application of the results

The method is fully automated, does not involve any fixed threshold, and requires no subjective adjustment for any of the more than 1200 images analyzed. The average poleward boundary across the bulge is a reasonably straight line, rather than being aligned with constant magnetic latitude (MLAT). The ILat position of the poleward boundary shows the plasma is proceeding toward the pole, as in a plasmoid traveling tailward.

This normalization method requires only an auroral image (such as Fig. 11.7) for which the entire bulge is identifiable and the time of the image in normalized substorm time. The pixel locations of the image can then be normalized to positions in the average substorm. Without these normalization steps, the results of any superimposed epoch analysis would give smeared and/or misleading results.

The typical bulge expanded to 74degrees ILat, all events expanded to at least 70 degrees and none beyond 80 degrees. The rates of expansion are highly variable, with the bulge ends and poleward boundary at the onset location exploding out of the onset point, but then slowing down considerably during the last half of the expansion period. For the first half of the expansion period, the average expansion rates are about 1.14 and 1.30 km/s, respectively, for the west and east ends, and about 0.66 km/s poleward from the onset point. The fastest expansion is the eastward end at about 2.0 km/s.

Ieda et al. (2008) have also shown that the flow depends strongly on the dawn-dusk (Y) location. They show GEOTAIL observations of tailward flows between 0 and 5 min after auroral breakups at locations on the equatorial plane for 65 breakup events. Most tailward flows were observed in the range $-5 \leq Y < 10\ R_E$, outside of which tailward flows were rare.

11.2.6   Validation of the Akasofu model

From this and other independent studies it can be concluded that the characteristics of auroral substorms are those foreseen by Akasofu (1964).

> *On the basis of this analysis we made the following conclusions. The normalization technique is highly efficient in maintaining the key features in the individual auroral emission patterns, even though the individual events varied significantly in intensity, size, position, and lifetime. Thus our normalization results quantitatively validate the Akasofu (1964) assumption that key auroral features exist in the bulge-type auroral substorm. After the substorm onset the auroral oval becomes clearly bifurcated consisting of two components: the oval aurora in the latitude range of the preonset oval and expanding primarily eastward postmidnight, and the bulge aurora, which emerges out of the oval, expanding poleward and both east and west in MLT.*
>
> **Gjerloev et al. (2008)**

Gjerloev et al. (2008) concluded that the optical emissions separate during the expansion phase of a substorm into two components that are quasi-independent.

- The bulge aurora starts with a localized onset, that is, a new region of emissions within the preonset oval, whereas the oval aurora brightens simultaneously within the preonset oval over some hours in local time.

- The bulge aurora expands poleward out of the presubstorm oval, westward as well as eastward, from the premidnight onset location. This results in a very asymmetric total emission pattern around midnight, the westward surge and omega band auroras.
- The oval aurora emissions at lower latitudes decay more rapidly than the bulge emissions, nearly settling to preonset levels by $T=2.0$ (about 30 min into the recovery phase).
- While the two emission regions are parts of the same phenomenon, the precipitation and energization mechanisms of electrons producing the emissions must be different.

These different temporal and spatial behaviors of the bifurcated auroral substorm have implications regarding the sources of electrons and ions that produce the emissions. The onset within the preonset oval implies that two different precipitation and energization mechanisms exist initially in the same region of the magnetosphere. The bifurcation of the oval during the expansion phase indicates a separation of the precipitation/energization mechanisms. The separate time constants of the decays emphasize the independence of the two sources.

11.3   Two models as apparent alternatives

Nearly six decades have passed since the concept of the auroral substorm was hypothesized by Akasofu (1964). Many ideas have been proposed for an explanation, but most of them have fallen by the wayside (see McPherron, 1995, p. 400).

> *Auroral substorms are the result of a complex process involving coupling between the solar wind, magnetosphere, and ionosphere. [Our] objective is to understand the processes that precede what is perhaps the most fundamental aspect, the brightening of a preexisting arc at onset. The notion that this brightening provides an indicator of substorm onset remains one of the few aspects of substorms that is not disputed.*
>
> **Lessard et al. (2007)**

The two substorm models of Fig. 11.10, the cross-field current disruption (CD) model (Lui, 1996, 2004), and the near Earth neutral line (NENL) model (e.g., Hones, 1984; Baker et al., 1996; Shiokawa et al., 1997), are in question now. They emphasize different aspects of substorm processes, setting the two as *apparent* alternatives.

In the inside-out model of Lui (2004) (Fig. 11.10A), current disruption takes place in the near-Earth region; this launches a rarefaction wave in the tailward direction producing a more stretched local magnetic configuration in the midtail. This in turn

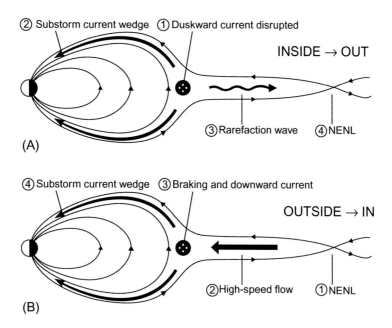

② Substorm current wedge ① Duskward current disrupted

INSIDE → OUT

③ Rarefaction wave   ④ NENL

(A)

④ Substorm current wedge ③ Braking and downward current

OUTSIDE → IN

② High-speed flow   ① NENL

(B)

Fig. 11.10 (A) The cross-field current disruption model of Lui (1996). (B) The near Earth neutral line or reconnection model of Hones (1984). Accurate timing of observed signatures of reconnection, current disruption, and the auroral breakup might clarify the causal relationship between the two processes and the substorm onset. From Ohtani, S., 2004. Flow bursts in the plasma sheet and auroral substorm onset: observational constraints on connection between mid-tail and near-Earth substorm processes. Space Sci. Rev. 113, 79.

sets up a favorable condition for a near Earth neutral line to form (Cheng, 2004; Ohtani et al., 2006). This view is supported by a cross-correlation analysis of Cluster magnetotail data and onset times from Substorm Onsets and Phases from Indices of the Electrojet (SOPHIE). Coxon et al. (2018) found that the magnetic energy density increased approximately linearly in the hour preceding onset and decreased at a similar rate afterward. The timing and magnitude of these changes varied with downtail distance, with observations from the midtail showing larger changes in the magnetic energy density that occur about 20 min after changes in the near-tail. The decrease in energy density in the near-tail region is observed before the ground onset identified by SOPHIE, implying that the substorm is driven from the magnetotail and propagates into the ionosphere.

The outside-in model of the latter group (Fig. 11.10B) is based on the idea that the near Earth neutral line is formed in the midtail region (Hones, 1984; Baker et al., 1996). This model describes depolarization as a pileup of the magnetic flux carried by the earthward plasma flow ejected from the near Earth neutral line.

As a result, the radial profile of the equatorial magnetic field becomes more gradual corresponding to the local reduction of the tail current. While each model has found its own observational support, they are still controversial; the interpretation is often inconclusive (Lin et al., 2009).

An alternative consideration for the substorm onset has been suggested by Song (2003), Song and Lysak (2001), and Lysak et al. (2009). It emphasizes that the substorm onset is a result of global Alfvénic interactions (GAI) in the current system including the tail and magnetopause current sheets as well as the auroral field-aligned current (FAC) system. This alternative global Alfvénic interaction scenario suggests that the substorm onset results from coupled wave-related and convection-related dynamical processes covering a broad range of temporal and spatial scales in a varying driven system. Therefore the substorm onset should follow a more complicated temporal sequence, requiring a new approach to analyze the timing relationship between onset signatures in different regions in the tail and auroral activity during substorms.

Results from a statistical study of 11 substorms (Lin et al., 2009) show that signatures in the midtail ($x \sim 15$–$25\ R_E$) typically occur before the ground signatures and those in the near-Earth tail ($x \sim 10\ R_E$) and that signatures in the midtail region observed prior to the substorm onset often occur at a time that was shorter than that expected from MHD wave propagation between the different regions. This suggests that the disturbance onsets in different active regions do not seem to have a simple causal relationship between them, as described by the reconnection or current disruption models of substorms. These results to some extent are consistent with suggested global Alfvénic interaction considerations, in which the substorm onset is the result of Alfvénic interaction in the global current systems.

11.3.1 Kinetic Alfvén waves and auroral beads

Recent work by Kalmoni et al. (2019) has shown that shear Alfvén waves of short perpendicular extent in the near-tail region may be responsible for substorm onset. Periodic auroral features which emerge and propagate along the arc (Fig. 11.11A), called auroral beads (after Henderson, 1994), are a common feature of the substorm onset arc. They have been observed simultaneously in the northern and southern hemisphere, suggesting a magnetospheric source (Motoba et al., 2012). The high time resolution of images from the Multi-spectral Observatory Of Sensitive EM-CCDs (MOOSE) all-sky imagers at Poker Flat in Alaska allows for the calculation of an observational dispersion relation from images of the bead's evolution over time. The authors used a

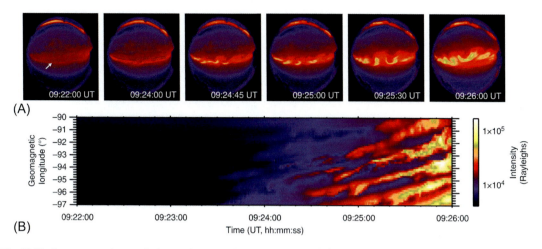

Fig. 11.11 A summary of auroral observations of substorm instability. (A) Snapshots of auroral intensities as a function of latitude and longitude in the 557.7 nm wavelength through the substorm. (B) An along-arc intensity profile (keogram) as a function of geomagnetic longitude and time. From Kalmoni, N.M.E., Rae, I.J., Watt, C.E.J., Murphy, K.R., Samara, M., Michell, R.G., et al., 2019. A diagnosis of the plasma waves responsible for the explosive energy release of substorm onset. Nat. Commun. 9, 3. https://doi.org/10.1038/s41467-018-07086-0.

kinetic theory of high-beta plasma to demonstrate that the shear Alfvén wave dispersion relation bears a remarkable similarity to the auroral dispersion relation. Such shear Alfvén waves in a high-beta environment are often called kinetic Alfvén waves. The spatial and temporal scales of the waves can be visually identified from the variability of auroral beads shown in Fig. 11.11B. The high time resolution of the auroral imagers allows measurements of wave frequencies up to the Nyquist frequency of $\omega_r = 11$ rad/s; however, no waves were seen above 0.8 rad/s. In contrast to prevailing theories of substorm initiation, Kalmoni et al. (2019) contend that the auroral beads seen during substorm onsets are likely the signature of kinetic Alfvén waves driven unstable in the high-beta magnetotail.

11.4 Substorm disturbance onsets

There are many observations of disturbance onsets but we will describe only two.

11.4.1 Auroral onset brightenings

Frank and Sigwarth (2000) reported a simultaneous depolarization and auroral breakup event. High-resolution global images of Earth's auroras in the visible emissions of atomic oxygen at

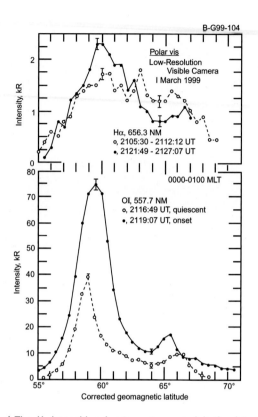

B-G99-104

Fig. 11.12 (Top) The *Hα* intensities due to protons precipitating into the upper atmosphere before and after substorm onset as observed with the Low-Resolution Visible Camera onboard the Polar spacecraft. These intensities are shown as functions of corrected geomagnetic latitude in the 0000–0100 MLT sector. (Bottom) Observations of the O I 557.7-nm emissions, which show the dramatic brightening of the equatorward arc at a latitude of 59 degrees. From Frank, L.A., Sigwarth, J., 2000. Findings concerning the positions of substorm onsets with auroral images from the Polar spacecraft. J. Geophys. Res. 105(A6), 12759.

557.7 nm with a camera onboard the Polar spacecraft were used to determine the MLAT of the auroral arc, which brightens at the onset of a substorm. Fig. 11.12 shows their results.

Six events were analyzed: an intensification and an onset on December 17, 1997, three substorms on January 6, 1998, and a substorm on March 1, 1999. The Tsyganenko (1989) model for the global magnetic field was used to map the field lines threading the center of the auroral arc to equatorial distances. During the period of relative magnetic quiescence on December 17 the equatorial distances were ∼9 R_E. During the magnetically disturbed periods of January 6 and March 1 the magnetic field lines for the auroral onset *brightenings were mapped to equatorial distances in the range of 5–7 R_E, and thus in the region of the extraterrestrial ring current. The equatorial position of the field lines was independently verified for the substorm on March 1 by observing the atmospheric footprint of proton precipitation from the ring current. This footprint was seen in the emissions of atomic hydrogen at 656.3 nm. The onset arc was located near the earthward edge of the ring current at magnetic shell parameter* $L = 4$ R_E.

Frank and Sigwarth (2000)

Fig. 11.12 shows 2 maxima, in line with the work discussed next.

Milan et al. (2019) identified two classes of substorm onset using observations of field-aligned currents by the Active Magnetosphere and Planetary Electrodynamics Response Experiment (AMPERE), which uses magnetic field data from the commercial Iridium satellite constellation. They classified high-latitude onsets (above 65 degrees magnetic latitude) as weak, and low-latitude onsets as intense. Weak onsets can develop into steady magnetospheric convection if the interplanetary magnetic field remains southward for a prolonged period following onset. Intense onsets develop a poleward-expanding auroral bulge and experience convection braking.

11.4.2 Two classes of auroral power

Shue et al. (2008) studied earthward plasma sheet fast flows from GEOTAIL plasma and magnetic fields with a criterion of $V_{\perp x} > 300$ km/s, where $V_{\perp x}$ is the X component of the plasma flow perpendicular to the ambient magnetic field. They estimated rates of change of the nightside auroral power over the courses of the fast flows using Polar UVI auroral images. They found that fast flows observed at $|Y| < 6$ R_E during 1997–98 can be classified into two classes (Fig. 11.13).

One class of the earthward fast flows (Class I) was often observed near X $= \sim - 10$ R_E *and the other class (Class II) was found at* X $< - 15$ R_E. *The auroral power rates of change of the fast flows in Class I in terms of time are found to be high. … The auroral power rates for most of the earthward fast flows in Class II are low. The auroral features, such as poleward boundary intensifications and pseudobreakups, are found to be associated with these fast flows.*

Shue et al. (2008)

It is essential to break [observations] down into two categories: those that take place in the region of H_β *emissions, and hence near the inner edge of the plasma sheet, and those that take place at the*

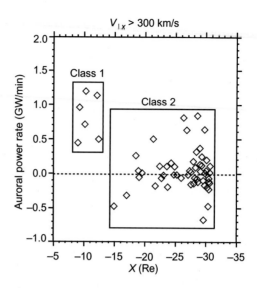

Fig. 11.13 Auroral power rates of change estimated over the courses of earthward fast flows ($V_{\perp x} > 300$ km/s) in terms of X in the GSM coordinate system. The auroral power is integrated over a region of 60–80 degrees MLAT and 2000–0400 MLT. From Shue, J.-H., Ieda, A., Lui, A.T.Y., Parks, G.K., Mukai, T., Ohtani, S., 2008. Two classes of earthward fast flows in the plasma sheet. J. Geophys. Res. 113, 5, A02205. https://doi.org/10. 1029/2007JA012456.

poleward border of the auroral oval. The former can be viewed as either legitimate EP [expansion phase] onsets or pseudobreakups. The latter must be viewed in a different light and take place in a completely different region of space farther out in the magnetotail … Each disturbance, whether it occurs in the poleward or equatorward region of the oval, is accompanied by the appearance of discrete arcs and surge activity together with enhanced westward electrojet current along the oval. The current near the equatorward edge can behave in a totally different fashion than the current at the poleward edge.

Rostoker (2002)

11.4.3 Action versus reaction

An important limitation in the models of Fig. 11.10 (CD and NENL) is related to a proper account of Newton's third law (for every action, there is an equal and opposite reaction). The direction of the force on the first object (the escaping plasmoid) is opposite to the direction of the force on the second object (the inner plasma sheet and the ionosphere). Forces always come in pairs: *equal and opposite action-reaction force pairs.* Rocket propulsion depends on this fact.

A plasmoid begins to travel tailward, somehow connected to the outward meander of the cross-tail current. If the speed is sufficiently fast there may be a shock wave. There is a force on the plasmoid $\mathbf{J} \times \mathbf{B}_Z$ necessary to overcome the resistance to travel tailward on closed field lines. Newton's third law states that there must also be a reaction, of equal magnitude, as suggested by the earthward meander on the left. These two results (Figs. 11.12 and 11.13) are clearly the reaction to the escaping plasmoid required by Newton's third law.

11.5   Substorm transfer event

Heikkila and Pellinen (1977, Pellinen and Heikkila, 1978) have proposed a different model for the substorm phenomenon shown in Figs. 11.14 and 11.15. It uses three-dimensional time-dependent considerations right from the start; using the *real* electric field (not the convection electric field). It is similar to a plasma transfer event (PTE) discussed in Chapter 6, but its details here are quite different. We shall call it a *substorm transfer event* (STE) following current thinning event (CTE) suggested in Chapter 10. The STE has at least three major features listed in the introduction of this chapter that deserve attention, discussed now.

11.5.1   Electromotive force to tap magnetic energy

An induction electric field is governed by Faraday's law,

$$\nabla \times \mathbf{E} = -\partial \mathbf{B}/\partial t \tag{11.1}$$

In the integral form the line integral of $\mathbf{E}$ around a closed contour generates the electromotive force (emf)

$$\varepsilon = \oint \mathbf{E} \cdot \mathbf{dl} = -d\Phi^M/dt \tag{11.2}$$

Only by this emf can we tap the energy stored in the magnetic field by Maxwell's equations. The electric field that is spatially constant, as assumed in theoretical work on standard magnetic reconnection (SMR), has zero curl and zero emf.

The electrostatic field without a curl cannot affect the curl of the induction electric field, nor the electromotive force. The charges flowing along the magnetic field lines in an attempt to reduce the parallel component of the electric field due to

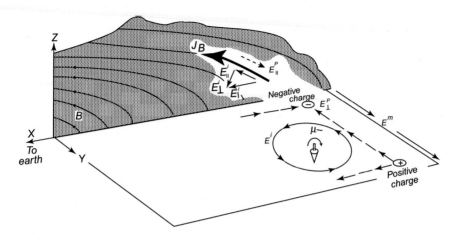

Fig. 11.14 The topological difference between the two types of electric fields may be appreciated by inspection of this figure. The charges flowing along the magnetic field lines in an attempt to reduce the parallel component of the electric field due to induction (shown by the *full arrows*) will *enhance* the dawn-to-dusk part of the perturbation electric field *(the dashed arrows)*. The electrostatic field without a curl cannot affect the curl of the induction electric field, nor the electromotive force. From Heikkila, W.J., Pellinen, R.J., 1977. Localized induced electric field within the magnetotail. J. Geophys. Res. 82, 1613.

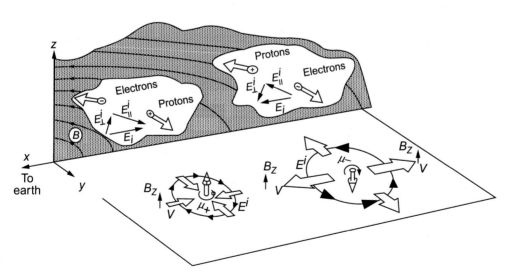

Fig. 11.15 The two possible choices for the current meanders in the equatorial plane are outward (shown by the magnetic moment $\mu-$) or inward (shown by μ_+). The outward meander is explosive in that the $\mathbf{E} \times \mathbf{B}$ direction is everywhere out of the disturbance. The induction electric field has only x and y components everywhere, due to its source, which is assumed to be the current meander in the equatorial plane. Because of the tail-like geometry this means that the electric field has a large component parallel to the magnetic field above and below the current plane. This leads to field-aligned forces on charged particles due to induction. From Heikkila, W.J., Pellinen, R.J., 1977. Localized induced electric field within the magnetotail. J. Geophys. Res. 82, 1613.

induction (shown by the full arrows in Fig. 11.14) will *enhance* the dawn-to-dusk part of the perturbation electric field (dashed arrows).

11.5.2 The first response

Pellinen and Heikkila (1978) constructed an ad hoc mathematical model involving linear growth in the perturbation current magnitude as well as in the size of the perturbed region. The activity starts with an instability, most likely with the ions because of their much larger gyration radius. The immediate action is to cause a current meander, as shown in Fig. 11.14. The meander can be to the right or left in the equatorial plane; the primary action should be to the right as the local plasma is seeking to escape. This meander is in the tailward sense; since the current is locally increasing, Lenz's law implies that the inductive electric field opposes the increase so that $\mathbf{E} \cdot \mathbf{J} < 0$. This action is hidden from us since it is a dynamo, with no precipitation (explained in Chapter 10).

On the other side the current density weakens and the result is an electrical load with $\mathbf{E} \cdot \mathbf{J} > 0$. The electron energy is increased, and so *the arc brightens*. As the current meander grows, it will deform the total magnetic field in the tail. Typical test particles were followed in these time-dependent fields, taking the full relativistic equation of motion.

11.5.3 Negative and positive meanders

The cross-tail current can *meander*, either to the left or to the right producing an induction field with opposite signs. The resulting induction electric field is solenoidal, opposing the perturbation current, by Lenz's law. The line integral around the current perturbation is the *emf* associated with the perturbation. The parallel component of the induced electric field in the tail lobes will accelerate protons and electrons on the dawn side and dusk side of the perturbation (only the dawn side is shown, the directions being reversed on the dusk side). Therefore we have two parts of the substorm to consider, labeled as negative and positive meanders in line with Newton's third law. They are parts of the same phenomenon, action versus reaction.

The results demonstrate clearly the capabilities of a rotational electric field, which are quite different from those of an electrostatic field. To study field-aligned acceleration, particles were launched slightly above the neutral sheet in the regions where the field lines were connected to the Earth (Fig. 11.16). The picture of precipitation at the substorm onset bore out Fukunishi's (1975)

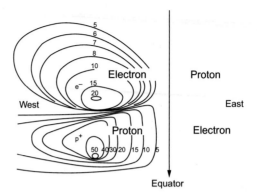

Fig. 11.16 Energy contours for electrons and protons heading toward the Earth. The energized electron flux at high-latitudes would produce the pure electron excited aurora in the westward traveling surge in an auroral breakup. A cyclic pattern of precipitation would result, as is observed during substorms. From Pellinen, R.J., Heikkila, W.J., 1978. Energization of charged particles to high energies by an induced substorm electric field within the magnetotail. J. Geophys. Res. 83, 1548.

finding (Fig. 11.17). In a typical substorm, the estimated area of electron precipitation exceeding 5 keV at the ionospheric level on the evening side is about 400–200 km^2, which would almost fill an all-sky photograph (diameter around 500 km). This agrees well with observations of auroral breakup.

11.6   Ion dynamics

The development of an intense current during the growth phase implies that magnetic energy is being stored in the magnetotail. A lower-energy state for the added cross-tail current is a filament or set of filaments. We note that such filaments are connected to auroral arcs during the growth phase. There are usually several arcs present but only one which "breaks up," usually the equatorward arc. Akasofu et al. (2010) have made this argument again, publishing several excellent examples.

We must have a breakup arc, an essential part of the substorm. Therefore the current carriers must include electrons and ions at small pitch angles, particles in the loss cone; these particles, both electrons and ions, gain energy from curvature drift in the plasma sheet as shown in Fig. 10.10 of Chapter 10.

At the beginning of the growth phase both electrons and ions are able to follow the field lines because of their low energy, a few 100 eV for electrons but greater for the ions (more than 1 keV). As they drift inward they gain energy adiabatically, conserving their magnetic moments, electrons to several keV, and ions to more

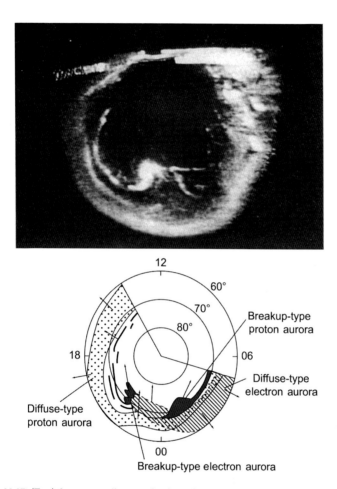

Fig. 11.17 (Top) An outstanding result of a substorm was recorded on December 22, 1971 by ISIS-2. The auroral oval and a well-developed substorm was in progress. (Bottom) A diagram (Fukunishi, 1975) showing a cyclic pattern. From Pellinen, R.J., Heikkila, W.J., 1984. Inductive electric fields in the magnetotail and their relation to auroral and substorm phenomena. Space Sci. Rev. 37, 58. https://doi.org/10.1007/BF00213957.

than 10 keV. Near the end of the growth phase near the region of high curvature, energization reaches the point where the ions (being massive, have much larger gyroradii, and also higher energy) cannot negotiate the sharp turn in the field-reversal region. This happens locally, at some local time near midnight, probably on the equatorward auroral arc (Akasofu, 1964) since the ions are more energetic farther in. The critical orbit is shown in Fig. 11.18A (Delcourt and Belmont, 1998a,b).

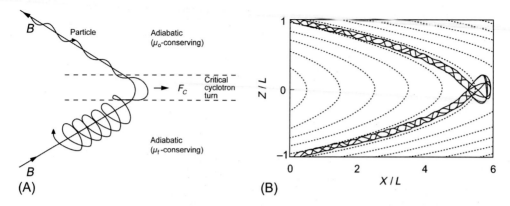

Fig. 11.18 (A) Schematic representation of the centrifugal impulse model. B denotes the magnetic field, Fc the impulsive force, and μ_o and μ_f the magnetic moments prior to and after crossing of the field reversal, respectively. (B) A value for $K=0.315$. The ion overshoots the midplane, forming currents above and below, a bifurcated current. From Heikkila, W.J., Chen, T., Liu, Z.X., Pu, Z.Y., Pellinen, R.J., Pulkkinen, T.I., 2001. Near Earth Current Meander (NECM) model of substorms. Space Sci. Rev. 95, 408.

11.6.1 Critical cyclotron turn

The nonadiabatic particle behavior can be characterized by the parameter κ defined as the square root of the minimum curvature radius to the maximum Larmor radius (Büchner and Zelenyi, 1989). Near the critical cyclotron turn the orbits become chaotic, and the energization becomes nonadiabatic with $\kappa < 1$. Closer to the Earth, the motion is again adiabatic where the magnetic field is greater, and also the curvature of the field lines is diminished, $\kappa \geq 3$.

The calculated linear current density during the growth phase is about 50 mA/m (Delcourt and Belmont, 1998b); over 10 R_E that would amount to 3×10^6 A, which is of the correct magnitude as deduced by Pulkkinen et al. (1991). In 1 R_E (as above) the total current would be 3×10^5 A; the local current density might even be higher due to the localization associated with a filamentary structure.

11.6.2 Bifurcated neutral sheet current

The magnetic field lines control the particle motions in adiabatic drift. As the curvature becomes higher and the particle gains energy, somewhere some particles will reach the critical moment where they overshoot. Fig. 11.18B shows the simulation result: they overshoot, some going northward and some southward depending on pitch angle. These are the current carriers of the cross-tail current; they also produce the auroral emissions. A value for $k = 0.315$ (Delcourt and Belmont, 1998b) is shown.

The ion overshoots the midplane, forming currents above and below, a twin current. This might be the cause of the bifurcation of the neutral current shown in Fig. 11.19. As pointed out by Heikkila et al. (2001), this initial jump is not small, approaching 1 R_E. Furthermore, the induction electric field is explosive, with the local plasma going everywhere out of the "bubble."

Examples have been found, for example, by Nakamura et al. (2006), Runov et al. (2005), and Shen et al. (2003, 2008). Runov et al. (2005) analyzed the vertical structure of the magnetotail current sheet for a few intervals during which Cluster repeatedly crossed the neutral sheet due to fast flapping motion. On October

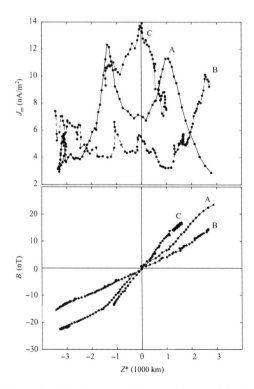

Fig. 11.19 On October 8, 2001, Cluster was located at $[-15.1, 9.4, 0.1]$ R_E. At the first crossing of the neutral sheet interval (*A*) the current is definitely bifurcated, with two similar peaks quasisymmetrical with respect to $Z^* = 0$, and a pronounced broad valley in-between. The crossing *A* displays the best example of the bifurcated current sheet. From Runov, A., Sergeev, V.A., Nakamura, R., Baumjohann, W., Zhang, T.L., Asano, Y., et al., 2005. Reconstruction of the magnetotail current sheet structure using multi-point cluster measurements. Planet. Space Sci. 53, 242.

8, 2001, Cluster was located at $[-15.1, 9.4, 0.1]$ R_E. The Alaska magnetometer chain detected a set of deflections with a peak of 500 nT during 1000–1600 UT. Three crossings of the neutral sheet, marked as A, B, and C in Fig. 11.19 were selected for the analysis. The timing analysis for crossing A shows that the boundary moved mostly duskward with velocity of 130–150 km/s. The analysis shows that during the 5-min-long interval the current sheet structure changed drastically from crossing to crossing. Fig. 11.19 shows the first crossing interval (A), where the current is definitely bifurcated. There are two similar peaks quasisymmetrical with respect to $Z^* = 0$, and a pronounced broad valley in-between. The next crossing (B) shows an incoherent structure of the current sheet with larger current density at the periphery than in the center, and the third interval (C) is more intensive and has a pronounced peak at $Z^* = 0$.

Reconstruction of the spatial profile of the magnetic field and the electric current in the time-varying plasma sheet can be done if spacecraft repeatedly cross the current sheet. Such crossings should be fast enough to be sure that the parameters of the sheet and its orientation remain the same within the crossing. This occurs during large amplitude vertical oscillation of the current sheet (flapping). For the presented studies we select several repeating crossings with a characteristic duration $30 \leq \tau \leq 120 s$ and a jump of B_X exceeding 15 nT.

Runov et al. (2005)

Multisatellite observations with the THEMIS mission have established that this flapping motion is not always a global process, but instead can be confined to a region of tailward flow. The observations indicated that the flapping motion in the tailward flow could have a different generation mechanism with that in the earthward flow (Wu et al., 2016b).

11.6.3 Plasma conditions in a thin sheet

Zhou et al. (2009) studied the structure of a thin current sheet on the basis of THEMIS observations in Fig. 11.20; this was prior to the expansion onset of a substorm event that occurred on February 26, 2008 in the near-Earth magnetotail. During this time interval, the ion distribution showed mushroom-shaped structures with clear nongyrotropic features. This indicates that the warmer component of the ions was unmagnetized, which becomes possible only if the gyroradii of these ions are comparable with the current sheet thickness. By comparing the observations with the model proposed by Sitnov et al. (2003), which is a modification

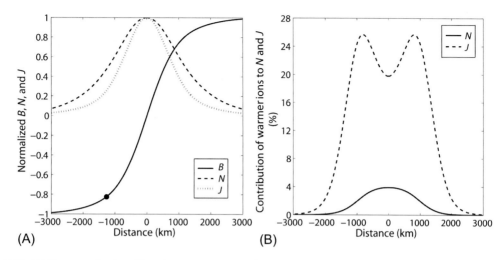

Fig. 11.20 (A) The normalized profiles of the magnetic field *(solid line)*, the plasma density *(dashed line)*, and the electric current density *(dotted line)* within the modeled current sheet. The *heavy dot* represents the location of the satellite. (B) The percentage of the warmer ion density to the local plasma density *(solid line)* and the percentage of the electric current carried by the warmer ion component to the local current density *(dashed line)* at each location of the current sheet. From Zhou, X.-Z., Angelopoulos, V., Runov, A., Sitnov, M.I., Coroniti, F., Pritchett, P., et al., 2009. Thin current sheet in the substorm late growth phase: modeling of THEMIS observations. J. Geophys. Res. 114, 7, A03223.

of the Harris model, and considering the effect of the meandering ions in thin current sheets, they reconstructed the current sheet structure in the late growth phase of the substorm. Warmer ions, mostly following meandering orbits across the neutral sheet, are found to remarkably alter the current sheet profiles and therefore play an important role in the formation of the thin current sheet. Sitnov et al. (2019b) used data mining techniques with a number of missions including IMP-8, Geotail, Polar, GOES, Cluster, THEMIS, Van Allen Probes, and MMS to find that thin current sheets may become as thin as $0.2\,R_E$, comparable to the thermal ion gyroradius, during the growth phase.

In summary, by taking into account the quasiadiabatic properties of the ion meandering motion in the distribution function, the current sheet model with nongyrotropic features is slightly modified to consider both the colder and the warmer ion components, and the modified model is validated by a best-fit procedure to compare with the observational data. Both the strong diamagnetic drift velocity of the warmer ion component and the meandering motion of these ions are found to play important roles in producing the nongyrotropic feature of the ion distribution, and to reorganize the structure of the thin current sheet.

11.7 Westward traveling surge

Undoubtedly the most spectacular auroral display is the westward traveling surge (WTS) that initiates the expansion phase of substorms. This was shown in Fig. 10.21 of Chapter 10, and Figs. 11.36 and 11.38 of this chapter. It is involved in a grand current diversion from the cross-tail current to field-aligned currents and closure in the ionosphere as shown in Fig. 11.21. These are the changes to be expected during a substorm expansion. Like a solar flare it signals a new phase of activity, not a continuation of the old. We must be brave and not be caught up in old habits: "they encourage linear thinking and discourage the factoring in of surprises, discontinuities and disasters" (Harries, 2005).

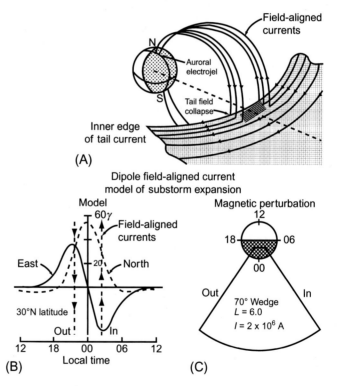

Fig. 11.21 Simple line current model of substorm expansion. (A) Perspective view of a diversion of the inner edge of the tail current. (C) An equivalent current system to be added to the tail current to model this diversion. (B) Midlatitude local time profiles of the magnetic perturbation due to this equivalent current. These are the changes to be expected during a substorm expansion. From Clauer, C.R., McPherron, R.L., 1974. Mapping the local time-universal time development of magnetospheric substorms using mid-latitude magnetic observations. J. Geophys. Res. 79, 2816.

11.7.1 Substorm current diversion

The magnetic perturbations observed on the Earth during the development of the midlatitude positive bay are the result of a current system similar to that shown at the top of Fig. 11.14.

This model suggests that at the beginning of a magnetospheric substorm expansion, a section of the cross-tail current is diverted; that is, it flows down magnetic field lines and then westward through the auroral ionosphere and returns along magnetic field lines. As the expansion phase progresses, the field-aligned currents expand further into the tail, and the returning field-aligned currents move westward. In the auroral zone this is observed as the northward expansion and the westward surge.

Clauer and McPherron (1974)

An all-sky image has been mapped onto a rectangular grid in Fig. 11.22 (Kosch et al., 2000). Plasma flow vectors point away from white dots, and the clockwise flow above the intense arc indicates negative charge, but counterclockwise at higher latitudes indicates positive charge. The surge is very localized, and *poleward of the breakup arc*; thus the substorm current diversion may be more complicated than simply a diversion of the neutral sheet current. Within minutes after the onset the z-component of the magnetic field increases in the plasma sheet with the field becoming more dipolar on lower L-shells. The earthward ion flow activates at the same time (Baumjohann et al., 1991) at these close distances ($<20\ R_E$).

Fig. 11.22 shows the midlatitude signature of such a current wedge. The essential characteristics of this signature are (1) a symmetric positive peak in the north-south component centered about the central meridian of the current wedge, and (2) an asymmetric perturbation in the east-west component with positive perturbation at earlier local time, negative perturbation at later local time, and a crossover on the central meridian.

11.7.2 THEMIS All Sky Imager observations

Liang et al. (2008) investigated a substorm auroral breakup on February 22, 2006 with the THEMIS All Sky Imager (ASI) as shown in Fig. 11.23. Their event database was selected from the THEMIS ASI observations based on good viewing conditions and tractability of auroral brightening and expansions. They are all characterized with noticeable poleward auroral expansion and definitive substorm features from corroborating ground-based and in-situ measurements (magnetometer, riometer, GOES, etc.). In

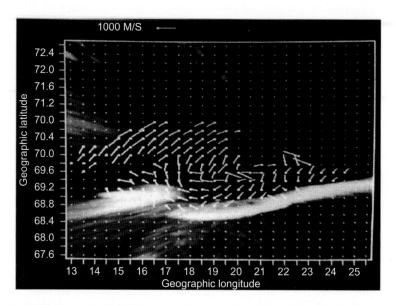

Fig. 11.22 The white-light TV image of the east-west aligned arc after it had brightened, taken at 2135 UT on January 15, 1980. The all-sky image has been mapped onto a rectilinear geographic grid, at an altitude of 100 km, to remove the all-sky distortion. Superimposed is a grid of Scandinavian Twin Auroral Radar Experiment (STARE) plasma flow vectors, with the direction pointing away from each *white dot.* Locations with no vectors correspond to no radar backscatter. From Kosch, M., Amm, O., Scourfield, M., 2000. A plasma vortex revisited: the importance of including ionospheric conductivity measurements. J. Geophys. Res. 105(A11), 24,891.

presentations of the images, they assume an emission height of 110 km, and use the altitude-adjusted corrected geomagnetic coordinate to define MLAT and magnetic longitude (MLON).

Prior to onset, the auroras are stationary and dark inside the field of view of the ASI, with a predominantly azimuthally aligned quiet arc above 66 degrees MLAT (Fig. 11.23). The arc is found to be stable with some slight equatorward motion for about 10 min during the late growth phase. Starting from 06:28:06 UT we see a few "spurs" sprouting. At 06:28:12 UT, another equatorward-extending "spur" begins to grow over the next couple of frames into an equatorward displaced arc segment (marked with arrow in the 06:28:18 frame) in which the initial signature of the auroral intensification developed.

The luminosity enhancement occurred both on the original arc and the new lower-latitude arc, whose formation over an extended azimuthal range becomes quite visible after 06:28:24. The overall

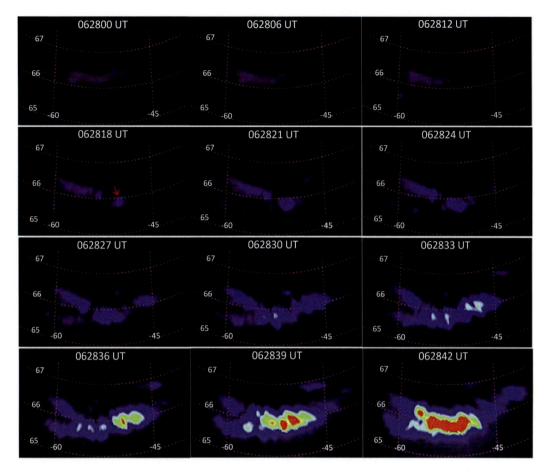

Fig. 11.23 A sequence of THEMIS ASI images showing a substorm auroral breakup on February 22, 2006. From Liang, J., Donovan, E.F., Liu, W.W., Jackel, B., Syrjäsuo, M., Mende, S.B., et al., 2008. Intensification of preexisting auroral arc at substorm expansion phase onset: wave-like disruption during the first tens of seconds. Geophys. Res. Lett. 35, 2, L17S19. https://doi.org/10. 1029/2008GL033666.

auroral intensity was strongly enhanced in the following frames, marking substorm auroral breakup. The onset time for this event is estimated to be 06:28:24±3s. A key feature constituting the core interest of this paper is that the auroral intensification developed in distinct forms of azimuthally-spaced structures. … After 06:28:39 noticeable poleward auroral expansion merged the original and new arcs into an overall latitudinally broad structure. The color scale saturates at 06:28:42 but the embedded intensity wave structures can be clearly seen later.

Liang et al. (2008)

The onset appears to occur over a longitudinally extended (10–15 degrees) arc segment and is characterized by distinct wave-like disruptions with a growth time scale of about 10 s. Examination of other THEMIS ASI onset events suggests that wave-like auroral activation is a quite common feature. High-resolution THEMIS ASI observations confirm that those azimuthal wave-like activations are intrinsic to the initial breakup process. The observations give strong hints of the existence of plasma instability waves in association with the substorm expansion phase onset.

The ballooning instability (Roux et al., 1991) was considered by a number of researchers (Lui, 1991; Cheng, 2004) as the underlying mechanism of the "explosive growth phase." Ohtani (2004) discussed rapid thinning of the near-Earth current sheet about a minute prior to expansion phase onset.

The "spurs" extending from the stable growth-phase arc about 10–20 s before onset may be signatures of fast earthward flows; consequently the development of the lower-latitude breakup arc was quite rapid. When the current sheet is severely thinned to a thickness comparable or less than the thermal ion gyroradius, it may become unstable to several modes of cross-field current instabilities that disrupt the cross-tail current. The potential role of preonset ballooning instability in leading to the cross-field current instability excitation was discussed by Cheng (2004).

Another key observation resulting from the high-resolution capability of THEMIS ASI is the very fast azimuthal spreading of the initial auroral activation over 10–15 degrees MLON within 10 s of the onset. For example, seconds after the onset, a few azimuthally discrete spots spanning from about −59 to −44 degrees MLON (marked in dashed lines in Fig. 11.24A) are evident in the 06:28:30 frame. The subsequent brightening seems more or less confined within this extent and more interestingly, in a partially standing mode fashion. After the ongoing current disruption has significantly changed the local conditions, the standing wave can no longer be sustained.

If the substorm onset stems from a point source which then expands in space, and covers a ∼10 degrees range in 10 s, it would map to an azimuthal speed of the order of ∼1000 km/s at the plasma sheet. A fast magnetosonic wave can barely reach this velocity in the local magnetosphere. The substorm current diversion of the three-dimensional current system is responsible for most of the effects seen by ground-based magnetometers at substorm onset.

A substorm current wedge to the ionosphere is formed as part of the auroral breakup. Kauristie et al. (1995) displayed radar data that showed the new ionization at the onset of the westward

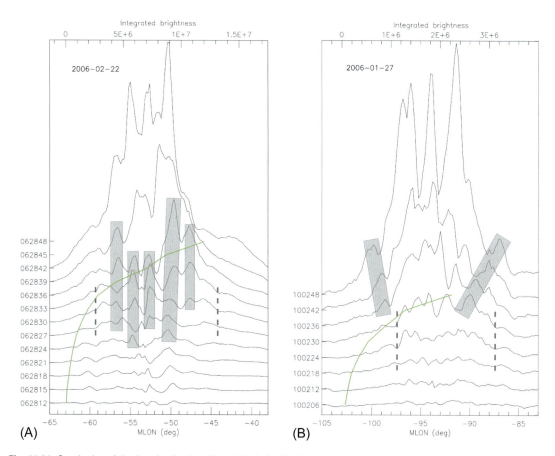

Fig. 11.24 Stack plot of the longitudinal profile of the latitudinally integrated auroral brightness around the expansion phase onsets for two events. Curves denote the temporal evolution of total auroral brightness (ticks on upper axis) with time (labeled on the left). Two solid curves are the growth of the emissions. From Liang, J., Donovan, E.F., Liu, W.W., Jackel, B., Syrjäsuo, M., Mende, S.B., et al., 2008. Intensification of preexisting auroral arc at substorm expansion phase onset: wave-like disruption during the first tens of seconds. Geophys. Res. Lett. 35, 4, L17S19. https://doi.org/10.1029/2008GL033666.

traveling surge was located below 90 km in altitude; the electron beam that suddenly appeared had energies approaching 100 keV.

- Magnetic perturbations caused by the DP-2 current system can be detected on the ground as the auroral electrojets by a chain of magnetic observatories.
- Diversion of the cross-tail current by the DP-1 current system starts at onset through the midnight ionosphere.
- The wedge shape of the projected equivalent current accounts for the name "substorm current wedge."

11.7.3   Auroral kilometric radiation

Morioka et al. (2007) examined the development of the auroral acceleration region during substorms by using high time resolution dynamic spectra of auroral kilometric radiation (AKR) provided by Polar Plasma Wave Instrument (PWI) observations. Two sources of AKR and their development prior to and during substorms were identified from electric field observations (Fig. 11.25). One source is a low-altitude source region corresponding to middle-frequency auroral kilometric radiation (MF-AKR), and the other is a high-altitude source region corresponding to low-frequency auroral kilometric radiation (LF-AKR). The former appears during the substorm growth phase in the altitude range of 4000–5000 km and is active both before and after substorm onset. A few minutes before the onset, the intensity of this source gradually increases, showing precursor-like behavior. It does not change drastically at the onset and is mostly insensitive to it.

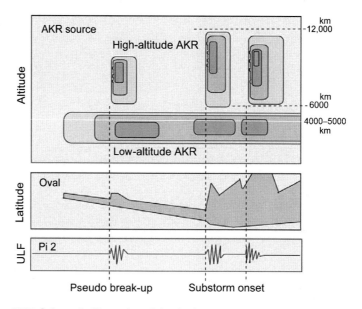

Fig. 11.25 Schematic illustration of the dual structure of auroral kilometric radiation (AKR) sources, showing (top) the dynamical distribution of AKR sources during the growth and expansion phase. AKR sources are composed of quasisteady low-altitude AKR sources and bursty high-altitude AKR sources. Also shown are (middle) the auroral oval evolution across latitudes during a substorm and (bottom) Pi 2 pulsations corresponding to substorm onsets. From Morioka, A., Miyoshi, Y., Tsuchiya, F., Misawa, H., Sakanoi, T., Yumoto, K., et al., 2007. Dual structure of auroral acceleration regions at substorm onsets as derived from auroral kilometric radiation spectra. J. Geophys. Res. 112, 10, A06245. https://doi.org/10.1029/2006JA012186.

*At Pi 2 onset, in contrast, high-altitude AKR appears abruptly with
intense power in a higher and wider altitude range of 6000 to
12,000 km. The increase in its power is explosive (increasing 1000
times within 20s), suggesting the abrupt growth of the parallel
electric fields that cause bursty auroral electron beams. The
statistically derived probability of both sources existing at substorm
onset is ∼70%, indicating that this duality of AKR sources is a
common feature of substorms.*

Morioka et al. (2007)

The intensity of the high-altitude AKR grew explosively up to
1000 times in about 10 s. The horizontal extent of this abruptly
built-up acceleration region is apparently restricted to a narrow
area on the auroral arc because extremely coherent AKR development in a short time-period can be achieved only when the
generation region is confined. Morioka et al. (2007) suggested
that the low-altitude AKR source is related to the large-scale
inverted-V acceleration region that would be generated
through the self-consistent distribution of the magnetospheric
plasma in the magnetosphere-ionosphere coupling region. The
high-altitude AKR source, which is an indicator of a substorm
onset, would be generated from the local field-aligned acceleration due to the current-driven instability or the Alfvénic
acceleration caused by substorm-associated short wavelength
Alfvén waves.

11.8   Bursty bulk flows

The plasma flow in the central plasma sheet of the Earth's magnetosphere includes short impulsive bursts (Baumjohann et al.,
1990; Angelopoulos et al., 1992). The bursty bulk flows (BBF) typically appear in 10 min sequences often consisting of several
∼1 min intensifications. The occurrence rate of high-speed flows,
with typical ion bulk velocities above 400 km/s, increases during
geomagnetically active periods. Since the bursts are almost always
directed earthward they are Class 1 of Fig. 11.13, the reaction to
the escaping plasmoid.

Superposed epoch analysis of the 1985–86 AMPTE/IRM database showed that bursty bulk flows are accompanied by transient
magnetic field dipolarizations and increases in ion temperature
and in the total pressure. A subset of events recorded entirely in
the inner plasma sheet reveals that the ion density decreases
within the bursty bulk flow (Angelopoulos, 1996). The bursty bulk
flow can account for a significant fraction of the earthward particle, energy, and magnetic flux transport. The longitudinal scale

sizes of bursty bulk flows are difficult to measure, but in the rare observations with two closely spaced spacecraft it has been estimated to be only a few R_E. The associated dipolarization fronts tend to move faster in the midtail (57% over 150 km/s as measured by MMS) than in the far-tail (35% over 150 km/s as measured by Cluster) (Schmid et al., 2016). More recent observations of bursty bulk flows, transient dipolarizations, and other explosive magnetotail activities have been reviewed by Sitnov et al. (2019a).

11.8.1 Auroral streamers

The evolution of north-south aligned auroral structures on a global scale and their role in the substorm process was first described using the Viking UV imager data (Rostoker et al., 1987; Henderson et al., 1998). In the study by Henderson et al. (1998), Viking auroral images for two different substorms in Fig. 11.26 illustrate the manner in which these structures develop. In Fig. 11.26A, a sequence of observations of the northern auroral region is shown acquired with the Viking UV imager on October 15, 1986. The 8-min sequence beginning at 1156:45 UT is striking! It confirms that they are Class 1 of Fig. 11.13 and are not related to the escaping plasmoid by Fig. 11.2 which already must be way poleward. The ionospheric effects of that are not obvious as they are in a dynamo with the raising mirror points.

In Fig. 11.26B the sequence of 10 images was acquired by Viking on December 1, 1986. In this case, multiple bright north-south aligned forms are ejected equatorward into the bulge from different locations along the poleward arc system. Note that the north-south aligned forms generated during this event tend to accumulate into bright diffuse patches at the equatorward edge of the bulge.

Auroras are probably so far the most extensively studied form of ionospheric phenomena related to BBFs. Two main kinds of auroral signatures have been observed during BBFs: pseudobreakups and auroral streamers (Nakamura et al., 2001a). Pseudobreakups show in global auroral images as a bright spot. The auroral brightness lasts only a few minutes and then fades, or brightens intermittently. Auroral streamers, on the other hand, are approximately north–south aligned, longitudinally narrow auroral forms that first appear at the poleward boundary of the auroral oval and from there expand equatorward. After reaching the equatorward boundary of the oval, they decay by evolving into a patch of diffuse or pulsating aurora. Streamers occur when the auroral oval is wide, both during substorm and nonsubstorm periods. Both substorm

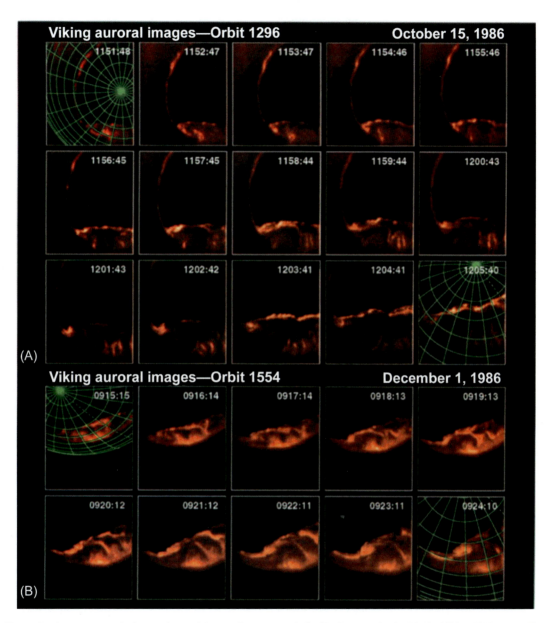

Fig. 11.26 A sequence of observations of the northern auroral distribution acquired with the Viking UV imager. The eccentric dipole coordinate system is superposed on the first and last images. (A) Between 1156:45 and 1205:40 UT on October 15, 1986, the poleward arc "bifurcates" and a spectacular array of north-south aligned auroral forms are rapidly ejected equatorward into the bulge from the poleward edge. (B) Multiple bright north-south aligned forms on December 1, 1986 are ejected equatorward into the bulge from different locations along the poleward arc system. From Henderson, M.G., Reeves, G.D., Murphree, J.S., 1998. Are north-south aligned auroral structures an ionospheric manifestation of bursty bulk flows? Geophys. Res. Lett. 25, 3739.

and nonsubstorm streamers frequently occur in several longitudinal locations of the oval simultaneously, and are often tilted from northwest to southeast. In the schematic picture (Fig. 11.27), the auroral streamer would be associated with the region of upward field-aligned current.

Juusola et al. (2009)

More recently, using the THEMIS all-sky-imagers, Lyons et al. (2018) found that substorm onset is frequently triggered by plasma sheet flow bursts that manifest in the ionosphere as auroral streamers. Substorm auroral onsets were defined in this study as a brightening followed by poleward expansion. The investigators were able to identify a detectable streamer heading to near the substorm onset location for all 60 cases studied, indicating that substorm onsets are very often triggered by the intrusion of plasma with lower entropy than the surrounding plasma to the onset region.

11.8.2 Bubble in plasma sheet

The relationship between the ionosphere and the magnetosphere during fast flows is essential to the understanding of the role of bursty bulk flows during substorms. The bubble model of Chen and Wolf (1993, 1999) describes how the narrow bursts are connected to the ionosphere. They suggested that a source of the low-pressure plasma in the near-Earth region could be plasma bubbles, that is, underpopulated flux tubes transported from the far tail (recall Fig. 11.15 and the explosive nature of the convection due to induction). The depleted plasma pressure is balanced by an increase of magnetic field, and the gradient and curvature drift currents inside the bubble are smaller than in the surroundings. Recent improvements in simulations such as the Rice Convection Model (RCM) to include inertial effects by correcting the expression for field-aligned currents allows it to better model some of the fluid dynamics of bursty bulk flows in the plasma sheet. A promising feature is that the injection of a low-entropy bubble produces braking oscillations and buoyancy waves that radiate away from the original bubble, forming multiple flow vortices, none of which could be produced by traditional RCM simulations (Yang et al., 2019).

Juusola et al. (2009) studied ionospheric equivalent currents related to bursty bulk flows using magnetic conjunctions between the four Cluster spacecraft and the International Monitor for Auroral Geomagnetic Effects (IMAGE) magnetometer network. For Cluster to be conjugate with IMAGE, it was required that the ionospheric footprint of Cluster, determined using the T89 model (Tsyganenko, 1989) with $K_p = 3$, fell within a box of 67–73 degrees in GEO latitude and 14–30 degrees in longitude. The relation of this box to the IMAGE magnetometers is illustrated in Fig. 11.28. Moreover, to ensure that Cluster was located close to the plasma sheet, it was required that $x < -10\,R_E$ and $|z| < 3\,R_E$. These requirements yielded 134 conjunction events with an approximate duration of about 1 h between 2001 and 2006. All conjunctions took place between the months of July and November between 16 and 03 UT. Owing to the location of IMAGE, this translated to ∼19–06 in MLT.

Fig. 11.27 illustrates schematically the plasma flow (black arrows) around a bubble in the plasma sheet and the corresponding convection pattern in the ionosphere. Such signatures were predicted in the schematic bubble model. Because of the charge accumulation at the bubble flanks, an additional polarization

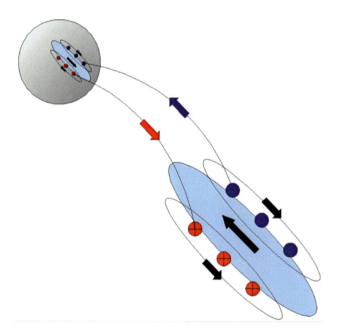

Fig. 11.27 Sketch of the plasma flow *(black arrows)* around a bubble in the plasma sheet and the corresponding convection pattern in the ionosphere. In the ionosphere, electric current in the opposite direction is associated with the convection. Corresponding to the convection pattern, there is upward field-aligned current *(red)* on the dusk side and downward field-aligned current *(blue)* on the dawn side of the bubble. From Juusola, L., Nakamura, R., Amm, O., Kauristie, K., 2009. Conjugate ionospheric equivalent currents during bursty bulk flows. J. Geophys. Res. 114, 2, A04313.

electric field forms inside and around the bubble. This convection pattern corresponds to a field-aligned current (FAC) system consisting of a pair of upward and downward currents, resembling thus a small substorm current wedge. The bubble theory provides a construct against which to compare observations in the plasma sheet.

> *These results support the idea that a dawn-to-dusk polarization electric field is created in the bubble to enhance the flows as predicted in the bubble model (Chen and Wolf, 1993). The result shows that not the entire flow burst region is possibly seen as auroras. It is therefore not simply a precipitation of the flow burst plasma that can be observed as auroras. The bright precipitation seems to be rather associated with upward field-aligned current centered duskward of the flow center as shown in (Fig. 11.27) and the precipitation therefore depends on field-aligned acceleration.*
>
> **Nakamura et al. (2001a)**

During the long-duration steady convection activity on December 11, 1998, the development of a few dozen auroral streamers was monitored by the Polar UVI instrument in the dark northern nightside ionosphere. On many occasions the DMSP spacecraft crossed the streamer-conjugate regions over the sunlit southern auroral oval, permitting the investigation of the characteristics of ion and electron precipitation, ionospheric convection, and FACs associated with the streamers. Sergeev et al. (2004) confirmed the conjugacy of streamer-associated precipitation. The observations display two basic types of streamer-associated precipitation. In its poleward most half, the streamer-associated (field-aligned) accelerated electron precipitation coincides with the strong (>2–$7\ \mu A/m^2$) upward FACs on the westward flank of the convection stream, sometimes accompanied by enhanced proton precipitation in the adjacent region. In the equatorward portion of the streamer, the enhanced precipitation includes both electrons and protons, often without indication of field-aligned acceleration. The convective streams in the ionosphere, when well resolved, had the maximal convection speeds ~ 0.5–$1 km/s$, total FACs of a few tenths of a MA, thicknesses of a few hundred kilometers and a potential drop of a few kV across the stream.

Hubert et al. (2007) studied the ionospheric signatures of one auroral streamer that was observed on December 7, 2000 between 2200 and 2206 UT during a substorm expansion phase (Fig. 11.29). An auroral streamer was observed to develop above Scandinavia with the IMAGE-FUV global imagers. The ionospheric equivalent current deduced from the IMAGE Scandinavian ground-based network of magnetometers (Fig. 11.28) is typical of a substorm-

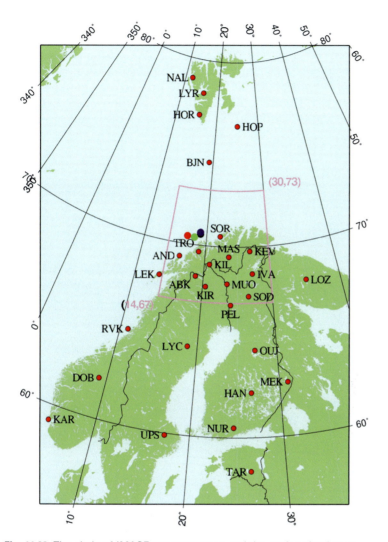

Fig. 11.28 The chain of IMAGE magnetometers and the conjunction box. From Juusola, L., Nakamura, R., Amm, O., Kauristie, K., 2009. Conjugate ionospheric equivalent currents during bursty bulk flows. J. Geophys. Res. 114, 3, A04313.

time streamer. Observations of the proton aurora using the SI12 imager onboard the IMAGE satellite (Mende et al., 2003) were combined with measurements of the ionospheric convection obtained by the SuperDARN radar network and the ionospheric electric field deduces by applying the method of Ruohoniemi and Baker (1998). The Wide Band Imaging Camera (WIC) and the Spectrographic Imager at 135.6 nm (SI13) instruments of the

IMAGE-FUV experiment, which are mostly sensitive to the emissions of the electron aurora, are also used to examine the morphology of the auroral features. We disagree with the conclusions of this study because the plasma is on closed field lines of the plasmoid by Fig. 11.29; instead we should note the sense of $\mathbf{E} \cdot \mathbf{J}$. There should be little precipitation where $\mathbf{E} \cdot \mathbf{J}$ is negative, at the top and bottom of the figure. This is the case on the

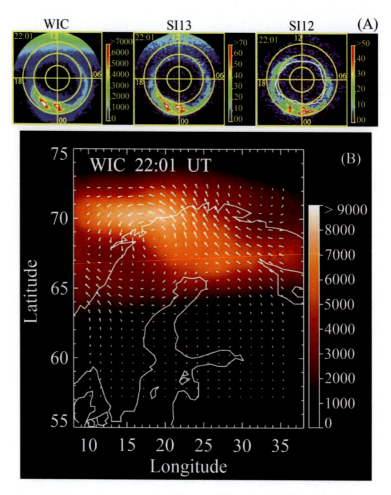

Fig. 11.29 (A) Polar view of the IMAGE-FUV WIC, SI13 and SI12 images (magnetic coordinates), with the open/closed field line boundary overlaid on the SI12 image at 2201 UT. (B) Map of the ionospheric equivalent currents above Scandinavia (geographic coordinates) in arbitrary units, with the auroral signal from the WIC image in AD units. From Hubert, B., Kauristie, K., Amm, O., Milan, S.E., Grocott, A., Cowley, S.W.H., et al., 2007. Auroral streamers and magnetic flux closure. Geophys. Res. Lett. 34, 2, L15105. https://doi.org/10.1029/2007GL030580.

two sides of the aurora in Fig. 11.29. "The ionospheric convection velocity, as measured by SuperDARN, appears to be reduced in the vicinity of the streamer."

11.8.3 Transverse **E** enhancement

We have repeatedly asserted in this book that current thinking forgets several important items. For one:

- The real electric field is $\mathbf{E} = -\partial\mathbf{A}/\partial t - \nabla\phi$, while the current analysis is based on the *convection electric field*: $\mathbf{E} + \mathbf{v} \times \mathbf{B} = \mathbf{R}$.

Here **R** is the complex term that represents the breakdown of ideal MHD. We are involved with *changes* in the cross-tail current (which is dawn-dusk by the magnetic topology) to find the induced electric field; these changes can result in the inductive electric field being either in the same sense as the net current (an electrical load) or the opposite (a dynamo).

The plasma is faced with the sudden onslaught of the induction electric field $-\partial\mathbf{A}/\partial t$ due to changes in plasma sheet convection; it tries to counter that with a redistribution of charge producing $-\nabla\phi$. However, it has an insurmountable problem: $-\nabla\phi$ has no curl. The plasma can only modify the field-aligned component (reduce in line with cause vs effect). In so doing, it must *enhance* the *transverse component* to preserve the curl (as noted by Fig. 11.14). Such enhancement "depends on field-aligned acceleration" reducing the parallel *voltage* (*not* the parallel *potential* drop!) due to the bursty bulk flow: the two are *linked together*. Whenever we see field-aligned acceleration due to $E_{\parallel}^{\text{ind}}$ (cause and effect), we know that there will be an increase in $E_{\perp}^{\text{ind}}$ (again cause and effect). The transverse component will be *increased*, and a bursty bulk flow results.

Zhao et al. (2020) used the P1 probe of ARTEMIS in lunar orbit to observe a tailward plasma flow and fluctuating magnetic field at about 54 R_E in the magnetotail on May 24, 2013. The wave activities of different frequencies in their observations were analyzed in detail. They found that the electron-scale whistler mode waves could be modulated by ion-scale waves, most likely kinetic Alfvén waves. The whistler mode waves periodically appeared at each cycle of the ion-scale waves. This shows that the dynamics of electrons and ions in the far magnetotail are coupled. The ion-scale waves are induced by streaming ion beams superposed on the ambient ion population. In this process, electrons are accelerated parallel to the magnetic field by the ion-scale wave, resulting in the deviation of the electron populations from equilibrium. The unstable electrons then release the excess energy by generating electron-scale whistler mode waves.

11.8.4 Auroral omega bands

Auroral forms in the recovery phase of substorms are periodic, wave-like undulations of the morningside diffuse aurora (Akasofu and Kimball, 1964). The auroral forms are drifting eastward with velocities ranging between 400 and 2000 m/s, with a tendency of increasing drift speed with time. These velocities have been found to agree closely with the local $\mathbf{E} \times \mathbf{B}$ drift velocity.

Amm et al. (2005a,b) studied ground-based data from the MIRACLE and BEAR ground networks and observations from the UVI and PIXIE instruments on the Polar satellite. They chose an event of an omega band over northern Scandinavia on June 26, 1998, which occurred close to the morningside edge of a substorm auroral bulge. Their analysis of the data concentrates on one omega band period from 03:18 to 03:27 UT, with a 1-min time resolution. In addition, the Assimilative Mapping of Ionospheric Electrodynamics (AMIE) method was used to derive global Hall conductance patterns. Their results show that zonally alternating regions of enhanced ionospheric conductances up to about 60 S and low conductance regions are associated with the omega bands. While they are moving coherently eastward with a velocity of about 770 m/s, the structures are not strictly stationary.

The current system of the omega band is a superposition of two parts: one consists of counter clockwise rotating Hall currents around the tongues, along with Pedersen currents with a negative divergence in their centers. The sign of this system is reversing in the low conductance areas. It causes the characteristic ground magnetic signature. The second part consists of zonally aligned current wedges of westward flowing Hall currents and is mostly magnetically invisible below the ionosphere. This system dominates the field-aligned current pattern and causes alternating upward and downward FAC at the flanks of the tongues with maximum upward FAC of about 25 μA/m^2. The total FAC of about 2 MA are comparable to the ones diverted inside a westward traveling surge. Throughout the event, the overwhelming parts of the FAC are associated with gradients of the ionospheric conductances, and 66%–84% of the FAC are connected with ionospheric Hall currents.

Fig. 11.30 summarizes the behavior of the substorm westward electrojet in the presence of omega bands. Instead of meandering along with the poleward boundary of the omega band tongues, as one could imagine from the structure of the equivalent currents, a large part of the substorm electrojet is interrupted intermittently by alternating sheets of upward and downward currents, located at the western and eastern flanks of the omega bands' tongues,

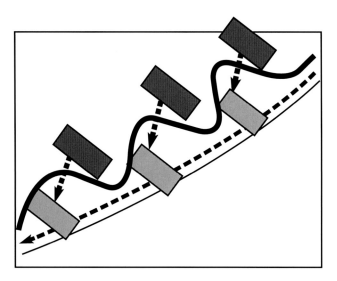

Fig. 11.30 Schematic sketch of the resulting geometry of the substorm electrojet current in the presence of omega bands. Areas with *darker shading* denote downward FAC (proton aurora), and areas with *lighter shading* denote upward FAC (electron aurora). From Amm, O., Aksnes, A., Stadsnes, J., Østgaard, N., Vondrak, R.R., Germany, G.A., et al., 2005. Mesoscale ionospheric electrodynamics of omega bands determined from ground-based electromagnetic and satellite optical observations. Ann. Geophys. 23(2), 339.

respectively. Only at the base of the omega bands is a continuous westward electrojet present.

It is likely that the result of the particle simulation of Fig. 3.12 and covered in Chapter 3 will be useful. The ions do gain energy as long as the parallel electric field is maintained, but they become bunched in the process. Some low-energy electrons circle the ion bunches in a form of Debye shielding (shown plainly in Fig. 3.12 of Chapter 3 showing the development of electron holes at an early stage). The ions (protons) carry the downward current in Fig. 11.30, while the electrons are the current carriers for the upward return current. How this fits the overall picture is not clear as yet.

11.8.5 Some other questions

There are questions that come to mind. The first one is by Fujii et al. (1994) in Fig. 11.31: What is the reason for the low conductivity and low electric field in the central part of the "auroral bulge"? Weimer et al. (1994), Fujii et al. (1994), and Hubert et al. (2007) find weak electric fields there. It is to be expected by Fig.

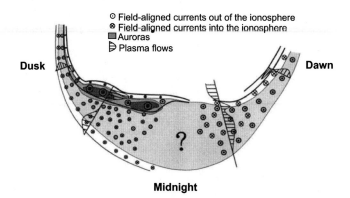

○ Field-aligned currents out of the ionosphere
⊗ Field-aligned currents into the ionosphere
▪ Auroras
▷ Plasma flows

Dusk

Dawn

Midnight

Fig. 11.31 The distributions of FACs, electric fields (plasma flows), and auroras associated with the generic aurora during a bulge-type substorm. At the poleward boundary of the bulge, a pair of upward and downward FACs is observed associated with a narrow eastward and/or antisunward plasma flow. From Fujii, R., Hoffman, R., et al., 1994. Electrodynamic parameters in the nighttime sector during auroral substorms. J. Geophys. Res 99(A4), 6108.

11.2; they must be forced to move the plasmoid out, and a dynamo with $\mathbf{E} \cdot \mathbf{J} < 0$. Recall that mirror points will be lifted, so there is no precipitation.

The steady-state current is from dawn to dusk, while the transient current can be decreasing or increasing. When it decreases it denotes an electrical load, so $\mathbf{E} \cdot \mathbf{J} > 0$ taking the sense of the inductive field into account; on the other hand, an increase counts as a dynamo and $\mathbf{E} \cdot \mathbf{J} < 0$. The result is a dynamo whenever the net current faces a dawnward sense.

Fig. 11.32 gives an example of a typical AE, AU, AL indices presentation covering a period of 72 h. The recovery phases of different substorm periods are indicated by shaded areas. The shortest substorm period is, in this case, on the order of 3 h. During this time interval, seven clear substorm activations can be identified. Given a constant electric field and a supply of plasma, substorms will occur with a time period of 1–3 h. The whole sequence of substorms ends in a so-called steady magnetospheric convection period lasting for about 16 h and having a continuous disturbance level of more than 500 nT.

To an engineer, it looks like the system is behaving like an oscillator: whenever we have an electric field in the magnetotail and a constant supply of plasma, there will be a response on a repetitive scale.

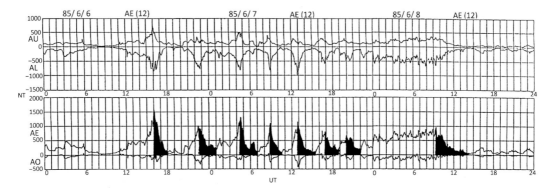

Fig. 11.32 AU, AL, and AE indices from June 6 to 8, 1985. The recovery phases of different substorm periods are indicated by *shaded areas*. The repetitive nature suggests the plasma sheet behavior is an oscillator; given a constant electric field and a supply of plasma, substorms will be present with a time period of 1–3 h. From Pellinen, R.J., Opgenoorth, H.J., Pulkkinen, T.I., 1992. Substorm recovery phase: relationship to next activation. In: Substorms 1, European Space Agency, SP-335, 469.

11.9   Observations of particle acceleration

The origin of energetic particles in the Earth's magnetosphere is still an unresolved mystery. It was suggested that the observed energetic particle bursts are related to magnetic reconnection and formation of a neutral line (e.g., Sarafopoulos and Sarris, 1988; Taylor et al., 2006) but that needs to be questioned or revised.

> *Using the four-spacecraft Cluster, we can clearly see that this large positive B_Z structure propagates in the earthward direction. Furthermore, we find that the energy spectrum of the energetic electrons becomes harder toward the downstream region. … To discuss the temporal and spatial profile of energetic electron acceleration in the magnetic reconnection region, we determined the spacecraft position in the temporally evolving magnetic structures of reconnection. Our observation clearly indicates second-step acceleration, in addition to X line acceleration, of energetic electrons in the downstream reconnection outflow region.*
>
> **Imada et al. (2007)**

Imada et al. (2007) used data from comprehensive measurements onboard the Cluster satellites, including high-energy electrons measured by RAPID (Research with Adaptive Particle Imaging Detectors), FGM (Fluxgate Magnetometer), and low-energy ions with the CIS (Cluster Ion Spectrometry) experiments. The observations were made during the crossing on the dusk side near-Earth magnetotail from 09:47 to 09:51 UT on October 1, 2001.

This event was identified by several previous studies as a crossing of the X-line from tailward to earthward (e.g., Runov et al., 2003; Cattell et al., 2005; Wygant et al., 2005).

Fig. 11.33 shows the two-dimensional energetic electron data from C3 in the spacecraft spin plane from 09:48 to 09:49 in the

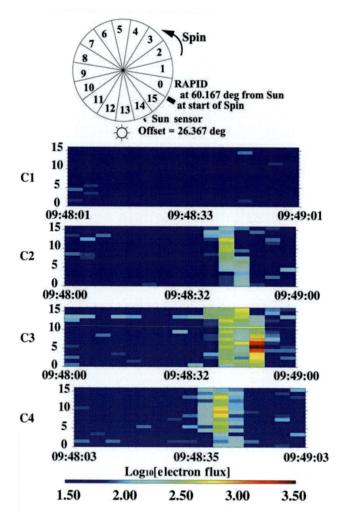

Fig. 11.33 Two-dimensional energetic electron flux (/(cm² str s keV)) behavior from 50.5 to 127.5 keV. The vertical axis shows azimuthal sectors, and the horizontal axis gives UT. Orientation of the RAPID azimuthal sectors relative to the Sun is also at the top of the figure. From Imada, S., Nakamura, R.P., Daly, W., Hoshino, M., Baumjohann, W., Mühlbachler, S., et al., 2007. Energetic electron acceleration in the downstream reconnection outflow region. J. Geophys. Res. 112, 4, A03202. https://doi.org/10.1029/2006JA011847.

energy range between 50.5 and 127.5 keV. The electron enhancements occur within a few spin periods. During this time interval, the three spacecraft C2, C3, and C4, are located in the plasma sheet with earthward flow. The electrons are assumed to be isotropic, although the flux varies with time. Fig. 11.34 shows the positions of three of the spacecraft at four times during the same 1-min interval as Fig. 11.33. C2 is located near the plasma sheet boundary layer on the earthward side during these observations. C3 is located in the central plasma sheet on the earthward side, and C4 is located very near to the X-line. It can be seen in Fig. 11.34 that the neutral line propagates tailward at roughly 100 km/s.

Imada et al. (2007) attempted to determine the acceleration region for energetic electrons in the structure of magnetic

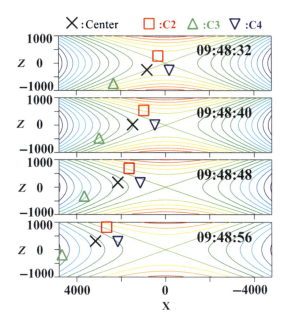

Fig. 11.34 Relative position of C2, C3, and C4 in the magnetic reconnection region. The vertical axis gives Z(km), and the horizontal axis gives X(km). The curves show contours of the magnetic vector potential. The results at (top to bottom) 09:48:32, 09:48:40, 09:48:48, and 09:48:56 are shown. C1 is located more than 1000 km away from the equatorial plane in the northern lobe region during this interval (not shown here). The *square, triangle, and inverted triangle* show the spacecraft position of C2, C3, and C4, respectively. From Imada, S., Nakamura, R.P., Daly, W., Hoshino, M., Baumjohann, W., Mühlbachler, S., et al., 2007. Energetic electron acceleration in the downstream reconnection outflow region. J. Geophys. Res. 112, 7, A03202. https://doi.org/10.1029/2006JA011847.

reconnection. The first enhancement of energetic electrons was observed at 09:48:41 by C4 as it moved to the inner central plasma sheet and observed a B_Z of about 5 nT. At the same time, C4 was located earthward and in the vicinity of the X-line as seen in the second plot of Fig. 11.34. The second energetic electron enhancement was observed at 09:48:43 by C2, while located earthward of the X-line and almost 500 km away from the equatorial plane (also second plot). The last and most energetic enhancement of electron fluxes was observed at 09:48:50 by C3, while it was located on the earthward side in the central plasma sheet and more than 3000 km away from the X-line. During the enhancement time, C3 observed a very strong B_Z of about 16 nT. This large normal magnetic field is 60% of the lobe magnetic field.

The value of energetic electrons observed by C1 is at approximately the background level during this time interval (not shown). The relative timing of the enhancement encounters start from the spacecraft nearest to the X point. The electrons are accelerated up to higher energy far from the X point toward the downstream region of the earthward side. The authors concluded that an acceleration region that has a large normal magnetic field B_Z was created near the X point and propagated to the downstream region while producing further high-energy electrons. Østgaard et al. (2009) have reached a similar but stronger conclusion.

> *Our main findings are: (1) The electron distribution from the reconnection region can generally not produce the auroral intensities observed in the ionosphere. ... Our results strongly indicate that acceleration in the reconnection region is usually marginal. (2) Although we can conclude that additional acceleration mechanism is needed, we are not able to identify the mechanism. (3) Reconnection is an expanding process observed along the poleward boundary of the aurora.*
>
> **Østgaard et al. (2009)**

Finally, Lui et al. (2007) present an analysis of what happened on August 22, 2001. The ion plasma measurements were taken by CIS. The proton plasma moments produced by the CODIF (Composition Distribution Function) sensor of the CIS instrument were used. The electron plasma measurements were made by PEACE (Plasma Electron and Current Experiment). The electric and magnetic field measurements were obtained by EFW (Electric Field and Wave) and FGM.

> *It is found that the breakdown occurred (1) in a low-density environment with moderate to large proton plasma flow and significant fluctuations in electric and magnetic fields, (2) in regions with predominantly dissipation but occasionally dynamo effect,*

and (3) at times simultaneously at two Cluster satellites separated by more than 1000 km in both X and Z directions. Evaluation of the terms in the generalized Ohm's law indicates that the anomalous resistivity contribution arising from field fluctuations during this event is the most significant, followed by the Hall, electron viscosity, and inertial contributions in descending order of importance. This result demonstrates for the first time from observations that anomalous resistivity from field fluctuations (implying kinetic instabilities) can play a substantial role in the breakdown of the frozen-in condition in the magnetotail during substorm expansions.

Lui et al. (2007)

The comparison of the B_Z component for these three satellites shown in Fig. 11.35 is very interesting. A noticeable feature is a transient bipolar B_Z signature from southward to northward near ~09:42:47 UT, which occurred in an earthward plasma flow environment. At nearly the same time, a similar deflection in the B_Z component was detected at C1 and C2 even though the sign of the B_Z component did not change by the deflection. Something remarkable is happening!

11.10   Acceleration of cold plasma

The energization of charged particles to high energies is a ubiquitous characteristic of plasmas. Observations have shown that electrons and protons are energized up to at least 1 MeV in the magnetotail during substorms.

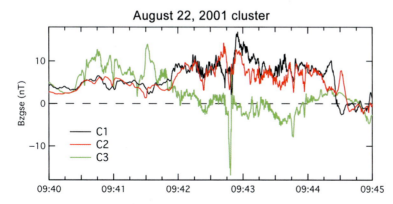

Fig. 11.35 Comparison of the dipolarization feature in the B_Z component seen by C1, C2, and C3 near an *X*-line. From Lui, A.T.Y., Zheng, Y., Rème, H., Dunlop, M.W., Gustafsson, G., Owen, C.J., 2007. Breakdown of the frozen-in condition in the Earth's magnetotail. J. Geophys. Res. 112, 7, A04215. https://doi.org/10.1029/2006JA012000.

Only the electric field is able to impart energy to charged particles. As its own energy is negligible, it acts as an intermediary, tapping energy either from an external source (static case) or from the magnetic field (inductive case). There are characteristic differences between the energization processes in the two cases.

$$\mathbf{E} = -\frac{\partial \mathbf{A}}{\partial t} - \nabla \phi \qquad (11.3)$$

We write the relation in this order to emphasize the point that the induction field, depending explicitly on time, is controlling. As electric potential fields are conservative, the acceleration that any individual particle can experience is limited by the potential difference transversed by its orbit, of the order of 50 keV corresponding to the total potential across the magnetosphere.

Electric induction fields are in several ways more efficient for particle energization. For one thing, the induction fields are stronger and hence allow greater acceleration. Another, and even more important, feature of induction fields is the ability to energize charged particles locally by betatron acceleration with the particles going around and around. This, too, has its limitations. One is that the ratio of final to initial energy is given by the ratio of final to initial magnetic field strength. Hence the largest relative energy increases are achievable near a magnetic neutral line where the initial field can be very small. Another way to say this is that, due to the dependence on the particle's magnetic moment (for relativistic particles the relevant quantity is instead the flux enclosed by the gyro-orbit), only particles with a substantial magnetic moment, that is, particles that already have a rather high energy, are sufficiently energized. In particular, betatron acceleration is ineffective on "cold" plasma particles.

Near a neutral line the difficulty of energizing "cold" plasma particles can be overcome by a two-step process (Pellinen and Heikkila, 1978); linear nonadiabatic acceleration along the neutral line provides sufficient momentum so that when the particle leaves the neutral line region and gets into a gyro-orbit in the surrounding magnetic field, it has a sufficient magnetic moment to benefit from the betatron acceleration (Pellinen, 1978). Therefore induction electric fields have several important consequences for substorm development, including that of energizing charged particles (Heikkila et al., 1979).

The two photographs in Fig. 11.36 were taken at onset about 20–30 s apart. Notice stars of the big dipper; these give a good sense of the direction (north to upper left), and the scale (beginning quite small). The left photo shows the first activation is slightly poleward of the existing arc. The photo on the right shows

Fig. 11.36 These two photos were taken about 30 s apart (notice stars of the big dipper). The photo on the left shows that the first activation is poleward of the existing arc. The photo on the right shows the burst of energized electrons (change in color). From the original film it is likely that the diffuse forms at the lower right could be the diffuse proton aurora. From Finnish Meteorological Institute.

the burst was produced by energized electrons deeper in the ionosphere (by the change in color). This figure (plus many others) will confirm that this new ionization was formed below 90 km. It is likely that the diffuse forms at the lower right could be the proton aurora (as had been observed by Fukunishi (1975)).

This is the significance of the sharp spike we saw in Fig. 11.35: the lightning strike, a mark of a discharge in a plasma! The zap in Fig. 11.37!

For reasonable values of the magnetic field, charged particles accompanying a substorm may be energized to MeV energies. The dependence of the mechanism of the neutral line region also illustrates how to achieve selective acceleration, rather than an equitable distribution of available energy among the whole particle population.

11.10.1 Discharge

Near a neutral line the difficulty of energizing "cold" plasma particles can be overcome by a two-step process. Linear nonadiabatic acceleration along the neutral line provides sufficient momentum so that when the particle leaves the neutral line region and gets into a gyro-orbit in the surrounding magnetic field, it has a sufficient magnetic moment to benefit from the betatron acceleration (Heikkila et al., 1979). For reasonable values of the magnetic field, charged particles accompanying a substorm may be energized to MeV energies, and quickly.

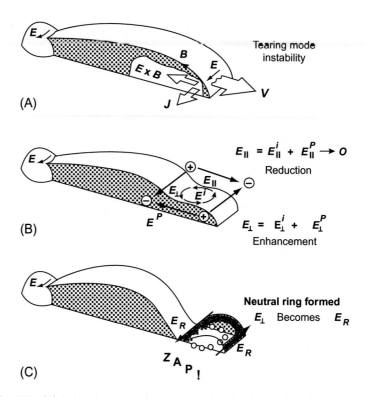

Fig. 11.37 (A) A thinning current can cause a localized meander of a cross-tail current filament. (B) This meander will produce an induction electric field with topology as shown, causing polarization of the plasma to reduce the parallel component and enhance the transverse component. (C) The enhanced transverse component will create a discharge along the neutral line. From Heikkila, W.J., 1983. The reason for magnetospheric substorms and solar flares. Solar Phys. 88, 329–336.

An electrical breakdown usually involves a large, abrupt rise in electric current in the presence of a small increase in electric voltage. Lightning is the most familiar example of breakdown. If the region of ionization bridges the gap between electrodes, thereby breaking down the insulation provided by the gas, the process is an ionization discharge. In a gas, such as the atmosphere, the potential gradient may become high enough to accelerate the ions and electrons to higher velocities. The conductivity along the X-line goes from essentially zero (Pedersen) to high (direct). When the conductivity along the X-line reaches a critical level, the current increases rapidly, as is apparent in Fig. 11.37.

The dependence of the neutral line region mechanism also illustrates how to achieve selective acceleration, rather than an equitable distribution of available energy among the whole particle population.

Fig. 11.38 Auroral photograph at about 1600 UT in Finland at the onset of the 911 substorm discussed in the next four figures. From Eklund and Heikkila.

In order to evaluate its capabilities a simple model of a localized, growing disturbance in the neutral sheet current was used to calculate perturbation magnetic and electric fields; the model includes the formation of X and O type neutral lines. Plasma sheet test particles were followed in these time-dependent fields by using the full relativistic equation of motion. The most efficient energizing mechanism is a two-step process, with an initial linear acceleration along a neutral line up to moderate energies, followed by betatron acceleration.

                                                                    Pellinen and Heikkila (1978)

11.10.2 Resonant diamagnetic acceleration

In Chapter 7 we presented a cyclotron resonant acceleration mechanism (CRAM) of the field with the charged particles. That same process may be the explanation of observations in the plasma sheet shown in Figs. 11.38–11.41 on September 11, 2002 called the 911 plasmoid. The photograph in Fig. 11.38 was taken just after dusk. Notice the bright emission to the left, lower in altitude; that is the beginning of the westward traveling surge we saw in Figs. 11.21 and 11.22. From there to the right is the region marked with a question mark in Fig. 11.31; it is where the escaping plasmoid is located. There must be a force to move the plasmoid out, a dynamo with $\mathbf{E} \cdot \mathbf{J} < 0$, and little precipitation.

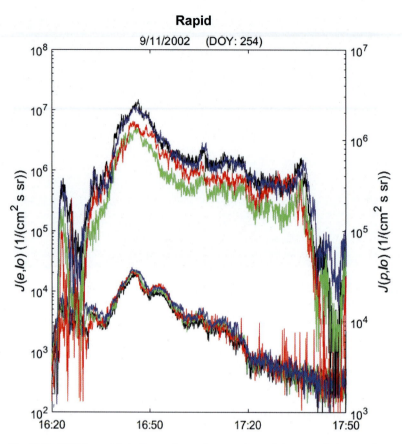

Fig. 11.39 RAPID provided observations of energetic particles at various energies. Inside the plasmoid, both the energetic particle intensity and the power spectral density of the magnetic fluctuations show increases by up to four orders of magnitude in comparison to an adjacent region. The spacecraft was in a plasmoid for an incredible 80 min with continuous evidence of energetic particles. From Zong, private communication.

As a reminder, the steady-state current in the magnetotail is from dawn to dusk, while the transient current depends on whether the current is decreasing or increasing. When it decreases it denotes an electrical load, so $\mathbf{E} \cdot \mathbf{J} > 0$. On the other hand, an increase in current counts as a dynamo by Lenz's law and $\mathbf{E} \cdot \mathbf{J} < 0$.

There is no doubt that upward portions in the photograph are in sunlight (just after dusk) but the auroral emissions must be caused by the primary electrons because of their narrow north-south extent. The violet color may indicate that the emissions are due to the presence of nitrogen in N_2 LBH band; they would

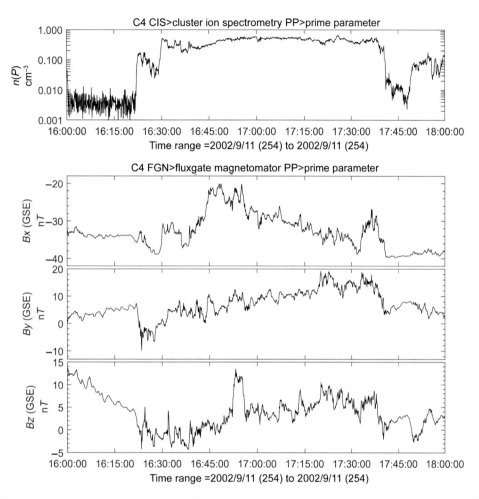

Fig. 11.40 Cluster was located at −19, +3, −1 GSM. The plasma density is very low until rising quickly at 1623 UT. This might be a shock wave as the magnetic field displays a sudden jump, and the beginning of dramatic plasma emission. From Zong, private communication.

be difficult see with the naked eye because of their ultraviolet nature. Lower down it would be the dominant green line of atomic oxygen at 558 nm.

Cluster was near apogee in midtail when a substorm took place, clearly seen by the magnetometer chain in Fig. 11.28. Onset of the expansion phase occurred at precisely 16:00 UT at Kevo, Finland. At that time, Cluster was located at a distance of −19, +3, −1 R_E GSM. The plasma density was very low before rising quickly at 16:23 UT as shown in Figs. 11.39–11.41. This might be a shock wave as the magnetic field displays a sudden jump, and

2002-09-11T16:00:00Z / 2002-09-11T18:00:00Z

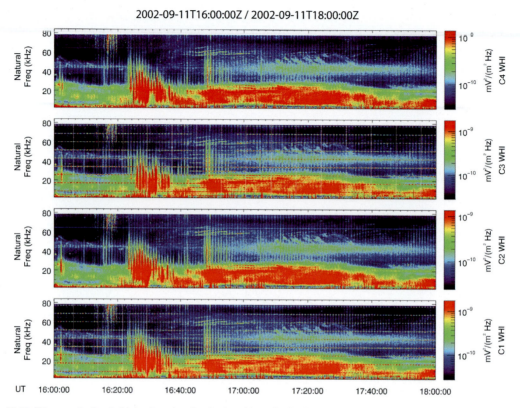

Fig. 11.41 Wave emission is high after the shock, then lower in the body of the plasmoid. Is it possible that these emissions were responsible for the high energies through cyclotron resonant acceleration? From Zong, private communication.

the beginning of dramatic plasma emission (Fig. 11.40). The spacecraft was in a plasmoid for an incredible 80 min, with continuous evidence of energetic particles. The average tailward speed of the plasmoid was ~ 50 km/s.

The following notes are from Chen (2008). The energization rate by the cyclotron resonant acceleration can be expressed analytically. The changing rate of the kinetic energy of a charged particle with velocity $\mathbf{v}$ in electric field $\mathbf{E}$ and magnetic field $\mathbf{B}$ is

$$dK/dt = \mathbf{F} \cdot \mathbf{v} \tag{11.4}$$

where $\mathbf{F} = q(\mathbf{E} + \mathbf{v} \times \mathbf{B})$ is the Lorentz force with q being the particle charge. Since $(\mathbf{v} \times \mathbf{B}) \cdot \mathbf{v} = 0$ (the magnetic force is always perpendicular to $\mathbf{v}$ and the latter term does no work on the particle), Eq. (11.4) is equivalent to

$$dK/dt = q\mathbf{E} \cdot \mathbf{v} \qquad (11.5)$$

Eq. (11.5) contains both perpendicular and parallel components. For particle cyclotron resonant acceleration, one only needs to consider the perpendicular component; that is,

$$dK_{\perp}/dt = qEv_{\perp} \qquad (11.6)$$

where E is the left-hand polarization-perpendicular electric field with frequency the same as the ion gyrofrequency, and $v_{\perp}$ is the particle's perpendicular velocity. For nonrelativistic ions,

$$v_{\perp} = (2K_{\perp}/m)^{1/2} \qquad (11.7)$$

In Chapter 7 it is shown that in the final result, over an ion gyroperiod (T_i), the energy increase due to ion cyclotron resonant acceleration (Chen, 2008) is:

$$\Delta K_{\perp} = (2K_{\perp}(0)/m)^{1/2}qET_i + q^2E^2T_i^2/(2m) \qquad (11.8)$$

The significance of this equation is that all terms on its right side are measurable and are independent of models and simulations. It indicates that an ion energy enhanced by the gyroresonant acceleration is a function of the initial perpendicular kinetic energy, charge/mass ratio, gyroperiod of the ion, and the left-hand polarization-perpendicular electric field.

Inside the plasmoid, both the energetic particle intensity and the power spectral density of the magnetic fluctuations show increases by up to four orders of magnitude in comparison to an adjacent region. Over the ULF range, the power spectral densities are dominated by the perpendicular component of the local magnetic field.

More research is needed to determine whether the measured left-hand polarization of the cusp electric field at ion gyrofrequencies can energize the local ions and electrons in the diamagnetic cavities and plasmoids; in particular, whether the cyclotron resonant acceleration can energize ions from keV to MeV in seconds.

11.11 Space weather implications

11.11.1 The Saint Patrick's Day Storm of 2015

One of the biggest geomagnetic storms in recent years began at 4:45 UT on March 17, 2015 with the arrival of a coronal mass ejection (CME) from the Sun. The sudden storm commencement (as measured by the symmetric ring current (SYM–H) index) coincided with a sharp jump in the solar wind speed and pressure,

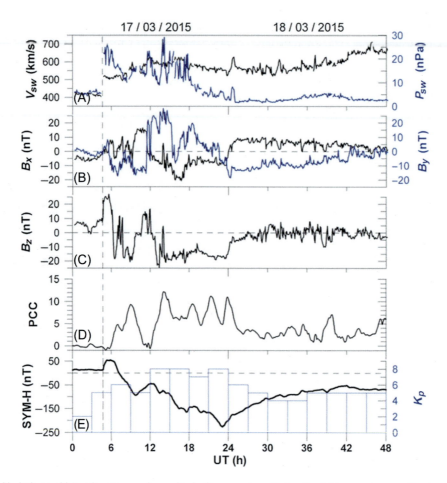

Fig. 11.42 Variations of interplanetary and geophysical parameters during the intense geomagnetic storm of March 17–18, 2015. (A) solar wind speed (V_{SW}, *black curve*) and solar wind ram pressure (P_{SW}, *blue curve*), (B) the IMF B_x *(black)* and B_y *(blue)* component, (C) the IMF B_z *(black)* component, (D) combined polar cap index, and (E) SYM-H *(black curve)* and K_p *(blue bars)* indices. All data are 5 min averaged, except for the PC index (1 min resolution) and the K_p index (3 h). The *vertical dashed line* shows the time of the SSC of 04L45 UT. For the data of interplanetary parameters, an additional shift of +15 min was applied in order to match the sudden storm commencement time. From Astafyeva, E., Zakharenkova, I., Förster, M., 2015. Ionospheric response to the 2015 St. Patrick's Day storm: a global multi-instrumental overview. J. Geophys. Res. 120, 9024. https://doi.org/10.1002/2015JA021629.

as seen in Fig. 11.42 (Astafyeva et al., 2015; Jacobsen and Andalsvik, 2016). It was the first G4 storm of solar cycle 24, according to the NOAA scale. The partial halo CME was seen by the SOHO spacecraft to erupt from the southwest portion of the Sun 2 days previously, at 1:48 UT on March 15 (Wu et al., 2016a). A sequence of images from this CME are shown in

0000UT 2015-03-15   0148UT 2015-03-15   0200UT 2015-03-15   0212UT 2015-03-15

0224UT 2015-03-15   0238UT 2015-03-15   0248UT 2015-03-15   0312UT 2015-03-15

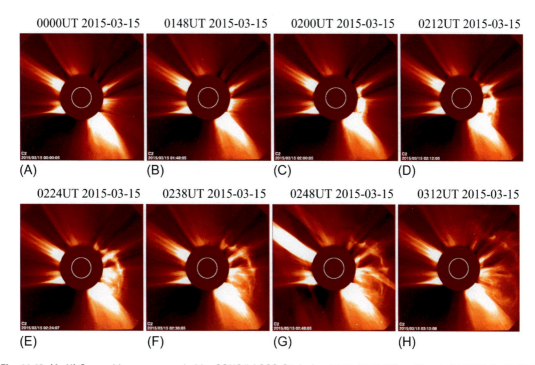

(A)   (B)   (C)   (D)

(E)   (F)   (G)   (H)

Fig. 11.43 (A–H) Coronal images recorded by SOHO/LASCO C2 during 00:00–03:12 UT on March 15, 2015. At 01:48 UT (B) C2 recorded a CME that erupted from the southwest. From Wu, C., Liou, K., Lepping, R.P., Hutting, L., Plunkett, S., Howard, R.A., Socker, D., 2016. The first super geomagnetic storm of solar cycle 24: "The St. Patrick's day event (17 March 2015)", Earth Planets Space 68, 151. http://doi.org/10.1186/s40623-016-0525-y, p. 3.

Fig. 11.43. Unlike previous studies (Liu et al., 2015; Astafyeva et al., 2015) that suggested that the storm was caused by the interaction of two successive CMEs, Wu et al. (2016a) found that this super-storm was caused by the strong negative B_z magnetic field in both the sheath (the plasma following the interplanetary shock at the leading edge of the CME), and a magnetic cloud that followed the sheath plasma, both of which are associated with the CME launched on March 15th.

In Fig. 11.42, the OMNI solar wind data has been retarded an additional 15 min beyond the normal propagation delay from the spacecraft to the magnetopause so that it coincides with the storm onset as seen in the SYM-H index (Fig. 11.42E). This additional delay time is within the error bars found by statistical studies (Case and Wild, 2012) and by cross-correlation estimations with ACE and Cluster observations. This delay is also partially accounted for by the propagation time from the bow shock nose down to ground-based magnetometers, which is estimated to be on the order of 5 min (Villante et al., 2004). The remaining time difference could be due to a breakdown of the "planar phase front" assumption at the solar wind CME shock front that is used for the OMNI data delay time estimates (Astafyeva et al., 2015).

The main phase of the storm began at about 7:30 UT when the IMF B_z turned southward in the sheath plasma and the SYM-H index began a gradual decrease (Fig. 11.42C and E). The magnetic cloud began at 10:36 UT (marked by the sharp change in B_x and B_y in Fig. 11.42B) and persisted with mostly strong southward B_z until 23:36 UT when all three components of the IMF abruptly changed again. The SYM-H index reached its minimum excursion of -233 nT at 22:45 UT before beginning a day-long recovery throughout March 18th. The planetary K_p index was at about its maximum of 8 from 12 to 24 UT on the 17th (Astafyeva et al., 2015; Wu et al., 2016a).

The ionospheric response to the St. Patrick's Day storm was complex and profound. The entirety of ionospheric variations induced by geomagnetic disturbances is commonly referred to as an "ionospheric storm." They are marked by largely reinforced auroral particle precipitation and enhanced high-latitude ionospheric currents and convection lasting for hours. The Joule heating, ion-drag forcing, and increased high-latitude ionization significantly affect the global dynamics and structure of the thermosphere and ionosphere. Heating and expansion of the thermosphere causes neutral winds and traveling atmospheric disturbances, which lead to global changes in the composition and dynamics of the thermosphere and also of the ionosphere. This is the source of increases (decreases) in electron plasma densities and total electron content, also known as positive (negative) storms. The largest positive ionospheric storm in this case occurred at low latitudes in the Earth's morning and postsunset sectors, with enhancements of 80%–150%. Analysis by Astafyeva et al. (2015) has shown that this region was most effected by thermospheric composition changes consisting of an increase in the O/N_2 ratio. At midlatitudes, there were inverse hemispheric asymmetries in different longitudinal regions. Such asymmetry is normally explained by seasonally driven thermospheric circulation; however, in this case the storm occurred at equinox. Obviously, other than seasonal factors are responsible for the asymmetries. These other factors may include the offset between geographic and magnetic poles, hemispheric differences in geomagnetic field strength, and storm time variation of the IMF B_y component (Astafyeva et al., 2015; Förster et al., 2011).

The complex nature of the storm onset meant that space weather agencies around the world failed to correctly predict its severity (Kamide and Kusano, 2015; Jacobsen and Andalsvik,

2016). Red auroras were seen in northern Japan for the first time in 11 years, along with many other low-latitude sightings in central Europe and the United States and Canada (Kamide and Kusano, 2015; Case and MacDonald, 2015). Higher latitudes such as Alaska witnessed bright and dynamic aurora (see cover photo). Auroral sightings differed significantly from the predictions of auroral models such as Newell et al. (2010) and Roble and Ridley (1987), with hundreds of verified public sightings reported through websites such as http://www.aurorasaurus.org (Case and MacDonald, 2015). Crowdsourced observations may become an important tool for improving our models and predictions for geomagnetic storms and their effects.

11.12 Discussion

A lot of commendable work over the past six decades has put space plasma physics on the road toward remarkable success. Figs. 11.2 and 11.44 show how the low-latitude boundary layer plays a key role in this development.

However, there are three reasons to be skeptical.

11.12.1 The real electric field

The generalized Ohm's law for a collisionless plasma has been considered essential with the inclusion of large amplitude and rapid fluctuations of electric and magnetic fields. This equation can be written:

$$\mathbf{E} + \mathbf{v} \times \mathbf{B} = \frac{1}{\varepsilon_o \omega_{pe}^2} \frac{d\mathbf{J}}{dt} + \frac{\mathbf{J} \times \mathbf{B}}{ne} - \frac{\nabla \cdot \mathbf{P}_e}{ne} \tag{11.9}$$

$$-\frac{1}{n} [\langle \delta \mathbf{E} \delta n \rangle + \langle \delta(n\mathbf{v}_e) \times \delta \mathbf{B} \rangle] \tag{11.10}$$

where **v** is the plasma bulk flow (Yoon and Lui, 2007). But, as we have repeatedly asserted in this book, the current thinking is based on the *convection electric field*:

$$\mathbf{E} + \mathbf{v} \times \mathbf{B} = \mathbf{R} \tag{11.11}$$

Here **R** is the complex term that represents the breakdown of ideal MHD. Hardly anyone has noted in space plasma physics that the *real electromagnetic field* (by Helmholtz' theorem) is:

$$\mathbf{E} = -\partial \mathbf{A}/\partial t - \nabla \phi \tag{11.12}$$

We must be able to display the magnetic topology and curl **E** to treat a given problem, in agreement with Faraday's law $\nabla \times \mathbf{E} = -\partial \mathbf{B}/\partial t$.

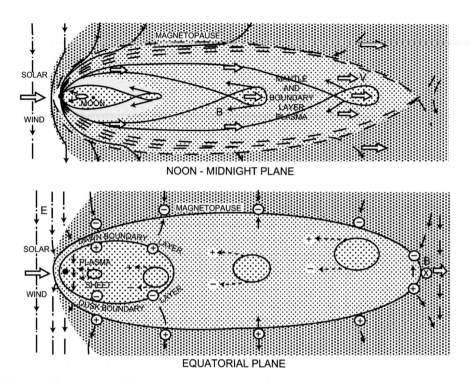

Fig. 11.44 A cut of the magnetotail in the noon-midnight as well as the equatorial planes. In this model, most of the distant magnetotail is composed of the combined dawn and dusk boundary layers. After it is formed in the near-Earth magnetotail within the plasma sheet, a plasmoid must burrow through a long region of closed field lines; during this time it must have an outward force that will be provided by a current loop back to the ionosphere. Auroral form disappears when the plasmoid reaches the end of the plasma sheet. From Heikkila, W.J., and Treilhou, J.-P., Proceedings of Results of the Arcad 3 Project, p. 281, Cepadues-Editions, Toulouse, France (1985).

In the integral form, the line integral of **E** around a closed contour generates the emf.

$$\varepsilon = \oint \mathbf{E} \cdot \mathbf{dl} = -d\Phi^M/dt \tag{11.13}$$

where Φ^M is the magnetic flux through the surface considered. Only by this emf can we tap the energy stored in the magnetic field, by Maxwell's equations, despite the claim of standard magnetic reconnection (SMR) (Vasyliunas, 1975).

11.12.2 Three dimensions

Dungey (1958, p. 4) has said; "The existence of the third dimension can radically alter a physical situation, particularly in regard to stability. A two-dimensional tight-rope walker confined to the vertical plane containing the rope could never fall off." Yet most work, including the presentation of results (see Fig. 11.10), use 2D. Birn et al. (2001) stated: "The conclusions of this study pertain explicitly to the 2D system." With 2D it is not easy to consider matters as action versus reaction using Newton's third law. With 2D we cannot consider generalized magnetic reconnection (GMR). With 2D in the noon-midnight plane we cannot consider the low-latitude boundary layer.

11.12.3 Cause versus effect

Paschmann et al. (2003, p. 360) stated: "The microphysics of the auroral acceleration regions is often treated without reference to the source of energy. In fact, the latter one is one of the most neglected elements in [space plasma physics]."

An electrical load means dissipation with $\mathbf{E} \cdot \mathbf{J} > 0$; SMR ignores the dynamo $\mathbf{E} \cdot \mathbf{J} < 0$, thus without reference to the source of energy. By this definition, it does not address the source of magnetic energy. SMR is suited only to discuss the dissipation region, which is only half the problem.

The real electromagnetic field permits the use of GMR. It depends on two electric fields (electrostatic and inductive), not one (convection electric field). It is time to use time dependence in three dimensions in the analysis of magnetic reconnection. This was initiated by Schindler et al. (1988), but seldom followed by anyone else. GMR involves magnetic helicity on the dawn and dusk sides during the growth and expansion phases.

With 3D it is possible to visualize plasmoids escaping the magnetotail. The tailward motion must be balanced by an earthward response, by Newton's third law. The *action* of this escape (the plasmoid) is a *reaction*, an inward meander (see Fig. 11.15), something not recognized before in space plasma physics.

11.13 Summary

A substorm is a time-dependent process in three dimensions involving four outstanding research topics in basic plasma physics: magnetic reconnection, current disruption, turbulence, and particle acceleration.
- Validation of the Akasofu model

A variety of independent studies have shown that the characteristics of auroral substorms are that foreseen by Akasofu (1964) at the beginning of the space age. The brightening of a small element of an existing auroral arc can be taken as the *defined onset* of a substorm. After the substorm onset the auroral oval becomes clearly bifurcated consisting of two components: the oval aurora in the latitude range of the preonset oval, and the bulge aurora that emerges out of the oval. The bulge aurora expands poleward out of the presubstorm oval, westward as well as eastward, from the premidnight onset location. An asymmetric emission pattern occurs, the premidnight westward surge toward dusk and omega band auroras postmidnight near dawn. The entire substorm behaves as a single component system but with many different aspects.

- Two classes of fast flows

The two substorm models of Fig. 11.10, the near-Earth neutral line or reconnection model (e.g., Hones, 1984; Baker et al., 1996) and the cross-field current disruption model (Lui, 1996, 2004), are in question at the present time. They emphasize different aspects of substorm processes and set the two as *apparent* alternatives. Lin et al. (2009) might imply another process, that of global Alfvénic interactions, proposed by Song and Lysak (2001), Lysak et al. (2009).

Shue et al., (2008) studied earthward plasma sheet fast flows from GEOTAIL plasma. They found that earthward fast flows can be classified into two classes, Class I was often observed near $X \sim -10$ and the other, Class II, was found at $X < -15 \ R_E$.

An important limitation in the models of Fig. 11.10 is that there is no mention of Newton's third law; that for every action, there is an equal and opposite reaction. The direction of the force on the first object (the escaping plasmoid) is opposite to the direction of the force on the second object (the inner plasma sheet and the ionosphere). Forces always come in pairs: *equal and opposite action-reaction force pairs*. The result of Shue et al. (2008) can be explained by this action-reaction.

- Substorm transfer event (STE)

We have proposed a model for the substorm phenomenon, one which uses time-dependent three-dimensional considerations right from the start. It is basically a PTE discussed in Chapter 6, but its details here are quite different mostly because of Newton's third law on reaction. We call it an STE following CTE, suggested in Chapter 10.

There are only two ways that the current can wander in the equatorial plane, to the left or to the right; these are associated with clockwise or counterclockwise perturbation current meanders. When the current is increasing, the induction electric field opposes the increase so that $\mathbf{E} \cdot \mathbf{J} < 0$; this local region is a dynamo. In contrast, the plasma gains energy where the current is decreasing and $\mathbf{E} \cdot \mathbf{J} > 0$, an electrical load.

- Westward traveling surge

The most spectacular auroral display is the westward traveling surge that initiates the expansion phase of substorms. The most reliable indicator of substorm onset is the sudden brightening of a small portion of an existing auroral arc due to increased particle precipitation. The substorm current diversion of the three-dimensional current system is responsible for most of the effects that can be seen by ground-based magnetometers at substorm onset. Magnetic perturbations caused by the type 2 current system can be detected on the ground as the auroral electrojets by a chain of magnetic observatories. Diversion of the cross-tail current by the type 1 current system starts at onset through the midnight ionosphere. The wedge shape of the projected equivalent current accounts for the name "substorm current wedge."

- Particle observation and simulation

The Cluster 3 spacecraft made an exceptional high-resolution measurement of a beam of electrons with low energies up to 400 keV (Taylor et al., 2006). Its pitch angle distribution evolved from antiparallel, through counterstreaming to parallel over a period of 20 s. Unique to Cluster 3, at the highest energies ($\geq$47.2 keV), RAPID observed an evolution of the pitch angle, moving from an initial antiparallel through to bidirectional and finally being more field aligned.

These observations are in agreement with the particle simulations in Chapter 3. The effect of the plasma on the applied electric field is not nearly as drastic as has been believed. The induction electric field is responsible for the energization; the plasma tries to counter that by charge separation to create an opposing electric field; this is highly restricted because it is electrostatic.

Morioka et al. (2007) examined the development of the auroral acceleration region during substorms. Two sources of auroral kilometric radiation (AKR) and their development prior to and during substorms were identified from high time resolution spectrograms provided by Polar/PWI electric field observations. One source is a low-altitude source region corresponding to middle frequency (MF-AKR); the other is a high-altitude source region corresponding to low frequency (LF-AKR).

- Bursty bulk flows (BBFs)

The plasma flow in the central plasma sheet of the Earth's magnetosphere includes short impulsive bursts (Baumjohann et al., 1990; Angelopoulos et al., 1992). The BBFs typically appear in 10 min sequences often consisting of several $\sim$1 min intensifications. The occurrence rate of high-speed flows, with typical ion bulk velocities above 400 km/s, increases during geomagnetically active periods. The bursts are almost always directed earthward.

Two main kinds of auroral signatures have been observed during bursty bulk flows: pseudobreakups and auroral streamers (Nakamura et al., 2001a; Juusola et al., 2009). Auroral streamers are approximately north-south aligned, longitudinally narrow auroral forms that first appear at the poleward boundary of the auroral oval and from there expand equatorward. After reaching the equatorward boundary of the oval, they decay by evolving into a patch of diffuse or pulsating aurora.

These results support the idea that a dawn-to-dusk polarization electric field is created in the bubble to enhance the flows as predicted in the bubble model (Chen and Wolf, 1993). The plasma is faced with the sudden onslaught of the induction electric field $-\partial \mathbf{A}/\partial t$ due to changes in plasma sheet convection; it tries to counter that with a redistribution of charge producing $-\nabla\phi$. However, it has an insurmountable problem: $-\nabla\phi$ has no curl ($\nabla \times \nabla\phi \equiv 0$). The plasma can only modify (reduce) the field-aligned component; in so doing, it must *enhance* the *transverse component* to preserve the curl, as noted by Fig. 11.14.

- Two methods of acceleration to high energies

The origin of energetic particles in the Earth's magnetosphere is still an unresolved mystery. In early satellite observations, energetic ion and electron bursts in the range of several 100 keV to 1 MeV (Kirsch et al. (1977); Sarafopoulos and Sarris (1988)) were observed in the magnetotail. They suggested that the observed energetic particle bursts were responding to an inductive electric field parallel to the magnetic field lines as in simulations, and as observed by Taylor et al. (2006) using Cluster.

An important feature of induction fields is the ability to energize charged particles locally by betatron acceleration. Only particles with a substantial magnetic moment, that is, particles that already have a rather high energy, are effectively energized. Near a neutral line, the problem of energizing "cold" plasma particles can be overcome by a two-step process. Linear nonadiabatic acceleration along the neutral line provides sufficient momentum so that when the particle leaves the neutral line region and gets into a gyro-orbit in the surrounding magnetic field, it has a sufficient magnetic moment to benefit from the betatron acceleration (Heikkila et al., 1979).

A different process of achieving high energies is by a cyclotron resonant interaction of the field with the charged particles, as pointed out in Chapter 7 (Chen, 2008). The energization rate by the cyclotron resonant acceleration can be expressed analytically. For particle cyclotron resonant acceleration, one only need consider the perpendicular component; that is, $dK_\perp/dt = qEv_\perp$ where E is the left-hand polarization perpendicular electric field with frequency the same as the ion gyrofrequency, and $\nu_\perp$ is the particle's perpendicular velocity. The energy increase is:

$$\Delta K_\perp = (2K_\perp(0)/m)^{1/2} qET_i + q^2 E^2 T_i^2/(2m) \qquad (11.14)$$

The significance of this equation is that over the ULF range, the power spectral densities are dominated by the perpendicular component of the local magnetic field. The measured left-hand polarization of the cusp electric field at ion gyrofrequencies demonstrates that the cyclotron resonant acceleration occurs in the diamagnetic cavities. It can energize from keV to MeV in seconds.

11.14 Problems

11.1. During Cluster's plasmoid encounter in the magnetotail on September 11, 2002, all four spacecraft were traveling in approximately the $-z$ direction so that their distances down-tail were not changing. How large would you estimate for the size of the plasmoid in the x dimension?

11.2. If the total current across a 7 R_E span of the magnetotail is 2×10^6 A, what is the linear current density?

11.3. The length of a magnetic field line segment going from the Earth to the subsolar point of the magnetopause can be roughly modeled as a half-circle. Given a delay time of 5 min and a typical value for the location of the magnetopause, what would be the propagation speed along the field line from the subsolar point to the Earth's surface?

11.4. The Earth camera on the Polar spacecraft has an angular resolution of 0.08 degrees with an overall field-of-view of $20° \times 20°$. What is the linear pixel size on the Earth when the spacecraft is near perigee at 0.65 R_E? At what altitude does the Earth just fit within its field of view?

11.5. Calculate the energy increase due to ion cyclotron resonant acceleration of H^+ if the initial velocity is 400 km/s, the gyroperiod is 3 s, and the perpendicular electric field is 30 mV/m.

References

Akasofu, S.-I., 1964. The development of the auroral substorm. Planet. Space Sci. 12 (4), 273–282. https://doi.org/10.1016/0032-0633(64)90151-5.

Akasofu, S.-I., Kimball, D.S., 1964. The dynamics of the aurora, 1. Instabilities of the aurora. J. Atmos. Terr. Phys. 26, 205.

Akasofu, S.-I., Lui, A.T.Y., Meng, C.-I., 2010. Importance of auroral features in the search for substorm onset processes. J. Geophys. Res.. 115, A08218 https://doi.org/10.1029/2009JA014960.

Amm, O., Donovan, E.F., Frey, H., Lester, M., Nakamura, R., Wild, J.A., et al., 2005a. Coordinated studies of the geospace environment using Cluster, satellite and ground-based data: an interim review. Ann. Geophys. 23, 2129–2170.

Amm, O., Aksnes, A., Stadsnes, J., Østgaard, N., Vondrak, R.R., Germany, G.A., et al., 2005b. Mesoscale ionospheric electrodynamics of omega bands determined from ground-based electromagnetic and satellite optical observations. Ann. Geophys. 23 (2), 325–342.

Angelopoulos, V., 1996. The role of impulsive particle acceleration in magnetotail circulation. In: Third International Conference on Substorms (ICS-3). ESA, Noordwijk, The Netherlands, p. 17.

Angelopoulos, V., Baumjohann, W., Kennel, C.F., Coroniti, F.V., Kivelson, M.G., Pellat, R., et al., 1992. Bursty bulk flow in the inner central plasma sheet. J. Geophys. Res. 97, 4027.

Astafyeva, E., Zakharenkova, I., Förster, M., 2015. Ionospheric response to the 2015 St. Patrick's Day storm: a global multi-instrumental overview. J. Geophys. Res. 120, 9023–9037. https://doi.org/10.1002/2015JA021629.

Baker, D.N., Pulkkinen, T.I., Angelopoulos, V., Baumjohann, W., McPherron, R.L., 1996. Neutral line model of substorms: past results and present view. J. Geophys. Res. 101 (A6), 2975–13010. https://doi.org/10.1029/95JA03753.

Baumjohann, W., Paschmann, G., Lühr, H., 1990. Characteristics of high-speed ion flows in the plasma sheet. J. Geophys. Res. 95 (A4), 3801–3809. https://doi.org/10.1029/JA095iA04p03801.

Baumjohann, W., Paschmann, G., Nagai, T., Lühr, H., 1991. Superposed epoch analysis of the substorm plasma sheet. J. Geophys. Res. 96 (A7), 11605–11608. https://doi.org/10.1029/91JA00775.

Birkeland, K., 1908. The Norwegian Aurora Polaris Expedition 1902–1903, vol. 1: On the Cause of Magnetic Storms and the Origin of Terrestrial Magnetism. H. Aschehoug and Co., Christiania, Denmark

Birn, J., Drake, J.F., Shay, M.A., Rogers, B.N., Denton, R.E., Hesse, M., et al., 2001. Geospace environmental modeling (GEM) magnetic reconnection challenge. J. Geophys. Res. 106, 3715.

Blake, J.B., Mueller-Mellin, R., Davies, J.A., Li, X., Baker, D.N., 2005. Global observations of energetic electrons around the time of a substorm on 27 August 2001. J. Geophys. Res.. 110, A06214 https://doi.org/10.1029/2004JA010971.

Büchner, J., Zelenyi, L.M., 1989. Regular and chaotic charged particle motion in magnetotaillike field reversals, 1. Basic theory of trapped motion. J. Geophys. Res. 94 (A9), 11821–11842. https://doi.org/10.1029/JA094iA09p11821.

Case, N.A., MacDonald, E.A., 2015. Aurorasaurus and the St Patrick's Day storm. Astron. Geophys. 56 (3), 3.13–3.14. https://doi.org/10.1093/astrogeo/atv089.

Case, N.A., Wild, J.A., 2012. A statistical comparison of solar wind propagation delays derived from multispacecraft techniques. J. Geophys. Res.. 117, A02101 https://doi.org/10.1029/2011JA016946.

Cattell, C., Dombeck, J., Wygant, J., Drake, J.F., Swisdak, M., Goldstein, M.L., et al., 2005. Cluster observations of electron holes in association with magnetotail reconnection and comparison to simulations. J. Geophys. Res. 110, A01211/1–A0121116. https://doi.org/10.1029/2004JA010519.

Chen, J., 2008. Evidence for particle acceleration in the magnetospheric cusp. Ann. Geophys. 26, 1993–1997.

Chen, C.X., Wolf, R.A., 1993. Interpretation of high-speed flows in the plasma sheet. J. Geophys. Res. 98, 21409–21419.

Chen, C.X., Wolf, R.A., 1999. Theory of thin-filament motion in Earth's magnetotail and its application to bursty bulk flows. J. Geophys. Res. 104 (A7), 14613–14626.

Cheng, C.Z., 2004. Physics of substorm growth phase, onset, and depolarization. Space Sci. Rev. 113 (1–2), 207–270. https://doi.org/10.1023/B:SPAC.0000042943.59976.0e.

Clauer, C.R., McPherron, R.L., 1974. Mapping the local time-universal time development of magnetospheric substorms using mid-latitude magnetic observations. J. Geophys. Res. 79, 2811–2820.

Coxon, J.C., Freeman, M.P., Jackman, C.M., Forsyth, C., Rae, I.J., Fear, R.C., 2018. Tailward propagation of magnetic energy density variations with respect to substorm onset times. J. Geophys. Res. 123, 4741–4754. https://doi.org/10.1029/2017JA025147.

Delcourt, D.C., Belmont, G., 1998a. Particle dynamics in the near-earth magnetotail and macroscopic consequences. In: Nishida, A, Baker, D.N., Cowley, S.W.H. (Eds.), New Perspectives on the Earth's Magnetotail. AGU Geophysical Monograph 105. American Geophysical Union, Washington, DC, pp. 193–210.

Delcourt, D.C., Belmont, G., 1998b. Ion dynamics at the earthward termination of the magnetotail current sheet. J. Geomagn. Geoelectr. 103, 4605.

Dungey, J.W., 1958. Cosmical Electrodynamics. Cambridge University Press, New York, NY.

Förster, M., Haaland, S.E., Doornbos, E., 2011. Thermospheric vorticity at high geomagnetic latitudes from CHAMP data and its IMF dependence. Ann. Geophys. 29 (1), 181–186. https://doi.org/10.5194/angeo-29-181-2011.

Frank, L.A., Sigwarth, J., 2000. Findings concerning the positions of substorm onsets with auroral images from the Polar spacecraft. J. Geophys. Res. 105 (A6), 12747–12761.

Frank, L.A., Sigwarth, J.B., Craven, J.D., Cravens, J.P., Dolans, J.S., Dvorsky, M.R., Hardebeck, P.K., Harvey, J.D., Muller, D.W., 1995. The visible imaging-system (VIS) for the Polar spacecraft. Space Sci. Rev. 71, 297.

Fujii, R., Hoffman, R., et al., 1994. Electrodynamic parameters in the nighttime sector during auroral substorms. J. Geophys. Res. 99 (A4), 6093–6112.

Fukunishi, H., 1975. Dynamic relationship between proton and electron auroral substorms. J. Geophys. Res. 80 (4), 553–574. https://doi.org/10.1029/JA080i004p00553.

Galperin, Y.I., Feldstein, Y.I., 1991. Auroral luminosity and its relationship to magnetospheric plasma domains. In: Meng, C.-I., Rycroft, M.J., Frank, L.A. (Eds.), Auroral Physics. Cambridge University Press, New York, NY, pp. 223–239.

Gjerloev, J.W., Hoffman, R.A., Sigwarth, J.B., Frank, L.A., 2007. Statistical description of the bulge-type auroral substorm in the far ultraviolet. J. Geophys. Res.. 112, A07213 https://doi.org/10.1029/2006JA012189.

Gjerloev, J.W., Hoffman, R.A., Sigwarth, J.B., Frank, L.A., Baker, J.B.H., 2008. Typical auroral substorm: a bifurcated oval. J. Geophys. Res.. 113, A03211 https://doi.org/10.1029/2007JA012431.

Harries, O., 2005. Suffer the intellectuals. The American Interest. 1(1).

Heikkila, W.J., Pellinen, R.J., 1977. Localized induced electric field within the magnetotail. J. Geophys. Res. 82, 1610–1614.

Heikkila, W.J., Pellinen, R.J., Fälthammar, C.-G., Block, L.P., 1979. Potential and induction electric fields in the magnetosphere during auroras. Planet. Space Sci. 27, 1383.

Heikkila, W.J., Chen, T., Liu, Z.X., Pu, Z.Y., Pellinen, R.J., Pulkkinen, T.I., 2001. Near Earth Current Meander (NECM) model of substorms. Space Sci. Rev. 95, 399–414.

Henderson, M.G., 1994. Implications of Viking Imager Results for Substorm Models. (Unpublished doctoral thesis)University of Calgary, Calgary, AB https://doi.org/10.11575/PRISM/22487.

Henderson, M.G., Reeves, G.D., Murphree, J.S., 1998. Are north-south aligned auroral structures an ionospheric manifestation of bursty bulk flows? Geophys. Res. Lett. 25, 3737–3740.

Hones Jr., E.W., 1984. Plasma sheet behavior during substorms. In: Hones, E.W (Ed.), Magnetic Reconnection in Space and Laboratory Plasmas. AGU

Geophysical Monograph 30. American Geophysical Union, Washington, DC, pp. 178 184.

Hubert, B., Kauristie, K., Amm, O., Milan, S.E., Grocott, A., Cowley, S.W.H., et al., 2007. Auroral streamers and magnetic flux closure. Geophys. Res. Lett.. 34, L15105 https://doi.org/10.1029/2007GL030580.

Ieda, A., Fairfield, D.H., Slavin, J.A., Liou, K., Meng, C.-I., Machida, S., et al., 2008. Longitudinal association between magnetotail reconnection and auroral breakup based on Geotail and Polar observations. J. Geophys. Res.. 113, A08207 https://doi.org/10.1029/2008JA013127.

Imada, S., Nakamura, R.P., Daly, W., Hoshino, M., Baumjohann, W., Mühlbachler, S., et al., 2007. Energetic electron acceleration in the downstream reconnection outflow region. J. Geophys. Res.. 112, A03202 https://doi.org/10.1029/2006JA011847.

Jacobsen, K.S., Andalsvik, Y.L., 2016. Overview of the 2015 St. Patrick's day storm and its consequences for RTK and PPP positioning in Norway. J. Space Weather Space Clim.. 6(A9) https://doi.org/10.1051/swsc/2016004.

Juusola, L., Nakamura, R., Amm, O., Kauristie, K., 2009. Conjugate ionospheric equivalent currents during bursty bulk flows. J. Geophys. Res.. 114, A04313.

Kalmoni, N.M.E., Rae, I.J., Watt, C.E.J., Murphy, K.R., Samara, M., Michell, R.G., Grubbs, G., Forsyth, C., 2019. A diagnosis of the plasma waves responsible for the explosive energy release of substorm onset. Nat. Commun. 9, 4806. https://doi.org/10.1038/s41467-018-07086-0.

Kamide, Y., Kusano, K., 2015. No major solar flares but the largest geomagnetic storm in the present solar cycle. Space Weather 13, 365–367. https://doi.org/10.1002/2015SW001213.

Kauristie, K., Pulkkinen, T.I., Pellinen, R.J., Janhunen, P., Huuskonen, A., Viljanen, A., et al., 1995. Analysis of the substorm trigger phase using multiple ground-based instrumentation. Geophys. Res. Lett. 22 (15), 2065–2068. https://doi.org/10.1029/95GL01800.

Kirsch, E., Krimigis, S.M., Sarris, E.T., Lepping, R.P., Armstrong, T.P., 1977. Evidence for an electric field in the magnetotail from observations of oppositely directed anisotropics of energetic ions and elections. Geophys. Res. Lett. 4, 137.

Kisabeth, J., Rostoker, G., 1974. The expansive phase of magnetospheric substorms, 1. Development of the auroral electrojets and auroral arc configuration during a substorm. J. Geophys. Res. 79 (7), 972–984.

Kosch, M., Amm, O., Scourfield, M., 2000. A plasma vortex revisited: the importance of including ionospheric conductivity measurements. J. Geophys. Res. 105 (A11), 24889.

Lessard, M.R., Lotko, W., LaBelle, J., Peria, W., Carlson, C.W., Creutzberg, F., et al., 2007. Ground and satellite observations of the evolution of growth phase auroral arcs. J. Geophys. Res.. 112, A09304 https://doi.org/10.1029/2006JA011794.

Liang, J., Donovan, E.F., Liu, W.W., Jackel, B., Syrjäsuo, M., Mende, S.B., et al., 2008. Intensification of preexisting auroral arc at substorm expansion phase onset: wave-like disruption during the first tens of seconds. Geophys. Res. Lett.. 35, L17S19 https://doi.org/10.1029/2008GL033666.

Lin, N., Frey, H.U., Mende, S.B., Mozer, F.S., Lysak, R.L., Song, Y., et al., 2009. Statistical study of substorm timing sequence. J. Geophys. Res.. 114, A12204 https://doi.org/10.1029/2009JA014381.

Liou, K., Meng, C.-I., Newell, P.T., Takahashi, K., Ohtani, S.-I., Lui, A.T.Y., 2000. Evaluation of low-latitude Pi2 pulsations as indicators of substorm onset using polar ultraviolet imagery. J. Geophys. Res. 105, 2495–2505.

Liou, K., Newell, P.T., Sibeck, D.G., Meng, C.-I., Brittnacher, M., Parks, G., 2001. Observation of IMF and seasonal effects in the location of auroral substorm onset. J. Geophys. Res. 106, 5799–5810.

Liou, K., Meng, C.-I., Lui, A.T.Y., Newell, P.T., Wing, S., 2002. Magnetic dipolarization with substorm expansion onset. J. Geophys. Res. 107, 1428. https://doi.org/10.1029/2001JA00179.

Liu, Y.D., Hu, H., Wang, R., Yang, Z., Zhu, B., Liu, Y.A., Luhmann, J.G., Richardson, J.D., 2015. Plasma and magnetic field characteristics of solar coronal mass ejections in relation to geomagnetic storm intensity and variability. Astrophys. J. Lett.. 809(2) https://doi.org/10.1088/2041-8205/809/2/L34.

Lui, A.T.Y., 1991. A synthesis of magnetospheric substorm models. J. Geophys. Res. 96, 1849–1856. https://doi.org/10.1029/90JA02430.

Lui, A.T.Y., 1996. Current disruption in the Earth's magnetosphere: observations and models. J. Geophys. Res. 101, 13067–13088.

Lui, A.T.Y., 2004. Potential plasma instabilities for substorm expansion onset. Space Sci. Rev. 113, 127–206.

Lui, A.T.Y., Zheng, Y., Rème, H., Dunlop, M.W., Gustafsson, G., Owen, C.J., 2007. Breakdown of the frozen-in condition in the Earth's magnetotail. J. Geophys. Res.. 112, A04215 https://doi.org/10.1029/2006JA012000.

Lyons, L.R., Zou, Y., Nishimura, Y., Gallardo-Lacourt, B., Angelopulos, V., Donovan, E.F., 2018. Stormtime substorm onsets: occurrence and flow channel triggering. Earth Planets Space 70 (81). https://doi.org/10.1186/s40623-018-0857-x.

Lysak, R.L., Song, Y., Jones, T.W., 2009. Propagation of Alfvén waves in the magnetotail during substorms. Ann. Geophys. 27, 2237–2246.

McPherron, R.L., 1970. Growth phase of magnetospheric substorms. J. Geophys. Res. 75, 5592–5599. https://doi.org/10.1029/JA075i028p05592.

McPherron, R.L., 1995. Magnetospheric dynamics. In: Kivelson, M.G., Russell, C.T. (Eds.), Introduction to Space Physics. Cambridge University Press, New York, NY, p. 400.

Mende, S.B., Carlson, C.W., Frey, H.U., Peticolas, L.M., Østgaard, N., 2003. FAST and IMAGE-FUV observations of a substorm onset. J. Geophys. Res. 108, 1344.

Meng, C.-I., Liou, K., 2004. Substorm timings and timescales: a new aspect. Space Sci. Rev. 113, 41.

Milan, S.E., Walach, M.-T., Carter, J.A., Sangha, H., Anderson, B.J., 2019. Substorm onset latitude and the steadiness of magnetospheric convection. J. Geophys. Res. 124, 1738–1752. https://doi.org/10.1029/2018JA025969.

Morioka, A., Miyoshi, Y., Tsuchiya, F., Misawa, H., Sakanoi, T., Yumoto, K., et al., 2007. Dual structure of auroral acceleration regions at substorm onsets as derived from auroral kilometric radiation spectra. J. Geophys. Res.. 112, A06245 https://doi.org/10.1029/2006JA012186.

Motoba, T., Hosokawa, K., Kadokura, A., Sato, N., 2012. Magnetic conjugacy of northern and southern auroral beads. Geophys. Res. Lett.. 39, L08108 https://doi.org/10.1029/2012GL051599.

Nakamura, R., Baumjohann, W., Brittnacher, M., Sergeev, V., Kubyshkina, M., Mukai, T., et al., 2001a. Flow bursts and auroral activations: onset timing and foot point location. J. Geophys. Res. 106 (A6), 10777–10789.

Nakamura, R., Baumjohann, W., Asano, Y., Runov, A., Balogh, A., Owen, C.J., et al., 2006. Dynamics of thin current sheets associated with magnetotail reconnection. J. Geophys. Res. 111, A11206. https://doi.org/10.1029/2006JA011706.

Nagai, T., Fujimoto, M., Saito, Y., Machida, S., Terasawa, T., Nakamura, R., 1998. Structure and dynamics of magnetic reconnection for substorm onsets with geotail observations. J. Geophys. Res. 103, 4419–4440.

Newell, P.T., Liou, K., Sotirelis, T., Meng, C.-I., 2001. Polar Ultraviolet Imager observations of global auroral power as a function of polar cap size and magnetotail stretching. J. Geophys. Res. 106 (A4), 5895–5905.

Newell, P.T., Sotirelis, T., Wing, S., 2010. Seasonal variations in diffuse, monoenergetic, and broadband aurora. J. Geophys. Res.. 115, A03216 https://doi.org/10.1029/2009JA014805.

Ohtani, S., 2004. Flow bursts in the plasma sheet and auroral substorm onset: observational constraints on connection between mid-tail and near-Earth substorm processes. Space Sci. Rev. 113, 77–96.

Ohtani, S., Singer, H.J., Mukai, T., 2006. Effects of the fast plasma sheet flow on the geosynchronous magnetic configuration: Geotail and GOES coordinated study. J. Geophys. Res.. 111, A01204 https://doi.org/10.1029/2005JA011383.

Østgaard, N., Snekvik, K., et al., 2009. Can magnetotail reconnection produce the auroral intensities observed in the conjugate ionosphere? J. Geophys. Res.. 114, A06204 https://doi.org/10.1029/2009JA014185.

Paschmann, G., Haaland, S., Treumann, R. (Eds.), 2003. Auroral Plasma Physics. Kluwer, Norwell, MA (Reprinted from *Space Science Reviews, 103*: 1–485).

Pellinen, R.J., 1978. Analytical model to study the behaviour of auroral particles in a local region of the magnetotail at the onset of a magnetospheric substorm. Geophysica 15 (1), 41–63.

Pellinen, R.J., Heikkila, W.J., 1978. Energization of charged particles to high energies by an induced substorm electric field within the magnetotail. J. Geophys. Res. 83, 1544.

Pellinen, R.J., Baumjohann, W., Heikkila, W.J., Sergeev, V.A., Yahnin, A.G., Marklund, G., et al., 1982. Event study on pre-substorm phases and their relation to the energy coupling between the solar wind and magnetosphere. Planet. Space Sci. 30, 371.

Pulkkinen, T.I., Baker, D.N., Fairfield, D.H., Pellinen, R.J., Murphree, J.S., Elphinstone, R.J., et al., 1991. Modeling the growth phase of a substorm using the Tsyganenko model and multi-spacecraft observations: CDAW-9. Geophys. Res. Lett. 18 (11), 1963–1966.

Roble, R.G., Ridley, E.C., 1987. An auroral model for the NCAR thermospheric general circulation model (TGCM). Ann. Geophys. 5, 369.

Rostoker, G., 2002. Identification of substorm expansive phase onsets. J. Geophys. Res. 107 (A7), 1137. https://doi.org/10.1029/2001JA003504.

Rostoker, G., Lui, A.T.Y., Anger, C.D., Murphree, J.S., 1987. North south structures in the midnight sector auroras as viewed by the Viking imager. Geophys. Res. Lett. 14 (4), 407–410.

Roux, A., Perraut, S., Robert, P., Morane, A., Pedersen, A., Korth, A., et al., 1991. Plasma sheet instability related to the westward traveling surge. J. Geophys. Res. 96 (A10), 17697–17714. https://doi.org/10.1029/91JA01106.

Runov, A., Nakamura, R., Baumjohann, W., Treumann, R.A., Zhang, T.L., Volwerk, M., et al., 2003. Current sheet structure near magnetic X-line observed by cluster. Geophys. Res. Lett. 30 (11), 1579. https://doi.org/10.1029/2002GL016730.

Runov, A., Sergeev, V.A., Nakamura, R., Baumjohann, W., Zhang, T.L., Asano, Y., et al., 2005. Reconstruction of the magnetotail current sheet structure using multi-point cluster measurements. Planet. Space Sci. 53, 237–243.

Ruohoniemi, J.M., Baker, K.B., 1998. Large-scale imaging of high-latitude convection with SuperDARN HF radar observations. J. Geophys. Res. 103, 20797–20811. https://doi.org/10.1029/98JA01288.

Sarafopoulos, D.V., Sarris, E.T., 1988. Inverse velocity dispersion of energetic particle bursts inside the plasma sheet. Planet. Space Sci. 36, 1181.

Schindler, K., Hesse, M., Birn, J., 1988. General magnetic reconnection, parallel electric fields, and helicity. J. Geophys. Res. 93, 5547–5557. https://doi.org/10.1029/JA093iA06p05547.

Schmid, D., Nakamura, R., Volwerk, M., Plaschke, F., Narita, Y., Baumjohann, W., Magnes, W., Fischer, D., Eichelberger, H.U., Torbert, R.B., Russell, C.T., Strangeway, R.J., Leinweber, H.K., Le, G., Bromund, K.R., Anderson, B.J.,

Slavin, J.A., Kepko, E.L., 2016. A comparative study of dipolarization fronts at MMS and Cluster. Geophys. Res. Lett. 43, 6012–6019. https://doi.org/10.1002/2016GL069520.

Sergeev, V.A., Liou, K., Newell, P.T., Ohtani, S.-I., Hairston, M.R., Rich, F., 2004. Auroral streamers: characteristics of associated precipitation, convection and field-aligned currents. Ann. Geophys. 22, 537–548.

Shen, C., Li, X., Dunlop, M., Liu, Z.X., Balogh, A., Baker, D.N., et al., 2003. Analyses on the geometrical structure of magnetic field in the current sheet based on cluster measurements. J. Geophys. Res. 108 (A5), 1168.

Shen, C., Liu, Z.X., Li, X., Dunlop, M., Lucek, E., Rong, Z.J., et al., 2008. Flattened current sheet and its evolution in substorms. J. Geophys. Res.. 113, A07S21 https://doi.org/10.1029/2007JA012812.

Shiokawa, K., Baumjohann, W., Haerendel, G., 1997. Braking of high-speed flows in the near-earth tail. Geophys. Res. Lett. 24 (10), 1179–1182.

Shue, J.-H., Ieda, A., Lui, A.T.Y., Parks, G.K., Mukai, T., Ohtani, S., 2008. Two classes of earthward fast flows in the plasma sheet. J. Geophys. Res.. 113, A02205 https://doi.org/10.1029/2007JA012456.

Sitnov, M.I., Guzdar, P.N., Swisdak, M., 2003. A model of the bifurcated current sheet. Geophys. Res. Lett. 30, 1712. https://doi.org/10.1029/2003GL017218.

Sitnov, M., Birn, J., Ferdousi, B., Gordeev, E., Khotyaintsev, Y., Markin, V., Motoba, T., Otto, A., Panov, E., Pritchett, P., Pucci, F., Raeder, J., Runov, A., Sergeev, V., Velli, M., Zhou, X., 2019a. Explosive magnetotail activity. Space Sci. Rev. 215, 31. https://doi.org/10.1007/s11214-019-0599-5.

Sitnov, M.I., Stephens, G.K., Tsyganenko, N.A., Miyashita, Y., Merkin, V.G., Motoba, T., Ohtani, S., Genestreti, K.J., 2019b. Signatures of nonideal plasma evolution during substorms obtained by mining multimission magnetometer data. J. Geophys. Res.. 124https://doi.org/10.1029/2019JA027037.

Song, Y., 2003. Challenge to the magnetic reconnection hypothesis. In: Lundin, R., McGregor, R. (Eds.), Proceedings of the Magnetic Reconnection Meeting, IRF Scientific Report 280. Swedish Institute of Space Physics, Kiruna, Sweden, pp. 25–35.

Song, Y., Lysak, R.L., 2001. Towards a new paradigm: from a quasi-steady description to a dynamical description of the magnetosphere. Space Sci. Rev. 95, 273–292. https://doi.org/10.1023/A:1005288420253.

Taylor, M.G.G.T., Reeves, G.D., Friedel, R.H.W., Thomsen, M.F., Elphic, R.C., Davies, J.A., et al., 2006. Cluster encounter with an energetic electron beam during a substorm. J. Geophys. Res.. 111, A11203 https://doi.org/10.1029/2006JA011666.

Torr, M.R., Torr, D.G., Zukic, M., Johnson, R.B., Ajello, J., Banks, P., et al., 1995. A far ultraviolet imager for the international solar-terrestrial physics mission. Space Sci. Rev. 71, 329.

Tsyganenko, N.A., 1989. A magnetospheric magnetic field model with a warped tail current sheet. Planet. Space Sci. 37, 5–20.

Vasyliunas, V.M., 1975. Theoretical models of magnetic field line merging, 1. Rev. Geophys. Space Phys. 13, 303–336.

Villante, U., Lepidi, S., Francia, P., Bruno, T., 2004. Some aspects of the interaction of interplanetary shocks with the Earth's magnetosphere: an estimate of the propagation time through the magnetosheath. J. Atmos. Sol. Terr. Phys. 66, 337–341. https://doi.org/10.1016/j.jastp.2004.01.003.

Weimer, D.R., Craven, J.D., Frank, L.A., Hanson, W.B., Maynard, N.C., Hoffman, R.A., et al., 1994. Satellite measurements through the center of a substorm surge. J. Geophys. Res. 99 (A12), 23639–23649. https://doi.org/10.1029/94JA01976.

Wu, C., Liou, K., Lepping, R.P., Hutting, L., Plunkett, S., Howard, R.A., Socker, D., 2016a. The first super geomagnetic storm of solar cycle 24: "The St. Patrick's day event (17 March 2015)" Earth Planets Space 68, 151. https://doi.org/10.1186/s40623-016-0525-y.

Wu, M.Y., Lu, Q., Volwerk, M., Vörös, Z., Ma, X., Wang, S., 2016b. Current sheet flapping motions in the tailward flow of magnetic reconnection. J. Geophys. Res. 121, 7817–7827. https://doi.org/10.1002/2016JA022819.

Wygant, J.R., Cattell, C.A., Lysak, R., Song, Y., Dombeck, J., McFadden, J., et al., 2005. Cluster observations of an intense normal component of the electric field at a thin reconnecting current sheet in the tail and its role in the shock-like acceleration of the ion fluid into the separatrix region. J. Geophys. Res.. 110, A09206 https://doi.org/10.1029/2004JA010708.

Yang, J., Wolf, R., Toffoletto, F., Sazykin, S., Wang, W., Cui, J., 2019. The inertialized rice convection model. J. Geophys. Res. 124, 10294–10317. https://doi.org/10.1029/2019JA026811.

Yoon, P., Lui, A., 2007. Anomalous resistivity by fluctuation in the lower-hybrid frequency range. J. Geophys. Res.. 112(A6), A06207.

Zhao, D., Fu, S., Parks, G.K., Chen, L., Liu, X., Tong, Y., et al., 2020. Modulation of whistler mode waves by ion-scale waves observed in the distant magnetotail. J. Geophys. Res. 125, e2019JA027278. https://doi.org/10.1029/2019JA027278.

Zhou, X.-Z., Angelopoulos, V., Runov, A., Sitnov, M.I., Coroniti, F., Pritchett, P., et al., 2009. Thin current sheet in the substorm late growth phase: modeling of THEMIS observations. J. Geophys. Res.. 114, A03223.

12

Epilogue

Chapter outline

> *Fluid dynamicists were divided into hydraulic engineers who observed what could not be explained, and mathematicians who explained things that could not be observed.*

Sir Cyril Hinshelwood, quoted by Lighthill in Nature (1956)

Earth's Magnetosphere. https://doi.org/10.1016/B978-0-12-818160-7.00012-0

12.1 Introduction

The space plasma physics community is active in extending the knowledge of plasma physics to the Sun and other objects in astrophysics. The NASA panel on Sun-Solar System Connection 2005–2035 Roadmap devoted to that aim suggested four fundamental processes as the critical immediate steps: magnetic reconnection, particle acceleration, the physics of plasma and neutral interactions, and the generation and variability of magnetic fields with their coupling to structures throughout the heliosphere. Each of these research focus areas involves the universal themes of energy conversion and transport, cross-scale coupling, turbulence, and nonlinear physics—concepts that are fundamental to the understanding of space and planetary systems.

Understanding the origins and manifestations of solar variability and the influence of the Sun on the Earth's atmosphere and geospace is a scientific pursuit driven both by intellectual inquiry and by societal needs. In the mid-20th century, the first space-age satellites were launched and space physics forever changed our view of geospace. Today, the once-separate fields of *solar physics* and *space physics* now strive jointly to elucidate how changes in the Sun and the solar wind cause changes in the ionospheres, thermospheres, and magnetospheres surrounding Earth and other planets, as well as the heliosphere that marks the boundaries of our solar system. While changes in the total solar radiative output are small, the energetic photons (gamma rays, x-rays, and extreme ultraviolet light) and the highly variable particles and fields that escape the Sun can have profound effects on the space environment, or even on terrestrial weather (Frederick et al., 2019). Together, solar and space physics offers a pathway to the critical understanding of this highly coupled natural system.

The interdisciplinary, interagency, and international enterprise of modern solar and space physics involves observing platforms in space, in the atmosphere, and on the ground that support a range of research, from fundamental, curiosity-driven investigations to predictive space weather applications. The field draws on the expertise of research physicists, instrument builders, technologists, and spacecraft engineers. In the United States, NASA and the National Science Foundation (NSF) support the field's space- and ground-based research, while the National Oceanic and Atmospheric Administration (NOAA) applies these and other resources to deliver space weather forecasts for users in the government and private sector. Research teams often include international contributions, with foreign instruments mounted on

US research platforms and vice versa. By drawing on this large set of talent and resources, the research community is able to make progress in understanding the complex and highly nonlinear phenomena explored by solar and space physics.

The investigation of solar system plasmas as coupled non-linear systems requires synergy between observational and theoretical initiatives, and between basic research and targeted research programs. NASA recognized this in establishing (and continually updating) a fleet of missions covering a wide range of locations and capabilities. Currently, three are located at the "L1" point between the Earth and the Sun:

- Advanced Composition Explorer (ACE)
- Solar and Heliospheric Observatory (SOHO)
- Wind
 Two are operating in heliocentric orbits:
- Solar Terrestrial Relations Observatory (STEREO)
- Parker Solar Probe
 The remainder in various geocentric orbits:
- Cluster
- Reuven Ramaty High-Energy Solar Spectroscopic Imager (RHESSI)
- Geotail
- Thermosphere Ionosphere Mesosphere Energetics and Dynamics (TIMED)
- Time History of Events and Macroscale Interaction during Substorms (THEMIS), two of which are now operating as Acceleration, Reconnection, Turbulence and Electrodynamics of the Moon's Interaction with the Sun (ARTEMIS)
- Aeronomy of Ice in the Mesosphere (AIM)
- Hinode
- Interstellar Boundary Explorer (IBEX)
- Interface Region Imaging Spectrograph (IRIS)
- Magnetospheric Multiscale (MMS)
- Solar Dynamics Observatory (SDO)
- Two Wide-angle Imaging Neutral-atom Spectrometers (TWINS)
- Van Allen Probes

There are also the Voyager probes near the heliopause. NASA's Heliophysics Division calls this spacecraft fleet the Heliophysics System Observatory (HSO). Together, these scientific platforms are able to observe the variable outputs of the Sun, provide advance information on the disturbances traveling to Earth, observe the resulting space weather responses in the geospace environment, and measures responses in the distant heliosphere. The Heliophysics System Observatory enables the coordinated investigation of space plasmas as complex coupled systems driven from the Sun to Earth and the heliosphere, as demonstrated by the observation and analysis of the famous "Halloween" solar storms of 2003 and the "Saint Patrick's Day" storm of 2015.

The roadmap referenced earlier clearly recognized the mission synergies and balance that underpin the Heliophysics System Observatory approach: mission selections and priorities in the Integrated Research Strategy were chosen in order to most efficiently maintain and augment such a research approach. The roadmap anticipated a midterm period where Hinode (Solar-B), Solar Terrestrial Relations Observatory (STEREO), and Solar Dynamics Observatory (SDO) would be providing more detailed information on the propagation of solar disturbances to Earth, while the Magnetospheric Multiscale (MMS) mission, the Van Allan Probes (Radiation Belt Storm Probes), the Ionosphere-Thermosphere Storm Probes (I-TSP), and the Geospace Electrodynamic Connections (GEC) mission would be providing the information needed to understand how the disturbances were processed within the geospace system. Throughout this period the Explorer program was expected to play a pivotal role in addressing new inquiries and strategic initiatives that would emerge during the conduct of the program. A revitalized sounding rocket program would also continue to provide unique capabilities associated with access to critical regions and small-scale features. Two of these missions, I-TSP and GEC, did not fly in the timeframe expected, a situation all too familiar to anyone working in the space sciences.

> *Unfortunately, very little of the recommended NASA program priorities from the decadal survey's Integrated Research Strategy will be realized during the period (2004–2013) covered by the survey. Mission cost growth, reordering of survey mission priorities, and unrealized budget assumptions have delayed or deferred nearly all of the NASA spacecraft missions recommended in the survey. As a result, the status of the Integrated Research Strategy going forward is in jeopardy, and the loss of synergistic capabilities in space will constitute a serious impediment to future progress.*

National Research Council (2009)

In addition, the present book shows that we desperately need to revise our analysis. There are the obvious problems that everyone should realize, although some will not necessarily wish to discuss them in public (e.g., Poynting's theorem). To draw upon an analogy with sailing, there are issues that represent possible shoals, submerged reefs that you can run into quite suddenly.

Presence of rocks would be catastrophic; one must be careful about these hazards, as a matter of course.

> *Given that a critical attitude is one of the guiding principles of science, it is natural to suppose that scientists will also adopt a critical attitude towards the way in which science is carried out and everything connected with this. Science in every age has experienced its own problems, pitfalls, and risks, but people have not necessarily wished to speak about these for fear that the discussion might undermine the credibility of science itself. In fact, exactly the opposite is true. The ability to look at itself in a mirror is one of the fundamental requirements for any science that strives to achieve impartial truth.*

Saarnisto (2008)

This was the subject for the Inaugural Symposium in November 23, 2007, for the Centenary Year of the Finnish Academy of Science and Letters. Four themes were considered:
- **Theme 1**: Ever tighter competition; chairman: Jorma Sipila
- **Theme 2**: Tensions within the academic community; chairman: Sirpa Jalkanen
- **Theme 3**: Pressures from outside; chairman: Ilkka Hanski
- **Theme 4**: Is science becoming isolated from its public? chairman: Ulla-Maija Kulonen

The academic world has always been one of competition in which scholars, schools of thought, and nations have competed among themselves in matters of reputation, honor, and financial success. The competition has become more severe than ever in recent times. Most research nowadays is carried out at least in an atmosphere of competition for external funding in which the universities attempt to augment their finances by producing more doctorates than ever before, and all manner of quantitative measures are derived to determine and compare with their "productivity figures."

12.1.1 Magnetic reconnection

Magnetic reconnection is said to occur in highly localized regions when interacting magnetic fields "snap" to a new, lower-energy configuration. Magnetic reconnection is supposed to release vast amounts of stored energy responsible for solar flares, CMEs, and geospace storms. We have developed an initial picture of where reconnection may occur and the apparent results; the detailed physical mechanisms, the role of large-scale topology, and the microphysical processes are still not understood.

12.1.2 Particle acceleration

By far the most distinguishing characteristic of plasmas is that they are efficient particle accelerators. A detailed understanding of the particle energization processes, the regions in which these processes operate, and the boundary conditions that control them, is crucial to the exploration of space. Radiation can be produced almost instantaneously through explosive processes, but also built up stepwise by processes acting under more benign conditions.

12.1.3 Plasma-neutral interactions

The Sun-Solar System Connection requires understanding of the fundamental physics of plasma and neutral particle coupling. This coupling encompasses a variety of mechanisms and regions from turbulence and charge exchange in the solar wind to gravity waves and chemical/collisional interactions in planetary atmospheres. Space plasmas are often in a nonequilibrium state and they can be a highly nonlinear medium.

12.1.4 Magnetic dynamos

Understanding the variations of the magnetic fields of the Sun and planets on both long and short time-scales is the key element of the Sun-Solar System Connection. The magnetic dynamo problem remains one of the outstanding problems in physics. How dynamos operate in such widely disparate systems, from stellar interiors to planetary cores, is poorly understood. Dynamos determine the characteristics of the solar activity cycle. The Sun's magnetic field controls the structure of the heliosphere and, thus, regulates the entry of galactic cosmic rays into the solar system. We must understand how dynamos are created and sustained, how they affect the nearby space environment, how to predict their variations, and ultimately their demise.

12.2 Main arguments in this book

This book presents a new way of looking at problems of solar wind interaction with the magnetosphere, and in space plasma physics in general. Several new or revised ideas are discussed, as well as reminders to the reader on matters that have apparently been ignored or forgotten.

12.2.1 Chapter 1: Historical introduction

A review of observations and theories during the International Geophysical Year of 1957–58 and the International Magnetospheric Study of 1976–79.

12.2.2 Chapter 2: Kirchhoff's laws

Dynamo $\mathbf{E} \cdot \mathbf{J} < 0$ versus electrical load $\mathbf{E} \cdot \mathbf{J} > 0$. This is well known to the engineering community. The circuit problem illustrates correctly the idea of cause and effect by the sign of $\mathbf{E} \cdot \mathbf{J}$.

> *From the rigorous starting point of the fundamental laws [Newton's and Maxwell's equations], it [is] found that for circuits which are small compared with wavelength, this exact approach leads directly to the familiar circuit ideas based upon Kirchhoff's laws, and the concepts of lumped inductances and capacitances are sufficient for analysis.*
>
> **Ramo and Whinnery (1953, p. 207)**

12.2.3 Chapter 3: Helmholtz's theorem

An efficient two-step acceleration process for magnetospheric substorms and solar flares.

> *Despite decades of observations in X-rays and gamma-rays, the mechanism for particle acceleration remains an enigma.*
>
> **Emslie and Miller (2003)**

A promising concept for this energization is demonstrated by 1D particle simulation of Omura et al. (2003). We note that the electric field is composed of two parts,

$$\mathbf{E} = -\nabla \phi - \partial \mathbf{A} / \partial t \qquad (12.1)$$

This allowed us to impose an inductive field-aligned electric field on a stable plasma; we found out, quite unexpectedly, how the plasma responds. This reaction is complex as it depends on the energy range of interest and the time of importance. These results apply directly to the substorm problem as follows:

- A transient parallel electric field $\mathbf{E}$ causes electrons and ions to be accelerated in a plasma almost as in free space despite some instabilities. This has been observed, notably by Taylor et al. (2006).

- At the onset of a magnetospheric substorm, electrons are abruptly energized to 100 keV. This is the westward traveling surge, quite similar to solar flares.
- Ions are accelerated also but these are accompanied by lower-energy electrons in a form of Debye shielding.
- Still higher energies exceeding 1 MeV result from the sudden breakdown at the X-line with low magnetic field, followed by an adiabatic process to higher values of **B** with the appropriate energies increased by the ratio of the magnetic field strength.

12.2.4 Chapter 4: Magnetohydrodynamic equations

The ionized plasma in near-Earth space and in the Solar Wind can be treated as a conducting fluid, for which the differential fluid dynamic and electromagnetic equations that govern magnetohydrodynamics can be applied. Since its introduction by Hannes Alfvén in 1942, MHD in space physics has become a highly specialized field utilizing the latest computing power to run simulations of ever-increasing complexity. Even so, assumptions and simplifications must be made, and any conclusions derived from the simulations must take those into account. As is often the case, the most interesting and important physics tends to occur when the idealized conditions break down.

There is no general rule for telling apart numerical and physical phenomena. In fact the division is artificial, as all phenomena in a numerical simulation are products of numerical calculations and in that sense of numerical origin. ... Thus classifying something as physical is not a statement about correspondence between the simulation and nature—it only says that within the reference frame of the physical model the phenomenon is real. Comparing the model to observations is then another question.

Laitinen et al. (2007, p. 114)

12.2.5 Chapter 5: Poynting's energy conservation theorem

Poynting's theorem shows just three choices for energy conservation. Poynting's theorem is hardly used in space plasma physics (except as student exercises). It clearly expresses a choice between three processes for solar wind interaction with the magnetosphere:
- A surface integral draws energy from an external source.
- Two time-dependent volume integrals tap stored electric and magnetic energies.

For the steady state the energy is supplied by an external dynamo; to understand this we must use 3D and/or circuit analysis. This is the term used in *standard magnetic reconnection* (SMR) involving an *X*-line in the magnetic topology, in 2D; no mention is made of the required dynamo. Time dependence and 3D is used only in *generalized magnetic reconnection* (GMR), in line with Faraday (1832). GMR should (must) be used in impulsive events, for example, in the so-called *flux transfer event* (FTE). Unfortunately, FTE is based on SMR, without a dynamo.

This point is so simple and compelling that one wonders how it was missed at the very beginning of the debate six decades ago. It is still missed as evidenced by the workshop in Kiruna in September 2002. The goal was to analyze the existing findings critically and to find a means of discriminating magnetic reconnection from other possible plasma physical processes. That workshop was intended to be at another location, but the referees for that institution voted against providing support. To wit, "There is a risk that a ruling paradigm may gain so much momentum that it becomes obvious and thus beyond criticism" (Lundin, 2003).

12.2.6 Chapter 6: Magnetopause

The key question was raised by the Cluster team:

The existence of a low-latitude boundary layer demonstrates that magnetosheath plasma can penetrate into the magnetosphere. How this actually happens has remained an unresolved question in magnetospheric physics, despite numerous studies based on single-spacecraft observations.

De Keyser (2005)

A PTE was observed by Cluster 3 on March 8, 2003, beginning at 0707 UT. For over a minute CIS saw, inside the magnetopause, a burst of solar wind plasma with a density up to $0.8 \, cm^{-3}$, a factor of 4 higher than was observed before and after. PEACE showed field-aligned fluxes at energies up to 500 eV. At higher energies from 1 to 40 keV the fluxes had a pancake distribution indicating closed field lines surrounding the event. The electric field was observed by EFW to vary between − 5 and 5 mV/m. WHISPER recorded strong plasma oscillations mostly in two bursts coincident with the fluxes recorded by PEACE.

Heikkila et al. (2006)

A plasma transfer event (PTE) allows:

- Transfer of solar wind plasma flow across the magnetopause
- Because both electric and magnetic fields reverse on the two sides, $\mathbf{E} \times \mathbf{B}$ is unidirectional; a plasma cloud can cross into the low-latitude boundary layer ...
- ... even onto closed field lines
- A finite B_n is crucial to a PTE, thus a dependence on the interplanetary magnetic field.

The other three Cluster spacecraft were in the magnetosheath; the data were used by Sonnerup et al. (2004) as evidence for an FTE, a major question but a separate issue from the PTE.

- GMR was advocated by Schindler et al. (1988) as a process affecting the topology of the magnetic field in three dimensions when $\mathbf{B}$ is finite (not an X-line); GMR is based on the breakdown of magnetic connection; the key issue is that there be an electric field parallel to the magnetic field lines in some region D_R; this is just what happens with a PTE.

12.2.7 Chapter 7: High-altitude cusps

The high-altitude cusp is a powerful source of energetic particles, a surprise to many. The high-altitude cusps have remained among the major unexplored regions of the Earth's magnetosphere, yet they are of major importance to solar wind mass and energy flow into the magnetosphere. This is true whether the interplanetary magnetic field is southward or northward, despite the great difference in topology. An unexpected finding is that particles, both electrons and ions, can attain MeV energies in the cusps on a regular basis (Fritz and Fung, 2005).

New evidence reveals that the charged particles can be energized locally in the magnetospheric cusp. The power spectral density of the cusp magnetic fluctuations shows increases by up to four orders of magnitude in comparison to an adjacent region. Large fluctuations of the cusp electric fields have been observed with an amplitude of up to 350 mV/m. The measured left-hand polarization of the cusp electric field at ion gyro-frequencies indicates that the cyclotron resonant acceleration mechanism is working in this region. The cyclotron resonant acceleration can energize ions from keV to MeV in seconds.

Chen (2008)

12.2.8   Chapter 8: Inner magnetosphere

The inner magnetosphere is dominated by the Van Allen Radiation Belts, zones of energetic charged particles trapped by the geomagnetic field at equatorial distances from about 1 to 10 Earth radii. The study of its dynamics has mostly focused on the response to magnetospheric substorms. The typical storm response was found to be a rapid decrease in electron fluxes during the main phase followed by reintensification. Two primary methods for electron acceleration were proposed to explain the early observations, radial diffusion and local wave-particle interactions. Recent analysis of data from the Van Allen Probe mission supports local electron acceleration via waves as the primary mechanism. Much remains to be learned about the details of this acceleration mechanism and the waves that enable it, but recent progress has been very encouraging.

The transient penetration of magnetosheath particles along field lines to the ionosphere can cause auroral emissions. Spacecraft measurements indicate that such events can be caused by magnetic disturbances in the solar wind. Studying the ionosphere in winter has the advantage of prolonged darkness, aiding in the study of auroral emissions due to the lack of ionization from sunlight. This also aids in the ability to map a large portion of the convection patterns within the ionosphere and polar cap. It has been found that light ions are dominant down to much lower altitudes in the winter polar ionosphere than in the sunlight. The polar cap in darkness is very different than when sunlit; there is almost no precipitation, no ionization, no excitation, and no auroras in the polar cap during southward IMF. Observations of the ionospheric convection signature at high latitudes during periods of prolonged northward IMF show that a four-cell convection pattern can frequently be observed. This exists in a region that is displaced to the sunward side of the dawn-dusk meridian regardless of season.

12.2.9   Chapter 9: Low-latitude boundary layer

The low-latitude boundary layer (LLBL) is a charged dipolar layer; it differs fundamentally from a charged layer. With a dipolar layer the electrostatic field interior to the layer can be strong in view of the antisunward flow; it has little effect on the external field allowing its transfer from one side to the other, using the superposition principle.

- Inadequacy of fluid theory
  Although the fluid theory of plasma is based on conservation of mass, momentum, and energy, it cannot do justice for many problems in space physics in part because it is not focused on the divergence and curl of **E**.

  > *It is rather fortunate, therefore, to have found that these fictitious quantities obey some quasihydrodynamic equations: an equation of continuity and an equation of motion. This is about as far as the analogy can be stretched. Unfortunately these equations do not suffice to provide solutions for the great number of unknown quantities.*

  Schmidt (1979, p. 57)

- The particle approach:
  Since the total charge is zero (negative equals the positive at the edges), the electric field in the magnetosheath is transmitted to the other (the plasma sheet) without much change. Thus:
  - The LLBL supplies both the electric field …
  - … and the plasma into the plasma sheet.

  The plasma must work hard in maintaining the electric field to conserve the momentum of the plasma flowing tailward, acting as a dynamo.

  > *It should not be thought, however, that the plasma can move indefinitely across a magnetic field. Particles drifting out into the surface layer experience a smaller electric field and hence a smaller drift than the bulk of the plasma and are consequently left behind. Since the total electric field in the plasma cannot change [because of the anti-sunward velocity], the particles lost from the surface layer are continuously replaced from the plasma interior. Thus the plasma velocity remains unchanged while the mass is gradually decreasing.*

  Schmidt (1979)

12.2.10 Chapter 10: Driving the plasma sheet

Practically all auroral theories treat only part of the total problem.

The essential element of the growth phase is the formation of elongated narrow auroral arcs during the growth phase.

*The microphysics of the auroral acceleration regions is often treated
without reference to the source of energy. In fact, the latter one is one
of the most neglected elements in the context of auroral arc
formation. Even less work has been done on the impact of arc
formation on the source plasma. In short, it is fair to say that we are
still far from a self-consistent theory, even on the macroscopic scale.*

Paschmann et al. (2003, p. 360)

Thin current sheet formation is a slow process with gradual
thinning during the growth phase, typically over 30–60 min. They
play a major role in controlling the magnetospheric dynamics.

- Raising and lowering mirror points:

Let us be reminded of a key concept of plasma theory: the first
adiabatic invariant μ_M, the magnetic moment, is conserved to a
remarkable degree when the particles move. If a plasma sheet particle gains or loses energy between points 1 and 2, then at the mirror points

$$\mu_M = \frac{\frac{1}{2}mv_1^2}{B_{MP1}} = \frac{\frac{1}{2}mv_2^2}{B_{MP2}} \qquad (12.2)$$

The mirror point comes down for an increase in energy; conversely, the mirror point is raised when the particle loses energy.
Precipitation will occur with an electrical load with $\mathbf{E} \cdot \mathbf{J} > 0$, resulting in aurora. A dynamo given by $\mathbf{E} \cdot \mathbf{J} < 0$ involves no precipitation;
the sky is black.

- Current thinning event (CTE):

Simulations suggest that the magnetotail flow is diverted
around the thin current sheet to the flanks in the inner part of
the tail (Pulkkinen et al., 2006). A CTE is similar to a PTE at the
magnetopause; again, charge separation cannot remove the electromotive force due to the induction. Newton's third law requires
that the slow action (tailward meander in the dynamo) requires a
reaction to the left (earthward meander) evident as the growth-
phase arc.

- The dark sky poleward is the dynamo with $\mathbf{E} \cdot \mathbf{J} < 0$.
- The growth-phase arc is located where $\mathbf{E} \cdot \mathbf{J} > 0$.

- Auroral current circuit:

The bright night-time aurora is caused by electrons accelerated toward Earth by an upward-pointing electric field. The return
current is carried by downgoing ions or by electrons accelerated
away from the auroral ionosphere. The ionospheric closure

current and the magnetospheric generator form the complete, and closed, auroral current circuit.

- Second low-altitude voltage generator:
  There can be a region above the arc where the potential is a maximum.
  – Secondary electrons are trapped by the electric mirror.
  – They excite the 630 nm emission of oxygen to cause the type-A red aurora.

12.2.11 Chapter 11: Magnetospheric substorms

Heikkila and Pellinen (1977) have proposed a model for the substorm phenomenon, one which uses time-dependent and three-dimensional considerations right from the start: the Substorm transfer event (STE). It is basically a PTE, but its details here are quite different mostly because of Newton's third law.

There are only two ways that the current can wander in the equatorial plane, to the left or to the right. These are associated with a clockwise or counterclockwise perturbation current meander, the action, and the reaction. The full electromagnetic field is $\mathbf{E} = -\nabla\phi - \partial\mathbf{A}/\partial t$. The action is a consequence of the appearance of the solenoidal induction electric field $-\partial\mathbf{A}/\partial t$, the reaction is the response by the plasma $-\nabla\phi$. This is quite limited because the electrostatic field has zero curl.

The easy transfer to the ionosphere in the CTE is not possible after onset of the expansion phase because everything happens in less than 1 min. The Alfvén velocity is only some 100 km/s, taking a few minutes to cover the distance to the ionosphere. The only alternative is to transfer the reaction directly to lower L-shells.

12.3 Substorm transfer event

A substorm is a time-dependent process in three dimensions. We must be able to display the magnetic topology and curl $\mathbf{E}$, in agreement with Faraday's law $\nabla \times \mathbf{E} = -\partial\mathbf{B}/\partial t$. In the integral form, the line integral of $\mathbf{E}$ around a closed contour generates the electromotive force (emf)

$$\varepsilon = \oint \mathbf{E} \cdot \mathbf{dl} = -d\Phi^M/dt \qquad (12.3)$$

Only this emf can tap the energy stored in the magnetic field, by Maxwell's equations. Such an emf is encountered with transfer events, both a dynamo $\mathbf{E} \cdot \mathbf{J} < 0$ and an electrical load $\mathbf{E} \cdot \mathbf{J} > 0$.

- PTE is a blast of solar wind plasma across the magnetopause and into the LLBL.
- CTE is thinning of the current in the magnetotail, with the reaction readily transferred to the ionosphere because of the slowness.
- STE is the expulsion of a plasmoid, the fast reaction being transferred to lower L-shells, and eventually to the polar caps.

All three processes are expressions of plasma convection due to an electromotive force around a continuous curve. The reaction, by Newton's third law, must be borne by the local plasma, on lower L-shells, or transferred to the ionosphere if time permits.

12.4 Four fundamental processes reexamined

12.4.1 Magnetic reconnection

Standard magnetic reconnection (SMR) treats only the problem of the electrical load; the source of energy, the dynamo, is ignored. It has many other problems, for example, with the so-called anomalous resistivity. General magnetic reconnection (GMR) depends on two electric fields (inductive and electrostatic), not one (convection electric field). It is time to recognize this difference.

12.4.2 Particle acceleration

By far the most distinguishing characteristic of plasmas is that they are efficient particle accelerators. This outcome can be produced directly by using the electromagnetic field as discussed in Chapters 3, 7, and 11. The effect of the plasma is major, contrary to the popular view as being passive.

12.4.3 Plasma-neutral interactions

The aurora is an excellent indicator of fundamental physics of plasma and neutral particle coupling. The second low-altitude dynamo for the auroral return current is a good example.

12.4.4 Magnetic dynamos

The magnetic dynamo problem remains one of the outstanding problems in physics. Better progress is likely to be made using the E,J paradigm rather than B,V. Akasofu et al. (2010) have illustrated that very well; Fig. 1.33 in Chapter 1 shows that the existence of a dynamo is essential, expressing cause versus effect.

12.5 Final summary

New ideas have been advanced, based on a simple examination of Maxwell's equations and Newton's laws. No elaborate computations were needed (particle simulation being an exception). Plasma transfer has been clarified, and the LLBL is the direct result. The LLBL allows the chain of cause and effect relationships to be followed throughout the entire process of solar wind interaction with the magnetosphere. If the plasma sheet has a constant supply of plasma and a dawn-dusk electric field it becomes an oscillator, repeating its action with a period of 1 or 2 h.

12.5.1 Reminders

- Circuit analysis
- Superposition principle
- Newton's third law
- Liénard-Wiechert potentials
- Poynting's theorem
- Generalized magnetic reconnection
- Charged dipolar layer
- Lowering of mirror points
- Betatron acceleration is ineffective on "cold" particles
- Charge separation cannot affect electromotive force

12.5.2 Nine new ideas

- Efficient particle energization mechanisms
- Plasma transfer event (PTE)
- Current thinning event (CTE)
- Substorm transfer event (STE)
- Low-latitude boundary layer (LLBL): driver for the plasma sheet
- Low-altitude dynamo for the auroral return current
- Bursty bulk flow analysis
- Omega band and pulsating auroras
- Broadband Electrostatic Noise (BEN)

References

Akasofu, S.-I., Lui, A.T.Y., Meng, C.-I., 2010. Importance of auroral features in the search for substorm onset processes. J. Geophys. Res. 115, A08218. https://doi.org/10.1029/2009JA014960.

Chen, J., 2008. Evidence for particle acceleration in the magnetospheric cusp. Ann. Geophys. 26, 1993–1997.

De Keyser, J., 2005. The Earth's magnetopause: reconstruction of motion and structure. Space Sci. Rev. 121 (1), 225–235.

Emslie, A.G., Miller, J.A., 2003. Particle acceleration. In: Dwivedi, B.N. (Ed.), Dynamic Sun. Cambridge University Press, Cambridge, UK, pp. 262–287.

Faraday, M., 1832. Experimental researches in electricity. Philos. Trans. R. Soc. Lond. 122, 125–162.

Frederick, J.E., Tinsley, B.A., Zhou, L., 2019. Relationships between the solar wind magnetic field and ground-level longwave irradiance at high northern latitudes. J. Atmos. Sol. Terr. Phys. 193. https://doi.org/10.1016/j.jastp.2019.105063.

Fritz, T.A., Fung, S.F. (Eds.), 2005. The Magnetospheric Cusps. Springer, Dordrecht, The Netherlands.

Heikkila, W.J., Pellinen, R.J., 1977. Localized induced electric field within the magnetotail. J. Geophys. Res. 82, 1610–1614.

Heikkila, W.J., Canu, P., Dandouras, I., Keith, W., Khotyaintsev, Y., 2006. Plasma transfer event seen by cluster. In: Cluster and Double Star Symposium—5th Anniversary of Cluster in Space. ESA SP-598.

Laitinen, T.V., Palmroth, M., Pulkkinen, T.I., Janhunen, P., Koskinen, H.E.J., 2007. Continuous reconnection line and pressure-dependent energy conversion on the magnetopause in a global MHD model. J. Geophys. Res. 112, A11201. https://doi.org/10.1029/2007JA012352.

Lighthill, M.J., 1956. Introductory remarks to the meeting of the Physical Society on "The Physics of Gas Flow at Very High Speeds" Nature 178, 343.

Lundin, R., 2003. Preface. In: Lundin, R., McGregor, R. (Eds.), Proceedings of the Magnetic Reconnection Meeting, IRF Scientific Report 280. Swedish Institute of Space Physics, Kiruna, Sweden, p. v.

National Research Council, 2009. A Performance Assessment of NASA's Heliophysics Program. The National Academies Press, Washington, DC. https://doi.org/10.17226/12608.

Omura, Y., Heikkila, W.J., Umeda, T., Ninomiya, K., Matsumoto, H., 2003. Particle simulation of plasma response to an applied electric field parallel to magnetic field lines. J. Geophys. Res. 108 (A5), 1197. https://doi.org/10.1029/2002JA009573.

Paschmann, G., Haaland, S., Treumann, R. (Eds.), 2003. Auroral Plasma Physics. Kluwer, Norwell, MA (Reprinted from *Space Science Reviews, 103*: 1–485).

Pulkkinen, T.I., Palmroth, M., Tanskanen, E.I., Janhunen, P., Koskinen, H.E.J., Laitinen, T.V., 2006. New interpretation of magnetospheric energy circulation. Geophys. Res. Lett. 33, L07101. https://doi.org/10.1029/2005GL025457.

Ramo, S., Whinnery, J.R., 1953. Fields and Waves in Modern Radio, second ed. Wiley, New York, NY.

Saarnisto, M., 2008. Emergence history of the Karelian Isthmus. In: Karelian Isthmus – Stone Age Studies in 1998–2003. Iskos, 16. The Finnish Antiquarian Society, Helsinki, Finland, pp. 128–139.

Schindler, K., Hesse, M., Birn, J., 1988. General magnetic reconnection, parallel electric fields, and helicity. J. Geophys. Res. 93, 5547–5557. https://doi.org/10.1029/JA093iA06p05547.

Schmidt, G., 1979. Physics of High Temperature Plasmas, second ed. Academic Press, New York, NY.

Sonnerup, B.U.Ö., Hasegawa, H., Paschmann, G., 2004. Anatomy of a flux transfer event seen by cluster. Geophys. Res. Lett. 31, L11803. https://doi.org/10.1029/2004GL020134.

Taylor, M.G.G.T., Reeves, G.D., Friedel, R.H.W., Thomsen, M.F., Elphic, R.C., Davies, J.A., et al., 2006. Cluster encounter with an energetic electron beam during a substorm. J. Geophys. Res. 111, A11203. https://doi.org/10.1029/2006JA011666.

Index

Note: Page numbers followed by *f* indicate figures.